T0299305

ASTROMETRY AND ASTROPHYSICS IN THE GAIA SKY

IAU SYMPOSIUM 330

COVER ILLUSTRATION:

Picture adapted from the Symposium poster (credits : A. Titarenko and Service Communication OCA): photomontage showing the Promenade des Anglais and the Mediterranean sea in Nice. It includes a drawing of the Gaia satellite and of the Grande Coupole of the Observatoire de la Côte d'Azur historical site (built by Gustave Eiffel and Charles Garnier).

IAU SYMPOSIUM PROCEEDINGS SERIES

INTERNATIONAL ASTRONOMICAL UNION

UNION ASTRONOMIQUE INTERNATIONALE

ASTROMETRY AND ASTROHPYSICS IN THE GAIA SKY

PROCEEDINGS OF THE 330th SYMPOSIUM OF THE INTERNATIONAL ASTRONOMICAL UNION HELD IN NICE, FRANCE APRIL 24–28, 2017

Edited by

ALEJANDRA RECIO-BLANCO
Université Côte d'Azur, Observatoire de la Côte d'Azur, CNRS, Laboratoire Lagrange, France

PATRICK DE LAVERNY
Université Côte d'Azur, Observatoire de la Côte d'Azur, CNRS, Laboratoire Lagrange, France

ANTHONY G.A. BROWN
Leiden Observatory, Leiden University, Leiden, The Netherlands

and

TIMO PRUSTI
ESA/ESTEC

CAMBRIDGE
UNIVERSITY PRESS

University Printing House, Cambridge CB2 8BS, United Kingdom

One Liberty Plaza, 20th Floor, New York, NY 10006, USA

477 Williamstown Road, Port Melbourne, VIC 3207, Australia

314-321, 3rd Floor, Plot 3, Splendor Forum, Jasola District Centre, New Delhi - 110025, India

103 Penang Road, #05-06/07, Visioncrest Commercial, Singapore 238467

Cambridge University Press is part of the University of Cambridge.

It furthers the University's mission by disseminating knowledge in the pursuit of education, learning and research at the highest international levels of excellence.

www.cambridge.org
Information on this title: www.cambridge.org/9781107170087

First published 2018

A catalogue record for this publication is available from the British Library

ISBN 978-1-107-17008-7 Hardback

Table of Contents

Preface ... xiv

Conference Photograph ... xv

Dedication of the Symposium ... xvi

The Organizing Committee ... xviii

Participants ... xx

Tribute to François Mignard's research ... 1
J. Kovalevsky

The Gaia Sky

The Gaia mission status ... 7
T. Prusti

The Gaia sky: version 1.0 ... 13
A. G. A. Brown

Gaia: from proposal to GDR1 ... 23
G. Gilmore

Gaia Photometric Data: DR1 results and DR2 expectations ... 30
D. W. Evans, M. Riello, F. De Angeli, G. Busso, F. van Leeuwen, L. Eyer, C. Jordi, C. Fabricius, J. M. Carrasco, M. Weiler, P. Montegriffo, C. Cacciari & E. Pancino

The Gaia Archive ... 35
A. Mora, J. González-Núñez, D. Baines, J. Durán, R. Gutiérrez-Sanchéz, E. Racero, J. Salgado & J. C. Segovia

Astrometry and Fundamental Physics

The Tycho-Gaia Astrometric Solution ... 41
L. Lindegren

Astrometric surveys in the Gaia era ... 49
N. Zacharias

Multiply imaged quasars in the Gaia DR1 ... 59
C. Ducourant, L. Delchambre, F. Finet, L. Galluccio, A. Krone-Martins, J. F. Le Campion, F. Mignard, E. Slezak, J. Surdej, R. Teixeira & O. Wertz

Local tests of gravitation with Gaia observations of Solar System Objects ... 63
A. Hees, C. Le Poncin-Lafitte, D. Hestroffer & P. David

GaiaNIR – A future all-sky astrometry mission ... 67
D. Hobbs & E. Høg

Gaia DR1 compared to VLBI positions 71
F. Mignard & S. Klioner

The LQAC-4, last update of the Large Quasar Astrometric Catalogue 75
J. Souchay, A. H. Andrei, C. Barache, F. Taris, C. Gattano & B. Coelho

The Differential Astrometric Reference Frame on short timescales in the Gaia Era 79
U. Abbas, B. Bucciarelli, M. G. Lattanzi, M. Crosta, M. Gai, R. Smart, A. Sozzetti & A. Vecchiato

The PMA Catalogue as a realization of the extragalactic reference system in optical and near infrared wavelengths.. 81
V. S. Akhmetov, P. N. Fedorov & A. B. Velichko

New Astronomical Reduction of Old Observations (the NAROO project) 83
J.-E. Arlot, V. Robert, V. Lainey, C. Neiner & N. Thouvenin

Using Gaia as an Astrometric Tool for Deep Ground-based Surveys........... 85
D. I. Casetti-Dinescu, T. M. Girard & M. Schriefer

Remarks of Gaia DR1 magnitude using ground-based optical monitoring of QSOs 88
G. Damljanović, F. Taris & A. Andrei

Outline of Infrared Space Astrometry missions:JASMINE................... 90
N. Gouda & JASMINE working group

GIER: A Danish computer from 1961 with a role in the modern revolution of astronomy - II 92
E. Høg

Astrometry with A-Track Using Gaia DR1 Catalogue 94
Y. Kılıç, O. Erece & M. Kaplan

Astrometry for New Reductions: The ANR method 96
V. Robert & C. Le Poncin-Lafitte

Optimisation of JWST operations with the help of Gaia 98
J. Sahlmann, E. G. Nelan, P. Chayer, B. McLean & M. Lallo

Kinematics of our Galaxy from the PMA and TGAS catalogues.............. 100
A. B. Velichko, V. S. Akhmetov & P. N. Fedorov

Nano-JASMINE and small-JASMINE data analysis........................... 104
Y. Yamada, Y. Shirasaki & R. Nishi

Light propagation in the Solar System for astrometry on sub-micro-arcsecond level 106
S. Zschocke

Galactic Archaeology

Self-consistent modelling of our Galaxy with Gaia data..................... 111
J. Binney

Stellar clusters in the Gaia era 119
A. Bragaglia

Galaxy simulations in the Gaia era 127
I. Minchev

Galactic Surveys in the Gaia Era 136
R. F. G. Wyse

Close stellar encounters with the Sun from the first Gaia Data Release 144
C. A. L. Bailer-Jones

Gaia DR1 completeness within 250 pc & star formation history of the Solar neighbourhood 148
E. J. Bernard

Self-consistent Modelling of the Milky Way using Gaia data 152
D. R. Cole & J. Binney

Abundance ratios & ages of stellar populations in HARPS-GTO sample 156
E. D. Mena, M. Tsantaki, V. Zh. Adibekyan, S. G. Sousa, N. C. Santos, J. I. González Hernández & G. Israelian

The kinematics and surface mass density in the solar neighbourhood using TGASxRAVE 160
J. H. J. Hagen & A. Helmi

Dynamical effects of the spiral arms on the velocity distribution of disc stars ... 164
K. Hattori, N. Gouda, T. Yano, N. Sakai & H. Tagawa

The evolution history of the extended solar neighbourhood 168
A. Just, K. Sysoliatina & I. Koutsouridou

Metallicity distribution functions using Gaia-DR1 data 172
G. Kordopatis & RAVE collaboration

RAVE-Gaia and the impact on Galactic archeology 176
A. Kunder

Hypervelocity star candidates in *Gaia* DR1/TGAS 181
T. Marchetti, E. M. Rossi, G. Kordopatis, A. G. A. Brown, A. Rimoldi, E. Starkenburg, K. Youakim & R. Ashley

Search for Galactic warp signal in Gaia DR1 proper motions 185
E. Poggio R. Drimmel, R. L. Smart, A. Spagna & M. G. Lattanzi

Can we detect Galactic spiral arms? 3D dust distribution in the Milky Way.... 189
S. Rezaei Kh., C. A. L. Bailer-Jones, M. Fouesneau & R. Hanson

Galactic Disk Structure and Metallicity from Mono-age Stellar Populations of LAMOST 193
M. Xiang, X. Liu, J. Shi, H. Yuan, Y. Huang, B. Chen & C. Wang

Mapping young stellar populations towards Orion with *Gaia* DR1............ 197
E. Zari & A. G. A. Brown

The kinematics of the white dwarf population from the SDSS DR12 201
B. Anguiano, A. Rebassa-Mansergas, E. García-Berro, S. Torres, K. Freeman & T. Zwitter

Accurate atomic data for Galactic Surveys 203
M. T. Belmonte, J. C. Pickering, C. Clear, F. Liggings & A. P. Thorne

Age dependence of metallicity gradients in the Galactic disc from astrometry and asteroseismology 206
L. Casagrande

Galactic disk structure as revealed by LAMOST A stars 208
B.-Q. Chen, X.-W. Liu, H.-B. Yuan, Y. Huang, M.-S. Xiang, C. Wang, Z.-J. Tian & H.-W. Zhang

Using ground based data as a precursor for *Gaia* in getting proper motions of satellites 210
T. K. Fritz , S. T. Linden, P. Zivick, N. Kallivayalil & J. Bovy

What we learn from TGAS about the moving groups of the Solar neighbourhood 214
B. Goldman, E. Schilbach, S. Röser, P. Schöfer, A. Derekas, A. Moor, W. Brandner & T. Henning

The AMBRE Project: r-process element abundances in the Milky Way thin and thick discs 216
G. Guiglion, P. de Laverny, A. Recio-Blanco & C. C. Worley

The thick disc according to Gaia-ESO 218
L. M. Howes & T. Bensby

The Galactic mass distribution from the LAMOST Galactic spectroscopic surveys 220
Y. Huang, X.-W. Liu, H.-B. Yuan, H.-W. Zhang, M.-S. Xiang, B.-Q. Chen & C. Wang

Modelling the Milky Way with *Gaia*-TGAS 222
J. A. S. Hunt

Identification of binary and multiple systems in TGAS using the Virtual Observatory 225
F. Jiménez-Esteban & E. Solano

Open star clusters and Galactic structure 227
Y. C. Joshi

The time evolution of gaps in tidal streams in axisymmetric potentials 229
H. H. Koppelman & A. Helmi

On-sky verification of the 6-h periodic basic angle variations of the Gaia satellite 231
S. Liao, M. G. Lattanzi, A. Vecchiato, Z. Qi, M. Crosta & Z. Tang

Open Cluster Dynamics via Fundamental Plane 233
C.-C. Lin & X.-Y. Pang

A 3D-Study of the residual vector field 235
F. J. Marco, M. J. Martínez & J. A. López

Impact on the Hipparcos2-UCAC4 geometric relation from some physical properties of the stars 237
F. J. Marco, M. J. Martínez & J. A. López

How far away and how old are these stars? 239
P. J. McMillan & the RAVE Collaboration

CNO distributions in the Solar neighborhood with Gaia data................ 241
Š. Mikolaitis, G. Tautvaišienė, R. Ženovienė, A. Drazdauskas, E. Pakštienė, R. Janulis, V. Bagdonas & L. Klebonas

Developing Automated Spectral Analysis Tools for Interstellar Features Extraction to Support Construction of the 3D ISM Map 243
L. Puspitarini, R. Lallement, A. Monreal-Ibero, H.-C. Chen, H. L. Malasan, M. Aprilia, I. Arifyanto & M. Irfan

Finding evolved stars in the inner Galactic disk with *Gaia* 245
L. H. Quiroga-Nuñez, H. J. van Langevelde, Y. M. Pihlström, L. O. Sjouwerman & A. G. A. Brown

Magellanic Clouds Proper Motion and Rotation with Gaia DR1.............. 249
J. Sahlmann & R. van der Marel

Proper motions of stars in the globular clusters using WFI@2.2 m telescope.... 251
D. P. Sariya, I.-G. Jiang & R. K. S. Yadav

Finding the stars that reionized the Universe 253
M. Sharma, T. Theuns & C. Frenk

Determining the Local Dark Matter Density with SDSS G-dwarf data 255
H. Silverwood, S. Sivertsson, J. Read, G. Bertone & P. Steger

The relation between velocity dispersions and chemical abundances in RAVE giants 259
R. Smiljanic & R. Silva de Souza

HST Proper Motions of Distant Globular Clusters: Constraining the Formation & Mass of the Milky Way 261
S. T. Sohn, R. P. van der Marel, A. Deason, A. Bellini, G. Besla & L. Watkins

Chemo-dynamical signatures in simulated Milky Way-like galaxies............ 263
A. Spagna, A. Curir, M. Giammaria, M. G. Lattanzi, G. Murante & P. Re Fiorentin

Revisiting TW Hydrae association in light of Gaia-DR1 265
R. Teixeira, E. R. Gonoretzky, C. Ducourant, P. A. B. Galli & A. G. O. Krone-Martins

[Y/Mg] stellar dating calibration 267
A. Titarenko, A. Recio-Blanco, P. de Laverny, M. Hayden, G. Guiglion & C. Worley

Dynamics of the Oort Cloud In the Gaia Era I: Close Encounters 269
S. Torres, S. P. Zwart & A. G. A. Brown

Stellar parameters with FASMA: a new spectral synthesis package............ 271
M. Tsantaki, D. T. Andreasen, G. D. C. Teixeira, S. G. Sousa, N. C. Santos, E. Delgado-Mena & G. Bruzual

Complex study of the open cluster NGC 2281............................... 273
J. Velčovský & J. Janík

Unveiling the stellar halo with TGAS 275
J. Veljanoski, L. Posti, A. Helmi & M. A. Breddels

Herbig Ae/Be stars with TGAS parallaxes in the HR diagram 277
M. Vioque, R. D. Oudmaijer & D. Baines

The age-metallicity relation with RAVE and TGAS 279
J. Wojno, G. Kordopatis, M. Steinmetz, P. J. McMillan & the RAVE collaboration

Reanalysis of 24 Nearby Open Clusters using Gaia data 281
S. X. Yen, S. Reffert, S. Röser, E. Schilbach, N. V. Kharchenko & A. E. Piskunov

Spectroscopic and Photometric Survey of Northern Sky for the ESA PLATO space mission 283
R. Ženovienė, V. Bagdonas, A. Drazdauskas, R. Janulis, L. Klebonas, Š. Mikolaitis, E. Pakštienė & G. Tautvaišienė

Stellar Physics

Variable stars in the Gaia era: Mira, RR Lyrae, δ and Type-II Cepheids 287
M. A. T. Groenewegen

Wide Binaries in TGAS: Search Method and First Results 297
J. J. Andrews, J. Chanamé & M. A. Agüeros

The white dwarf mass-radius relation with Gaia, Hubble and FUSE 301
S. R. G. Joyce, M. A. Barstow, S. L. Casewell, J. B. Holberg & H. E. Bond

Optical interferometry and Gaia parallaxes for a robust calibration of the Cepheid distance scale 305
P. Kervella, A. Mérand, A. Gallenne, B. Trahin, S. Borgniet, G. Pietrzynski, N. Nardetto & W. Gieren

Gaia view of low-mass star formation 309
C. F. Manara, T. Prusti, J. Voirin & E. Zari

Calibration and characterisation of the Gaia Red Clump 313
L. Ruiz-Dern, C. Babusiaux, F. Arenou, C. Danielski, C. Turon & P. Sartoretti

White dwarfs in the Gaia era 317
P.-E. Tremblay, N. Gentile-Fusillo, J. Cummings, S. Jordan, B. T. Gänsicke & J. S. Kalirai

Runaway companions of supernova remnants with Gaia 321
D. Boubert, M. Fraser & N. W. Evans

The TGAS HR diagram of barium stars 323
A. Escorza, H. M. J. Boffin, A. Jorissen, L. Siess, S. Van Eck, S. Shetye, D. Pourbaix & H. Van Winckel

Dynamical masses of Cepheids from the GAIA parallaxes 325
A. Gallenne, P. Kervella, A. Mérand, N. R. Evans & C. Proffitt

Confronting the Gaia and NLTE spectroscopic parallaxes for the FGK stars . . . 327
T. Sitnova, L. Mashonkina & Y. Pakhomov

Double, triple and quadruple-line spectroscopic binary candidates within the Gaia-ESO Survey . . . 329
T. Merle, S. Van Eck, A. Jorissen, M. Van der Swaelmen, G. Traven & T. Zwitter

Stellar Parameters, Chemical composition and Models of chemical evolution . . . 331
T. Mishenina, M. Pignatari, B. Côté F.-K. Thielemann, C. Soubiran, N. Basak, T. Gorbaneva, S. A. Korotin, V. V. Kovtyukh, B. Wehmeyer, S. Bisterzo, C. Travaglio, B. K. Gibson, C. Jordan, A. Paul, C. Ritter & F. Herwig

Long term near infrared observation of very bright stars at Kagoshima University 333
T. Naagaya

The Baade-Wesselink p-factor of Cepheids in the Gaia area . . . 335
N. Nardetto

G-Band Period-Luminosity Relation For Galactic Cepheids Based on Gaia DR1 Measurements . . . 337
C.-C. Ngeow, A. Bhardwaj & S. M. Kanbur

The mass-ratio distribution of spectroscopic binaries along the main-sequence . . 339
H. M. J. Boffin & D. Pourbaix

OB stars towards NGC 6357 and NGC 6334 . . . 341
D. Russeil

Gaia observations of naked-eye stars: status update . . . 343
J. Sahlmann, A. Mora, J. M. Martín-Fleitas, A. Abreu, C. Crowley & M. Fink

The TGAS HR diagram of S-type stars . . . 345
S. Shetye, S. Van Eck, A. Jorissen, H. Van Winckel & L. Siess

Understanding Li enhancement in K giants and role of accurate parallaxes . . . 348
R. singh & B. E. Reddy

Detection of spectroscopic binaries: lessons from the Gaia-ESO survey . . . 350
M. Van der Swaelmen, T. Merle, S. Van Eck, A. Jorissen & T. Zwitter

S stars in the Gaia era: stellar parameters and nucleosynthesis . . . 352
S. Van Eck, D. Karinkuzhi, S. Shetye, A. Jorissen, S. Goriely, L. Siess, T. Merle & B. Plez

Observational Facilities of Sternberg Astronomical Institute for Ground-Based Photometric Study of Newly Identified GAIA Objects, — CV-candidates. . 354
I. Voloshina & V. Sementsov

The nearby triple star HIP 101955 . . . 356
F. Xia

A new method for orbit determination on the Gaia SB1s . . . 358
W. Xiaoli

Clarification of the formation process of the super massive black hole by Infrared astrometric satellite, Small-JASMINE 360
T. Yano, JASMINE-WG

Mathematical Assessment of Physical and Chemical Processes from the Middle B to the Early F Type Main Sequence Stars 362
K. Yüce & S. J. Adelman

Massive companions of binary systems........ 364
D. Jableka, S. Zola, B. Zakrzewski, J. M. Kreiner & W. Ogloza

Solar system and Exoplanets

Characterisation of exoplanet host stars: A window into planet formation........ 369
N. C. Santos

Exploring the Solar System using stellar occultations 377
B. Sicardy

Prediction of stellar occultations by distant solar system bodies in the Gaia era. 382
J. Desmars, J. Camargo, B. Sicardy, F. Braga-Ribas, R. Vieira-Martins, M. Assafin, D. Bérard & G. Benedetti-Rossi

Prospects for asteroid mass determination from close encounters between asteroids: ESA's Gaia space mission and beyond 386
A. Ivantsov, D. Hestroffer & S. Eggl

T_c-trend and terrestrial planet formation: The case of Zeta Reticuli........ 391
V. Adibekyano, E. Delgado-Mena, N. C. Santos, S. G. Sousa & P. Figueira

Shape and spin of asteroid 967 Helionape 393
G. Apostolovska, A. Kostov, Z. Donchev, E. V. Bebekovska & O. Kuzmanovska

Preliminary Results of Low Dispersion Asteroid Spectroscopy Survey at NAO Rozhen 395
E. V. Bebekovska, G. Borisov, Z. Donchev & G. Apostolovska

Solar system astrometry, Gaia, and the large surveys – a huge step ahead to stellar occultations by distant small solar system bodies 397
J. I. B. Camargo, M. V. Banda-Huarca, R. L. Ogando, J. Desmars, F. Braga-Ribas, R. Vieira-Martins, M. Assafin, B. Sicardy, D. Bérard, G. Benedetti-Rossi, L. A. N. da Costa, M. A. G. Maia, M. Carrasco-Kind & A. Drlica-Wagner

Using Gaia spectrophotometric data for the purposes of asteroid taxonomy 399
A. Cellino, P. Tanga, M. Delbo, L. Galluccio, P. Bendjoya & F. De Angeli

Follow-up studies of Gaia transients at the Terskol Observatory........ 401
V. Godunova, V. Reshetnyk, A. Simon, S. Velichko, O. Sergeev & V. Taradii

Alerting observations of asteroids at the SBG telescope of the Kourovka Astronomical Observatory in the Gaia-FUN-SSO Network 403
E. Kuznetsov, D. Glamazda, G. Kaiser & Y. Wiebe

Searching for planets around eclipsing binary stars using timing method: NSVS 14256825 405
I. Nasiroglu, K. Goźiewski, A. Słowikowska, K. Krzeszowski, M. Żejmo, S. Zola & H. Er

Observations of the satellites of the major planets at Pulkovo Observatory: history and present 407
N. A. Shakht, A. V. Devyatkin, D. L. Gorshanov & M. S. Chubey

Precise CCD positions of Triton in 2014-2016 from the Gaia DR1 409
N. Wang, Q. Y. Peng, H. W. Peng & Q. F. Zhang

Astrometric Reduction of Cassini ISS Images of Enceladus in 2015 Based on Gaia DR1 411
Q. F. Zhang, V. Lainey, A. Vienne, N. J. Cooper, Q. Y. Peng & N. Wang

Astrometry and Spectra Classification of Near Earth Asteroids with Lijiang 2.4 m Telescope 413
X. L. Zhang, B. Yang & J. M. Bai

Author index 415

Preface

Astrometry has historically been fundamental to all the fields of astronomy, driving many revolutionary scientific results. Keplers laws, deduced after analysing the observations of Tycho Brahe, are an outstanding example of this. Four centuries later, the ESA Gaia mission is astrometrically, photometrically and spectroscopically surveying the full sky since July 2014. This survey will be complete to magnitude 20 for the astrometry and photometry and to magnitude 16 for the spectroscopy (about 1 billion and 150 million sources expected, respectively). The Gaia astrometry allows stellar distance and age estimations with unprecedented accuracy, and with the complement of radial velocities, it will provide the full kinematic information of the targets. Moreover, the photometric and spectroscopic data will be used to classify objects and astrophysically characterize stars.

The International Astronomical Union Symposium 330 has been the occasion to review the first 2.5 years of the Gaia activities and to present and discuss the first scientific results derived from the first Gaia Data Release (GDR1), seven months after its delivery in September 2016. The most significant illustration of the high impact of Gaia is probably the large involvement of the international astronomical community in this symposium: 276 participants from 35 different countries were present in Nice for this one-week symposium. Already from its first data release Gaia is therefore undoubtedly changing not only our understanding of the Galaxy and its planetary and stellar components, but also our way of working.

On the one hand, the significant increase in the precision of the astrometric measurements is sharpening our view of the Milky Way, but also of the physical processes involved in the stellar and galactic evolution. This implies an enhanced synergy between different communities (astrometry, stellar and galactic physics), and a refinement of the models and simulations that are now better constrained thanks to Gaia data.

On the other hand, the increasing number of available data has a clear impact on the analysis approaches that have to be adopted by the astronomical community, imposing robust statistical treatments, opening the path to unsupervised classification techniques and generally enriching our knowledge of the detailed physics at play in the studied astronomical objects (the Milky Way, the stars, the Solar System objects...). Moreover, from the point of view of the models and simulations, the increasing number of constraints even for low probability processes or rare objects is already acting as the catalyst of a new era in our understanding of the Galaxy, its stars, and the Solar System.

In summary, the extremely important improvement in the astrometric precision and in the number of studied objects is leading to a transformation comparable to the impact of the telescope invention about four centuries ago. IAU Symposium 330 has confirmed the start of the announced Gaia revolution and the articles in this volume testify of it. We hope that you will enjoy reading them as much as we have done.

Alejandra Recio-Blanco, Anthony Brown and Timo Prusti, co-chairs SOC,
Patrick de Laverny, chair LOC
Nice, July 2017

CONFERENCE PHOTOGRAPH

Dedication of the Symposium

Photo 1. François Mignard, expert in astrometry and reference frames, and former chair of the Gaia Data Processing and Analysis Consortium, to whom this symposium is dedicated

The organizers wish to dedicate the IAU Symposium 330 to François Mignard for his outstanding contribution to the Gaia mission. François Mignard is an expert in astrometry and reference frames, and was the chair of the Gaia Data Processing and Analysis Consortium (DPAC) since its formation until the end of 2012.

François Mignard graduated in Physics from the Ecole Normale Supérieure in 1974 and he moved to Nice at the newly created CERGA (*Centre d'études et recherches géodynamiques et astronomiques*) in Grasse. His doctoral thesis, under the supervision of Jean Kovalevsky, was dedicated to the study of tidal phenomena in the evolution of planet-satellite systems. After a one year post-doc in Cornell University with J.A. Burns and one year professorship in Marseille, he returned to CERGA. During that period, he worked on radiation pressure and dust particle dynamics, on grain dynamics in rings, the rotation of Hyperion, the dynamics of Oort cloud bodies and of binary asteroids, among many other subjects. In addition, one of his constant interests was General Relativity for which the advent of space astrometry opened the way to new practical applications.

Since 1985, he became a major member of the FAST Consortium in charge of the Hipparcos data reduction. In particular, François worked on the precision and the accuracy of the Hipparcos photometric catalogue and the light-curves of variable stars, showing their powerful scientific value.

In 1991, he became member of the Hipparcos Science Team and replaced Jean Kovalevsky as the the FAST Consortium coordinator. In addition, he became responsible for a working group of the Hipparcos Science Team in charge of constructing the final catalogue of double stars.

From 1993, the concept of what later will be called Gaia was presented. Since then, François was advocating it and working on its feasibility and the expected science. In 1997, a first draft of a scientific proposal was presented to ESA and François Mignard was member of the Scientific Advisory group created after this. Finally, as a recognition of his major role in the preparation of the data treatment, he became chair of the Gaia DPAC in 2006. He tireless worked on DPAC activities since then, even after stepping down as DPAC head at the end of 2012.

In addition to his international responsibilities, François had many local and national management activities. In particular, he was director of the CERGA laboratory from 1993 to 2004, principle investigator for the French *Action Spécifique* Gaia, and founder of the Gaia group at the Observatoire de la Côte d'Azur. Thanks to his national and local investments, a long list of researchers and engineers got permanent positions in France to work on Gaia science. In Nice, a whole new Gaia generation emerged thanks to him.

If, as François says, science is a learned mixture of *savoir faire* and *savoir dire,* François himself is an outstanding example of it. But, more than that, he is a universalist capable to explain to you during an informal conversation the dates of the swift migrations, the physical details of the Corsica island vision from Nice, the different types of spiders, the exact antipodes of your birth place, the trips of the ancient explorers of the world, the rugby matches of the French national team, etc... People like us, having the wonderful opportunity to work with him, can testify that François is not only a remarkable scientist, but also a remarkable human being.

On behalf of the SOC and the LOC, Alejandra Recio-Blanco, Anthony Brown, Timo Prusti and Patrick de Laverny

THE ORGANIZING COMMITTEE

Scientific

Anthony Brown (**co-chair**, Netherlands)
Katia Cunha (Brazil)
Gayandhi De Silva (Australia)
Gerry Gilmore (UK)
Naoteru Gouda (Japan)
Sergei Klioner (Germany)
Tamara V. Mishenina (Ukraine)
Timo Prusti (**co-chair**, ESA, Netherlands)
Alejandra Recio-Blanco (**chair**, France)
Bacham Reddy (India)
Gonzalo Tancredi (Uruguay)
Paolo Tanga (France)
David Vokrouhlicky (Czech Republic)
Patricia A. Whitelock (South Africa)
Norbert Zacharias (USA)
Manuela Zoccali (Chile)

Local

Patrick de Laverny (**chair**)
Marco Delbo
Vanessa Hill
Thierry Lanz
François Mignard
Sophie Rousset
Mathias Schultheis

Acknowledgements

The IAU Symposium 330 Astrometry and Astrophysics in the Gaia sky was supported by IAU Division A (Fundamental Astronomy), Division F (Planetary Systems and Bioastronomy), Division G (Stars and Stellar Physics), Division H (Interstellar matter and Local Universe), Commission 4 (Ephemerides), Commission 7 (Celestial Mechanics and Dynamical Astronomy), Commission 8 (Astrometry), Commission 29 (Stellar Spectra), Commission 30 (Radial Velocities), Commission 33 (Structure and Dynamics of the Galactic System), Commission 35 (Stellar Constitution), Commission 36 (Theory of Stellar Atmospheres), Commission 47 (Cosmology) and Commission 53 (Exoplanets).

The SOC and LOC chairs sincerely thank the Lagrange Laboratory and the Observatoire de la Côte d'Azur for their help and their involvement in this symposium organisation.

The SOC and LOC acknowledge financial support from several institutions: The symposium has been strongly supported by the Centre National d'Etudes Spatiales (CNES), the European Space Agency (ESA), the AIRBUS company and the University of Côte d'Azur (UCA, through the Agence Nationale de la Recherche UCAJEDI project number ANR-15-IDEX-01).

Additional sponsors are the Observatoire de la Côte d'Azur (OCA), Ville de Nice, Provence Alpes Côte d'Azur Région (PACA), Université of Nice – Sophia Antipolis (UNS), Société Française d'Astronomie et d'Astrophysique (SF2A), INSU/CNRS through its National Programs of Stellar Physics (PNPS), Galaxies and Cosmologie (PNCG) and Planetology (PNP) and its Action Spécifique Gravitation, Références, Astronomie et Métrologie (ASGram).

The majority of the articles in this volume made use of data from the European Space Agency (ESA) mission Gaia (https://www.cosmos.esa.int/gaia), processed by the Gaia Data Processing and Analysis Consortium (DPAC, https://www.cosmos.esa.int/web/gaia/dpac/consortium). Funding for the DPAC has been provided by national institutions, in particular the institutions participating in the Gaia Multilateral Agreement.

Participants

First name	Last name	Email address	Organisation	Town/ City	Country
Ummi	Abbas	abbas@oato.inaf.it	INAF - Osservatorio Astrofisico di Torino	Pino Torinese	ITALIE
Carlos	Abia	cabia@ugr.es	Carlos Abia	Granada	ESPAGNE
Vardan	Adibekyan	vadibekyan@astro.up.pt	Centro de Investigaçao em Astronomia/Astrofisica da Universidade do Porto	Porto	PORTUGAL
Carlos	Allende Prieto	callende@iac.es	Instituto de Astrofísica de Canarias	San Cristóbal de La Laguna	ESPAGNE
Jeff	Andrews	andrews@physics.uoc.gr	University of Crete	Heraklion	GRECE
Borja	Anguiano	astrobaj@gmail.com	Department of Astronomy University of Virginia	Charlottesville	ETATS-UNIS
Frederic	Arenou	Frederic.Arenou@obspm.fr	CNRS UMR 8111 / GEPI, Observatoire de Paris	Meudon	FRANCE
Jean-Eudes	Arlot	jean-eudes.arlot@obspm.fr	CNRS UMR8028 IMCCE	paris	FRANCE
Chrysa	Avdellidou	chrysa.avdellidou@esa.int	ESTEC	AZ Noordwijk, Zuid Holland	PAYS-BAS
Coryn	Bailer-Jones	calj@mpia.de	Max Planck Institute for Astronomy	Heidelberg	ALLEMAGNE
Ulrich	Bastian	bastian@ari.uni-heidelberg.de	ARI/ZAH, Heidelberg University	Heidelberg	ALLEMAGNE
Rachael	Beaton	rbeaton@obs.carnegiescience.edu	Rachael Beaton	PASADENA	ETATS-UNIS
Charles	Beichman	chas@ipac.caltech.edu	NASA Exolqanet Science Institute/IPAC	Pasadena	ETATS-UNIS
Maria Teresa	Belmonte	m.belmonte-sainz-ezquerra@imperial.ac.uk	Maria Teresa Belmonte	London	ROYAUME-UNI
Diane	Berard	diane.berard@obspm.fr	Observatoire de Paris	Paris	FRANCE
Edouard	Bernard	ebernard@oca.eu	Observatoire de la Côte d'Azur	NICE Cedex 4	FRANCE
Filip	Berski	filip.berski@amu.edu.pl	Adam Mickiewicz University	Poznam	POLOGNE
Olivier	Bienaymé	olivier.bienayme@unistra.fr	Observatoire de Strasbourg	Strasbourg	FRANCE
Albert	Bijaoui	albert-bijaoui@orange.fr	Observatoire de la Côte d'Azur	NICE Cedex 4	FRANCE
James	Binney	binney@physics.ox.ac.uk	University of Oxford	Oxford	ROYAUME-UNI
Alex	Bombrun	alex.bombrun@sciops.esa.int	Camino Bajo del Castillo	Villanueva de la Cañada, Madrid	ESPAGNE
Robin	Bonannini	robin.bonannini@oca.eu	Patrick de Laverny - Observatoire de la Côte d'Azur	NICE Cedex 4	FRANCE
Douglas	Boubert	d.boubert@ast.cam.ac.uk	Institute of Astronomy	Cambridge	ROYAUME-UNI
Jo	Bovy	bovy@astro.utoronto.ca	Department of Astronomy and Astrophysics – University of Toronto	TORONTO	CANADA
Angela	Bragaglia	angela.bragaglia@oabo.inaf.it	Bologna Observatory	Bologna	ITALIE
Danielle	Briot	danielle.briot@obspm.fr	Danielle Briot	Paris	FRANCE
Anthony	Brown	brown@strw.leidenuniv.nl	Anthony Brown	Leiderdorp	PAYS-BAS
Beatrice	Bucciarelli	bucciarelli@oato.inaf.it	Torino Observatory	Pino Torinese	ITALIE
Julio Ignacio	Bueno de Camargo	camargo@on.br	Julio Ignacio Bueno de Camargo	Rio de Janeiro	BRESIL
Nicole	Capitaine	nicole.capitaine@obspm.fr	Nicole Capitaine	Meudon	FRANCE
Raymond	Carlberg	raymond.carlberg@utoronto.ca	Raymond Carlberg	Toronto	CANADA

First name	Last name	Email address	Organisation	Town/ City	Country
Luca	Casagrande	luca.casagrande@anu.edu.au	Luca Casagrande	Weston Creek	AUSTRALIE
Dana	Casetti	danacasetti@gmail.com	Dana I. Casetti	Hamden	ETATS-UNIS
Alberto	Cellino	cellino@oato.inaf.it	INAF, Torino Observatory	Cantarana (AT)	ITALIE
Corinne	Charbonnel	Corinne.Charbonnel@unige.ch	Département d'Astronomie - Université de Genève	Versoix	SUISSE
Patrick	Charlot	patrick.charlot@u-bordeaux.fr	Laboratoire d'Astrophysique de Bordeaux - Université de Bordeaux	Pessac	FRANCE
Alain	Chelli	Alain.Chelli@oca.eu	Laboratoire Lagrange, OCA	Nice	FRANCE
Bingqiu	Chen	bchen@pku.edu.cn	Peking University	Beijing	CHINE
Andrea	Chiavassa	andrea.chiavassa@oca.eu	Observatoire de la Côte d'Azur	Nice	FRANCE
David	Cole	david.cole@physics.ox.ac.uk	David R Cole	Stamford	ROYAUME-UNI
Maria	Cordero	mjcorde@ari.uni-heidelberg.de	Astronomisches Rechen-Institut (Heidelberg University)	Heidelberg	ALLEMAGNE
Johanna	Coronado	coronado@mpia-hd.mpg.de	MPIA Heidelberg	Heidelberg	ALLEMAGNE
Philippe	Crane	philippe.crane@dartmouth.edu	Philippe Crane	Belmont	ETATS-UNIS
Orlagh	Creevey	ocreevey@oca.eu	Observatoire de la Côte d'Azur	NICE Cedex 4	FRANCE
Katia	Cunha	kcunha@on.br	NOAO	Tucson	ETATS-UNIS
Goran	Damljanovic	gdamljanovic@aob.rs	Goran Damljanovic	Belgrade	SERBIE
Patrick	de Laverny	laverny@oca.eu	Observatoire de la Côte d'Azur	NICE Cedex 4	FRANCE
Alis	Deason	alisdeason@gmail.com	Institute for Computational Cosmology, Department of Physics, Durham University,	Durham	ROYAUME-UNI
Marco	Delbo	marco.delbo@oca.eu	Observatoire de la Côte d'Azur	NICE Cedex 4	FRANCE
Elisa	Delgado Mena	elisa.delgado@astro.up.pt	Centro de Investigacao Astronomia/Astrofisica da Universidade do Porto	Porto	PORTUGAL
Michel	Dennefeld	dennefel@iap.fr	CNRS/IAP	Paris	FRANCE
Josselin	Desmars	josselin.desmars@obspm.fr	Observatoire de Meudon	Meudon	FRANCE
Paola	Di Matteo	paola.dimatteo@obspm.fr	Observatoire de Meudon	Meudon	FRANCE
Ronald	Drimmel	drimmel@oato.inaf.it	INAF - Osservatorio Astrofisico di Torino	Pino Torinese	ITALIE
Christine	Ducourant	christine.ducourant@u-bordeaux.fr	laboratoire d'Astrophysique de Bordeaux	Pessac Cedex	FRANCE
Josef	Durech	durech@sirrah.troja.mff.cuni.cz	Josef Durech	Prague	TCHEQUE, REPUBLIQUE
Iakov	Elyashev	eluashev-jak@yandex.ru	Iakov Elyashev	Moscow	RUSSIE, FEDERATION DE
Chris	Engelbrecht	engelbrecht.chris@gmail.com	University of Johannesburg	Johannesburg	AFRIQUE DU SUD
Ana	Escorza	ana.escorza@kuleuven.be	Ana Escorza	Leuven	BELGIQUE
Bacham	Eswar Reddy	ereddy@iiap.res.in	Indian institute of Astrophysics	bengaluru	INDE
Dafydd Wyn	Evans	dwe@ast.cam.ac.uk	Institute of Astronomy	Cambridge	ROYAUME-UNI
Wyn	Evans	nwe@ast.cam.ac.uk	Wyn Evans	Cambridge	ROYAUME-UNI
Laurent	Eyer	Laurent.Eyer@unige.ch	Geneva Observatory, University of Geneva	Versoix	SUISSE

First name	Last name	Email address	Organisation	Town/ City	Country
Andressa	Ferreira	andressa.ferreira@ astro.up.pt	Centro de Investigacao em Astronomia/Astrofisica da Universidade do Porto	Porto	PORTUGAL
Diane	Feuillet	feuillet@mpia.de	Diane Feuillet	Corvallis	ETATS-UNIS
Tobias	Fritz	tkf4w@virginia.edu	Tobias Fritz	Charlottesville	ETATS-UNIS
Alexandre	Gallenne	agallenn@eso.org	European Southern Observatory	Santiago de Chile	CHILI
Laurent	Galluccio	laurent.galluccio@ oca.eu	Observatoire de la Côte d'Azur	NICE Cedex 4	FRANCE
Shuang	Gao	sgao@bnu.edu.cn	Beijing Normal University	Beijing	CHINE
Cesar	Gattano	cesar.gattano@ obspm.fr	SYRTE - Observatoire de Paris	Paris	FRANCE
Ralph	Gaume	rgaume@nsf.gov	National Science Foundation	Arlington, VA	ETATS-UNIS
Stephan	Geier	geier@astro.uni-tuebingen.de	Institute for Astronomy and Astrophysics, University of Tuebingen	Tuebingen	ALLEMAGNE
Gerry	Gilmore	gil@ast.cam.ac.uk	Gerard Gilmore	Cambridge	ROYAUME-UNI
Ian	Glass	isg@saao.ac.za	SAAO	Cape Town	AFRIQUE DU SUD
Vira	Godunova	V_Godunova@ bigmir.net	Vira Godunova	Kiev	UKRAINE
Bertrand	Goldman	goldman@mpia.de	Max Planck Institute for Astronomie	Heidelberg	ALLEMAGNE
Naoteru	Gouda	naoteru.gouda@ nao.ac.jp	Naoteru Gouda	Mitaka, Tokyo	JAPON
Carl	Grillmair	carl@ipac.caltech.edu	Caltech/IPAC	Pasadena	ETATS-UNIS
Martin	Groenewegen	martin.groenewegen @oma.be	M. Groenewegen/ Royal Observatory of Belgium	Brussel	BELGIQUE
Guillaume	Guiglion	guillaume.guiglion@ oca.eu	Leibniz-Institut für Astrophysik Potsdam	Potsdam	ALLEMAGNE
Difeng	Guo	difengguo.astro@ gmail.com	Difeng Guo	Amsterdam	PAYS-BAS
Jincheng	Guo	jincheng.guo@ pku.edu.cn	Beijing University	Haidian district	CHINE
Jorrit	Hagen	hagen@astro.rug.nl	Jorrit Hagen	Hoogezand	PAYS-BAS
Kohei	Hattori	khattori@umich.edu	Kohei Hattori	Ann Arbor	ETATS-UNIS
Keith	Hawkins	khawkins@astro. columbia.edu	Keith Hawkins	New York	ETATS-UNIS
Michael	Hayden	mhayden@oca.eu	Observatoire de la Cote d'Azur	Nice cedex4	FRANCE
Misha	Haywood	misha.haywood@ obspm.fr	Misha Haywood	Meudon	FRANCE
Amina	Helmi	ahelmi@astro.rug.nl	Kapteyn Astronomical Institute	AV Groningen	PAYS-BAS
Jose	Hernandez	Jose.Hernandez@ esa.int	Jose Hernandez	Villanueva de la Canada, Madrid	ESPAGNE
Daniel	Hestroffer	hestro@imcce.fr	IMCCE - CNRS D. Hestroffer	PARIS	FRANCE
Vanessa	Hill	Vanessa.Hill@oca.eu	Laboratoire Lagrange, Observatoire de la Cote d'Azur	NICE Cedex 4	FRANCE
David	Hobbs	david@astro.lu.se	Lund Observatory	Lund	SUEDE
Erik	Hoeg	ehoeg@hotmail.dk	Erik Hoeg	Bagsvaerd	DANEMARK
Louise	Howes	louise@astro.lu.se	Louise Howes	Lund	SUEDE
Yang	Huang	yanghuang@ pku.edu.cn	Yang Huang	Beijing	CHINE
Christian	Hummel	chummel@eso.org	Christian Hummel	Munich	ALLEMAGNE
Jason	Hunt	jason.hunt@ utoronto.ca	Jason Hunt	TORONTO	CANADA
Anatoliy	Ivantsov	an.ivantsov@ gmail.com	Akdeniz University	Antalya	TURQUIE
Christopher	Jacobs	Chris.Jacobs@ jpl.nasa.gov	Christopher Jacobs	Pasadena	ETATS-UNIS
Jan	Janik	honza@ physics.muni.cz	Masaryk University	Brno	TCHEQUE, REPUBLIQUE
Fran	Jimenez-Esteban	fran@cab.inta-csic.es	Centro de Astrobiologia (INTA-CSIC)	Villanueva de la Canada	ESPAGNE

First name	Last name	Email address	Organisation	Town/ City	Country
Yipeng	Jing	ypjing@sjtu.edu.cn	Shanghai Jiao Tong University	Shanghai	CHINE
Kenneth	Johnston	ktjohnston11@verizon.net	Naval Observatory	Alexandria	ETATS-UNIS
Stefan	Jordan	jordan@ari.uni-heidelberg.de	ARI/ZAH, University Heidelberg	Heidelberg	ALLEMAGNE
Alain	Jorissen	alain.jorissen@ulb.ac.be	Université Libre de Bruxelles	Bruxelles	BELGIQUE
Yogesh	Joshi	yogesh@aries.res.in	Aryabhatta Research Institute of Observational Sciences (ARIES)	Nainital	INDE
Simon	Joyce	srgj1@le.ac.uk	University of Leicester	Leicester	ROYAUME-UNI
Kumamoto	Jun	j.kumamoto@astr.tohoku.ac.jp	T-corpo #201 4-16-12 Osawa	Tokyo	JAPON
Andreas	Just	just@ari.uni-heidelberg.de	Astron. Rechen-Institut	Heidelberg	ALLEMAGNE
Yucel	Kilic	yucel.kilic@linux.org.tr	Akdeniz University	Antalya	TURQUIE
Pierre	Kervella	pierre.kervella@obspm.fr	Observatoire de Paris	Meudon cedex	FRANCE
Sergei	Klioner	Sergei.Klioner@tu-dresden.de	Lohrmann-Observatorium Technische Universitaet Dresden	Dresden	ALLEMAGNE
Helmer	Koppelman	h.h.koppelman@rug.nl	H.H. Koppelman	Groningen	PAYS-BAS
Georges	Kordopatis	gkordo@oca.eu	Observatoire de la Côte d'Azur	Nice Cedex 4	FRANCE
Jean	Kovalesky	laverny@obs-nice fr	Patrick de Laverny - Observatoire de la Côte d'Azur	NICE Cedex 4	FRANCE
Alberto	Krone-Martins	algol@sim.ul.pt	Dept. de Física, Faculdade de Ciências, Universidade de Lisboa	Lisboa	PORTUGAL
Andrea	Kunder	amkunder@gmail.com	Leibniz Institut fur Astrophysik	Potsdam	ALLEMAGNE
Eduard	Kuznetsov	eduard.kuznetsov@urfu.ru	Ural Federal University	Yekaterinburg	RUSSIE, FEDERATION DE
Jacques	Lepine	jacques.lepine@iag.usp.br	Universidade de Sao Paulo	Sao Paulo	BRESIL
Nadege	Lagarde	nadege.lagarde@utinam.cnrs.fr	Observaoire de Besançon	Besançon	FRANCE
Valery	Lainey	lainey@imcce.fr	IMCCE	Paris	FRANCE
Rosine	Lallement	rosine.lallement@obspm.fr	Observaoire de Paris	Meudon	FRANCE
Olivier	LaMarle	Olivier.LaMarle@cnes.fr	CNES	paris	FRANCE
Uwe	Lammers	uwe.lammers@sciops.esa.int	ESAC	Villanueva de la Canada	ESPAGNE
Thierry	Lanz	thierry.lanz@oca.eu	OCA	Nice	FRANCE
Christophe	Le Poncin-Lafitte	christophe. lep-oncin@obspm.fr	Observatoire de PARIS-CNRS	Meudon CEDEX	FRANCE
Haining	Li	lhn@nao.cas.cn	National Astronomical Observatories, CAS	Beijing	CHINE
Jingjing	Li	jjli@pmo.ac.cn	Purple Mountain Observatory	Nanjing	CHINE
Shilong	Liao	liao@oato.inaf.it	INAF-Astronomical Observatory of Turin (OATo)	Turin	ITALIE
Chien-Cheng	Lin	cclin@shao.ac.cn	Chien-Cheng Lin	Changhua	TAIWAN, PROVINCE DE CHINE
Lennart	Lindegren	lennart@astro.lu.se	Lund Observatory	Lund	SUEDE
Jia-Cheng	Liu	jcliu@nju.edu.cn	Nanjing University	Nanjing	CHINE
Niu	Liu	liuniu@smail.nju.edu.cn	Nanjing University	Nanjing	CHINE
Xiaowei	Liu	x.liu@pku.edu.cn	Xiaowei Liu	Beijing	CHINE

First name	Last name	Email address	Organisation	Town/ City	Country
Yujuan	Liu	lyj@bao.ac.cn	National Astronomical Observatories, CAS	Beijing	CHINE
Eugene	Magnier	eugene@ifa. hawaii.edu	Eugene Magnier	Honolulu	ETATS-UNIS
Carlo Felice	Manara	cmanara@ cosmos.esa.int	Carlo Felice Manara	Valkenburg ZH	PAYS-BAS
Tommaso	Marchetti	marchetti@mail. strw.leidenuniv.nl	Leiden Observatory, Leiden University	Leiden	PAYS-BAS
Francisco J	Marco	marco@mat.uji.es	Universidad Jaume I (Francisco J Marco Castillo)	Castellon	ESPAGNE
Maria J	Martinez	mjmartin@ mat.upv.es	Maria J Martinez Uso	Valencia	ESPAGNE
Paul	McMillan	paul@astro.lu.se	Paul McMillan	Lund	SUEDE
Andrew	McWilliam	andy.ociw@ gmail.com	Andrew McWilliam	Castaic, California	ETATS-UNIS
Thibault	Merle	tmerle@ulb.ac.be	Thibault Merle	Bruxelles	BELGIQUE
Areg	Mickaelian	aregmick@ yahoo.com	Areg Mickaelian	Yerevan	ARMENIE
Francois	Mignard	francois.mignard@ oca.eu	Observatoire de la Côte d'Azur	NICE Cedex 4	FRANCE
Sarunas	Mikolaitis	Sarunas.Mikolaitis@ tfai.vu.lt	Sarunas Mikolaitis	Vilnius	LITUANIE
Ivan	Minchev	iminchev1@ gmail.com	Ivan Minchev	Berlin	ALLEMAGNE
Alexey	Mints	mints@mps.mpg.de	Max Planck Institute for Solar System Research	Goettingen	ALLEMAGNE
Felix	Mirabel	felix.mirabel@cea.fr	CEA/SACLAY	Gif-sur-Yvette	FRANCE
Tamara	Mishenina	tmishenina@ukr.net	Astronomical Observatory Odessa National University	Odessa	UKRAINE
Maria	Monguio	m.monguio@ herts.ac.uk	University of Hertfordshire	Hatfield	ROYAUME-UNI
David	Montes	dmontes@ucm.es	David Montes	Madrid	ESPAGNE
Alcione	Mora	alcione.mora@ esa.int	Aurora Technology BV	Lisse	PAYS-BAS
Bruno Eduardo	Morgado	brunomorgado@ on.br	Bruno Eduardo Morgado	Paris	FRANCE
Takahiro	Nagayama	nagayama@sci. kagoshima-u.ac.jp	Department of Physics and Astronomy, Graduate school of Science, Kagoshima University	Kagoshima	JAPON
Govind	Nandakumar	govind.nandakumar @oca.eu	Observatoire de la Côte d'Azur	NICE Cedex 4	FRANCE
Nicolas	Nardetto	Nicolas.Nardetto@ oca.eu	Observatoire de la Côte d'Azur	NICE Cedex 4	FRANCE
Chow-Choong	Ngeow	cngeow@astro. ncu.edu.tw	Graduate Institution of Astronomy, National Central University	Taoyuan City	TAIWAN
Birgitta	Nordstrom	birgitta@nbi.ku.dk	Niels Bohr Institute	Copenhagen	DANEMARK
Jurgen	Oberst	Juergen.Oberst@ dlr.de	Jurgen Oberst	Berlin	ALLEMAGNE
Karen	O'Flaherty	koflaher@esa.int	Karen O'Flaherty	Oegstgeest	PAYS-BAS
Go	Ogiya	Go.Ogiya@oca.eu	Observatoire de la Côte d'Azur	NICE Cedex 4	FRANCE
Christophe	Ordenovic	christophe. ordenovic@oca.eu	Observatoire de la Côte d'Azur	NICE Cedex 4	FRANCE
Rene	Oudmaijer	r.d.oudmaijer@ leeds.ac.uk	Rene D Oudmaijer	Leeds	ROYAUME-UNI
Thierry	Pauwels	thierry.pauwels@ oma.be	Thierry Pauwels	Brussel	BELGIQUE
Qingyu	Peng	tpengqy@jnu.edu.cn	Jinan University	Guangzhou	CHINE
Ruth	Peterson	peterson@ ucolick.org	Ruth Peterson	Palo Alto	ETATS-UNIS
Jean-Marc	Petit	Jean-Marc.Petit@ normalesup.org	Institut Utinam CNRS UMR6213	Besançon cedex	FRANCE

First name	Last name	Email address	Organisation	Town/ City	Country
Leonid	Petrov	Leonid.Petrov@ lpetrov.net	ADNET Systems Inc	Falls Church	ETATS-UNIS
Marc	Pinsonneault	pinsonneault.1@ osu.edu	Marc Pinsonneault	Columbus	ETATS-UNIS
Giampaolo	Piotto	giampaolo.piotto@ unipd.it	Padova Observatory	Padova	ITALIE
Eloisa	Poggio	poggio.eloisa@ gmail.com	Eloisa Poggio	Torino	ITALIE
Ennio	Poretti	ennio.poretti@ brera.inaf.it	Ennio Poretti – Osservatorio	Merate	ITALIE
Dimitri	Pourbaix	pourbaix@astro. ulb.ac.be	Université Libre de Bruxelles	Bruxelles	BELGIQUE
Paresh	Prema	paresh.prema@ ukho.gov.uk	UK Hydrographic Office	Taunton	ROYAUME-UNI
Timo	Prusti	tprusti@cosmos. esa.int	Timo Prusti	Oegstgeest	PAYS-BAS
Lucky	Puspitarini	lucky.puspitarini@ gmail.com	FMIPA ITB, Bosscha Observatory	Bandung	INDONESIE
Luis Henry	Quiroga Nunez	quiroganunez@ strw.leidenuniv.nl	Leiden Observatory	Leiden	PAYS-BAS
Altair	Ramos Gomes Junior	altair08@ astro.ufrj.br	Observatório do Valongo	Centro – Rio de Janeiro/RJ	BRESIL
Paola	Re Fiorentin	re_fiorentin@ oato.inaf.it	INAF - Osservatorio Astrofisico di Torino	Pino Torinese (TO)	ITALIE
Alejandra	Recio-Blanco	arecio@oca.eu	Observatoire de la Côte d'Azur	NICE Cedex 4	FRANCE
Nicole	Reindl	nr152@le.ac.uk	Nicole Reindl	Leicester	ROYAUME-UNI
Celine	Reyle	celine@obs-besancon.fr	Institut Utinam CNRS UMR6213	Besançon cedex	FRANCE
Sara	Rezaei-khoshbakht	sara@ mpia.de	Max Planck institute for Astronomy	Heidelberg	ALLEMAGNE
Robert	Rich	rmr@astro.ucla.edu	Robert Rich	Los Angeles	ETATS-UNIS
Vincent	Robert	vincent.robert@ obspm.fr	IMCCE / OBSPM	Paris	FRANCE
Annie	Robin	annie.robin@obs-besancon.fr	Institut Utinam CNRS UMR6213	Besançon cedex	FRANCE
Brigitte	Rocca-Volmerange	rocca@iap.fr	Institut d'Astro-physique de Paris	Paris	FRANCE
Tineke	Roegiers	troegiers@cosmos. esa.int	HE Space Operations B.V.	Noordwijk	PAYS-BAS
Alvaro	Rojas-Arriagada	alvaro.rojas. astrono-mia@gmail.com	Alvaro Rojas-Arriagada	Santiago	CHILI
Daniel	Rouan	daniel.rouan@ obspm.fr	LESIA – Observatoire de Paris	Meudon cedex	FRANCE
Laura	Ruiz Dern	laura.ruiz-dern@ obspm.fr	LAURA RUIZ DERN	Meudon	FRANCE
Delphine	Russeil	delphine.russeil@ lam.fr	LAM	Marseille	FRANCE
Johannes	Sahlmann	jsahlmann@ stsci.edu	Space Telescope Science Institute	Baltimore, MD	ETATS-UNIS
Nuno	Santos	nuno.santos@ astro.up.pt	Centro de Investigacao em Astrono-mia/Astrofisica da Universidade do Porto	Porto	PORTUGAL
Devesh P.	Sariya	deveshpath@ gmail.com	Devesh P. Sariya	Hsinchu	TAIWAN, PROVINCE DE CHINE
Mathias	Schultheis	mathias. schultheis@oca.eu	Observatoire de la Côte d'Azur	NICE Cedex 4	FRANCE
Branimir	Sesar	bsesar@mpia.de	Max Planck Institute for Astronomy	Heidelberg	ALLEMAGNE
Nataliia	Shakht	natalia.shakht@ yandex.ru	Nataliia Shakht	St Petersburg	RUSSIE, FEDERATION DE
Mahavir	Sharma	mahavir.44@ gmail.com	Durham University	Durham	ROYAUME-UNI

First name	Last name	Email address	Organisation	Town/ City	Country
Shreeya	Shetye	shreeyashetye15@gmail.com	Institute of Astronomy and Astrophysics, Université Libre de Bruxelles	Brussels	BELGIQUE
Bruno	Sicardy	bruno.sicardy@obspm.fr	Bruno Sicardy	Chambourcy	FRANCE
Hamish	Silverwood	hamish.silverwood@gmail.com	Hamish Silverwood	Christchurch	NOUVELLE-ZELANDE
Tatyana	Sitnova	sitnova@inasan.ru	INASAN	Moscow	RUSSIE, FEDERATION DE
Rodolfo	Smiljanic	rsmiljanic@camk.edu.pl	Nicolaus Copernicus Astronomical Center - NIP 525-000-89-56	Warsaw	POLOGNE
Keith	Smith	ksmith@science-int.co.uk	Keith Smith	Cambridge	ROYAUME-UNI
Martin	Smith	dr.mcsmith@me.com	The Shieling, Old Back Lane, Wiswell,	Clitheroe	ROYAUME-UNI
Tony	Sohn	tsohn@stsci.edu	Space Telescope Science Institute	Baltimore	ETATS-UNIS
Enrique	Solano	esm@cab.inta-csic.es	Instituto Nacional de Tecnica Aeroespacial (INTA)	Torrejon de Ardoz	ESPAGNE
Caroline	Soubiran	caroline.soubiran@u-bordeaux.fr	Laboratoire d'Astrophysique de Bordeaux	PESSAC	FRANCE
Jean	Souchay	Jean.Souchay@obspm.fr	Jean SOUCHAY	Paris	FRANCE
Alessandro	Spagna	spagna@oato.inaf.it	Osservatorio Astrofisico di Torino	Pino Torinese	ITALIE
Alain	Spang	alain.spang@oca.eu	Observatoire de la Côte d'Azur	NICE Cedex 4	FRANCE
Federica	Spoto	fspoto@oca.eu	Observatoire de la Côte d'Azur	Nice Cedex 4	FRANCE
Philippe	Stee	Philippe.Stee@oca.eu	Observatoire de la Côte d'Azur	NICE Cedex 4	FRANCE
Susan	Stewart	susan.stewart@usno.navy.mil	US Naval Observatory	Washington, DC	ETATS-UNIS
Nikolay	Stroilov	n.stroilov@gmail.com	Optico-physical department Space Research Institute of the Russian Academy of Sciences	Moscow	RUSSIE
Laszlo	Szabados	szabados@konkoly.hu	MTA CSFK	SOPRON	HONGRIE
Gustav Andreas	Tammann	G-A.Tammann@unibas.ch	Gustav Andreas Tammann	Basel	SUISSE
Paolo	Tanga	Paolo.Tanga@oca.eu	Observatoire de la Côte d'Azur	NICE Cedex 4	FRANCE
Francois	Taris	francois.taris@obspm.fr	Observatoire de Paris - SYRTE	Paris	FRANCE
Ramachrisna	Teixeira	rama.teixeira@iag.usp.br	Ramachrisna Teixeira	SÃ£o Paulo	BRESIL
William	Thuillot	William.Thuillot@obspm.fr	CNRS - SCTD 0195	Vandoeuvre-les-Nancy Cedex	FRANCE
Chris	Tinney	c.tinney@unsw.edu.au	UNSW Sydney	UNSW	AUSTRALIE
Patrick	Tisserand	tisserand@iap.fr	CNRS, Institut d'Astrophysique de Paris	Paris	FRANCE
Anastasia	Titarenko	atitaren@oca.eu	Observatoire de la Côte d'Azur	NICE Cedex 4	FRANCE
Santiago	Torres	storres@strw.leidenuniv.nl	Santiago Torres	NULL	PAYS-BAS
Boris	Trahin	boris.trahin@obspm.fr	LESIA Observatoire de Paris, Section de Meudon	Meudon CEDEX	FRANCE
Pier-Emmanuel	Tremblay	P-E.Tremblay@warwick.ac.uk	University of Warwick	Coventry	ROYAUME-UNI

First name	Last name	Email address	Organisation	Town/ City	Country
Wilma	Trick	trick@ mpia.de	Max-Planck-Institut fÃ1/4 r Astronomie	Heidelberg	ALLEMAGNE
Maria	Tsantaki	m.tsantaki@crya.unam.mx	Instituto de Radioastronomia y Astrofisica (UNAM)	Morelia	MEXIQUE
Takuji	Tsujimoto	taku.tsujimoto@nao.ac.jp	Takuji Tsujimoto	Tokyo	JAPON
Catherine	Turon	catherine.turon@obspm.fr	GEPI, Observatoire de Paris	Meudon	FRANCE
Gerard	van Belle	gerard@lowell.edu	Gerard van Belle	Flagstaff	ETATS-UNIS
Mathieu	Van der Swaelmen	mathieu.van.der.swaelmen@ulb.ac.be	Mathieu Van der Swaelmen	Bruxelles	BELGIQUE
Sophie	Van Eck	svaneck@astro.ulb.ac.be	Université Libre de Bruxelles	Bruxelles	BELGIQUE
Huib	van Langevelde	langevelde@jive.eu	JIVE ERIC	Dwingeloo	PAYS-BAS
Gerard	Vauclair	gerard.vauclair@irap.omp.eu	Gerard Vauclair	Toulouse	FRANCE
Sylvie	Vauclair	sylvie.vauclair@irap.omp.eu	Sylvie Vauclair	Toulouse	FRANCE
Elena	Vchkova Bebekovska	elenavckova@gmail.com	Elena Vchkova Bebekovska	Skopje	MACEDOINE, L'EX-REPUBLIQUE YOUGOSLAVE DE
Jaroslav	Velcovsky	375641@mail.muni.cz	Masaryk University	Brno	TCHEQUE, REPUBLIQUE
Anna	Velichko	astronomo@mail.ru	Anna Velichko	Kharkiv	UKRAINE
Jovan	Veljanoski	jovan@astro.rug.nl	Jovan Veljanoski	Groningen	PAYS-BAS
Miguel	Vioque	miguel.vioque@gmail.com	Miguel Vioque	Leeds	ROYAUME-UNI
David	Vokrouhlicky	vokrouhl@cesnet.cz	Institute of Astronomy, Charles University, Prague	Prague 8	TCHEQUE, REPUBLIQUE
Irina	Voloshina	voloshina.ira@gmail.com	Sternberg Astronomical Institute	Moscow	RUSSIE, FEDERATION DE
Joachim	Wambsganss	jkw@uni-hd.de	Zentrum fÃ1/4 r Astronomie der Universitaet Heidelberg	Heidelberg	ALLEMAGNE
Chun	Wang	wchun@pku.edu.cn	Department of Astronomy Peking University	Beijing	CHINE
Na	Wang	twangna@jnu.edu.cn	Jinan university	Guangzhou	CHINE
Shu	Wang	shuwang@pku.edu.cn	Peking University	Beijing	CHINE
Xiaoli	Wang	wangxl@ynao.ac.cn	Yunnan Observatories,CAS	Kunming	CHINE
Jennifer	Wojno	jwojno@aip.de	Leibniz Institute for Astrophysics Potsdam	Potsdam	ALLEMAGNE
Rosemary	Wyse	wyse@jhu.edu	Rosemary Wyse	Baltimore	ETATS-UNIS
Fang	Xia	xf@pmo.ac.cn	Fang Xia	Nanjing	CHINE
Maosheng	Xiang	msxiang@nao.cas.cn	Maosheng Xiang	Beijing	CHINE
Ye	Xu	xuye@pmo.ac.cn	Purple Mountain Observatory	Nanjing	CHINE
Yoshiyuki	Yamada	yamada@amesh.org	Kyoto University	Kyoto	JAPON
Taihei	Yano	yano.t@nao.ac.jp	National Astronomical Observatory of Japan	Tokyo	JAPON
Steffi	Yen	syen@lsw.uni-heidelberg.de	Landessternwarte, ZAH, University Heidelberg	Heidelberg	ALLEMAGNE
Bin	Yu	tlmrobin@163.com	Beijing Normal University	Beijing	CHINE
Haibo	Yuan	yuanhb@bnu.edu.cn	Beijing Normal University	Beijing	CHINE
Kutluay	Yuce	kyuce@ankara.edu.tr	Ankara University	Ankara	TURQUIE
Marion	Zacharias	ma_no@verizon.net	Marion Zacharias	Edgewater	ETATS-UNIS

First name	Last name	Email address	Organisation	Town/ City	Country
Norbert	Zacharias	nzIAUc8@gmail.com	Norbert Zacharias	Edgewater	ETATS-UNIS
Eleonora	Zari	zariem@strw.leidenuniv.nl	Leiden Observatory	Leiden	PAYS-BAS
Renata	Zenoviene	renata.zenoviene@tfai.vu.lt	Vilnius University	Vilnius	LITUANIE
Fupeng	Zhang	zhangfp7@mail.sysu.edu.cn	Sun Yat-Sen University	Guangzhou	CHINE
Huawei	Zhang	zhanghuawei100@hotmail.com	Huawei Zhang	Beijing	CHINE
Qingfeng	Zhang	tqfz@jnu.edu.cn	Jinan university	Guangzhou	CHINE
Xiliang	Zhang	zhangxiliang@ynao.ac.cn	Yunnan Observatories,CAS	Kunming	CHINE
Huo	Zhiying	zhiyinghuo@bao.ac.cn	Zhiying Huo	Beijing	CHINE
Stanislaw	Zola	szola@oa.uj.edu.pl	Jagiellonian University	Krakow	POLOGNE
Sven	Zschocke	sven.zschocke@tu-dresden.de	Technical University Dresden	Dresden	ALLEMAGNE

Astronomy and Astrophysics in the Gaia sky
Proceedings IAU Symposium No. 330, 2017
A. Recio-Blanco, P. de Laverny, A.G.A. Brown & T. Prusti, eds.

doi:10.1017/S1743921317006743

Tribute to François Mignard's research

Jean Kovalevsky

Observatoire de la Côte d'Azur and French National Academy of Sciences

It is a great pleasure for me to report about your colleague and my friend François Mignard. I know him since now 43 years and shared with him some scientific adventures. If, because of my age, I had to leave the Gaia project, I have followed the tremendous work done since by François Mignard and noticed that, despite his masterly played role as the conductor of the Gaia orchestra, he kept an important scientific activity. Actually, Gaia is far from being all his scientific achievements. And this is what I shall try to convince you.

In 1974, François Mignard graduated in Physics from the Ecole Normale Supérieure, a renown University level establishment in Paris which, in U.K. and USA would be called *College*. Then, he came to Nice with an assistantship at the University and decided to start his scientific career at the newly created CERGA in Grasse (Centre d'études et recherches géodynamiques et astronomiques). Progressing in his university third cycle curriculum, he had to make an original scientific research. I proposed him to work on the motion of a high eccentricity satellite and apply his results to the motion of Nereid. I thought that it would be a good preparation for a doctoral thesis, but the resulting paper astonishingly solved most of the difficulties. The tables derived from his theory were used by the international Ephemerides. This is how I discovered that François was an exceptionally gifted researcher.

He later proposed himself the subject of his doctoral thesis *Tidal phenomena in the evolution of planet-satellite system*, a very difficult and a new subject at that time,. The result was a remarkable piece of work whose application to the evolution of the lunar orbit was described in three articles in *The Moon and Planets* from 1979 to 1981. Using the same algorithms, he also studied the problem of the evolution of the motion of the satellites of Mars. He published a paper on the subject in the Monthly Notices. There he showed that the Laplace plane, used as an intermediary reference, allowed to understand the long-term evolution of the orbit of Phobos. He used a similar approach in the case of Pluto and Charon. In addition, in the frame of his duties in the observing service, he observed with an astrolabe, favouring the observation of planets (Mars and Jupiter).

In 1981, a new period of his activity started with a one year post-doc in Cornell University with J.A. Burns and one year professorship in Marseille. Then, he returned to CERGA. He was interested in many aspects of dynamical planetology, general relativity, data analysis, reference systems and frames. During the ten year period (1982-1992), he published on a large variety of subjects. Let us present some of them.

- He was particularly interested in radiation pressure and dust particle dynamics in the interplanetary medium as well as in the higher atmosphere with emphasis on LAGEOS which was then a major tool of geodesy.
- He studied the dynamics of grains in rings under radiation pressure force. Realizing that the direction of this force varies with time, he gave an integrable solution.
- He worked on the rotation of Hyperion, showing that its oblong shape, together with the high eccentricity of its orbit and the one to one resonance between the orbital and rotational periods, lead to a chaotic rotation.

• Studying the dynamics of the bodies in the Oort cloud, he found that if a star approaches the Solar system by less than one parsec, chances are that the perihelions of some of the bodies become sufficiently low, so that they may be captured by the major planets.

• He also got interested in the dynamics of binary asteroids, although there were no such example at that time.

• One of his constant interests was General Relativity to which the advent of space astrometry opened the way to new practical applications.

Most of these subjects were later, and some even until now, the object of original studies and publications or reports in conferences. This large variety of subjects, encompassing most of Astrometry and Dynamical Astronomy, to which one may add his great interest and knowledge in history of astronomy, made François Mignard a rare example of an astronomer with a wide knowledge and personal achievements. It is with such a background that he entered into space astrometry.

I dragged him into it in 1985, when I realized that some members of the FAST Consortium preparing the Hipparcos data reduction were not in a situation to perform some of the tasks in time. These were photometry and double star reduction. François was the only man in my environment I could trust to redress the situation, and he did it perfectly. He took up the tasks, directed people engaged, but actually did most of the work himself.

In photometry, he showed how to compare the variable amplitude of signals and to establish their eventual periodicities so as to recognize eclipsing variables, Cepheids, RR Lyr, etc.. While the HIPPARCOS mission was originally foreseen to be strictly astrometric, the work of François Mignard showed that each observation gave a mean uncertainty of 0.01 magnitude and the final calibrated combination presented an accuracy of 0.001mag. So, the resulting photometric catalogue and the light-curves of variable stars appeared to be a major improvement to earlier expectations.

In his approach to double stars, he faced a great variety of situations, making the work particularly difficult. He discovered a specific property of photon counts that opened the way to the resolution of binary systems. This led him, using all the observations accumulated during the mission, to obtain the relative astrometry (separation and direction) together with the magnitude difference. This work presented many difficulties, and F. Mignard wrote a very large and complex software that took into account all the possible situations. In this way, he treated 25,000 objects, discovering 10,000 new double stars out of which 70% astrometric data were obtained, and for 85% of them, the difference of magnitudes. Of course, the final catalogue takes into account the results obtained by both reduction consortia, but it is not exaggerating to state that François Mignard was the main contributor. Actually, he continued to work on the data, and extended the results, using ground-based astrometry, so as to determine the parallax of systems and obtain masses and absolute brightnesses of the components of some thirty double star systems.

Progressively he got interested in all the reduction procedure and became a major member of the Consortium FAST. For instance, the FAST software on astrometry had some difficulties with stars that happened to be far from their actual position in the Input Catalogue. François wrote a different and independent software that proved to be very useful in those difficult cases and, in addition, it contributed globally to the improvement of the solution. Among his involvement in other tasks contributing to the final catalogue, one may mention the astrometry of minor planets and the efforts to render the catalogue absolute.

In 1991, he became member of the Hipparcos Science Team and, later, he progressively replaced me as the coordinator of the FAST consortium. He became responsible of a working group of the Hipparcos Science Team in charge of constructing the final catalogue of double stars. In addition to the specific articles on photometry and double stars, the presence of his name in several publications describing the intermediate and final reduction procedures, as well as in the presentation of the final catalogue, prove his wide views of the mission. In addition, he was responsible of seven chapters of the volume describing the methods used in both consortia to construct the catalogue.

Once the Hipparcos catalogue was published, he naturally started to work on the astronomical implications of such rich and precise data. In particular, he made a very deep and detailed analysis of the local galactic kinematics using proper motions from the Hipparcos catalogues.

Let me say that Hipparcos was not the only activity of François Mignard during this time: he continued to work in the fields of atmospheric drag and radiation pressure and published several papers on these subjects. In addition, his long lasting interest in General Relativity led him to be associated to a proposal of an ensemble of four helio-synchronous satellites to test the principle of equivalence. This first proposal was not accepted by ESA. It was however presented again in an improved configuration and, under the new name MICROSCOPE, successfully launched last year and works now very satisfactorily.

Finally, let me mention that, when I retired from my administrative duties, he became director of CERGA in 1993 until 2004. However, at that time for François, Hipparcos was already the past. As early as 1993, together with a few colleagues, the concept of what could be Gaia was presented. Since then, he was advocating it and participating in various groups working on the feasibility and the expected science of this mission. In 1997, a first draft of a scientific proposal was presented by this group to ESA. Interested, ESA created a Scientific Advisory group that included François Mignard, while first technical studies, based on this report, were initiated by ESA.

At that time, NASA was also considering to enter space astrometry and François was called in 1996 to become a member of the Science Working Group on SIM, a space interferometer with a limited number of stars, but very precise even at high magnitudes. I mention this fact to emphasize that, already at that time, he gained a world-wide recognition as a scientist.

During the hectic period in the beginning of this century, during which, the actual future of Gaia was questioned, he strongly defended the scientific quality of the mission, as President of the ESA Gaia Data Analysis Coordination Committee. His major role in the preparation of the treatment of the data was recognized so that when, in 2006, the DPAC (Data Processing and Analysis Consortium) was created, he was naturally chosen as the chairman of its executive Committee for two three years mandates.

During this period ranging from 1997 to 2006, his scientific activities did not weaken but were essentially directed towards General Relativity and reference frames to which Gaia was expected to bring completely new results. Among these, he studied the relativistic effects of Jupiter, in the frame of classical as well as quantum gravity. More generally, he analysed the various aspects of fundamental physics in the framework of Gaia mission. He also examined the best way to attach the Gaia catalogue to an absolute reference frame, the objective being to achieve better than 0.01 microarcsecond per year. He was also interested in the science that could be obtained from the observations of asteroids by Gaia: for instance the determination of their masses or the eventual detection of the Yarkovsky effect in the motion of close to Earth asteroids.

His activity as chairman of DPAC was fantastic. He was present in all the general meetings of the consortium and many of the thematic ones. Actually, he was President of two of them (Solar system and Relativity and reference frames). In this way, he coordinated, and was the driving force of the whole group. This was not a simple task because on one side he had to be the voice of the scientific community within ESA and, on the other side, to ensure within DPAC that every one has a part of responsibilities in the most favourable conditions for the success of the mission. Later, he was still very much involved in the preparation and in the follow up of the launch. Most of you were witnesses of his formidable energy in this context. This hectic activity of François is, I am sure, partly responsible of the major health problems he suffered.

During this meeting, the first remarkable results of Gaia much better than I could expect at first will be presented. It appears to be only a glimpse in comparison with what François will present you at the end of the colloquium and for which he has had a major share. You will judge by yourselves.

So, I shall stop here and hope that I convinced you that François Mignard is one of the most gifted and hard working astronomer of our time.

The Gaia Sky

The Gaia Project Scientist, Timo Prusti.

Alejandra Recio-Blanco opening the IAU symposium 330, as chair of the SOC.

Astronomy and Astrophysics in the Gaia sky
Proceedings IAU Symposium No. 330, 2017
A. Recio-Blanco, P. de Laverny, A.G.A. Brown & T. Prusti, eds.

doi:10.1017/S1743921317005816

The Gaia mission status

T. Prusti

European Space Agency
ESTEC
Noordwijk
The Netherlands
email: tprusti@cosmos.esa.int

Abstract. Gaia is an ESA cornerstone mission conducting a full sky survey over its 5 year operational period. Gaia performs astrometric, photometric and spectroscopic measurements. The data processing is entrusted to scientists and engineers who have formed the Gaia Data Processing and Analysis Consortium (DPAC). The photometric science alerts started in 2014. The first intermediate data release (Gaia DR1) took place 14 September 2016 and it has been extensively used by the community. Gaia DR2 is scheduled for April 2018. Gaia is expected to be able to continue observations roughly for another 5 years after the nominal phase. The procedure to grant funding for the extension period has been initiated. In case funding is granted, the total operational time of Gaia may be 10 years.

Keywords. space vehicles, catalogs, surveys, astrometry, techniques: photometric, techniques: radial velocities

1. Introduction

Gaia is an ESA cornerstone mission building on the heritage of the Hipparcos mission as detailed in Gaia Collaboration *et al.* (2016b). Gaia covers the full sky making an astrometric, photometric, and spectroscopic survey. The spacecraft, including its payload, was built by industry with Airbus DS as the prime contractor. ESA has the overall management role as well as the spacecraft operations. The scientific community participation is through the Data Processing and Analysis Consortium (DPAC), which has been selected by ESA to produce the scientific catalogues for the community.

Gaia was launched 19 December 2013 from Kourou with a Soyuz rocket. Its nominal mission includes 5 years of operations and 3 years of post-operations to finalise the catalogues. The photometric science alerts, commenced 2014, were the first products provided to the community. The first intermediate release (Gaia DR1) took place 14 September 2016 (Gaia Collaboration *et al.* 2016a). The Solar system alerts for new asteroids also started 2016. The next data release, Gaia DR2, is planned for April 2018.

2. Operations

After a commissioning period lasting about half a year, Gaia has been conducting routine observations. On average Gaia detects and measures 70 million objects per day. On top days, when the scanning is parallel to the Galactic Plane, the count can exceed 300 million. In total, at the moment of this Symposium (24 April 2017), Gaia has observed 70 billion transits. Overall the Gaia operations are nominal.

Astrometry. By 24 April 2017 Gaia has made 688 billion astrometric measurements. Gaia has an automatic on-board detection to decide whether an object is celestial, point-like and brighter than 20.7 mag. The faint limit is not precise as the on-board magnitude

Figure 1. Gaia spacecraft. Copyright ESA

Figure 2. Gaia payload module at the time of vibration testing. Copyright Airbus DS

estimate for the faintest stars is of the order of 0.3 mag. The bright limit of Gaia is between 2 and 3 magnitudes. Brighter objects cause such a large saturation area on the detector, that Gaia cannot anymore determine the object as point like. The bright limit is not fixed as there are sensitivity variations across the CCDs and the brighter stars can be seen if they pass through a less sensitive part of a detector. In order to achieve full completeness, special observation are scheduled for the very brightest stars.

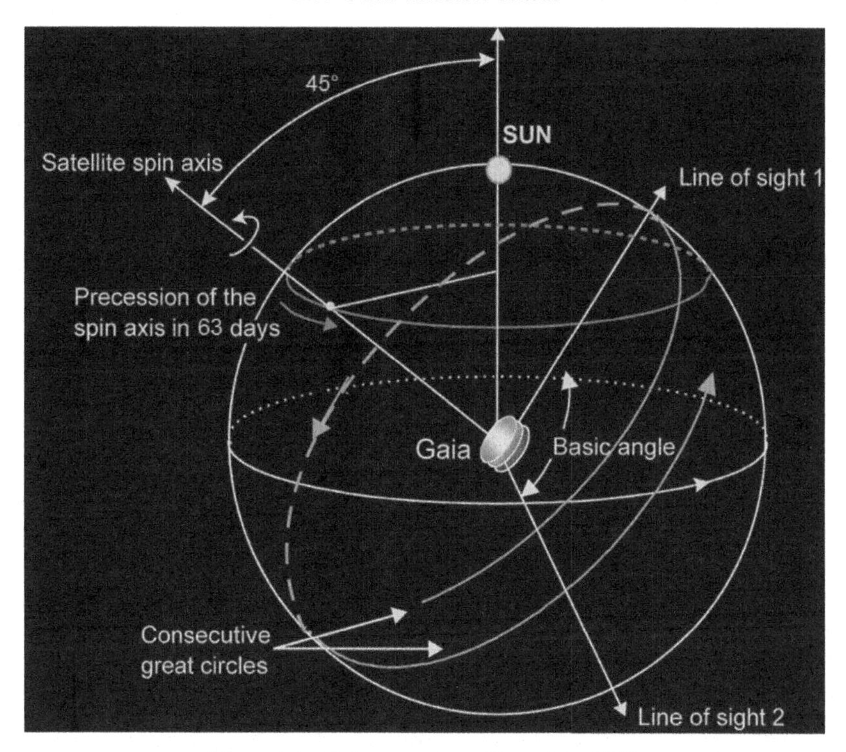

Figure 3. Gaia scanning. Copyright ESA

The ultimate accuracy of those measurements is not yet known, but will be of lower quality than measurements made in the regular way. In addition, selected dense regions are imaged with a special Gaia mode. This is done to compensate for the fact that in the most crowded parts of the sky, the Gaia on-board resources cannot cope with the stellar densities. By producing the images, some astrometric information can be gathered also for the faintest objects in these regions.

Photometry. All astrometric measurements of Gaia are also photometrically calibrated. Therefore for every astrometric measurements there is a corresponding photometric measurement. In addition Gaia has dedicated prisms and CCDs to record spectrophotometry of all objects detected by Gaia. The spectrophotometry is split into two ranges. BP covers wavelengths from 330 to 680 nm and RP from 640 to 1050 nm. The number of spectrophotometric measurements by 24 April 2017 is 147 billion. While the dispersed spectra can be used as narrow-band photometry, the BP and RP channels can also be integrated to provide more precise magnitudes for the wavelength ranges covered. Although every astrometric measurement has accompanying spectrophotometric measurement, it is important to note that in crowded regions, where the spectra overlap, deblending methods are needed to disentangle close-by sources from each other. In very crowded regions this leads to lower quality spectrophotometry than in nominal source density areas.

Spectroscopy. The spectrometer on-board Gaia is called Radial Velocity Spectrometer (RVS). The main task is to enable deducing radial velocities. The spectra cover wavelength range from 845 to 872 nm with resolution of the order of 11,000. In the on-board detection algorithm the limiting magnitude for gathering RVS spectrum is set to 16.2 in the RVS wavelength range. By 24 April 2017 about 13.7 billion spectra have been collected. For most of the objects the only astrophysical quantity that can be deduced is the radial velocity. However, for stars brighter than about 11 to 14 magnitudes also

Figure 4. Gaia focal plane assembly. Groups from left to right: 12 RVS CCDs, 7 RP CCDs, 7 BP CCDs, 62 astrometric CCDs and 14 sky mapper CCDs. The two most extreme CCDs in the corner are for the Basic Angle Monitor and the one out of the row of astrometric CCDs together with the one most to the right in the middle of the focal plane are for wavefront censors (the wavefront censor is missing still in this picture). Copyright Airbus DS

spectroscopy can be done (the brighter the star, the better the signal to noise and the more spectroscopic analyses are possible).

3. Intermediate data releases

Gaia DR1 took place 14 September 2016. The number of publications went very quickly up and many of the published (or close to publication) results can be found from these proceedings.

Gaia DR2 is scheduled for April 2018. The expected contents and precisions are detailed to some level in these proceedings and especially the scientific expectations are listed in many contributions.

4. Mission extension

The nominal Gaia mission will be completed by mid-2019. At that moment Gaia has been scanning the sky for 5 years. The operations are smooth and can continue beyond the nominal life time. If nothing unexpected happens, then the Gaia mission life time will be limited by the consumable, cold gas, for the micro propulsion system. The cold gas is used to keep the spin matching exactly the readout speed of the CCDs. The consumption is very regular, but with an expectation, that the consumption will increase as the satellite surfaces age and the Solar wind causes more torque on Gaia. The current best estimate of cold gas exhaustion is mid-2024. After cold gas exhaustion the spin control is not anymore sufficient for precision astrometry work.

In order to continue operations beyond the nominal end of mission to the real, functional end of the mission, additional funding is required for the 5 years from mid-2019 till mid-2024. As mission extensions are not unusual for ESA spacecrafts, there is a procedure that is followed up within the science programme. Every two years there is a decision made whether to continue or to stop. In addition a forward look is done for years 3 and 4 ahead. As Gaia is in the nominal phase till mid-2019, at this stage only the issue of forward look to the second half of 2019 and 2020 has been on the table. Extension decision for the second half of 2019 and 2020 will take place toward the end of 2018 in combination with a forward look to 2021 and 2022.

The nominal 5 year mission life time is has played a fundamental role in Gaia design. Various technical and operational aspects are tuned for completing a homogeneous survey in 5 years. After mid-2019, the end of the nominal mission, the estimated end of life in mid-2024 results to another 5 year operational period. Given the technical and operational characteristics of Gaia, the anticipated 5 year extended life time and the survey nature of the mission, it was decided to prepare a science case for an extra 5 year period although administratively the extensions to missions are funded in the above mentioned two year cycles.

It is clear that more Gaia data, 10 years in total instead of the nominal 5 years, will give better science. However, in some fields the gain is more impressive than in others. In all Gaia areas, astrometry, photometry and spectroscopy, the basic performance for bright objects is limited by calibration accuracy, while for the faint objects signal to noise increase will improve the deduced astronomical parameters. As most Gaia sources are faint (due to faint sources outnumbering the bright ones), the gain from an extension affects the majority of Gaia detected objects. For bright stars the gains are in more complex systems. In astrometry these are related to kinematics and dynamics while in photometry and spectroscopy variable objects will benefit.

In this contribution two case studies with significant science improvement due to Gaia extension are presented. These examples concern exoplanets and reference frames.

Exoplanets. Gaia is able to detect exoplanets both by photometry and astrometry. It is specifically the astrometric method, which is of special interest as Gaia is more or less the only facility able to provide this complementary information. Perryman *et al.* (2014) estimated that based on data from the nominal mission, Gaia will detect about 20,000 exoplanets. By extending the operational period to 10 years will result to 70,000 exoplanet detections. The method is the same as for unresolved double stars where a wobble of the stellar light centroid around its joint, with the exoplanet, barycentre can be detected on top of the system parallax and proper motion. Clearly this is easier for massive planet detection as the wobble is larger and Gaia is mostly detecting 'Jupiters'. Nevertheless, the huge advantage is that with astrometry also the inclination of the system may be determined. This allows exoplanet mass determination without the sin(inclination) ambiguity. A longer Gaia life time not only allows more exoplanets to be detected, but specifically Jupiter mass objects further away from their host star in orbits longer than 5 years. If we believe that a giant planet in distant orbit is useful in 'guarding' possible exoplanets closer by a star, then Gaia is the way to find these systems.

Reference frame. One of the important results from Gaia will be the establishment of an accurate and dense optical reference frame. This can also be aligned with the radio reference frame based on quasars as Gaia is able to detect a sufficient number of quasars. The advantage of a dense optical reference frame is that practically all astronomical images contain Gaia objects in the field, allowing good derivation of the astrometry in that field. However, the density of objects is achieved with stars and stars have proper motions with errors assigned to this quantity. This means that in the course of time the optical reference frame will degrade. The strength of Gaia is that the proper motion errors are typically very small and therefore the Gaia defined optical reference frame will be sufficiently accurate for decades. Nevertheless, the evolution in e.g. ground based facilities continues with the extremely large telescopes. For some instruments on the future giant telescopes the field of view can be extremely small. Therefore a very accurate optical reference frame is a necessity. Gaia with a 10 year operational period can provide this, as the proper motion errors scale with time to exponent 1.5 giving a factor of 2.8 better astrometry with respect to 5 year nominal mission. With a 10 year Gaia all extremely large telescopes can count on having a sufficiently accurate optical reference frame at least till 2050.

5. Conclusions

Gaia mission operations are nominal and on average 70 million transits are recorded daily. Gaia DR1 has triggered studies in many areas of astronomy and raised the expectations toward Gaia DR2. We can conclude that Gaia is well on its way to fulfil 'The Promise of Gaia'. The best way to enjoy that and to get ready for Gaia DR2 is — to use Gaia DR1.

Acknowledgement

This work has made use of data from the European Space Agency (ESA) mission *Gaia* (https://www.cosmos.esa.int/gaia), processed by the *Gaia* Data Processing and Analysis Consortium (DPAC, https://www.cosmos.esa.int/web/gaia/dpac/consortium). Funding for the DPAC has been provided by national institutions, in particular the institutions participating in the *Gaia* Multilateral Agreement.

References

Gaia Collaboration, Brown, A. G. A., Vallenari, A., *et al.* 2016, *A&A*, 595, A2
Gaia Collaboration, Prusti, T., de Bruijne, J. H. J., *et al.* 2016, *A&A*, 595, A1
Perryman, M., Hartman, J., Bakos, G.Á., & Lindegren, L., 2014, *ApJ*, 797, 14

Astronomy and Astrophysics in the Gaia sky
Proceedings IAU Symposium No. 330, 2017
A. Recio-Blanco, P. de Laverny, A.G.A. Brown & T. Prusti, eds.

doi:10.1017/S1743921317006664

The Gaia sky: version 1.0

Anthony G. A. Brown

Sterrewacht Leiden, Leiden University, Niels Bohrweg 2, 2333 CA, Leiden, Netherlands
email: brown@strw.leidenuniv.nl

Abstract. In this contribution I provide a brief summary of the contents of *Gaia* DR1. This is followed by a discussion of studies in the literature that attempt to characterize the quality of the *Tycho-Gaia* Astrometric Solution parallaxes in *Gaia* DR1, and I point out a misconception about the handling of the known systematic errors in the *Gaia* DR1 parallaxes. I highlight some of the more unexpected uses of the *Gaia* DR1 data and close with a look ahead at the next *Gaia* data releases, with *Gaia* DR2 coming up in April 2018.

Keywords. catalogs, surveys, astrometry

---

1. Overview of *Gaia* DR1

With the announcement on September 14 2016 of the first data release from the ESA *Gaia* mission (*Gaia* DR1) the astronomical community truly entered the *Gaia* era. This data release is the culmination of over 10 years of effort by ESA and the members of the *Gaia* Data Processing and Analysis Consortium (DPAC), the community of European astronomers responsible for the data processing for the *Gaia* mission. The *Gaia* satellite was launched in December 2013 to collect data that will allow the determination of highly accurate positions, parallaxes, and proper motions for over one billion sources brighter than magnitude $G = 20.7$ in the *Gaia* white-light photometric band. The astrometry is complemented by multi-colour photometry, measured for all sources observed by *Gaia*, and radial velocities which are collected for stars brighter than $G_{\rm RVS} \sim 16$ in the pass-band of Gaia's radial velocity spectrograph. The scientific goals of the mission and the scientific instruments on board *Gaia* are summarised in Gaia Collaboration, *et al.* (2016a). The raw data collected during the first 14 months of the mission were processed by the DPAC, involving some 450 astronomers and IT specialists, and turned into the first version of the *Gaia* catalogue of the sky (Gaia Collaboration, *et al.* 2016).

The bulk of *Gaia* DR1 consists of celestial positions (α, δ) and G-band magnitudes for about 1.1 billion sources. The distribution of the *Gaia* DR1 sources in magnitude is show in Fig. 1. With median positional accuracies of 2.3 milli-arcsec (mas) and a spatial resolution comparable to the Hubble Space Telescope, the *Gaia* DR1 catalogue represents the most accurate map of the sky to date, including the most precise and homogeneous all-sky photometry, ranging from milli-magnitude uncertainty at the bright end of the *Gaia* survey to 0.03 magnitude uncertainty at the faint end. In addition the combination of *Gaia* data and the positions from the *Hipparcos* and *Tycho-2* catalogues allowed the derivation of highly precise proper motions and parallaxes for the 2 million brightest sources in *Gaia* DR1 (the so-called *Tycho-Gaia* Astrometric Solution or TGAS; Michalik *et al.* 2015, Lindegren *et al.* 2016). The typical parallax uncertainty is 0.3 mas, while the proper motion uncertainties are about 1 mas yr^{-1} for the stars from the *Tycho-2* catalogue and as small as 0.06 mas yr^{-1} for the stars from the *Hipparcos* catalogue. *Gaia* DR1 in addition contains light curves and variable type classifications for a modest sample of some 2600 RR Lyrae and 600 Cepheid variables (Clementini *et al.* 2016) as

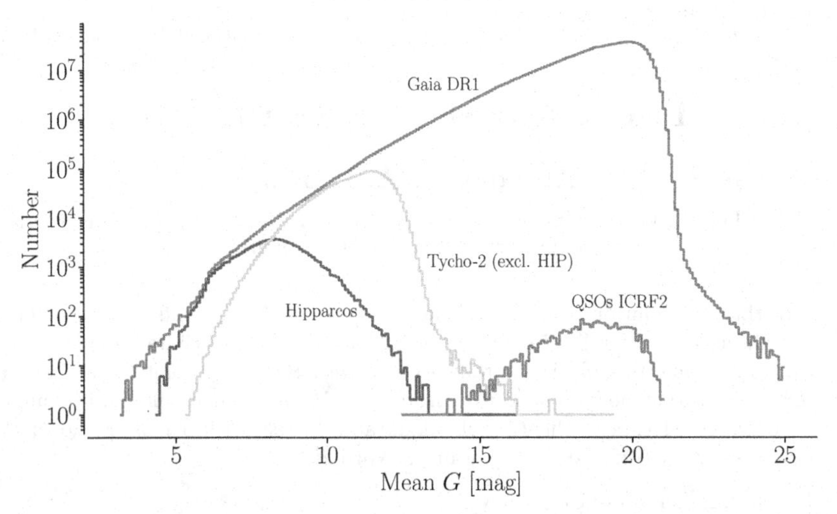

Figure 1. Magnitude distribution of the sources in the *Gaia* DR1 catalogue. The TGAS magnitude distribution is split into the *Hipparcos* stars and the stars from the *Tycho-2* catalogue (excluding *Hipparcos*). The distribution of the magnitude of the $\sim$ 2000 ICRF2 QSOs is also shown separately.

well as the optical positions of about 2000 ICRF2 sources (Mignard *et al.* 2016). More details can be found in Gaia Collaboration, *et al.* (2016).

2. On the quality of the TGAS parallaxes

Since the publication of *Gaia* DR1 various papers have treated the topic of the quality of the TGAS parallaxes, in particular focusing on the possibility of systematic offsets with respect to independent parallax measurements or distance estimates. The *Gaia* DR1 catalogue validation done by DPAC (Arenou *et al.* 2017) confirms the estimate by Lindegren *et al.* (2016) that the global parallax zero-point for TGAS is at the ± 0.1 mas level (an average of ~ -0.04 mas for various comparison samples was found, see table 2 in Arenou *et al.* 2017). The validation effort also confirmed the conclusion by Lindegren *et al.* (2016) that locally additional offsets at the ± 0.2 mas level can exist (see for example figure 24 in Arenou *et al.* 2017). The latter are spatially correlated and colour-dependent. Analyses of the period-luminosity relations of local Cepheids and RR Lyrae revealed no global offset in the TGAS parallaxes to $\sim$0.02 and $\sim$0.05 mas precision, respectively (Casertano *et al.* 2016, Sesar *et al.* 2017). In contrast Jao *et al.* (2016) and Stassun & Torres (2016) find rather large systematic offsets of 0.24 and 0.25 mas, where the Gaia parallaxes are too small by these amounts compared to independently measured or predicted parallaxes for a sample of nearby stars and a sample of eclipsing binaries, respectively.

De Ridder *et al.* (2016) analyzed a sample of nearby (parallaxes larger than $\sim$5 mas) dwarfs and sub-giants as well as a sample of more distant red giants (parallaxes less than $\sim$3 mas), for which asteroseismic distances are available based on data from the Kepler mission (Borucki *et al.* 2010). In both cases the asteroseismically derived distances were converted to parallaxes and a linear relationship, $\varpi_{\rm predicted} = \alpha + \beta\varpi_{\rm TGAS}$, was fit between these parallaxes and the TGAS values. For the dwarfs and sub-giants the one-to-one relation between the parallax values, with a zero offset, was found to be a plausible model, while for the red giants De Ridder *et al.* (2016) find $\alpha \sim 0.3$ mas and $\beta \sim 0.75$, implying that the TGAS parallaxes are too small. Davies *et al.* (2017) used a sample of Red Clump (RC) stars in the Kepler field to assess the TGAS parallaxes. They predicted the parallaxes from the apparent magnitudes of the RC stars and the assumed mean absolute magnitude for the RC, using the K_s band. Fitting a linear relation they find $\alpha \sim 0.24$ and $\beta \sim 1.64$, implying that the TGAS parallaxes are too large. Finally, Huber *et al.* (2017) present an analysis of 2200 stars in the Kepler field, from the main sequence to the giant branch. They conclude that the offsets between the TGAS parallaxes and distances derived from the properties of eclipsing binaries and asteroseismic samples have been overestimated for parallaxes in the 5–10 mas range and find a significantly smaller deviation than De Ridder *et al.* (2016) for smaller parallaxes. They also find that the remaining differences can be partially compensated by adopting a hotter $T_{\rm eff}$ scale, leaving differences between TGAS parallaxes and asteroseismic distances at the $\sim$2 per cent level.

These apparently contradictory conclusions on the differences between TGAS parallaxes and independent distance indicators merit a couple of remarks:

- Spurious differences between two sets of parallax measurements can occur due to the effects of truncating the sample on the value of the parallax or that of the relative parallax uncertainty. For example, when comparing parallax measurements that differ in precision and truncating the sample on the value of the lower precision parallaxes it can be shown (assuming no systematic errors in either sample and Gaussian errors) that for a case similar to the Jao *et al.* (2016) study (using only parallaxes larger than 40 mas in the comparison drawn from an underlying sample reaching to 25 mas in parallax) the mean difference in the parallaxes is expected to be ~ -0.1 mas, in the sense of high minus low precision parallax values. This is of the order of the offsets claimed and can be understood when considering that truncation on the low precision parallaxes combined with the steep increase in stars toward smaller true parallaxes, leads to an excess of stars in the low precision sample having overestimated parallaxes. It should be stressed that Jao *et al.* (2016) are aware of this issue and have made the comparison for various subsamples and also by truncating on the value of the high precision (TGAS) parallaxes. The differences they find cannot readily be explained away by the sample truncation effect only.
- Similarly a truncation of the sample on apparent magnitude can introduce a spurious difference between two sets of parallax estimates. This issue could play a role in the studies that use standard candles, such as RC stars, to estimate parallaxes from the apparent magnitude. In the study of Davies *et al.* (2017) many of the RC stars have apparent magnitudes around the completeness limit of TGAS (this is also pointed out in the work by Gontcharov & Mosenkov, 2017). This will lead to favouring the intrinsically brighter RC stars for which the parallaxes will be underestimated when using the mean RC absolute magnitude to calculate the parallax. This specific effect is probably not very important in the Davies *et al.* (2017) study if the intrinsic spread in RC star absolute magnitudes is as small as derived in Hawkins *et al.* (2017). In general the properties of a selected sample, the properties of the parent population it is drawn from, as well as the survey selection functions (in both TGAS and other surveys the samples are drawn from) need to be well understood in order to properly interpret any differences between different sets of parallaxes.
- Offsets between TGAS and other parallax estimates that increase with the size of the parallax could be indicative of a scaling error in the non-astrometric parallaxes. Silva Aguirre *et al.* (2017), Huber *et al.* (2017), and Yildiz *et al.* (2017) point out the possibility that the $T_{\rm eff}$ scale used in the calculation of asteroseismic distances (which scale as $T_{\rm eff}^{2.5}$) could partly explain the offsets between TGAS and their distances estimates. Likewise an underestimate of the absolute magnitude in the case standard candles are used could explain the trend seen in the study of Davies *et al.* (2017), where the ratio of estimated to

true parallax scales as $10^{0.2\Delta M}$, with ΔM the difference between the estimated and true absolute magnitude. However $\Delta M \sim 1$ is required to explain the effect found by Davies *et al.* (2017), and it is highly unlikely that the existing Red Clump absolute magnitude calibrations are in error by that amount. An overestimate of the extinction has a similar effect but again an implausibly large overestimate of the extinction toward the Kepler field would be required.

• Comparisons of TGAS parallaxes to independent parallax measurements or to parallaxes estimated by other means should always consider that systematic effects can occur in either set of parallaxes. Independently estimated distances that are converted to parallaxes (using $\varpi = 1/d$) will suffer from the same non-linear transformation problems as when calculating distances as $1/\varpi$. Depending on the relative distance error, the error on the predicted parallax will be highly non-Gaussian and the mean can be biased away from the true value.

• The actual properties of the parallax uncertainties in both TGAS and the independent parallax measurements/estimates play an important role. All studies above implicitly assume normally distributed errors for which the values are correct. However Lindegren *et al.* (2016) show that the distribution of the normalized TGAS-Hipparcos parallax differences contains exponential tails, while its width hints at uncertainties being underestimated in one or both data sets. The differences between TGAS and previously published trigonometric parallaxes are modelled as Lorentzians by Jao *et al.* (2016) in order to accommodate for extended wings, again hinting at partly non-Gaussian errors in the parallaxes of one or both samples. The parallaxes estimated from astrophysical properties of a particular sample of stars can likewise suffer from non-Gaussianity and/or under- or overestimation of the errors. Errors that deviate significantly from normal behaviour will amplify the effects discussed above (sample truncation, scale errors in distance estimators).

• A number of studies rely on stars from the Kepler field only and extrapolating the results to the entire TGAS catalogue is a dubious undertaking. The special validation solutions for TGAS discussed in appendix E of Lindegren *et al.* (2016), as well as the QSO analysis in Arenou *et al.* (2017) show that there are regional systematics on the sky. Stars distributed over the Kepler field may well suffer from similar parallax systematics as suggested by figure E.1 in Lindegren *et al.* (2016). However the offsets between TGAS and independent parallax estimates for the Kepler field should not blindly be extrapolated to other regions on the sky.

The studies of TGAS parallaxes carried out so far have mostly not or only partly addressed the above issues. This makes it difficult to come to clear interpretation of the offsets seen between TGAS parallaxes and alternative parallax measurements or estimates. I conclude that there is no strong reason to revise the estimates of the level of systematic errors in TGAS parallaxes as presented in Lindegren *et al.* (2016) and Arenou *et al.* (2017).

2.1. *TGAS parallax uncertainties, treatment of (spatially correlated) systematics*

As described in Lindegren *et al.* (2016) an inflation factor was applied to the formal uncertainties on the astrometric parameters (as determined in the astrometric data processing) to arrive at the uncertainties quoted in the *Gaia* DR1 catalogue. This inflation factor was derived from a comparison between the TGAS and the Hipparcos parallaxes. There are indications in various studies that the inflated parallax uncertainties may be overestimated at the 10–20% level (Casertano *et al.* 2016, Gould *et al.* 2016, Sesar *et al.* 2017). If desired the inflation factor applied in Lindegren *et al.* (2016) can be undone (see their section 4.1) but then this factor should be re-estimated as part of the data analysis (see Sesar *et al.* 2017, for an example).

Since the publication of *Gaia* DR1 there has been some confusion in the astronomical community (also reflected in the literature) as to how to deal with the systematic uncertainties known to be present in the TGAS astrometry, in particular for the parallaxes. This was partly caused by a misleading statement in the paper describing *Gaia* DR1 (Gaia Collaboration *et al.* 2016b) in which it is recommended to 'consider the quoted uncertainties on the parallaxes as $\varpi \pm \sigma_\varpi$ (random) ± 0.3 mas (systematic)'. This creates the impression that the typical 0.3 mas systematic uncertainty should be added in quadrature to the uncertainty quoted in the *Gaia* DR1 catalogue. It should be stressed here that this *should not be done.* The reason is that the calibration of the TGAS parallax uncertainties by comparison to the Hipparcos parallaxes automatically leads to the inclusion of the local systematics in the quoted uncertainty. There is no simple recipe to account for the systematic uncertainties. The advice is to proceed with one's analysis of the *Gaia* DR1 data using the uncertainties quoted in the catalogue, but to keep the systematics in mind when interpreting the results of the data analysis.

As illustrated in Arenou *et al.* (2017) and Lindegren *et al.* (2016) the systematic uncertainties on the parallaxes vary over the sky and are spatially correlated in the sense that the systematics over small patches of the sky tend to be in the same direction. No attempt was made during the *Gaia* DR1 processing and validation to derive a correlation length scale. Zinn *et al.* (2017) made use of the precise asteroseismic distances for stars in the Kepler field to calibrate the spatial correlation length scale of systematic parallax uncertainties in *Gaia* DR1. They also provide a model of the spatial correlations that can be used to construct a covariance matrix for data analyses that involve TGAS parallaxes. It is not obvious that this finding for the Kepler field holds for the entire sky, but it does provide a useful estimate of the local correlations and perhaps the appropriate values of the model parameters for other sky regions can be estimated as part of the data analysis.

3. Science from *Gaia* DR1

Notwithstanding the complexity of dealing with its error characteristics, the scientific exploitation of the first *Gaia* data release has been taken up enthusiastically by the world-wide astronomical community, as evidenced by the numerous workshops organized to collectively work on the analysis of *Gaia* data†, and the over 300 papers that have appeared in the literature since September 14 2016 which are based on or make use of the *Gaia* DR1 data.

The Gaia Collaboration has published two performance verification papers that provide a new inventory of the nearby open clusters (Gaia Collaboration *et al.* 2017a), and a test of the TGAS parallaxes through a thorough study of the local Cepheid and RR Lyrae populations (Gaia Collaboration *et al.* 2017b), where the K-band period luminosity relations show a substantial improvement in the TGAS parallaxes compared to the *Hipparcos* values. I highlight below a few of the more creative and unexpected analyses of the *Gaia* DR1 data.

Mapping the structure of the Magellanic clouds. Belokurov *et al.* (2017) describe a very clever method for tracking down variable stars in *Gaia* DR1, even though their light curves have not been published and no explicit indication is included on the possible variability of catalogue sources (keep in mind that light curves and variable star charac-

† For example the 'Gaia Sprints' (`http://gaia.lol/`), and the Gaia 2016 Data Workshop (`https://www.cosmos.esa.int/web/gaia-2016-data-workshop/home`).

terizations were included in *Gaia* DR1 only for a very modest sample of 2600 RR Lyrae and 600 Cepheid variables, see Clementini *et al.* 2016). Belokurov *et al.* (2017) make use of the fact that the photometric uncertainties quoted in the *Gaia* DR1 catalogue reflect the scatter in the individual observations made for each source. This leads to overestimates of the uncertainty on the mean G band value for variable sources, making these stars stand out in a diagram of the uncertainty in G vs. the value of G. By calibrating against samples of known variable stars Belokurov *et al.* (2017) were able to identify candidate RR Lyrae stars in a field covering the Magellanic clouds. These candidate RR Lyrae beautifully outline the LMC and SMC and in particular reveal the bridge of old stars between the two Milky Way companions. A combination of *Gaia* DR1 with *GALEX* (Bianchi *et al.* 2014) data revealed the existence of a bridge of younger stars, offset from the bridge of old stars and coincident with the known HI bridge between the LMC and SMC. The technique to find variable stars in *Gaia* DR1 was also applied by Deason *et al.* (2017a) in order to map the structure of the Magellanic system through Mira variables.

A cluster hiding near Sirius. The power of a high spatial resolution, high dynamic range, all-sky star map was demonstrated nicely in the paper by Koposov *et al.* (2017). Although *Gaia* observations of bright sources suffer from CCD saturation effects, there is no need to avoid the vicinity of even the brightest stars on the sky and hence *Gaia* can observe sources very near such stars. Koposov *et al.* (2017) made use of this by creating an all-sky map of potential source over-densities and in that way discovered a hitherto unknown star cluster very near the brightest star in the sky, Sirius. The reality of the *Gaia* 1 cluster was confirmed by combining the *Gaia* DR1 information with the photometry form the *2MASS* (Skrutskie *et al.* 2006), *WISE* (Wright *et al.* 2010), and Pan-Starrs1 (Chambers *et al.* 2016) surveys. Simpson *et al.* (2017) carried out spectroscopic follow-up observations and concluded that *Gaia* 1 is an intermediate age (~3 Gyr) open cluster with a mass of roughly $10^4\ M_\odot$.

De-noising the TGAS colour magnitude diagram. A central goal of the *Gaia* mission is the establishment of a precise and accurate empirical description of the colour magnitude diagram, which opens the way to accurate luminosity calibrations of stars across the CMD and to an accurate calibration of the theoretical Hertzsprung-Russell diagram. The measurement of stellar distances plays a fundamental role in this endeavour but accurate distances cannot always be obtained through parallaxes alone. In particular for the more luminous and rarer stars near the bright end of the CMD, even the end of mission *Gaia* parallaxes may have relative errors above the level where one should not simply invert the parallax to obtain a distance (see also Bailer-Jones 2015). Hence it is imperative to combine multiple pieces of information to estimate accurate distances to stars. Two papers based on*Gaia* DR1 (Leistedt & Hogg 2017 and Anderson *et al.* 2017) present approaches in which the information contained in the photometry of stars (apparent brightness and colour) is combined with the TGAS parallax information to arrive at more precise representations of the CMD than can be obtained through TGAS parallaxes alone. In both cases a hierarchical Bayesian model is employed albeit with a different approach to constructing the prior on the distribution of stars in the CMD. Both studies successfully demonstrate how this type of modelling leads to shrinkage in the errors on the inferred distances (absolute magnitudes) of the stars, even if strictly speaking they only provide a more precise description of the contents of the TGAS CMD, rather than of the CMD per se (which would require folding in selection functions and considerations on the degree to which the solar neighbourhood is representative). The hope is that eventually this type of analysis of the *Gaia* data leads to a data-driven predictive models of stars which would very tightly constrain our physical models of stars.

The needle in the haystack. The work by Marchetti *et al.* (2017) shows how machine learning (in this case a neural network) can be applied to large and rich data sets such as *Gaia* DR1. The goal of this work was to find the very few hyper-velocity stars (which were ejected from the Galactic centre) expected to be present in the TGAS catalogue (a few hundred to a few thousand hyper-velocity stars are expected in the full billion star *Gaia* data set). This was done by training and artificial neural network on simulated data containing both the Milky Way and a population of hyper-velocity stars. The optimized neural network was then applied to the TGAS data which resulted in the isolation of 80 hyper-velocity star candidates purely on the basis of astrometric information. A careful follow-up of these candidates through the collection of radial velocity information and the assessment of whether the orbits of the stars imply that they come from the Galactic centre, resulted in one candidate hyper-velocity star that might be unbound from the Milky Way and 5 candidates that appear to be bound. The results of this study greatly strengthen the confidence that in future *Gaia* data releases (where the application of machine learning techniques will be more important) many more hyper-velocity stars can be uncovered.

Stellar occultations. Although this was not an unforeseen application of *Gaia* DR1 it is an excellent illustration of the benefits of an accurate star map. The accurate prediction of the path on the earth from where the occultation of a star by minor body in the solar system can be observed depends very much on the accuracy to which the orbit of the body is known and the accuracy to which the star's position at the observation epoch is known. The latter is greatly improved by the availability of *Gaia* DR1, with more improvements expected in future *Gaia* data releases when proper motions and parallaxes are available for all sources observed by *Gaia*. A taste of the possibilities was provided in the summer of 2016 through the exceptional early release of the Gaia position for a star that would be occulted by Pluto. The better knowledge of Pluto's ephemeris due to the New Horizons flyby was combined with the more accurate *Gaia* position for the star to enable a much more accurate prediction of the occultation path on earth†. The subsequent successful occultation campaign allowed to add a further observational point to the evolution of the atmospheric pressure on Pluto, showing a hint that the pressure increase seen since 1988 (despite Pluto's moving away from the Sun) is now coming to an end, perhaps indicating the start of Pluto's predicted atmospheric 'collapse' due to the lower solar flux.

The above examples are an illustration of the new and complementary ways in which astronomical science can be pursued in the era of large surveys. Creative 'playing' with the data can lead to significant discoveries and new understanding, while at the same time the hard work of developing statistical/numerical/data-driven methods that can efficiently deal with the large amount of information to be uncovered is indispensable.

Finally, it should be noted that *Gaia* DR1 has quickly become the standard against which other surveys are calibrated both astrometrically and photometrically. An example of *Gaia* DR1 serving as the astrometric standard for another large survey is provided by SMASH (Nidever *et al.* 2017) for which the astrometry was re-reduced to the *Gaia* DR1 reference frame. A number of proper motion catalogues have been constructed from the *Gaia* DR1 positions in combination with other surveys. The 'Hot Stuff for One Year' proper motion catalogue (Altmann *et al.* 2017) combines *Gaia* DR1 with the PPMXL (Röser *et al.* 2010) positions in order to derive proper motions for over 500 million stars. The combination of Pan-Starrs1 and *Gaia* DR1 led to the GPS1 proper motion catalogue, covering three quarters of the sky (Tian *et al.* 2017), and Deason *et al.* (2017b) make

† `https://www.cosmos.esa.int/web/gaia/iow_20160914`

use of a proper motion catalogue derived by combining *Gaia* DR1 and SDSS (York *et al.* 2000). Finally the UCAC series of proper motion catalogues was extended with the creation of UCAC5 by re-reducing the existing UCAC observations to the *Gaia* DR1 reference frame and then combining them with the *Gaia* DR1 positions to to derive new proper motions (Zacharias *et al.* 2017).

4. Looking ahead

Although the exploitation of the *Gaia* DR1 data is still in full swing the next data release will arrive soon, in April 2018. *Gaia* DR2 will be based on 22 months of input data and allow for a *Gaia* stand-alone astrometric solution (so the *Hipparcos/Tycho-2* positions will no longer be used), including parallaxes and proper motions for a much larger number (of order one billion) sources. The larger amount of data, the improvements in the various instrument calibrations, and the introduction of colour terms in the astrometric solution will lead to large reductions of the astrometric uncertainties. A major difference between *Gaia* DR2 and*Gaia* DR1 will be the presence of radial velocities for stars brighter than $G_{\rm RVS} = 12$ and the availability of a broad-band colour, $(G_{\rm BP} - G_{\rm RP})$, for all stars on an all-sky homogeneous photometric system. Perhaps these two elements represent the biggest advance from *Gaia* DR1 to *Gaia* DR2. In addition for stars brighter than $G = 17$ the effective temperature and extinction will be determined from the broad-band photometry in combination with the parallaxes, and a major extension of the variable star catalogue is foreseen, including an all-sky RR Lyrae survey. Finally, the epoch astrometry for a pre-selected list of about 10 000 asteroids will be released.

In connection with *Gaia* DR2 it is important to be aware of the following issue concerning the traceability of sources from *Gaia* DR1 to *Gaia* DR2. The data processing leading up to a data release starts with a process that groups individual Gaia observations and links them to sources on the sky. This leads to a working catalogue of sources ('the source list') and their corresponding observations, which forms the basis for the subsequent data processing. The algorithm that carries out the grouping and linking had been much improved before the start of the *Gaia* DR2 processing and this led to many changes in these groups.

When using the Gaia data one should thus be aware that the source list for *Gaia* DR2 should be treated as independent from *Gaia* DR1. Although the majority of sources in *Gaia* DR1 can be identified with the same source in *Gaia* DR2 through the Gaia source identifier, the improved source list will lead to the following changes in the linking of the observations to the source identifiers for a substantial fraction of entries in the source list:

- The merging of groups of observations previously linked to more than one source will lead to a new source associated to the merged observations (with a new source identifier) and the disappearance of the original sources (along with their source identifiers).
- The splitting of groups of observations previously linked to one source will lead to new sources associated to the split groups of observations (with new source identifiers) and the disappearance of the original source (along with its source identifier).
- The list of observations linked to a source may change (and hence the source characteristics may change), while the source identifier remains the same.

A means to trace sources from *Gaia* DR1 to *Gaia* DR2 will be provided, but a one-to-one relation will not exist for all sources. It will then be up to the catalogue user to judge which *Gaia* DR2 source (best) matches a given *Gaia* DR1 source.

Beyond *Gaia* DR2 we can look forward to *Gaia* DR3, targeted for mid to late 2020, and *Gaia* DR4, targeted for the end of 2022. Details on the contents of these releases can be found on the *Gaia* web pages†. Note that *Gaia* DR4 will be the final release for the nominal (5 year) *Gaia* mission. Should the *Gaia* mission be extended, at least one additional data release is foreseen at the end of the extended mission operations. There is much more to come!

Acknowledgements

The early release of Gaia data and the scientific success of this IAU symposium have only been possible due to the excellence and the tireless efforts of the DPAC and ESA teams.

References

Altmann, M., Roeser, S., Demleitner, M., Bastian, U., & Schilbach, E., 2017, *A&A* 600 L4

Anderson, L., Hogg, D. W., Leistedt, B., Price-Whelan, A. M., & Bovy, J., 2017, arXiv:1706.0505

Arenou, F., Luri, X., Babusiaux, C., Fabricius, C., Helmi, A., *et al.*, 2017, *A&A* 599, A50

Bailer-Jones, C. A. L., 2015, *PASP* 127, 994

Belokurov, V., Erkal, D., Deason, A. J., Koposov, S. E., De Angeli, F., *et al.*, 2017, *MNRAS* 466, 4711

Bianchi, L., Conti, A. & Shiao, B., 2014, *Advances in Space Research* Vol. 53, 900

Borucki, W. J., Koch, D., Basri, G., Batalha, N., Brown, T., *et al.*, 2010, *Science* 327, 977

Casertano, S., Riess, A. G., Bucciarelli, B., & Lattanzi, M., 2016, *A&A* 599, A67

Chambers, K. C., Magnier, E. A., Metcalfe, N., Flewelling, H. A., Huber, M. E., *et al.*, 2016, arXiv:1612.05560

Davies, G. R., Lund, M. N., Miglio, A., Elsworth, Y., Kuszlewicz, J. S., *et al.*, 2017, *A&A* 598, L4

Deason, A. J., Belokurov, V., Erkal, D., Deason, A. J., Koposov, S. E., & Mackey, D., 2017, *MNRAS* 467, 2636

Deason, Alis J., Belokurov, V., Koposov, S. E., Gómez, F. A., Grand, R. J., *et al.*, 2017, *MNRAS* 470, 1259

De Ridder, J., Molenberghs, G., Aerts, C., & Eyer, L., 2016, *A&A* 595, L3

Clementini, G., Ripepi, V., Leccia, S., Mowlavi, N., Lecoeur-Taibi, I., *et al.*, 2016, *A&A* 595, A133

Gaia Collaboration, Prusti, T., de Bruijne, J. H. J., Brown, A. G. A., Vallenari, A., & Babusiaux, C., *et al.* 2016a, *A&A* 595, A1

Gaia Collaboration, Brown, A. G. A., Vallenari, A., Prusti, T., de Bruijne, J. H. J., & Mignard, F., *et al.* 2016b, *A&A* 595, A2

Gaia Collaborations, van Leeuwen, F., Vallenari, A., Jordi, C., Lindegren, L., Bastian, U., *et al.*, 2017a, *A&A* 601, A19

Gaia Collaborations, Clementini, G., Eyer, L., Ripepi, V., Marconi, M., Muraveva, T., *et al.*, 2017b, *A&A* in press, arXiv:1705.00688

Gontcharov, G., & Mosenkov, A., 2017, arXiv:1705.09063

Gould, A., Kollmeier, J. A., & Sesar, B., 2016, arXiv:1609.06315

Hawkins, K., Leistedt, B., Bovy, J., & Hogg, D. W., 2017, arXiv:1705.08988

Huber, D., Zinn, J., Bojsen-Hansen, M., Pinsonneault, M., Sahlholdt, C., *et al.*, 2017, arXiv:1705.04697

Jao, W.-C., Henry, T. J., Riedel, A. R., Winters, J. G., Slatten, K. J., & Gies, D. R., 2016, *ApJL* 832, L18

Kopsov, S. E., Belokurov, V., & Torrealba, G., 2017, *MNRAS* 470, 2702

Leistedt, B. & Hogg, D. W., 2017, arXiv:1703.08112

Lindegren, L., Lammers, U., Bastian, U., Hernández, J., Klioner, S., *et al.*, 2016, *A&A* 595, A4

† `https://www.cosmos.esa.int/web/gaia/release`

Marchetti, T., Rossi, E. M., Kordopatis, G., Brown, A. G. A., Rimoldi, A., *et al.*, 2017, *MNRAS* 470, 1388
Michalik, D., Lindegren, L., & Hobbs, D., 2015, *A&A* 574, A115
Mignard, F., Klioner, S., Lindegren, L., Bastian, U., Bombrun, A., *et al.*, 2016, *A&A* 595, A5
Nidever, D. L., Olsen, K., Walker, A. R., Vivas, A. K., Blum, R. D., *et al.*, 2017, arXiv:1701.00502
Röser, S., Demleitner, M., & Schilbach, E., 2010, *AJ* 139, 2440
Sesar, B., Fouesneau, M., Price-Whelan, A., Bailer-Jones, C. A. L., Gould, A., & Rix, H.-W., 2017, *ApJ* 838, 107
Silva Aguirre, V., Lund, M. N., Antia, H. M., Hall, W. B., Basu, S., *et al.*, 2017, *ApJ* 835, 173
Simpson, J. D., De Silva, G. M., Martell, S. L., Zucker, D. B., Ferguson, A. M. N., *et al.*, 2017, arXiv:1703.03823
Skrutskie, M. F., Cutri, R. M., Stiening, R., Weinberg, M. D., Schneider, S., *et al.*, 2006, *AJ* 131, 1163
Stassun, K. G. & Torres, G., 2016, *ApJL* 831, L6
Tian, H.-J., Gupta, P., Sesar, B., Rix, H.-W., & Martin, N. F., 2017, arXiv:1703.06278
Wright, E. L., Eisenhardt, P. R. M., Mainzer, A. K., Ressler, M. E., Cutri, R. M., *et al.*, 2010, *AJ* 140, 1868
Yildiz, M., Çelik Orhan, Z., Örtel. S., & Roth, M., 2017, arXiv:1705.08313
York, D. G., Adelman, J., Anderson, J. E., Jr., Anderson, S. F., Annis, J., *et al.*, 2000, *AJ* 120, 1579
Zacharias, N., Finch, C., & Frouard, J., 2017, *AJ* 153, 166
Zinn, J. C., Huber, D., Pinsonneault, M. H., & Stello, D., 2017, arXiv:1706.09416

Astronomy and Astrophysics in the Gaia sky
Proceedings IAU Symposium No. 330, 2017
A. Recio-Blanco, P. de Laverny, A.G.A. Brown & T. Prusti, eds.

doi:10.1017/S1743921317005798

Gaia: from proposal to GDR1

Gerard Gilmore
Institute of Astronomy, Madingley Road, Cambridge CB3 0HA, UK
email: gil@ast.cam.ac.uk

Abstract. In this concluding article I recall the early history of the Gaia mission, showing that the original science case and expectations of wide community interest in Gaia data have been met. The quarter-century long partnership involving some 1,000 scientists, engineers and managers in industry and academia is delivering a large, high-quality and unique data set which will underpin astrophysics across many sub-fields for years to come.

Keywords. stellar dynamics, Galaxy: kinematics and dynamics, (cosmology:) dark matter, history and philosophy of astronomy, space vehicles.

---

1. *Gaia*: origins of the ESA mission

The *Gaia* mission concept evolved naturally from the successful Hipparcos space astrometric mission. Hipparcos was launched in 1989, and operated until March 1993. It delived reliable parallaxes and proper motions for 118,000 stars, and was supplemented by a star-mapper flux and position catalogue, Tycho2, of 2.5million stars, being a nearly complete sample to magnitude 11. Hipparcos established that absolute parallaxes could indeed be determined from space observations, following an original suggestion by Pierre Lacroute. A brief overview of the Hipparcos mission, from its proposal to its completion, is provided by Perryman (2011).

Astrometry from space has unique advantages over ground-based observations. All-sky coverage is possible, removing the challenge, with risk of systematic errors, of cross-calibrating complementary hemispheric facilities. A relatively stable and temperature- and gravity-invariant operating environment is viable. Even more importantly, absolute astrometry is possible. Narrow field astrometry, from ground or with a typically-designed telescope, such as the Hubble Space Telescope, measures differential parallaxes between all objects in its narrow field of view. In a small field of view every object has similar angular distance from the Sun-Earth baseline, and so has similar parallactic angle. All parallactic ellipses are aligned, providing no information on absolute scale. One attempts to convert to absolute parallaxes either by modelling the distance (parallax) of distant stars in the field, or by comparing to zero-parallax extragalactic objects, with due compensation for their different energy distributions and/or image structure. The possibility of systematic errors is always present. Space astrometry introduces, through appropriate optical design, a large differential angle between stars which are separated by only a small angle on the detector. Thus precise small-angle measurement, together with knowledge of the large angle offset delivered by the spacecraft optical design, compares stars with very different but known parallax factors, and so allows absolute parallax measures. Hipparcos delivered the large angle with an optical system which projected fields separated on the sky by 58° through a modulating grid onto a single-pixel image dissector scanner detector. Gaia has two separate telescopes, with angular separation ("basic angle") of 106°, delivering two $0.^{o}7$ fields of view onto a single very large focal plane, made of an array of 106 CCDs. Manifestly, observing two widely separated fields of view simultaneously from the ground is more difficult, and inevitably involves different observing conditions

ROEMER

Proposal for the Third Medium Size ESA Mission (M3)

A mission dedicated to very accurate astrometric
and photometric measurement of 10^8 stars

U. Bastian, G. Gilmore, J.L. Halbwachs, E. Høg, J. Knude,
J. Kovalevsky, A. Labeyrie, F. van Leeuwen, L. Lindegren,
J.W. Pel, H. Schrijver, R. Stabell and P. Thejll

Edited by L. Lindegren

Lund 1993

Figure 1. The ROEMER proposal to ESA in 1993. The original is bright green.

for each line of sight, though radio-wavelength VLBI achieves precision. Space access is required in the optical. During the latter stages of the Hipparcos mission space astrometry had become a proven technique, so opportunities for successor missions to deliver the very wide science case thus enabled were investigated.

During the 1980's the scientific interest in Galactic Structure developed rapidly. The all-sky photographic surveys (Palomar POSS-II, ESO, UK Schmidt) were digitised leading to quantitative advances ranging from stellar luminosity functions, to discovery of the Galactic thick disk, to quantification of the local Dark Matter density, and to discovery of the Sgr dwarf galaxy and stellar streams in the Galactic halo, direct evidence that Galaxy evolution continued today, to name just some results of personal interest (Kroupa, Tout & Gilmore 1993; Gilmore & Reid 1983; Kuijken & Gilmore 1989; Ibata, Gilmore & Irwin 1994). This, together with rapid advances in understanding stellar evolution, provided the range of science cases and the large scientific community available to interpret and analyse the data which justified a much more ambitious determination of stellar distances, kinematics and chemical abundances.

Technology had also advanced, with availability of large-format 2-D CCD detectors ensuring major efficiency and precision gains. The other aspect of the context was ESA's strategic interest in interferometry. Various strategic studies led to ESA reports SP-1135 *"A Proposed Medium-Term Strategy for Optical Interferometry in Space"* and SP-354 *"Targets for Space-Based Interferometry"*, proposing global astrometry as a high-priority area for space interferometry.

In this context two proposals were submitted. The first (May 1993), to the ESA M3 Call (1993), was for the ROEMER concept (Fig. 1), to provide astrometric and photometric data with 100microarcsec precision for 10^8 stars, was highly rated. The second (October 1993) (Fig. 2) which introduced the acronym GAIA - Global Astrometric Interferometer for Astrophysics - was more ambitious for a Cornerstone mission under the Horizon 2000 programme, to observe 5.10^7 objects with accuracy 10microarcsec at magnitude 15. In an important Annex to the ROEMER proposal, J. Kovalevsky noted that use of CCDs additionally removed the need for an Input Source Catalogue, as Hipparcos required, so that the mission could be a true survey mission. These proposals led to an ESA-funded

Response to Call for Mission Concepts for Horizon 2000 Follow Up
Proposal for an astrometric interferometer as an ESA Cornerstone Mission

GAIA

Global Astrometric Interferometer for Astrophysics

L. Lindegren, M.A.C. Perryman, U. Bastian, J.C. Dainty, E. Høg, F. van Leeuwen, J. Kovalevsky, A. Labeyrie, F. Mignard, J.E. Noordam, R.S. Le Poole, P. Thejll and F. Vakili

Figure 2. The 1993 proposal for a Cornerstone interferometric astrometric mission.

ἤτοι μὲν πρώτιστα Χάος γένετ᾽· αὐτὰρ ἔπειτα
Γαῖ᾽ εὐρύστερνος, (...)

Γαῖα δέ τοι πρῶτον μὲν ἐγείνατο ἶσον ἑωυτῇ
Οὐρανὸν ἀστερόενθ᾽, (...)

Figure 3. *Gaia*, Hesiod, *Theogony 116/117 & 126/127* Loeb Classical Library, Hesiod 1, 2006.

industrial study complemented by an astronomy Science Advisory Group. This group was tasked to develop the science case for a Cornerstone mission. The group was Chaired by the ESA Project Scientist, Michael Perryman, with Project Manager Oscar Pace, and as members G. Gilmore, E. Hoeg, M. Lattanzi, L. Lindegren, F. Mignard, S. Roeser, and P.T. de Zeeuw. Roeser was later replaced by K. de Boer, and X. Luri joined.

Interestingly, rather early in the industry study it became apparent that interferometry was not the optimum technical solution to the scientific challenge. This forms yet another example of the common lesson that solutions looking for problems are rarely implemented. The initial acronym GAIA mutated into the name *Gaia* and survives. It turns out to be an appropriate name in its own right - and motivating the fairing logo (Fig 5) - after the ancient goddess as she appears, for example, in Hesiod's *Theogony 116/117 & 126/127* (Fig. 3) who came into being after *Chaos* and generated the starry sky. One interpretation of her coming into being is as a contrast to the unintelligible (*Chaos*, a gap, a wide opening) and as a generator of the explorable (the starry sky amongst many others).

The Study Team developed the *Gaia Concept and Technology Study Report* ESA-SCI(2000)4, commonly referred to as the mission "Red Book", although the printed cover is white, which is the full proposal for the Gaia mission. [The full document remains available via the ESA *Gaia* web site, while an accessible summary was published in Perryman, de Boer, Gilmore *et al.* (2001).] Following the study the *Gaia* mission was formally presented to a selection meeting of the ESA communities, together with the other proposed missions, on September 13, 2000, in Paris. Presentations made were

- Scientific Case: P.T. de Zeeuw
- Payload, Accuracy and Data Analysis: L. Lindegren
- Spacecraft and Mission Implementation: O. Pace
- Why, How and When? G. Gilmore

Following these presentations *Gaia* was adopted as a Cornerstone Mission. Detailed spacecraft design and cost-reduction-motivated redesign continued under an industry prime contract to Astrium (now Airbus Defence & Space), leading to the spacecraft successfully operating today.

2. The *Gaia* science case - then and now

The "Key Science Objectives" presented for approval of the *Gaia* mission addressed the top-level ambition to provide the data needed to describe the origin, formation and evolution of the Galaxy.

- Structure and Kinematics of our Galaxy
  - shape and rotation of the bulge, disk and halo
  - internal motions of star forming regions, clusters, etc
  - nature of spiral arms and the stellar warp
  - space motions of all Galactic satellite systems
- Stellar populations
  - physical characteristics of all Galactic components
  - initial mass function, binaries, chemical evolution
  - star formation histories
- Tests of Galaxy Formation
  - dynamical determination of dark matter distribution
  - reconstruction of merger and accretion history

Other science products of the survey mission included Stellar Astrophysics, from luminosity calibration of large samples, including distance scale calibration; studies of the Solar System, with unique capability to map potentially earth-crossing asteroids orbiting interior to 1AU; discovery of large volume-complete samples of extra-solar planets; important contributions to galaxies and quasars; and establishment of a dense high-precision reference frame. Dynamical determination of the Galactic gravitational potential to determine the distribution of Dark Matter was a major goal. Although the term "Dark Energy" was not in wide use at the time of the proposal, the science case included both precision distance calibration in cosmology and mapping orbits of nearby galaxies, from the Galactic halo into the Hubble flow. Particular emphasis was made on contributions to tests of General Relativity and the metric.This is an obvious science of interest for *Gaia*, since light-bending by the Sun (and planets) is the largest astrometric signal for most sources - Solar light bending is 4milliarcsec even at 90° from the Sun. Other GR effects include strong- and micro-lensing, and possible detection of gravitational waves. Indeed a special "experiment" is underway with *Gaia* to measure light-bending by Jupiter, a first.

The science case summary in the year 2000 was

- *Gaia* will determine
  - when the stars in the Milky Way formed
  - when and how the Milky Way was assembled
  - how dark matter in the Milky Way is distributed
- *Gaia* will also make substantial contributions to
  - stellar astrophysics
  - Solar System studies
  - extra-solar planetary science
  - cosmology
  - fundamental physics

This remains a topical summary of research activity in 2017. In part this is recognition that data of the type and accuracy and volume which *Gaia* is designed to deliver cannot be obtained in any other way. *Gaia* remains at the forefront of the field.

Type of Data		*Model Function*		*Physics*
Kinematics	$\Longleftrightarrow$	Dynamics	$\Rightarrow$	
radial velocities proper motions		phase space distribution function spatial distributions		gravitational potential dissipational history
Chemical Abundances	$\Longleftrightarrow$	Chemical Evolution	$\Rightarrow$	
line strengths, photometry		star formation history ISM history		stellar initial mass function gas flows, dissipation, SFR
Luminosity Profiles	$\Longleftrightarrow$	Galactic Structure	$\Rightarrow$	
colour–magnitude data surface brightness		spatial distribution function luminosity functions		stellar IMF, binarity, dissipation, SFR, ...

Proceedings of a joint RGO-ESA Workshop on Future Possibilities for Astrometry in Space, *Cambridge, UK, 19-21 June 1995* (ESA SP-379, September 1995)

Figure 4. A view of the approaches to data analysis from one of the *Gaia* preparation meetings.

2.1. *From science case to data analyses*

An important aspect of the *Gaia* proposal was to deliver early and regular data releases to ensure access to data by the wide community, in spite of the need to accumulate data over long times to enable precise astrometry. One special aspect of the proposal was what has become the Gaia transient science alerts (https://gaia.ac.uk). These are proving of both intrinsic scientific interest and an opportunity to involve the wider community, including amateur astronomers and schoolchildren, in the ground-based follow-up. More generally regular major data releases - GDR1 being just the first - are designed to get as much data published and available for analysis as early as possible. This approach is complemented by a very special feature of the *Gaia* mission which was an integral aspect of the original proposal case: *Gaia* data are made available freely to the entire astronomical community as soon as possible. There is no proprietary time, or proprietary science, retained for private analysis inside the data processing consortium. This approach was recommended by the original Study Team mentioned above because of both the volume of the data, and because of the very wide range of science analyses and applications possible. The enthusiastic and wide involvement of the community in GDR1 data shows the wisdom of this philosophy.

Nonetheless, reliable intepretation of astrometric data is not a trivial task: the data sets are large, the errors are not simple, correlations are everywhere. Simply inverting or combining parallaxes is not the thing to do. Prior to *Gaia* astrometric data analyses were a specialist topic with rather few practitioners. Thus the *Gaia* data processing team has been concerned to ensure the wider community has the tools and expertise to analyse the data robustly. One example is the tutorial papar on how to estimate distances from parallaxes (Bailer-Jones 2015). Another is the top-level analysis protocol methodology summarised in Fig. 4. The article by Binney in this Proceedings is an example of how the wider community has risen to the *Gaia* data challenge. Schools, tutorials, workshops, Challenges and so on are now widespread across the community.

3. Lessons from history

The GDR1 data release is the largest and highest accuracy and precision astrometric data set yet available. Almost daily papers using the data appear. It has certainly achieved its primary aim of introducing the astronomical community to large astrometric data sets. Nonetheless, GDR1 with its 2million TGAS parallaxes is tiny compared to future data

Figure 5. The People behind *Gaia*, and the fairing logo.

releases. GDR2 will be 3 orders of magnitude larger, and start to make available the wealth of photometric, spectroscopic, spectrophotometric and time-series data, as well as derived quantities, for the more than 2.5billion objects which Gaia observes.

Obtaining these data depends entirely on the superb *Gaia* spacecraft, designed, built and operated by very many talented and dedicated people in some 60 companies, under the leadership of the Prime contractor Astrium, now known as Airbus Defence & Space, and soon to become simply Airbus. A large team at ESA and ESOC manage and implement the mission. These efforts are supported by the several hundred active participants in the *Gaia* Data Processing and Analysis Consortium - DPAC. All these many people (Fig. 5) deserve our recognition and thanks. Over the quarter century of dedicated efforts since the first proposals for what became the *Gaia* mission and now, where we enjoy GDR1 and await GDR2 and its successors, there has been a close and productive partnership between ESA, the community and industry. We all benefit from that, and should acknowledge that we now stand on the shoulders of a giant team.

4. Acknowledgements

This work has made use of data from the European Space Agency (ESA) mission *Gaia* (https://www.cosmos.esa.int/gaia), processed by the *Gaia* Data Processing and Analysis Consortium (DPAC, https://www.cosmos.esa.int/web/gaia/dpac/consortium). Funding for the DPAC has been provided by national institutions, in particular the institutions participating in the *Gaia* Multilateral Agreement. Gilmore acknowledges partial support through the European Union FP7 programme grant ERC 320360, and the assistance of G. Pebody for Hesiod.

References

Bailer-Jones, C. A. L. 2015, *PASP*, 127, 994

Gilmore, G. & Reid, N., 1983, *MNRAS*, 202, 1025

Ibata, R., Gilmore, G., & Irwin, M., *Nature*, 370, 194

Kroupa, P., Tout, C. A., & Gilmore, G., 1993, *MNRAS*, 262, 545

Kuijken, K. & Gilmore G., 1989, *MNRAS*, 239, 605

Perryman, M. A. C., de Boer, K., Gilmore, G., Hoeg, E., Lattanzi, M. G., Luri, X., Mignard, F., Pace, O., & deZeeuw, P. T. 2001, *A&A*, 369, 339

Perryman, M. A. C. 2011, *Astron. Astrophys. Rev*, 19, 45

Astronomy and Astrophysics in the Gaia sky
Proceedings IAU Symposium No. 330, 2017
A. Recio-Blanco, P. de Laverny, A.G.A. Brown
& T. Prusti, eds.

doi:10.1017/S1743921317006329

Gaia Photometric Data: DR1 results and DR2 expectations

Dafydd Wyn Evans[1], Marco Riello[1], Francesca De Angeli[1], Giorgia Busso[1], Floor van Leeuwen[1], Laurent Eyer[2], Carme Jordi[3], Claus Fabricius[3], Josep Manel Carrasco[3], Michael Weiler[3], Paolo Montegriffo[4], Carla Cacciari[4] and Elena Pancino[5]

[1]Institute of Astronomy, University of Cambridge, Madingley Road, Cambridge CB3 0HA, UK

[2]Department of Astronomy, University of Geneva, Ch. des Maillettes 51, 1290 Versoix, Switzerland

[3]Institut de Ciències del Cosmos, Universitat de Barcelona (IEEC-UB), Martí Franquès 1, 08028 Barcelona, Spain

[4]INAF–Osservatorio Astronomico di Bologna, via Ranzani 1, 40127 Bologna, Italy

[5]INAF–Osservatorio Astrofisico di Arcetri, Largo Enrico Fermi 5, 50125 Firenze, Italy
email: dwe@ast.cam.ac.uk

Abstract. Gaia DR1 was released in September 2016 and contained a photometric catalogue of over 1 billion sources. At this stage, this only included mean G-band photometry and an estimate of the error. Even though this may sound limited in nature, interesting science can still be achieved with this data thanks to its quality. A high level overview of the photometric processing and some validation results will be presented. Additionally, epoch photometry in the G-band was released in Gaia DR1 for a small number of variable sources in the South Ecliptic Pole which covers the LMC. The second data release (Gaia DR2) is currently being prepared and, if available, some preliminary validation results will be presented. It is planned that this release will contain colour information in the form of integrated BP and RP photometry in addition to the latest G-band photometry.

Keywords. Astronomical data bases, Catalogues, Surveys, Instrumentation:photometers, Techniques:photometric, Galaxy:general

---

1. Introduction

The main photometric content of Gaia DR1 consists of mean fluxes and errors for over a billion sources in the G passband. The magnitude range over which the catalogue is likely to be complete is $6 < G < 20$. An overview of the photometry can be found in van Leeuwen *et al.* (2017). Figure 1 shows the magnitude distribution of the main catalogue along with its Tycho 2 and Hipparcos subsets.

In addition to the mean photometry, there is also epoch G-band photometry for a selected number of variable stars. This includes 599 Cepheids (43 new identifications) and 2595 RR Lyrae (343 new). Eyer *et al.* (2017) describes the general variability processing that was carried out and Clementini *et al.* (2016) the special RR Lyrae and Cepheid processing.

2. General principles of the calibration

The calibration of the photometry is carried out using internal standards. Initially, the reference fluxes for the standards are generated from the raw, uncalibrated, fluxes.

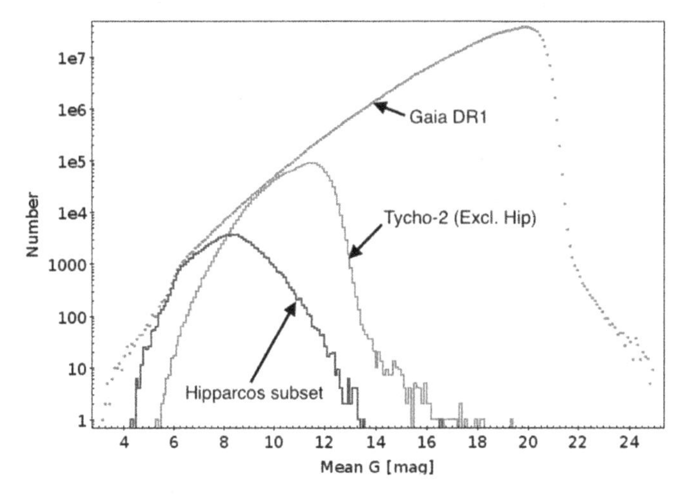

Figure 1. The magnitude distribution of Gaia DR1 along with its Tycho 2 and Hipparcos subsets.

Following a first calibration using these raw fluxes, a new set of reference fluxes is generated using these calibrations. This process is iterated until a stable photometric system is obtained. Care must be taken with this procedure since it will only work if sources are observed under many different conditions. In general, this is easily the case, but when this does not occur, special calibrations must be made.

Currently, the main part of the internal calibration consists of a large-scale and small-scale calibration. The large-scale calibration has a model with colour and across-scan position dependencies. Calibrations are carried out for each combination of the observing configurations (CCD, field-of-view, gate and window class) and approximately every day. The small-scale calibration is effectively a 1d flat field and is averaged out over 4 pixel columns. The model used for Gaia DR1 for the small-scale calibration was a zero point only one, but this is configurable for future processing cycles.

Following the launch of Gaia, a number of unexpected features were found which have an effect on the photometry and required new or improved calibrations. The largest two such complexities are:

(*a*) Contamination of the mirrors and CCDs by water originating from the spacecraft. This is mitigated by periodic heating (decontamination). This varying amount of contamination stresses the photometric initialization and requires special calibration procedures. Following a number of decontamination events, most of the contamination has now gone.

(*b*) A higher background level has been found on the CCDs. This is mainly caused by more straylight from the Sun finding its way to the focal plane. This has meant that the background-subtraction algorithms had to be improved. Even so, this has caused a loss of performance at the faint end due to the data now being sky limited.

Finally, in order to bring the photometry onto a more physical system, a small set of 94 spectrophotometric standard stars are used to provide an external calibration. These standard stars are described in Pancino *et al.* (2012). From this calibration, Vega and AB magnitude zeropoints are generated. While no specific passband information was provided in Gaia DR1, various colour-colour relationships are given in van Leeuwen *et al.* (2017).

For more details about the internal and external calibration, see Carrasco *et al.* (2016).

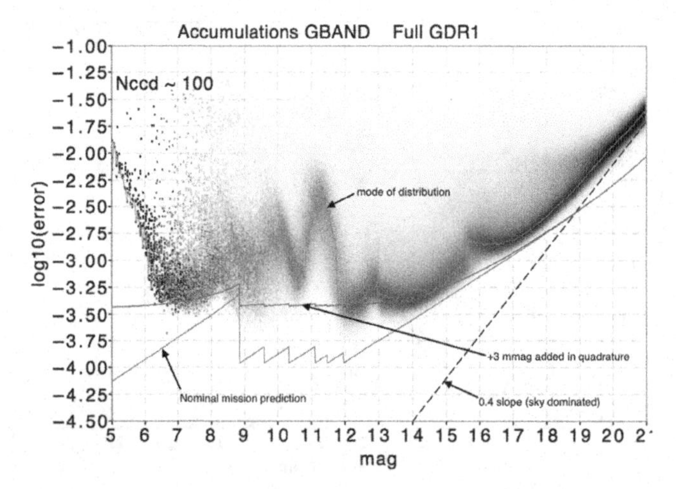

Figure 2. The error on the mean G flux as a function of magnitude for sources with about 100 CCD observations. The nominal mission prediction is shown along with a version with 3 mmag added in quadrature. Also shown is a line with a 0.4 slope indicating the effect of sky domination.

3. Validation of Gaia DR1

Figure 2 shows the distribution of the error on the mean flux as a function of magnitude. By restricting this plot to sources with the same number of observations (about 100 in this case), comparisons can be made with the expectations from the nominal mission. Various features can be seen:

- At the faint end, the error expectation is not achieved due to the data being sky limited. This is a consequence of the higher than expected sky background.
- At G=13 and 16, the higher errors seen are a consequence of changing configurations (window class) at these magnitudes. These performances will be improved with changes to the calibration models.
- The various bumps seen brighter than G=12, are caused by saturation effects which were not accounted for in Gaia DR1.

By adding 3 mmag in quadrature to the nominal error prediction it can be seen that for these calibrations a calibration floor of 3 mmag has been reached. This represents the order of magnitude of systematic errors remaining to be calibrated.

Comparisons with external catalogues can be made to validate the data, but great care must be taken since it will be uncertain as to the source of any differences seen. Additional information is usually needed to identify the source of the differences. Analysis shows that possible systematic effects as a function of magnitude of around 10 mmag exists in the photometric data of Gaia DR1. This is confirmed by an internal consistency check that was carried out (see Fig. 40 of Arenou *et al.*, 2017).

Further details on the photometric validation can be found in Evans *et al.* (2017).

4. Gaia DR2: contents and expectations

For the second data release, in addition to the mean G fluxes and errors, mean BP and RP fluxes (and their errors) will be available giving colour information for more than a billion sources. Additionally, epoch photometry and colours (G, BP and RP) will be released for tens of thousands of variable sources. Full passband information will also be available along with the Vega/AB zeropoints.

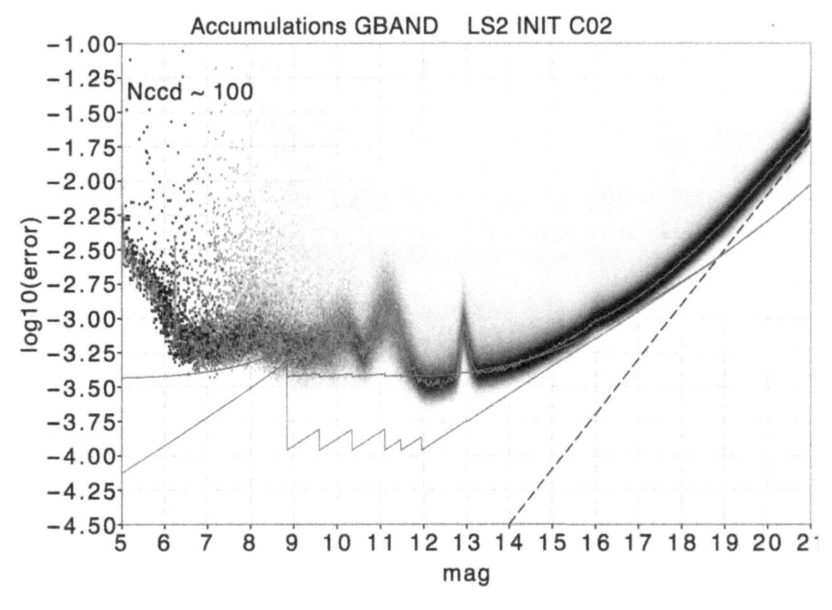

Figure 3. As for Figure 2, but for an early stage of the Cycle 2 reductions.

A number of improvements have been carried out to the calibration procedures:

• The Image Parameter Determination used for the fluxes has been improved due to the use of more sophisticated LSF/PSF models. Saturation is also partially taken into account.

• The cross match has been improved leading to fewer outlier observations.

• By using data from a low contamination period, the initialization of the photometric system has been improved.

• The handling of outliers has been improved leading to fewer anomalous magnitudes.

• More extreme-colour and bright sources will be included in the photometric catalogue.

Figure 3 shows the error on the mean, cf. Fig. 2, at an early stage of the reductions for Gaia DR2. Already the calibration floor of 3 mmag has been surpassed and it is expected that further improvements will be made. Also note that many of the bumps seen in Fig.2 have been reduced.

As a consequence of the changes to the processing of the data, two differences with respect to Gaia DR1 exist. The new photometric initialization implies a (slightly) different passband. This is due to using less contaminated data for the initialization. Also a new cross-match procedure means that some Source Identifiers will have changed. This is not a simple one-to-one relationship. The moral of these differences is that the user should treat Gaia DR2 as a totally new catalogue.

Acknowledgments: This work has made use of data from the European Space Agency (ESA) mission *Gaia* (`https://www.cosmos.esa.int/gaia`), processed by the *Gaia* Data Processing and Analysis Consortium (DPAC, `https://www.cosmos.esa.int/web/gaia/dpac/consortium`). Funding for the DPAC has been provided by national institutions, in particular the institutions participating in the *Gaia* Multilateral Agreement.

References

Arenou, F., Luri, X., Babusiaux, C., *et al.* 2017 *A&A*, 599, A50
Carrasco, J. M., Evans, D. W., Montegriffo, P., *et al.* 2016 *A&A*, 595, A7
Clementini, G., Ripepi, V., Leccia, S., *et al.* 2016 *A&A*, 595, A133
Evans, D. W., Riello, M., De Angeli, F., *et al.* 2017 *A&A*, 600, A51
Eyer, L., Mowlavi, N., Evans, D. W., *et al.* 2017, arXiv:1702.03295
van Leeuwen, F., Evans, D. W., De Angeli, F., *et al.* 2017, *A&A*, 599, A32
Pancino, E., Altavilla, G., Marinoni, S., *et al.* 2012 *MNRAS*, 426, 1767

Astronomy and Astrophysics in the Gaia sky
Proceedings IAU Symposium No. 330, 2017
A. Recio-Blanco, P. de Laverny, A.G.A. Brown & T. Prusti, eds.

doi:10.1017/S1743921317006548

The Gaia Archive

Alcione Mora[1], Juan González-Núñez[2], Deborah Baines[2], Javier Durán[2], Raúl Gutiérrez-Sanchéz[2], Elena Racero[2], Jesús Salgado[2] and Juan Carlos Segovia[2]

[1]ESA-ESAC Gaia Science Operations Centre, Camino Bajo del Castillo s/n, Urb. Villafranca del Castillo, 28692 Villanueva de la Cañada, Madrid, Spain
email: alcione.mora@esa.int

[2]ESA-ESAC Science Data Centre, Camino Bajo del Castillo s/n, Urb. Villafranca del Castillo, 28692 Villanueva de la Cañada, Madrid, Spain

Abstract. The Archive is the main Gaia data distribution hub. The contents of DR1 are briefly reviewed and the data structures discussed. The system architecture, based on Virtual Observatory standards, is also presented, together with the extensions that allow e.g. authenticated access, persistent uploads and table sharing. Finally some usage examples are provided.

Keywords. astronomical data bases: miscellaneous, catalogs, surveys, astrometry, stars: distances, Cepheids

---

1. Introduction

The main contents of Gaia Data Release 1 (DR1) are included in one big catalogue table: `gaia_source`, including astrometric parameters and average photometry for 1.14 billion sources. `tgas_source` contains the subsample of 2.06 million stars for which a good five parameters astrometric solution was obtained. Other data comprise a number of selected external catalogues, their corresponding pre-computed cross-matches and light curves for a sample of Cepheids and RR Lyrae stars. Sect. 2 provides further details on the DR1 contents.

Astronomical archives are challenged by the need to provide increasingly larger data sets. The Gaia mission is surveying more than a billion sources. The final data release volume will be in the scale of petabytes. New projects and missions will push this boundary by several orders of magnitude (e.g. Euclid, LSST, SKA). Consequently, one key objective in modern archives is to carry out as much processing as possible on the server side, according to the "code to the data" paradigm. The Gaia Archive can be considered a step in this direction. Its architecture is described in Sect. 3, while some usage examples are shown in Sect. 4.

2. Archive contents

For many science cases, DR1 can be considered just the big table `gaia_source`, providing astrometric and photometric information (columns) for each object (row). However, the Gaia Archive contains more information, which is presented below.

Light curves and detailed variability data for a sample of around 3000 pulsating variables (Cepheids and RR Lyrae) in the LMC, due to the observing geometry during the initial phase of the nominal mission, when the ecliptic poles were repeatedly observed. The data are provided in a collection of flat tables. For example the `rrlyrae` table includes a number of parameters for each RR Lyrae (period, variability amplitude, Fourier decomposition, ...). The light curves are included in `phot_variable_time_series_gfov`, which includes a row for each star and observing epoch. The latter structure is flat, simple and conveniente, but its applicability to the final Data Release is uncertain, because the table size would approach the trillion entries.

The Gaia data is most powerful when combined with additional information. A number of major surveys have been adapted and ingested into the Archive (2MASS, PPMXL, SDSS9 , UCAC4, URAT1, WISE). Pre-computed cross-matches have been generated for improved usability. For each catalogue, both the best neighbour and all possible matches in a given neighbourhood have been provided in dedicated pivot tables. The tables include one match per row, relating the Gaia to external catalogue IDs in different columns.

Finally, there is a number of additional catalogues which are considered immutable and are provided just for reference, such as Hipparcos, Tycho or IGSL. The latter is considered superseded by DR1 and should not be used for any future scientific analysis.

All tables considered, the total size of the Gaia Archive as of DR1 is 11.6 billion rows.

3. Architecture

Fig. 1 shows the main components of the Gaia Archive. The tables are stored in a PostgreSQL data base, and can be accessed through a VO TAP interface using the ADQL query language, a SQL dialect specifically designed for Astronomy. There are multiple ways to interact with the archive, including its own Graphical User Interface (GUI), external applications such as Topcat, command line tools and more recently a Python module within AstroPy.

Custom extensions on TAP, called TAP+, have been developed to provide additional functionalities outside the standard, most notably persistent uploads, table sharing and server cross-match. The key difference is that TAP is basically an anonymous protocol, while the extensions support user authentication to provide privacy. All jobs executed while logged in the system are associated to the user and kept until actively erased. User defined tables can be created either using the output of a job or uploading external data. Those tables can be used in queries exactly as if they were public data, but are only visible to their owner and those users with whom he decides to share them. This feature is very powerful. On the one hand, it provides the means to execute and refine complex queries in various steps. On the other hand, it allows the interactive analysis of preliminary results by a small number of people. Sharing is particularly useful during the validation campaigns before data releases. Query execution time and user space quotas can be defined on a per-user basis. Finally, an optimised geometrical cross-match service is offered, which allows users to carry out cone searches between tables with sky coordinate information.

The GUI can be accessed using any browser at the Archive address†. It offers the possibility to carry out basic and advanced queries and to quickly inspect the output. The data can be downloaded in a variety of formats, including VO table, fits and csv or directly exported to other VO tools via SAMP. Help is provided in different ways. A dedicated tab in the GUI contains general information, tutorials, explanations on ADQL custom functions, and instructions for command line and programmatic access. Information on each table and column is available from a tree view. Units are shown in the table preview and exported into output files. Links to external resources are also provided (e.g. DR1 documentation and data model detailed description).

† `http://archives.esac.esa.int/gaia`

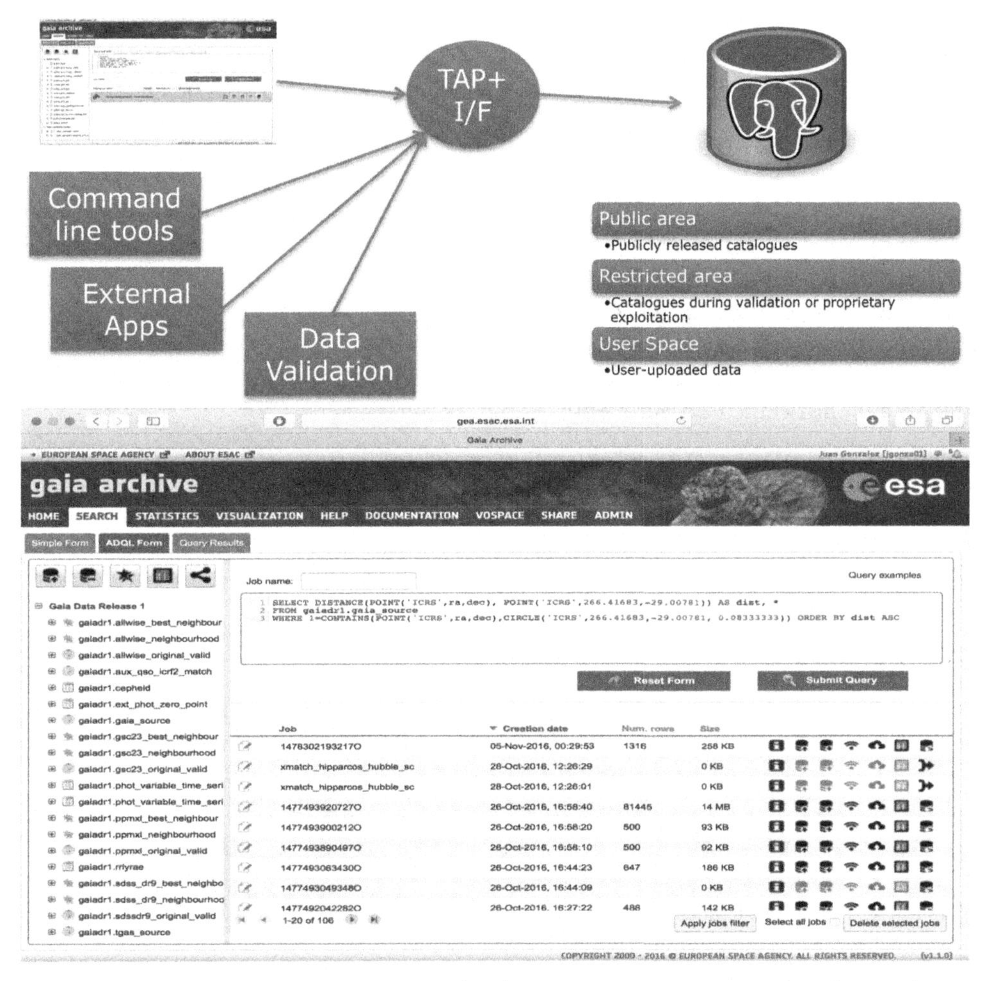

Figure 1. Gaia Archive architecture (top) and Graphical User Interface (bottom)

4. Examples

The use of ADQL to query the Gaia Archive has pros and cons. The two biggest advantages are probably flexibility and reproducibility. Complex operations can be defined within a query to retrieve derived quantities (e.g. absolute magnitudes, colours, variability phase, ...). Bute even more important is the ability to exactly reproduce the results of any previous work where the original ADQL queries are included.

Being a relatively new tool, learning ADQL can be an entrance barrier to advanced Archive users. The learning curve is relatively shallow, though, due to the reduced syntax and simplicity of the language. Fig. 2 shows a simple query used to retrieve candidate members of the Pleiades open cluster based on position on the sky (two degrees radius circle) and proper motion (box of size 20 mas/yr in RA and 15 mas/yr in DEC approximately centred on the average value). The selected objects in this region can be clearly separated from field stars in proper motion space.

Gaia Collaboration *et al.* (2016). provides several examples of ADQL queries illustrating the contents of DR1. Fig. 3 adapts two of them: the TGAS HR diagram and the phase folded light curve of an RR Lyrae star. The original HR diagram was a scatter plot containing more than a million $G-Ks$ colours and M_G absolute magnitudes. While this

```
SELECT *
FROM gaiadr1.tgas_source
WHERE CONTAINS(
    POINT('ICRS', ra, dec),
    CIRCLE('ICRS', 56.75, 24.1167, 2)
    )=1
AND pmra IS NOT NULL AND pmra != 0
AND pmdec IS NOT NULL AND pmdec != 0
AND abs(pmra_error/pmra) < 0.10
AND abs(pmdec_error/pmdec) < 0.10
AND pmra BETWEEN 15 AND 25
AND pmdec BETWEEN -55 AND -40
```

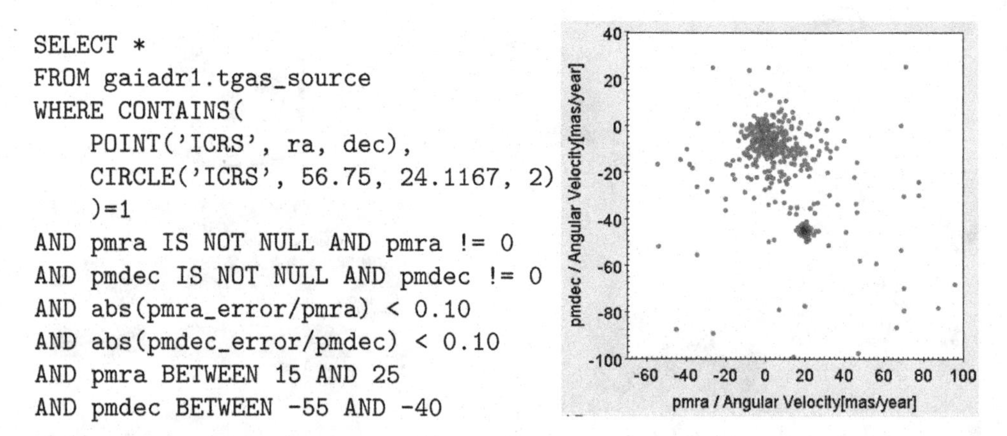

Figure 2. Simple selection of Pleiades member candidates. Left: a simple ADQL query shows how to explore a circular region on the sky adding extra constraints on the proper motion. Right: the candidate members (red) show a clearly distinct behaviour compared to field stars in proper motion space.

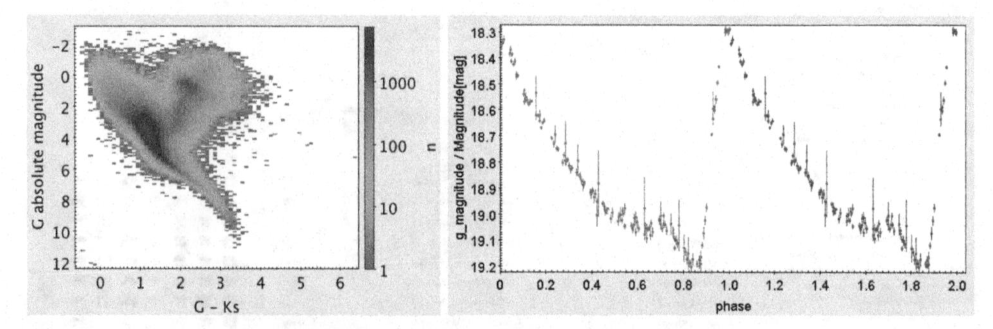

Figure 3. Example DR1 plots based on Gaia Collaboration *et al.* (2016). Left: TGAS HR diagram computed as a 2D histogram. Right: phase folded light curve for RR Lyrae with source_id 5284240582308398080.

is OK for DR1, representing a billion points in a diagram for DR2 would be very difficult and most probably an overkill. The figure shows how to provide the same information using a memory efficient 2D histogram. The RR Lyrae light curve has been folded, and the errors have been directly estimated using an ADQL query within the Archive.

Acknowledgements

This work has made use of data from the European Space Agency (ESA) mission *Gaia* (`https://www.cosmos.esa.int/gaia`), processed by the *Gaia* Data Processing and Analysis Consortium (DPAC, `https://www.cosmos.esa.int/web/gaia/dpac/consortium`). Funding for the DPAC has been provided by national institutions, in particular the institutions participating in the *Gaia* Multilateral Agreement.

References

Gaia Collaboration, Brown, A. G. A., Vallenari, A., Prusti, T. *et al.* 2016, *A&A*, 595A, 2G

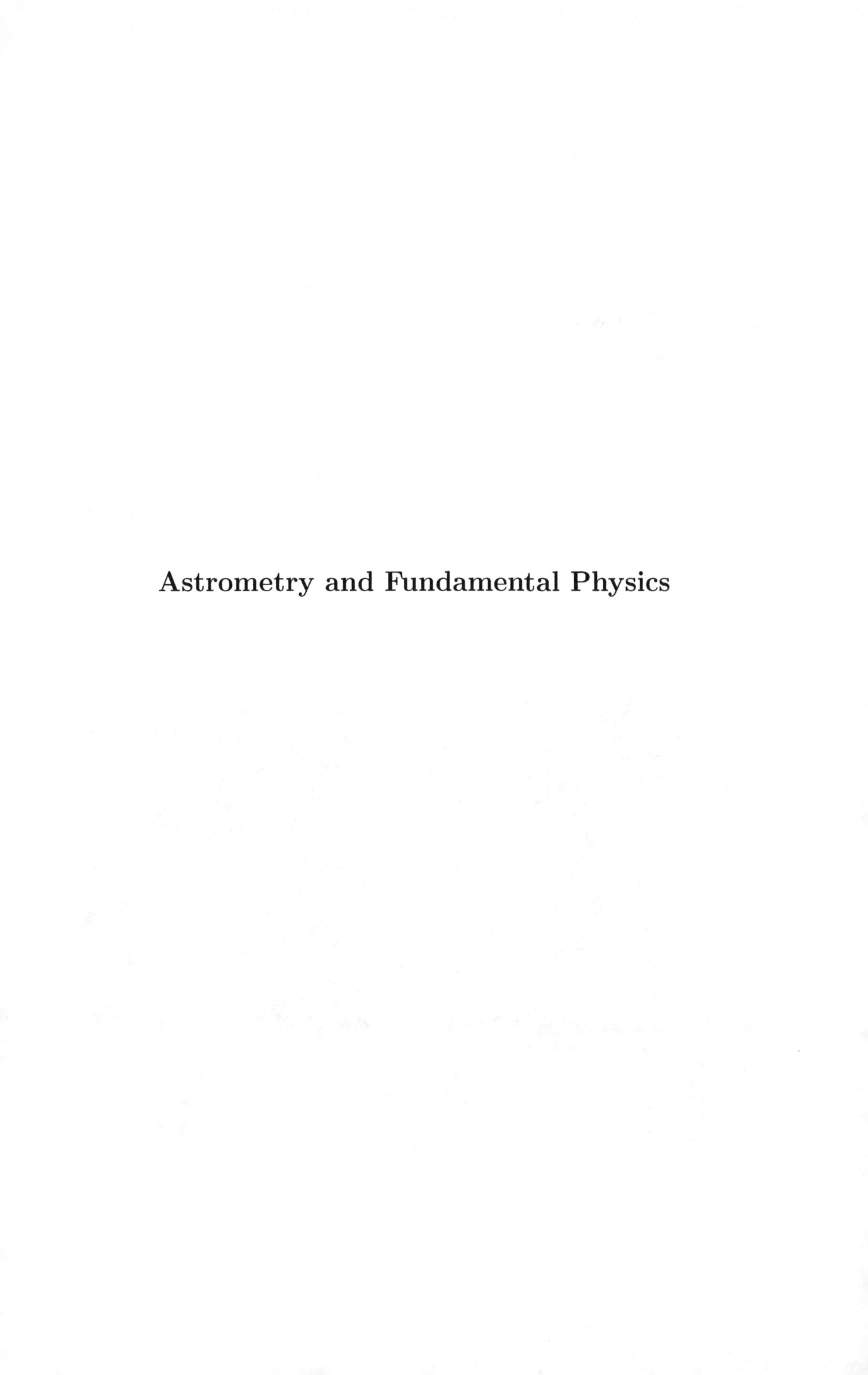

Astrometry and Fundamental Physics

Some of the Hipparcos satellite and Gaia mission leaders: (from left to right) Ulrich Bastian, Dafydd Wyn Evans, Lennart Lindegren, Jean Kovalevsky, François Mignard, Catherine Turon and Erik Hoeg.

François Mignard thanking Jean Kovalevsky after his tribute.

Astronomy and Astrophysics in the Gaia sky
Proceedings IAU Symposium No. 330, 2017
A. Recio-Blanco, P. de Laverny, A.G.A. Brown
& T. Prusti, eds.

doi:10.1017/S1743921317005919

The Tycho-Gaia Astrometric Solution

Lennart Lindegren
Lund Observatory, Department of Astronomy and Theoretical Physics
Lund University, Box 43, SE-22100 Lund, Sweden
email: lennart@astro.lu.se

Abstract. Gaia DR1 is based on the first 14 months of Gaia's observations. This is not long enough to reliably disentangle the parallax effect from proper motion. For most sources, therefore, only positions and magnitudes are given. Parallaxes and proper motions were nevertheless obtained for about two million of the brighter stars through the Tycho-Gaia astrometric solution (TGAS), combining the Gaia observations with the much earlier Hipparcos and Tycho-2 positions. In this review I focus on some important characteristics and limitations of TGAS, in particular the reference frame, astrometric uncertainties, correlations, and systematic errors.

Keywords. Catalogs, astrometry, reference systems, stars: distances, stars: general

---

1. Introduction

The *Tycho-Gaia astrometric solution* (TGAS) is the special astrometric processing applied to bright stars in Gaia Data Release 1 (DR1; Gaia Collaboration, Brown, *et al.* 2016) in order to provide full astrometric information (positions, parallaxes, and proper motions) for 2 057 050 sources. † For the remaining $\sim 1.14 \times 10^9$ sources in Gaia DR1, no parallaxes or proper motions are given. The theoretical background, contents, properties, and validation of TGAS are described in several papers (Michalik *et al.* 2015a; Lindegren *et al.* 2016; Arenou *et al.* 2017). Here I will focus on some properties of TGAS that were perhaps not sufficiently highlighted or explained in these papers.

2. Historical background and concept

TGAS dates back to an idea presented in 2009 by François Mignard to the ESA Gaia Science Team, called the *Hundred Thousand Proper Motions* project (HTPM; Mignard 2009). The basic idea is simple: to improve the proper motions of the ~100 000 Hipparcos stars roughly by an order of magnitude by combining the early Gaia astrometry acquired for these stars with the Hipparcos positions at an epoch more than 20 years earlier. Simulations by Mignard showed that HTPM would be possible with just six months of Gaia data – the minimum needed to cover the whole celestial sphere – by using the Hipparcos parallaxes to transform the Gaia observations to barycentric directions, as needed for the accurate calculation of proper motions.

The combination of astrometric catalogues from different space missions (Hipparcos, Gaia, NanoJASMINE) was studied in a research project started in 2009 by D. Michalik at Lund Observatory, with support from ESA (Dr. U. Lammers). By rigorously formulating HTPM in a Bayesian framework, it was demonstrated how the solution could improve both the proper motions and parallaxes of the Hipparcos stars (Michalik *et al.* 2014). Since simulations showed that only the positions were needed from the earlier catalogue,

† In Gaia terminology, a "source" is any approximately point-like object observed by Gaia. A TGAS source may be a single star, a component in a resolved double or multiple system, the photocentre of two or more stellar components, or a compact extragalactic object.

it was a very natural step to include also the Tycho-2 positions in the solution. This effectively led to TGAS (Michalik *et al.* 2015a), which then superseded HTPM.

The simplest kinematic model of a star assumes that it moves with uniform velocity relative to the Solar System Barycentre. Knowing the location and velocity of the observer, the apparent path of the source on the celestial sphere is then described by five astrometric parameters: the barycentric position (α, δ) at some chosen reference epoch (J2015.0 for TGAS), the parallax (ϖ), and the components of proper motion ($\mu_{\alpha *}$, μ_δ). (For a small number of nearby high-velocity stars, a spectroscopic radial velocity is also needed.) Successful estimation of all five parameters requires observations covering a time interval of at least a few years, with a suitable distribution in time to disentangle parallax from proper motion. With only 14 months of Gaia observations available for DR1, the astrometric solution would be highly degenerate for most sources, resulting in large uncertainties and strong correlations between the parallax and proper motion components. By incorporating positions from the Hipparcos and Tycho-2 catalogues at their much earlier epochs (around 1991.25), the proper motions are decorrelated from the other parameters, making a non-degenerate solution possible. In a Bayesian framework the early positions can be regarded as prior information for the estimation of the complete set of astrometric parameters.

3. Priors used in TGAS

Only the *positions* in the Hipparcos and Tycho-2 catalogues are used as priors in TGAS, leaving the parallaxes and proper motions in this solution statistically independent of their corresponding values in the Hipparcos catalogue.

For the Hipparcos sources in TGAS, positions at epoch J1991.25 were taken from the re-reduction by van Leeuwen (2007). For other TGAS sources, positions were taken from the Tycho-2 catalogue (Høg *et al.* 2000) using the "observed" coordinates at the original mean epochs roughly spanning 1990.9 to 1992.1. As shown in the left panel of Fig. 1 the two catalogues have very different positional uncertainties at these epochs, which makes it relevant to distinguish between the "Hip subset" (93 635 sources) and "Tyc subset" (1 963 415 sources) of TGAS. In the Gaia DR1 catalogue of the Gaia Archive† the flag `astrometric_priors_used` equals 3 for the Hip subset and 5 for the Tyc subset.

4. Reference frame of TGAS

The reference frame of TGAS is linked to the extragalactic radio frame ICRF2 (Fey *et al.* 2015) by means of the auxiliary quasar solution (Michalik & Lindegren 2016). The system of positions in TGAS is therefore accurately aligned (to < 0.1 mas) with ICRF2 at the reference epoch J2015.0. Since the Hipparcos catalogue was aligned with the radio frame to better than 0.6 mas at its epoch J1991.25, this makes the proper motion system of TGAS non-rotating with respect to the quasars to better than 0.03 mas yr^{-1}. TGAS thus supersedes the Hipparcos catalogue as the currently best optical realisation of the International Celestial Reference System (ICRS; Arias *et al.* 1995). When constructing TGAS it was found that the Hipparcos Reference Frame (HRF) rotates with respect to ICRS at a rate of 0.24 ± 0.03 mas yr^{-1}. This is still well within the stated uncertainty of the HRF, which is ± 0.25 mas yr^{-1} per axis (Kovalevsky *et al.* 1997).

If proper motions in the HRF (e.g. from the Hipparcos catalogue) are to be compared with TGAS (in the reference frame of Gaia DR1), they should first be corrected for the

† `http://gea.esac.esa.int/archive/`

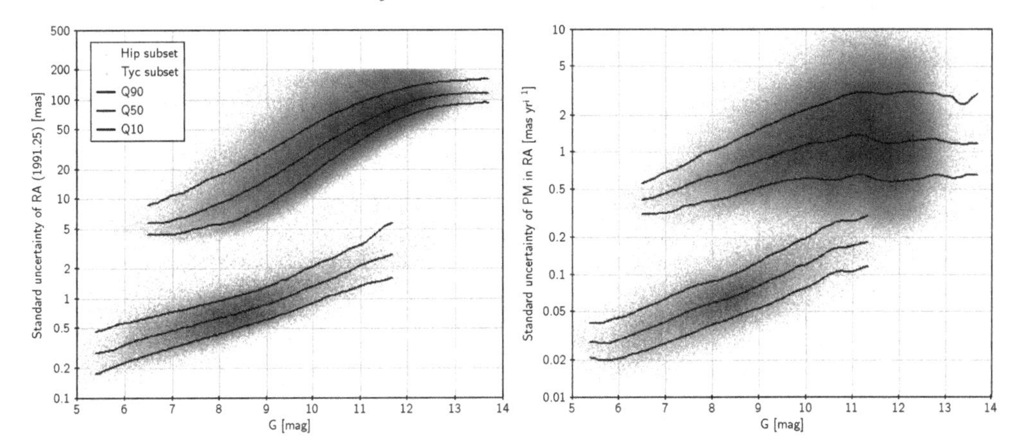

Figure 1. *Left:* Positional uncertainties at epochs around J1991.25 in the Hipparcos and Tycho-2 catalogues. These positions are used as priors in the TGAS solution. *Right*: Uncertainties of the proper motions in the TGAS catalogue. In both diagrams the lower cloud shows the ~93 000 Hipparcos sources in TGAS, the upper cloud the ~2 million Tycho-2 sources. The curves are quantiles at 10%, 50% (median), and 90%. Only the components in right ascension are shown.

rotation of the HRF using the expressions (in mas yr^{-1}):

$$\left.\begin{aligned} \mu_{\alpha*}^{\rm GRF1} &= \mu_{\alpha*}^{\rm HRF} - 0.126\sin\delta\cos\alpha + 0.185\sin\delta\sin\alpha - 0.076\cos\delta \\ \mu_{\delta}^{\rm GRF1} &= \mu_{\delta}^{\rm HRF} + 0.126\sin\alpha + 0.185\cos\alpha \end{aligned}\right\} \quad (4.1)$$

The 1 billion non-TGAS sources in Gaia DR1 provide a dense optical reference frame all the way down to 20th magnitude, which is on the same positional system (ICRS) as TGAS. It is accurate to within a few mas at the reference epoch J2015.0 but quickly deteriorates when moving away from this epoch due to the unspecified proper motions.

5. Astrometric uncertainties

The standard uncertainties of the five astrometric parameters are the main indicators of the quality of the solution for a TGAS source. In the Gaia Archive they are given in the fields `ra_error`, `dec_error`, `parallax_error` (in mas) and `pmra_error`, `pmdec_error` (in mas yr^{-1}). `ra_error` is $\sigma_{\alpha*} = \sigma_\alpha \cos\delta$ and `pmra_error` is $\sigma_{\mu\alpha*} = \sigma_{\mu\alpha}\cos\delta$. They should be interpreted as estimated standard deviations of the actual total (random plus systematic) errors, remembering that actual error distributions may be non-Gaussian with an excess of errors beyond several standard deviations. Only the uncertainties in parallax and proper motion are discussed here.

5.1. *Proper motions*

As expected, the uncertainties of the TGAS proper motions strongly depend on the precision of the old (~1991) positions used as priors, as shown in the right panel of Fig. 1. Superficially the diagram mirrors the prior uncertainties in the left panel, taking into account the epoch difference of ~24 years, but the range of uncertainties is actually a lot smaller for the proper motions (a factor ~250×) than it is for the prior positions (~1000×). This is discussed in Sect. 6.

5.2. *Parallaxes*

By contrast, as shown in Fig. 2, the uncertainties in parallax are not drastically different between the Hip and Tyc subsets. The overall median uncertainty is 0.28 mas for Hip

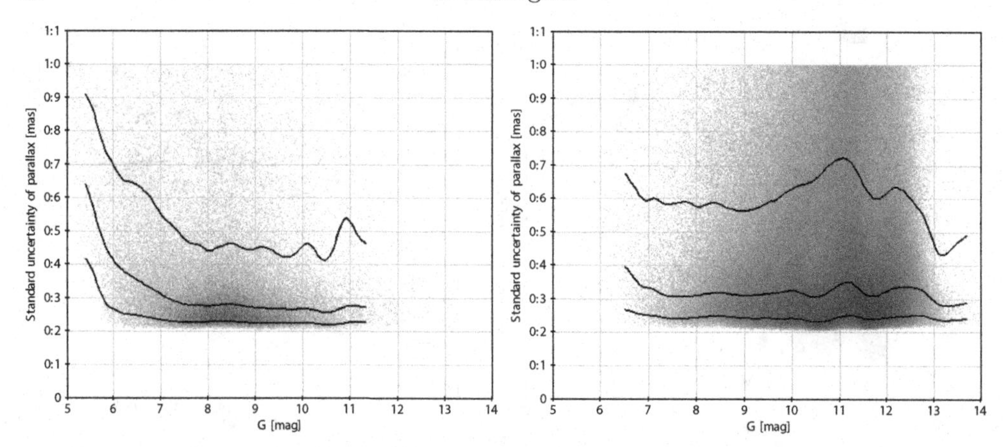

Figure 2. *Left:* Parallax uncertainties versus magnitude for the Hip subset of TGAS. *Right:* Same for the Tyc subset. The curves are quantiles at 10%, 50% (median), and 90%.

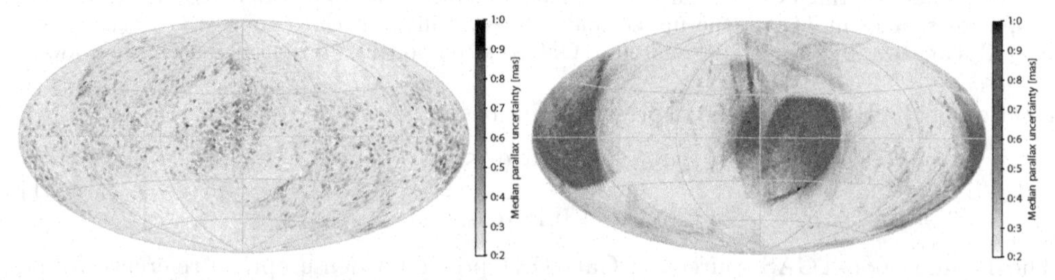

Figure 3. *Left:* Parallax uncertainty versus position for the Hip subset of TGAS. *Right*: Same for the Tyc subset. The maps use an equatorial projection with $\alpha = \delta = 0$ at the centre and α increasing from right to left. Median uncertainties are shown in cells of about 3.36 deg^2.

and 0.32 mas for Tyc, with relatively little variation with magnitude. One can conclude that the parallaxes in TGAS on the whole do not depend strongly on the prior information. Their uncertainties are dominated by errors in the Gaia data – mainly unmodelled calibration errors such as chromaticity (at all magnitudes) and saturation effects (for $G < 7$), both of which were not yet taken into account in the processing for Gaia DR1.

Figure 3 shows the median parallax uncertainty as a function of position for the Hip and Tyc subsets. The similarity in parallax uncertainty between the subsets does not extend to the whole celestial sphere. In the Tyc subset there are two big areas, each covering about 4000 deg^2, where the typical uncertainties are roughly twice as large as in comparable areas along the ecliptic. These areas do not generally have much fewer Gaia observations than other areas along the ecliptic; it is just that the particular temporal and directional distributions of the scans in these areas happen to be less favourable for disentangling the parallax from the proper motion. In the Hip subset this is not a problem because the proper motions are sufficiently well constrained by the priors.

6. Correlations

Correlations are an integral and essential part of the statistical description of the errors in an astrometric catalogue. Under the assumption of unbiased errors the covariance of any two parameters X and Y is $\text{Cov}(X, Y) = \text{E}(e_X e_Y) = \rho(X, Y)\sigma_X \sigma_Y$, where e_X, e_Y are the (unknown) errors in X and Y (deviations from the true values), σ_X, σ_Y their uncertainties, and $\rho(X, Y)$ their correlation, with $|\rho(X, Y)| < 1$.

Correlations matter whenever a calculation involves more than one astrometric parameter. To take a simple example, consider the calculation of the transverse velocity (in km s^{-1}) using the formula $v_{\mathrm{T}} = 4.7405\mu/\varpi$, where μ is the proper motion (in mas yr^{-1}) along α or δ, and ϖ the parallax (in mas). In the limit of high signal-to-noise ratios ($\varpi/\sigma_\varpi \gg 1$, $|\mu/\sigma_\mu| \gg 1$), classical error propagation gives

$$\left(\frac{\sigma_{v\mathrm{T}}}{v_{\mathrm{T}}}\right)^2 = \left(\frac{\sigma_\mu}{\mu}\right)^2 + \left(\frac{\sigma_\varpi}{\varpi}\right)^2 - 2\rho(\varpi,\mu)\left(\frac{\sigma_\mu}{\mu}\right)\left(\frac{\sigma_\varpi}{\varpi}\right). \tag{6.1}$$

Neglecting the last term, proportional to the correlation between ϖ and μ, could lead to $\sigma_{v\mathrm{T}}$ being either under- or over-estimated depending on the sign of $\rho(\varpi,\mu)$. More sophisticated tools to estimate astrophysically interesting quantities, e.g. using chi-square minimisation, maximum-likelihood estimation, or Bayesian methods, involve metrics like

$$\chi^2 = \begin{bmatrix} \Delta\varpi & \Delta\mu_{\alpha*} & \Delta\mu_\delta \end{bmatrix} \mathbf{C}^{-1} \begin{bmatrix} \Delta\varpi \\ \Delta\mu_{\alpha*} \\ \Delta\mu_\delta \end{bmatrix}, \tag{6.2}$$

where the Δ-quantities are model residuals in the astrometric data, and $\mathbf{C}$ the covariance matrix of the astrometric parameters. Ignoring correlations, by using only the main diagonal in $\mathbf{C}$, could lead to biased estimates and under- or over-estimated uncertainties.

In an astrometric catalogue there are two kinds of correlations to worry about: between-source correlations and within-source correlations. *Between-source correlations* are important when combining astrometric data from several sources. Typical examples include the estimation of integral properties of a group of stars, such as the mean distance, radial extent, bulk velocity, or internal velocity dispersion of a stellar cluster. In the absence of between-source correlations, the mean parallax or mean proper motion of a cluster would improve as $\sigma/\sqrt{n}$, if σ is the uncertainty for individual member stars and n the number of such stars. The parallaxes or proper motions in a small area are however expected to have some degree of positive between-source correlation, due to shared errors from the attitude and instrument calibration. The uncertainty of the mean then decreases less rapidly with n, approaching a floor $\simeq \sigma\sqrt{\langle\rho\rangle}$ for large n, where $\langle\rho\rangle > 0$ is the average correlation between all pairs of cluster stars. In TGAS positive between-source correlations are expected on spatial scales up to a few tens of degrees. They are however very difficult to estimate, and are not quantified in the Gaia Archive for Gaia DR1. For all practical purposes they look just like the systematic errors discussed in Sect. 7.

Within-source correlations are the correlations between the different astrometric parameters of the same source, such as between μ and ϖ in Eq. (6.1). In TGAS there are always exactly five astrometric parameters per source, and consequently ten within-source correlations per source. In the Gaia Archive they are given in the fields `ra_dec_corr` for $\rho(\alpha,\delta)$, etc. Maps of their median values are shown in Fig. 7 of Lindegren *et al.* (2016). See also Sect. 8.1 of the Gaia DR1 validation paper (Arenou *et al.* 2017).

The complete 5×5 covariance matrix for the astrometric parameters of a TGAS source can be computed from the five uncertainties and ten correlations given in the Gaia Archive. This matrix is needed in some applications, for example to propagate the uncertainties to an epoch different from the TGAS reference epoch J2015.0, or to a different reference system such as galactic coordinates; see Sects. 4.3.2 and 3.1.7 of the Gaia DR1 on-line documentation.†

Users who care about the errors in ϖ, $\mu_{\alpha*}$, and μ_δ should also be concerned about the three correlations $\rho(\varpi,\mu_{\alpha*})$, $\rho(\varpi,\mu_\delta)$, and $\rho(\mu_{\alpha*},\mu_\delta)$. For TGAS as a whole, roughly

† `https://gaia.esac.esa.int/documentation/GDR1/`

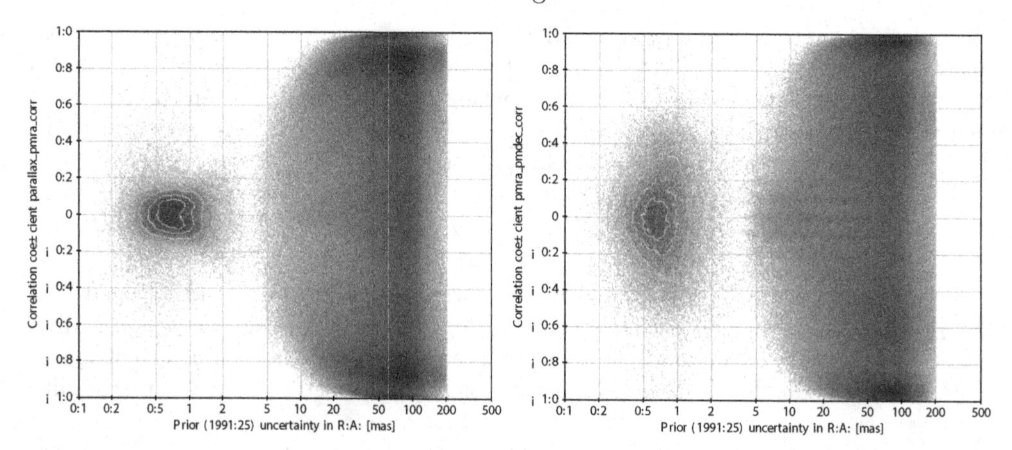

Figure 4. *Left:* Correlation coefficient $\rho(\varpi, \mu_{\alpha*})$ plotted against the positional prior in α at the Hipparcos/Tycho-2 epoch. *Right*: Same for $\rho(\mu_{\alpha*}, \mu_\delta)$. In both panels the (pink) cloud at prior $\lesssim 3$ mas is the Hip subset, the bigger (orange) cloud is the Tyc subset.

half of them exceed 0.5 in absolute value, and many are quite close to ± 1. For the Hip subset, however, only a few per cent exceed 0.5 in absolute value. This difference between the Hip and Tyc subsets is clearly related to the strength of the prior. In Fig. 4 two of the correlation coefficients are plotted against the prior uncertainty at 1991.25 (a plot of $\rho(\varpi, \mu_\delta)$ looks similar). The correlations are moderate for sources with a strong prior (positional prior uncertainty < 2 mas), but start to bifurcate towards larger positive and negative correlations for Hip sources with a weaker prior. The progression towards larger (positive or negative) correlations with increasing prior uncertainty continues in the Tyc subset, but the paucity of sources with intermediate prior uncertainty, around 3 mas, still makes it relevant to think about Hip and Tyc as distinct subsets.

The interpretation of Fig. 4 is straightforward. Recalling (Sect. 2) that the purpose of the prior was to decorrelate parallax from proper motion, we must conclude that this was not fully achieved for most Tyc sources due to a too weak prior position in Tycho-2. Effectively, the TGAS results for many sources in the Tyc subset are almost entirely based on the 14 months of Gaia observations covered by DR1. This also explains the reduced range of proper motion uncertainties seen in the right panel of Fig. 1.

7. Systematic errors

A systematic error (bias) may be understood as the part of an observational error that cannot be eliminated by averaging many measurements. This definition is not very useful in the present context, where there is just one "measurement" of each quantity, namely the particular value given in the catalogue. It is more useful to regard the bias as an (essentially) unknown function of several "circumstantial variables", such as the position, magnitude, and colour of the source. For example, if $e_i = \varpi_i^{\rm TGAS} - \varpi_i^{\rm true}$ is the (unknown) error of the parallax of source i, we imagine that it consists of two parts,

$$e_i = r_i + s(\alpha_i, \delta_i, G_i, C_i, p_i), \tag{7.1}$$

where r_i is the random error and s the systematic error as function of the circumstantial variables. The random errors are unbiased and uncorrelated between the sources.

This error model implies that several sources having similar circumstantial variables also have similar bias s. Averaging many such sources will not reduce the error below s. This behaviour is analogous to the effect of between-source correlations discussed in

Sect. 6. In practice there is no way to separate the between-source correlations from the systematic errors, and the discussion below applies to both.

It is useful to consider which circumstantial variables are believed to be most relevant for the systematics in TGAS, and why. The variables in Eq. (7.1) are as follows.

Position (α_i, δ_i) is important because the scanning geometry and temporal distribution of the scans across a source depend on its approximate position. Similar calibration errors affect the calculated astrometric parameters differently depending on these factors. The scanning law (Gaia Collaboration, Prusti, *et al.* 2016) imprints a complex pattern on the celestial sphere, where the number and distribution of observations may vary discontinuously on angular scales of a few arcmin. This spatial complexity can be seen in Fig. 3 and, on smaller scales, in high-resolution images of the full Gaia DR1 source density (e.g. Fig. 2 in Gaia Collaboration, Brown, *et al.* 2016). Even within apparently "smooth" areas of the sky the bias could vary significantly since the sources will have been observed in different parts of the field of view with different calibration errors.

Magnitude G_i: sources brighter than $G \simeq 12$ obtain shorter CCD integration times, to avoid detector saturation, through the use TDI gates (Gaia Collaboration, Prusti, *et al.* 2016). Each gate effectively means a different geometric instrument, requiring a dedicated geometric calibration with its own calibration errors. This results in different s as a function of G_i even for sources at practically the same position.

Colour C_i: the image centroiding in Gaia DR1 and the subsequent TGAS solution did not take into account that sources have different colours. The combination of wavelength-dependent diffraction and unavoidable optical aberrations makes the image centroid position sensitive to the colour. The colour-coefficient of the shift depends both on the time of the observation and on its location in the field of view. This so-called chromaticity can be calibrated, and thus eliminated, using stars of known colour index. In TGAS this was not yet possible, but it will be done for future data releases. The colour-dependent shifts of the individual CCD observations of a given source combine to create astrometric biases that depend not only on the colour of the source but also on the detailed scanning geometry and the times of the observations.

Prior p_i: in Eq. (7.1) this represents the strength of the prior position taken from the Hipparcos or Tycho-2 catalogue. As shown in Sect. 6 the TGAS astrometry for a source with a weak prior is almost unconstrained by the old position, and it is also expected to be more susceptible to systematic errors from the incomplete instrument calibration. This concerns mainly the Tyc subset and especially the faint part of it. The dependence on p_i will again be a complicated function of the detailed scanning geometry, etc.

On the other hand, the systematic errors are *not* expected to depend on the actual (true) values of the astrometric parameters at the sub-arcsec level. For example, while the systematic error in parallax may depend on the (approximate) position and magnitude of the source, it should not depend on the actual parallax, i.e. on the barycentric distance to the source, other factors being equal. The reason for this is the linearity of the astrometric model, where both the fitted astrometric parameters and their errors are linear combinations of the corresponding centroid shifts on the detectors.

From the discussion above it should be clear that it is very difficult to characterise the function s except in the broadest terms. For a limited number of sources some conclusions can be drawn from comparisons with external data. In their independent validation of the Gaia DR1 catalogue, Arenou *et al.* (2017) confirm the overall quality of the data but also find evidence for a global parallax zero point error of about -0.04 mas, as well as spatially varying offsets on the level of one or a few tenths of a mas.

As described in Appendix E of Lindegren *et al.* (2016), special validation solutions were computed by the Gaia astrometry team as part of their internal assessment of TGAS.

The differences between these solutions and the published catalogue give an idea about the likely sizes and spatial correlations of the systematic errors. In one experiment the input data were partitioned in two halves according to which part of the astrometric field of view was used. This is probably the closest one can get to having two independent Gaia instruments, as the resulting two solutions are expected to have very different $s(\alpha_i, \delta_i, G_i, C_i, p_i)$. A map of the median parallax differences at a resolution of a few degrees shows large-scale patterns with a standard deviation of about 0.1 mas and exceeding ± 0.3 mas in less than 1% of the sky. Other validation solutions give similar results, and these numbers are our best guess at the overall statistics of s at an angular resolution of a few degrees and averaged over the other circumstantial variables.

The parallax uncertainties in TGAS refer to the total error e_i, i.e. the σ_ϖ given in the Gaia Archive is our best estimate of the standard deviation of e_i. This follows from the way the TGAS parallax uncertainties were calibrated (and inflated) through a comparison with the independent Hipparcos catalogue (Appendix B in Lindegren *et al.* 2016).

Gaia Collaboration, Brown, *et al.* (2016) recommend that uncertainties on the parallaxes are quoted as $\varpi \pm \sigma_\varpi$ (random) $\pm$ 0.3 mas (systematic). In view of the previous discussion, this does not mean that the total (random + systematic) uncertainty is $\sqrt{\sigma_\varpi^2 + 0.3^2}$ mas. Since σ_ϖ already represents the total uncertainty, that would clearly be an overestimation. Furthermore, the ± 0.3 mas (systematic) should not be interpreted as an RMS value, but rather as an amplitude of what can reasonably be expected. The user should however always be aware of the possibility of even larger systematics in some areas, for the faint part of the Tyc subset, or sources with very blue or red colours.

Acknowledgements

This work is based on data from the European Space Agency (ESA) mission Gaia (`https://www.cosmos.esa.int/gaia`), processed by the Gaia Data Processing and Analysis Consortium (DPAC, `https://www.cosmos.esa.int/web/gaia/dpac/consortium`). Funding for the DPAC has been provided by national institutions, in particular the institutions participating in the Gaia Multilateral Agreement. The author gratefully acknowledges support from the Swedish National Space Board.

References

Arenou, F., Luri, X., Babusiaux, C., Fabricius, C., Helmi, A., *et al.* 2017, *A&A*, 599, A50

Arias, E. F., Charlot, P., Feissel, M., & Lestrade, J.-F. 1995, *A&A*, 303, 604

Fey, A. L., Gordon, D., Jacobs, C. S., Ma, C., Gaume, R. A., Arias, E. F., Bianco, G., Boboltz, D. A., *et al.* 2015, *AJ*, 150, 58

Gaia Collaboration, Brown, A. G. A.., Vallenari, A., Prusti, T., de Bruijne, J. H. J., Mignard, F., Drimmel, R., Babusiaux, C., *et al.* 2016, *A&A*, 595, A2

Gaia Collaboration, Prusti, T., de Bruijne, J. H. J., Brown, A. G. A., Vallenari, A., Babusiaux, C., Bailer-Jones, C. A. L., Bastian, U., *et al.* 2016, *A&A*, 595, A1

Høg, E., Fabricius, C., Makarov, V. V., *et al.* 2000, *A&A*, 355, L27

Kovalevsky, J., Lindegren, L., Perryman, M. A. C.., Hemenway, P. D., Johnston, K. J., Kislyuk, V. S., Lestrade, J. F., Morrison, L. V., *et al.* 1997, *A&A*, 323, 620

Lindegren, L., Lammers, U., Bastian, U., Hernández, J., Klioner, S., *et al.* 2016, *A&A*, 595, A4

Michalik, D., Lindegren, L., Hobbs, D., & Lammers, U., 2014, *A&A*, 571, A85

Michalik, D., Lindegren, L., & Hobbs, D., 2015a, *A&A*, 574, A115

Michalik, D. & Lindegren, L., 2016, *A&A*, 586, A26

Mignard, F., 2009, unpublished technical note GAIA-C3-TN-OCA-FM-040, http://www.rssd.esa.int/doc_fetch.php?id=2939272

van Leeuwen, F., 2007, *Hipparcos, the New Reduction of the Raw Data*, Astrophysics and Space Science Library, Vol. 350

Astronomy and Astrophysics in the Gaia sky
Proceedings IAU Symposium No. 330, 2017
A. Recio-Blanco, P. de Laverny, A.G.A. Brown
& T. Prusti, eds.

doi:10.1017/S1743921317005889

Astrometric surveys in the Gaia era

Norbert Zacharias
U.S. Naval Observatory, 3450 Mass.Ave.NW
Washington DC 20392, USA
email: nzIAUc8@gmail.com

Abstract. The Gaia first data release (DR1) already provides an almost error free optical reference frame on the milli-arcsecond (mas) level allowing significantly better calibration of ground-based astrometric data than ever before. Gaia DR1 provides positions, proper motions and trigonometric parallaxes for just over 2 million stars in the Tycho-2 catalog. For over 1.1 billion additional stars DR1 gives positions. Proper motions for these, mainly fainter stars ($G \geqslant 11.5$) are currently provided by several new projects which combine earlier epoch ground-based observations with Gaia DR1 positions. These data are very helpful in the interim period but will become obsolete with the second Gaia data release (DR2) expected in April 2018. The era of traditional, ground-based, wide-field astrometry with the goal to provide accurate reference stars has come to an end. Future ground-based astrometry will fill in some gaps (very bright stars, observations needed at many or specific epochs) and mainly will go fainter than the Gaia limit, like the PanSTARRS and the upcoming LSST surveys.

Keywords. astrometry, surveys, catalogs, reference systems, stars: kinematics.

---

1. Introduction

This paper is limited to a review of current and future ground-based, optical to near-IR, wide-field astrometric survey research in the Gaia era. There are 2 distinct phases. We are now in the first phase which will last until the second Gaia data release, DR2, scheduled for April 2018. In this period of time ground-based efforts concentrate on a) using DR1 data for almost error-free reference stars to re-calibrate CCD data, and b) derive proper motions for large numbers of stars, not available in DR1. The second phase is after DR2 when full astrometric solutions including proper motions and parallaxes will become available for almost any star and many compact extragalactic sources in the about 4 to 20.5 magnitude range at unprecedented accuracy.

2. Gaia DR1

The astrometric properties of Gaia DR1 data are described in detail by Lindegren *et al.* (2016). A short summary of the important numbers is provided in Table 1.

The Tycho-Gaia Astrometric Solution (TGAS) of DR1 uses the Hipparcos Catalogue (van Leeuwen 2007) and the Tycho-2 catalog (Høg *et al.* 2000) to resolve the degeneracy between proper motion and parallax over the short time span Gaia data. The major difference between the Hipparcos stars subset and the entire TGAS set of stars is the over 10 times smaller error in the Hipparcos set proper motions. This difference is not very important when using the Gaia DR1 data for reference stars in astrometic solutions of ground-based CCD data because there are few Hipparcos stars and most of them are too bright for the purpose.

The DR1 data do provide highly precise and accurate positions at a near current epoch (2015). However, of great importance here is the lack of proper motions of all

Table 1. Overview of Gaia DR1 astrometry. Errors are given for the 10 and 90 percentile according to Lindegren *et al.* (2016).

	Hipparcos		TGAS		secondary	
number of sources	93,635		2,057,050		1.14 billion	
main magnitude range	6 - 10		8 - 11		12 - 21	
position error [mas]	0.15	0.39	0.14	0.60	0.26	12.9
proper motion error [mas/yr]	0.03	0.13	0.44	2.67	—	—
parallax error [mas]	0.23	0.50	0.24	0.64	—	—

Table 2. Recent proper motion surveys utilizing Gaia DR1 data.

survey name	1st epoch date [year]	number of stars [million]	proper motion error [mas/yr]	based on what 1st epoch data
HSOY	1950 - 1975	583	1 – 5	PPMXL, USNO-B1, Schmidt plates
UCAC5	1998 - 2004	107	1 – 5	UCAC, CCD data, no photogr.plates
GPS1	multiple	350	1.5 – 3	PS1, SDSS, 2MASS
PMA	≈ 2000	421	2 – 10	2MASS

non-TGAS stars in DR1. This fact prompted a flurry of efforts to supplement Gaia with preliminary proper motions. These are better than anything offered by ground-based, global astrometry before. Of course, the life-span of these products will be short. The Gaia DR2 will make those products obsolete. However, in the meantime they provide much needed support for astrometric reductions of ground-based CCD imaging.

3. Phase 1: between DR1 and DR2

3.1. *New proper motions*

A summary of these new proper motion surveys and their properties is given in Table 2. All are using the Gaia DR1 (secondary sources) as their 2nd epoch and all are constructed to be on the International Celestial Reference Frame (ICRF). Most are claiming quasi absolute proper motions with a direct link to extragalactic sources. All are based on a single set of early epoch data, except the GPS1. All cover the entire sky, except for GPS1.

The first product of new proper motions to become available is the Hot Stuff for One Year (HSOY) catalog (Altmannn *et al.* 2017). It is based on PPMXL (Röser *et al.* 2010) using USNO-B1 Schmidt plate survey data (Monet *et al.* 2003) as first epoch. There is a big difference in epoch difference between the northern part (early Palomar data) and the remaining southern hemisphere area which was observed much later. The epoch difference of course propagates into the proper motion errors.

The UCAC5 (Zacharias, Finch & Frouard 2017) proper motions are based on a completely new reduction of the USNO CCD Astrograph Catalog (UCAC) observations using TGAS and then combine the about epoch 2000 data with DR1. This has the advantage of having both sets of data on the exact same reference frame to begin with and avoiding any photographic plate data which are often subject to large systematic errors depending on a combination of magnitude, color and location in the field. However, the UCAC data are not entirely free of these systematic errors (along right ascension) due to the poor charge transfer efficiency (Zacharias *et al.* 2004) of the particular detector used.

The GPS1 proper motion catalog (Tian *et al.* 2017) uses several early epoch data: the first PanSTARRS (panstarrs.stsci.edu) data release (PS1) (Chambers *et al.* 2017), Sloan Digital Sky Survey (SDSS) (Munn *et al.* 2004) and near-IR 2MASS data (Skrutskie *et al.* 2006). The GPS1 proper motion catalog covers about 3/4 of the sky (the PS1 area) and claims systematic errors below 0.3 mas/yr. Using more than just one early epoch to calculate new proper motions should result in realistic error estimates. However, the scatter of observed proper motions of extragalactic sources in GPS1 is significantly larger than expected from their formal errors.

The Proper Motion Absolute (PMA) catalog (Akhmetov *et al.* 2017) uses 2MASS data as first epoch. It was constructed to be independent of the ICRF and Hipparcos Celestial Reference Frame (HCRF) using extragalactic sources for the zero-point of its proper motion system and covers the large 8 to 21 magnitude range.

3.2. *New parallaxes*

Not directly linked to Gaia data but also a temporary product until the Gaia DR2 becomes available is the URAT Parallax Catalog (UPC) (Finch & Zacharias 2016) based on the USNO Robotic Astrometric Telescope (URAT) nothern hemisphere observations. UPC gives results for over 112,000 stars in the 6.5 to 17 mag range. This is the largest catalog of trigonometric parallaxes since the Hipparcos space mission.

The UPC parallaxes compare very well with previously published trigonometric parallaxes, however, the precision is limited with typical errors of about 3 to 10 mas, which is not surprising considering that the telescope used for these observations has a focal length of only 2.06 m. Selection criteria for stars in common with other, previously available data have been relaxed (e.g. number of required observations, formal error on parallax), while for stars without previously published trigonometric parallax the selection criteria are tighter to avoid contamination of the sample by false positives. Nevertheless UPC provides first published trigonometric parallaxes for over 53,000 nearby stars.

3.3. *Example: UCAC5*

How much do Gaia DR1 reference stars actually help in the re-reduction of earlier astrometric catalog data? As an example the UCAC5 data are used here (see also above). Over 250,000 CCD exposures were taken with the USNO "redlens" astrograph for the UCAC project between 1998 and 2004. Observing begun at Cerro Tololo, Chile, covering the sky from $\delta = -90°$ to about $\delta = +25°$. The rest of the northern sky was observed from Flagstaff, AZ. While the field of view of the astrograph lens is about 9° in diameter, only a 1.0 square degree area near the optical axis was used with a Kodak 4k by 4k CCD at a scale of 0.905 arcsec/pixel. At the time this was the largest CCD on the mountain. A single, fixed bandpass (579 to 643 nm) was used. The survey covers the about R = 8 to 16 mag range with a 2-fold overlap pattern and a long (100 to 150 sec) and a short exposure (1/5 of long) on each field.

Results based on Tycho-2 reference stars were published previously (UCAC4, Zacharias *et al.* 2013). Systematic errors of the CCD data were corrected as a function of magnitude, location in the field and subpixel phase using 2MASS data allowing for residuals covering the entire UCAC magnitude range. However, due to the poor charge transfer efficiency of the 4k CCD systematic errors mainly along RA remain on the 10 mas level.

Now TGAS was used as reference star catalog for a re-reduction of the UCAC exposures. These positions were published as UCAC5 together with new proper motions in combination with Gaia DR1 data. Both UCAC4 and UCAC5 astrometric solutions use a 6-parameter, linear "plate" model in a weighted least-squares adjustment. The same systematic error correction model as function of magnitude (CTE) was adopted for UCAC5

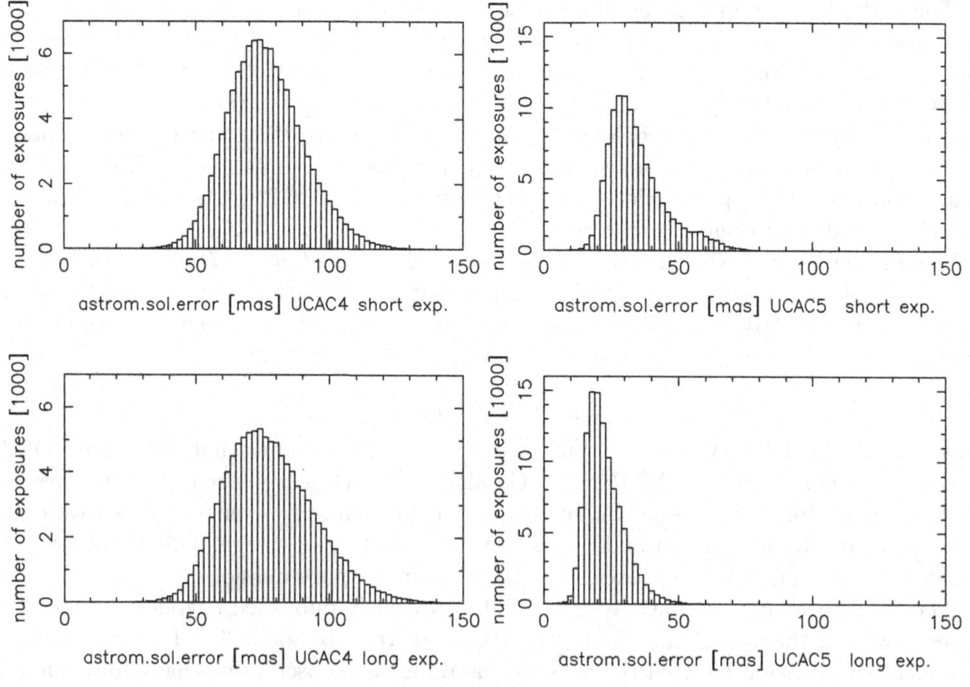

Figure 1. Astrometric solution error for UCAC4 (left) and UCAC5 (right).

as was used with UCAC4. The field distortion pattern and sub-pixel corrections were updated for UCAC5. Deriving new corrections as a function of magnitude (most needed here) could not be accomplished due to the limited magnitude range of TGAS, ending at about V = 11.5. The distribution of the astrometric solution errors for these about 250,000 individual exposures is shown in Fig. 1 with the same binning, separately for the long and short exposures.

A dramatic decrease in astrometric solution error (about factor of 3) is seen when using Gaia data instead of Tycho-2 reference stars. Interesting to note is that the mean astrometric solution error for UCAC4 is slightly smaller for short than for long exposures. The opposite, with significant difference is true for the UCAC5 data. This is because the error contribution from the turbulent atmosphere is much smaller for the long than the short exposures. However, for the UCAC4 data the long exposures are impacted by larger than average Tycho-2 reference star errors while the more precise, brighter reference stars are saturated and not used in the reductions. The short exposures have the benefit of using more, in particular brighter, more accurate Tycho-2 reference stars. These 2 factors (atmospheric turbulence and reference star errors) cancel each other about out for UCAC4. For UCAC5 however, the reference star errors (TGAS) are very small for all stars including the faint end and "plate" solutions of the long exposures are not significantly degraded. Thus for UCAC5 the benefit of the longer integration times (smaller errors from the atmopshere) becomes dominant.

But how much does the Gaia data help for the field star positions obtained from the UCAC exposures? This is a matter of error propagation of the "plate" parameters. Over 50 years ago a monumental paper was published on this topic (Eichhorn & Williams 1963). The answer to our question strongly depends on the complexity of the "plate" model. It is highly recommended for anyone dealing with astrometric reductions of CCD image data to read this paper. Often a complex model, like a full 2nd or even 3rd order

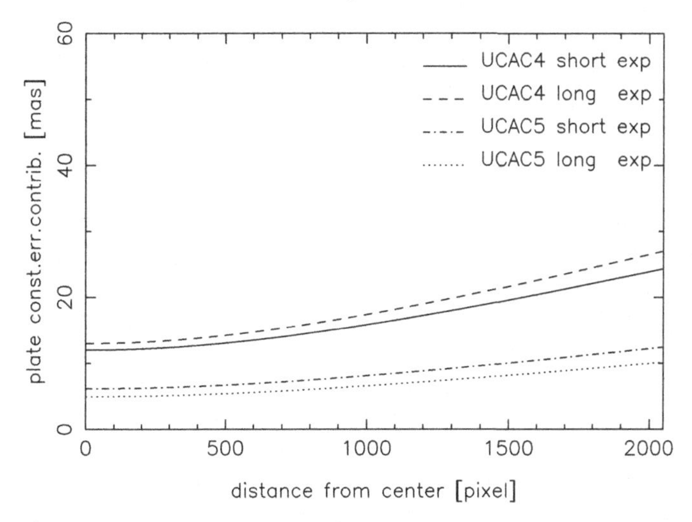

Figure 2. Contribution from plate parameter error propagation.

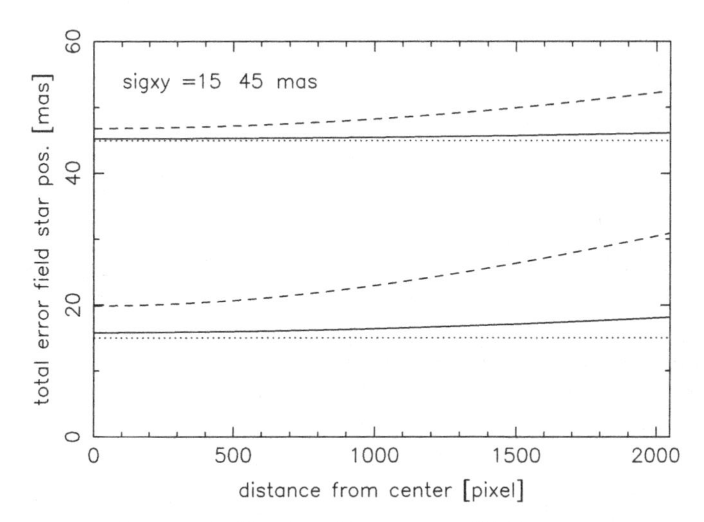

Figure 3. Total position errors of field stars.

polynomial is used to "make sure" possible systematic effects in the focal plane mapping are covered, without noticing what harm is done by doing so. Although the residuals of such a reduction look "very good" the parameters bear large errors which are propagated into field star positions.

For the UCAC4 / UCAC5 example this fortunately is benign due to the simple linear "plate" model. Figure 2 shows the "plate" parameter error contribution for field star positions as a function of distance from the CCD center, separately for long/short exposures and UCAC4 / UCAC5 case. As can be seen the increase in error contribution toward the edge of the field is moderate. These error contributions are about a factor of 2 smaller for UCAC5 data (TGAS reference stars) than for UCAC4 data (Tycho-2 stars).

Going a step further, Fig. 3 shows the root-sum-square (RSS) of image center position errors (sigxy) and the "plate" parameter error propagation contribution to field star positions. Results for 2 cases are shown for sigxy = 15 and 45 mas (lower and upper set of lines, respectively). For each case the larger error line is for UCAC4 data, the lower for UCAC5 and the dashed line for the limiting case with sigxy only and no "plate" parameter error contribution. The smallest image center error in UCAC4 data has sigxy = 9 mas, for a bright star just under saturation of high quality data. The sigxy = 15 mas case is a typical star of high S/N ratio and the 45 mas case is an average field star at S/N = 20.

As can be seen the improvement in positions obtained from UCAC data using Gaia reference stars is largest for high S/N, bright targets, while for most field stars the improvement near the center of the CCD is minimal and not much larger at the edge of the field. This just means that in the case of UCAC (1 square degree field, with typically about 25 to 100 Tycho-2 reference stars) the astrometric solution is already very good despite the relatively large errors of individual Tycho-2 star positions.

More important than the small random errors are the very small systematic errors of the Gaia DR1 data. Re-reduction of ground-based survey data will be on the Gaia coordinate system with much smaller local systematic errors than previously achieved. As for random errors, the big improvement of TGAS over Tycho-2 reference stars is not so important if enough reference stars are in the field. The quality of the CCD data (image center positions) now becomes the limiting factor for the overall result. Major improvements for re-calibration of ground-based astrometric survey data will likely have to wait until the Gaia DR2, when a much wider magnitude range becomes available with very small position errors, allowing to eliminate the critical systematic errors as function of magnitude and combinations of magnitude with other parameters.

4. Phase 2: after DR2

The Gaia DR2 will likely have an even bigger impact on optical astrometry than DR1. DR2 will provide a full astrometic solution (position, proper motion and parallax) for most stars (and compact, extragalactic sources) in the 4 to 20.7 mag range on the 0.1 to 1 mas level (depending on magnitude). All above mentioned, preliminary proper motion catalogs based on Gaia DR1 and earlier, ground-based data will become obsolete. In fact, all traditional, optical, ground-based surveys like CMC, GSC, NPM, SPM, PPM, PPMXL, UCAC, URAT, USNO-B, and others will become obsolete for astrometric data. At this point the era of optical, ground-based, wide-field astrometric surveys which begun with the Astrographic Catalogue project in the 1880s (Eichhorn 1974) is coming to an end. No optical reference frame densification effort will be needed, Gaia DR2 and later releases will provide the reference frame and the most dense realization possible (all objects to a limiting magnitude) at the same time.

So, is there anything left for ground-based astrometry, is there life after Gaia? Yes, there is. Ground-based, wide-field astrometry has already and will in future branch out into the few remaining areas not covered by Gaia. These are:

(*a*) Very bright stars ($G \leqslant 4$), until near final Gaia DR, likely after 2020.
(*b*) Faint objects ($G \geqslant 20.7$).
(*c*) Other than optical bandpass, like near-IR.
(*d*) Complex motion cases, or variable center objects.

Each of these areas will be described in more detail below. Astrometry will continue to support astrophysics in this new era. It is important not to forget the mindset, methods and skills needed for astrometry in this transition. Many of the ideas and tools of past astrometric projects are still applicable today and in the future. A big help for students in this respect is the recent textbook (van Altena 2013) on this subject.

Table 3. Current and future, deep, wide-field surveys.

survey name	first light	survey begin	telescope aperture	bandpass	camera size Mpx	field of view sq.deg	R mag range
SDSS	1998	2000	2.5 m	u,g,r,i,z	120	1.5	14-23
Skymapper	2008	2014	1.3 m	u,v,g,r,i,z	268	5.7	13-22
PanSTARRS	2008	2009	1.8 m	g,r,i,z,y	1000	7.0	14-23
CFHT	2003		3.6 m	u,g,r,i,z +	340	1.0	15-24
DECam	2012	2013	4.0 m	u,g,r,i,z,Y	520	3.0	15-24
ZTF	2017	2018	1.2 m	g, R	576	47.0	14-21
LSST	2020	2022	8.4 m	U,G,R,I,Z,Y	3000	9.6	18-26
Vista	2009	2010	4.1 m	near-IR	67	0.6	

Note: Limiting magnitudes given here are for stacked images except for ZTF. The saturation limit is approximate, strongly depending on exposure time and seeing conditions.

4.1. *Very bright stars*

Accurate positions of nacked-eye stars are important for navigation. The currently best data are still based on Hipparcos observations (ESA 1997) and Tycho-2 proper motions, which are often more reliable than the Hipparcos proper motions due to the much longer time baseline (about 100 years vs. 3.5 years). Northern Hemisphere URAT observations are limited to stars of about magnitude 6 and fainter, while the Gaia DR1 lists only few stars brighter than 5th magnitude. Positions of stars brighter than the DR1 limit are only on the 30 mas level at current epochs with a good fraction of stars showing much larger inconsistencies likely due to unresolved multiplicity issues.

The UBAD program (Subasavage, private com.) uses the 61in telescope at the Naval Observatory Flagstaff Station (NOFS) with very dense neutral density spots of 9 and 12.5 magnitude attenuation to observe bright stars. URAT, operated from CTIO uses a neutral density spot with about 4 mag attenuation in combination with an objective grating, targeting all stars from Sirius to R = 4.5 mag within about $-89° \leqslant \delta \leqslant +23°$. For most targets several exposures of 30 or 60 sec are taken per night, while the handful of brightest stars are observed with about a dozen 5 or 10 sec exposures. This results in mean positions of 10 mas precision per night. Each target star is observed multiple times during a year. These observations begun in October 2015 and are still ongoing. Results from both programs will be published soon.

4.2. *Faint objects*

Going deeper than Gaia is the main area of research for wide-field, ground-based, astrometric surveys these days and in the coming years. However, astrometry is only a part of these programs with photometry being another major objective. These programs fall under the umbrella of "time domain astronomy" with high cadence observations to achieve multiple science goals. A main driver for photometric time domain observations is the discovery and characterization of rare or transient phenomena. The astrometric relevance of these high cadence observations is described in more detail in the following section. Table 3 provides an overview of current and future, deep, ground-based surveys with relevance to astrometry.

The Sloan Digital Sky Survey (SDSS) project was the first of its kind (Gunn *et al.* 2006) aiming at a deep, wide-field survey. Beyond imaging SDSS also can obtain spectra of thousands of objects via fibers. SDSS is operated from Apache Point, New Mexico and was used for various projects, now being past its 13th data release.

Skymapper (Keller *et al.* 2007) is the southern hemisphere counterpart to SDSS operating from Siding Spring, Australia. An early data release was in May 2016 with a first full-sky data release expected in 2017.

The Panchromatic Survey Telescope and Rapid Response System (PanSTARRS) operates from Haleakala, Hawaii with now 2 of the planned 4 telescopes. The focal plane consists of 60 orthogonal-transfer CCDs. PanSTARRS had its first data release (PS1) in Dec. 2016. PS1 consists of several surveys with the 3π steradian survey covering all of the sky north of $\delta = -30°$. A second data release is scheduled for early 2018.

The Canadian-Frensh-Hawaii Telescope (CFHT) MegaCam instrument was used for deep, wide-field imaging from its Mauna Kea location on Hawaii. It is used for many projects and the telescope also has other instruments than the imaging camera, including the near-IR WIRCam.

The Dark Energy Camera (DECam) operates at the prime focus of the Blanco telescope at Cerro Tololo, Chile (Abbott *et al.* 2016). It is currently working on a 5-year survey of a 5000 sq.deg area in the southern sky.

The Zwicky Transient Facility (ZTF) (Bellm 2014) is the follow-up project to the successful Palomar Transient Factory (PTF), both using the 48in Palomar Schmidt telescope. While the PTF had a relatively small focal plane detector, the ZTF utilizes the full field of view of this survey telescope which undertook the famous Palomar Schmidt Surveys in the 1950s and 1970s on photographic plates.

The Large Synoptic Survey Telescope (LSST) will operate from Chile. Construction begun in August 2014. Full survey operation is expected to start in 2022 for 10 years. The telescope has a primary mirror of 8.4 m diameter but a very large central obstruction resulting in an effective collecting area equivalent to a 6.7 m diameter, unobstructed mirror. LSST will take 15 sec exposures every 20 sec on 189 CCDs with 0.2 arcsec/px sampling. The technological and data reduction challenges, and its budget is comparable to a mid-size space mission.

The Visible and Infrared Survey Telescope for Astronomy (VISTA) currently has a single, large, near-IR imaging instrument and is operated from Paranal, Chile. A variety of surveys have been undertaken with different sky coverage and depth. A multi-object spectrograph is under development.

4.3. *Complex motion, variable center location*

Understanding the dynamics of natural satellites in our solar system and deriving accurate ephemerides is an example where a very long time span of moderately accurate observations is more important than a short time span with very accurate observations. Digitizing photographic plates taken over many decades and reducing these data with modern reference stars provided significantly better results than Hipparcos observations alone can do (Robert *et al.* 2011).

These are cases where the motion of targets is complex. Although Gaia will observe each target on average about 50 to 100 times, the cadence is pre-determined by the scanning law and observations are limited to a relatively short lifetime of the mission. Astrometric objectives which require a target specific cadence, more observations or a longer timespan will resort to ground-based observations in addition to Gaia data.

Other examples of supporting ground-based observations are mass determination of asteroids from close encounters (Ivantsov 2017) and observations of binary star systems, where typically 50% or more of an orbital motion needs to be seen to obtain a high quality orbit solution. Things can become even more complex when the observed center shifts due to brightness variation or color induced motion (Makarov *et al.* 2017).

4.4. *Other than optical bandpass*

Using different wavelength or bandpass observations is mainly performed to sample the flux of a target, i.e. for photometry. However, in some cases there are also astrometric implications. A shift of the center of an object may be wavelength (frequency) dependent. Different astrophysical phenomena are seen at different wavelengths. An example even relevant for the definition of the reference frame itself are possible center shifts of active galactic nuclei (AGN) e.g. between optical and radio data due to source structure seen differently at different wavelengths (Petrov 2017).

Thus ground-based surveys performed in the near-IR can offer additional astrometric data beyond what Gaia data can provide. For most tagets (like ordinary stars) the difference between optical and near-IR positions will be very small. However, those objects which display a significant difference will be of astrophysical importance.

5. Summary

(*a*) The primary Gaia DR1 catalog (TGAS) with its 2 million stars is the new optical reference frame surpassing Hipparcos and Tycho-2 in accuracy. Zonal systematic errors in Tycho-2 proper motions are seen which are no longer present in TGAS.

(*b*) The main short-term ground-based, astrometric activities (until DR2) are:

- use TGAS data to reduce earlier epoch observations
- combine the DR1 position data of 1.14 billion stars with earlier deep surveys to obtain improved proper motions on the Gaia system
- derive trigonometric parallaxes from recent gound-based, high precision surveys, mainly the URAT parallax catalog

(*c*) All traditional, ground-based, wide-field, astrometric surveys in the 4 to 20.7 mag range will become obsolete with Gaia DR2 by about April 2018.

(*d*) Current and future ground-based, astrometric surveys go into areas of reseach not covered by Gaia:

- Very bright stars ($G \leqslant 4$) are currently observed with UBAD and URAT.
- Many surveys go deeper than Gaia, like PanSTARRS and LSST. This is a major area of reseach but not limited to astrometry, rather including also photometry and time-domain astronomy in general.
- Complex motions (solar system, masses of asteroids, multiple star systems, exoplanets, object center motion as function of variability) need many observing epochs, often at specific times or over very long periods of time, where Gaia plays a role but needs to be supplemented by additional ground-based observations.
- Astrometric observations at other than optical bandpasses (like near-IR and radio) are sometimes needed to investigate position shifts as function of wavelength, as for example seen in some AGN objects or double stars with different color components.

References

Abbott, T. *et al.* (Dark Energy Survey Collaboration) 2016, *arXiv:*1601.00329v3
Akhmetov, V. S., Federov, P. N., Velichko, A. B., & Shulga, V. M. 2017, in press with *MN*
Altmann, M. *et al.* 2017, *arXiv:*1701.02629
Bellm, E. C. 2014, *arXiv:* 1410.8185
Chambers, K. C., Magnier, E. A. *et al.* 2017, *arXiv:*1612.05560

Eichhorn, H. 1974, *Astronomy of Star Positions*, Frederick Ungar Publishing, New York
Eichhorn, H. & Williams, C. A. 1963, *AJ*, 68, 221
European Space Agency, 1997, SP 1200
Finch, C. T. & Zacharias, N. 2016, *AJ*, 151, 160
Gunn, J. E., *et al.* 2006, *AJ*, 131, 2332
Høg, E., Fabricius, C., Makarov, V. V. *et al.* 2000, *A&A*, 335, 27; Vizier catalog I/259
Ivantsov, A. 2017, *this proceedings* asteroid masses from close encounters
Keller, S. C. *et al.* 2007, *Pub. Asron. Soc. Australia*, 24, 1
Lindegren, L., Lammers, U., Bastian, U. *et al.* 2016, *A&A*, 595, A4
Makarov, V. Frouard, J., Berghea, C. T. *et al.* 2017, *ApJ*, 835, 30
Munn J. A., Monet, D. G., Levine, S. E. *et al.* 2004, *AJ*, 127, 3034
Monet, D. *et al.* 2003, *AJ*, 125, 984
Petrov, L. 2017, *this proceedings*
Robert, V., De Cuyper, J. P., Arlot, J. E. *et al.* 2011, *MNRAS*, 415, 701
Roeser, S., Demleitner, M., & Schilback, E. 2010, *AJ*, 139, 2440
Skrutskie, M. F., Cutri, R. M., Stiening, R. *et al.* 2006, *AJ*, 131, 1163
Tian, H.-J. *et al.* 2017, *arXiv*:1703.06278
van Altena, W. F. (Editor) 2013, *Astrometry for Astrophysics*, Cambridge University Press
van Leeuwen, F. 2007, *A&A* 474, 653; Vixier catalog I/311
Zacharias, N., Urban, S. E., Zacharias, M. I. *et al.* 2004, *AJ*, 127, 3043
Zacharias, N., Finch, C. T., Girard, T. M. *et al.* 2013, *AJ*, 145, 44
Zacharias, N., Finch, C., & Frouard, J. 2017, *AJ*, 153, 166

Discussion

?: What is the percentage of double stars causing problems? Most stars are multiple, what is the impact on these catalog data?

ZACHARIAS: This is not very well known at this point. The impact on wide-field astrometric surveys strongly depends on the type of double star, i.e. depending on their separation and magnitude difference. An approximate estimate is 10 to 20% of stars are seen as "astrometric problematic" in many surveys.

HØG: What is the status of LSST, when will it become online?

ZACHARIAS: LSST is on schedule. The ground has been broken in Chile and constructions are underway. Major parts of the telscope have already been completed. First light is expected in 2020. Routine survey observing will begin about 2 years after first light.

Astronomy and Astrophysics in the Gaia sky
Proceedings IAU Symposium No. 330, 2017
A. Recio-Blanco, P. de Laverny, A.G.A. Brown
& T. Prusti, eds.

doi:10.1017/S1743921317006445

Multiply imaged quasars in the Gaia DR1

C. Ducourant[1], L. Delchambre[2], F. Finet[2], L. Galluccio[3], A. Krone-Martins[4], J.F. Le Campion[1], F. Mignard[3], E. Slezak[3], J. Surdej[2], R. Teixeira[5] and O. Wertz[6]

[1]Laboratoire d'Astrophysique de Bordeaux, Univ. Bordeaux, CNRS, France
email: christine.ducourant@u-bordeaux.fr

[2]University of Liège, Belgium

[3]Observatoire de Côte d'Azur - OCA, France

[4]Universidade de Lisboa (CENTRA/SIM/FCUL), Portugal

[5]Universidade de São Paulo, IAG, São Paulo, Brazil

[6]Bonn University, Germany

Abstract. Because of to its exceptional resolving power, Gaia should detect a few thousands gravitational lensed systems. These consist in multiple images of background quasars. The estimated number of lens phenomena in the sky, however, depends on the cosmological model considered. By taking into account the observational bias that will restrict the detection of lensed quasars, identification of these up to a given limiting magnitude will constrain the cosmological parameters.

We have investigated the known gravitationally lensed quasars present in the Gaia DR1, and found that a significant number of components of these systems have been measured and are present in the Gaia DR1 catalogue although quasi none of them have all their components detected. We additionally examined the immediate surroundings of QSOs from the large Quasar catalogue, LQAC3, and detected several configurations compatible with gravitational lensing phenomena. A more global strategy to systematically detect the potential candidates in the various releases of the Gaia catalogue is presented.

Keywords. astrometry, gravitational lensing, quasars: general

---

1. Introduction

Although it has been designed for the exploration of the Milky Way, the ESA - Gaia satellite (Perryman *et al.* 2001, Gaia Collaboration 2016b) is observing many extragalactic objects. It is expected to detect about one million of galaxies (De Souza *et al.* 2014) and 0.5 million quasars (Mignard 2012). These distant objects may be subject to lensing phenomena and in some cases multiple images of quasars will be detected by the satellite (Finet and Surdej 2016). Depending on the model of Universe considered and on the deflecting population of galaxies, thousands quasar gravitational lenses (GL) could be present in the Gaia catalogue. These ones would be for most composed of 2 images but several with 3 and 4 images should also be detectable. Counts of GL will bring independent constraints on the cosmological parameters (H_0, Ω_0, Λ_0) (Finet and Surdej 2016).

Up to now, quasar gravitational lensing phenomena have been discovered from ground with the limitation of poor resolving power and only a few hundreds systems are known today. They were either discovered in quasar studies or by mining essentially the Sloan Digital Sky Survey (Inada *et al.* 2012, More *et al.* 2016, etc.) which is not an all-sky survey. In this respect Gaia will be the first all-sky survey of gravitational lenses.

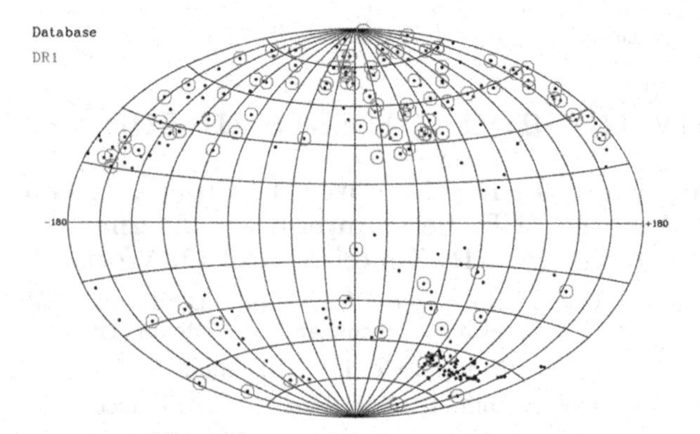

Figure 1. Distribution of the known multiply imaged quasars in galactic coordinates. Circles indicate the detection by Gaia (as given in DR1) of at least one of the images of the quasars.

Moreover, since Gaia observes from space its resolving power is exceptional and should allow to disentangle multiple images that would be merged into a single image viewed from ground and that would thus remain undetectable without Gaia.

We present here the efforts that we dedicate to detect GL from the various data releases of Gaia (Gaia Collaboration 2016). These efforts are in three directions : 1- search for known GL, 2- explore the surrounding of known quasars for neighbours (i.e. potential multiple images of the central quasar) and 3- blind search the Gaia data releases for lensing configurations.

2. The known quasar gravitational lenses

To evaluate the potentiality of Gaia in terms of detection of gravitational lens systems we searched the litterature to set up an updated database of multiply imaged quasars. The main source of data comes from the Gravitational Lens Database of Castles (Munoz *et al.* 1999) (see https://www.cfa. harvard.edu/castles/) which unfortunately seems not to have been updated since several years. We completed these data essentially with the SQLS - SDSS Quasar Lens Survey (Inada *et al.* 2012) that contains resolved and unresolved GL candidates and with the SDSS - III BOSS quasar lens survey (More *et al.* 2016). Our database comprises 272 GL systems from which 20 have 3 or more images. We present in Fig. 1 the locations on the sky of the gravitational lenses from our database together with their detection by Gaia as published in the Gaia DR1 (Gaia Collaboration 2016a). Most of the known multiply imaged quasars are partially present in the Gaia DR1 but most of these do not have all their components published. The selection of the objects to be published in this first release is at the origin of the absence of many of these components and the situation should be much improved in the Gaia DR2 with this respect.

3. Automatic search around known quasars

To detect new multiply imaged quasars, we searched the Gaia DR1 for companions around known quasars. We thus considered the $\sim$ 322 000 sources of the Large Quasars Astrometric Catalogue LQAC3 (Souchay *et al.* 2015) and searched the Gaia DR1 in a radius of 3". In the case of about 209 000 quasars, a single counterpart is found. But in 1300 cases, 2 sources (including the LQAC3 quasar) are found, in 70 cases 3 sources and in 10 cases 4 and more sources. Most of these cases correspond to optical

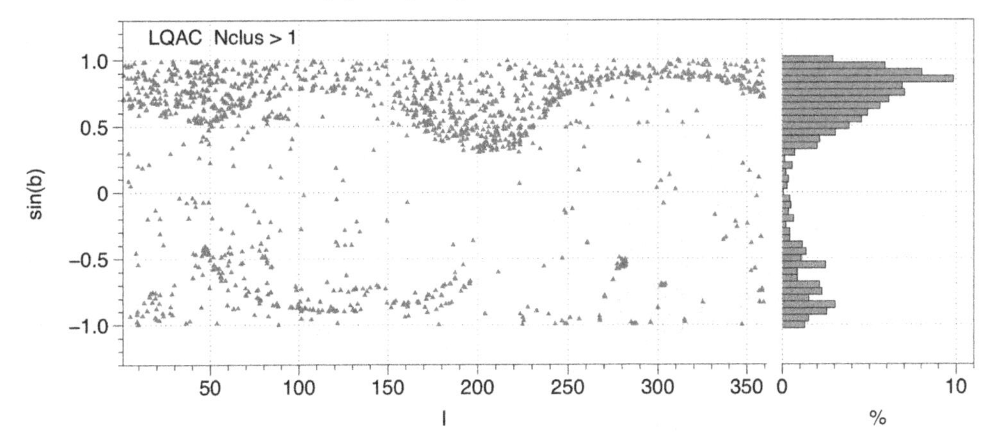

Figure 2. Distribution of clusters of objects including a LQAC3 quasar in Gaia DR1 (Galactic coordinates). On right part frequency of clusters as function of galactic latitude.

projections of stars in the direction of the quasar but several of these configurations may be gravitational lens systems. We notice that some well known gravitational lenses are recovered by this method; the unknown configurations need serious analysis and additional observations for validation. We present in Fig. 2 the locations on the sky of the clusters of objects around LQAC3 quasars that comprise 2 or more sources found in Gaia DR1. The inhomogeneous repartition of candidates clearly reproduces the footprint of the LQAC3 catalogue (mostly SDSS footprint).

4. Automatic blind detection of Gravitational Lenses in Gaia DR1

To systematically detect the gravitational lens systems in Gaia releases we developed the SELenA (Systematic Exploration of Lenses from Astrometry) pipeline. This is an *intelligent* and adaptable framework designed to search lenses in astrometric catalogues and to model them. It may incorporate additional informations and constraints such as photometry, spectroscopy, time-series, etc. The Gaia DR1 is obviously an excellent benchmark for this framework. At first the full Gaia DR1 is handled and clusters of objects in the (RA,Dec) space are isolated and their similarity in other spaces such as G magnitudes, color, proper motions, etc... is also tested (see Fig. 3-*Left* for the known properties of images of lensed quasars). Then a machine learning based classifier is applied to exclude the low probability configurations. The classifier is beforehand trained with simulations. Finally a forward model implemented in OpenCL (that may run in CPU, GPU or FPGA) performs an efficient sampling of the parameter space of the lens system (deflector : axis ratio, major axis, Einstein radius (H0+Mass+redshift), position angle, shear terms – background quasar : angular distance to the lens, position angle) for the Bayesian inference process (see for example Fig. 3-*Right* for the results of one modelling). This model is able to test millions of lens configurations per second. We applied the first step of SELenA to the whole Gaia DR1, and could derive a list of a few thousands candidates GL with 3 and 4 images outside the galactic plane ($| b | \geqslant 50$), with differences of positions smaller than 3" and differences in magnitude less than 3 magnitudes. Indeed the availability of proper motions and BP/RP photometry in the next data release will greatly help to reduce the number of candidates. The candidates are under analysis in order to propose the most promising ones for validation via multi colour imaging.

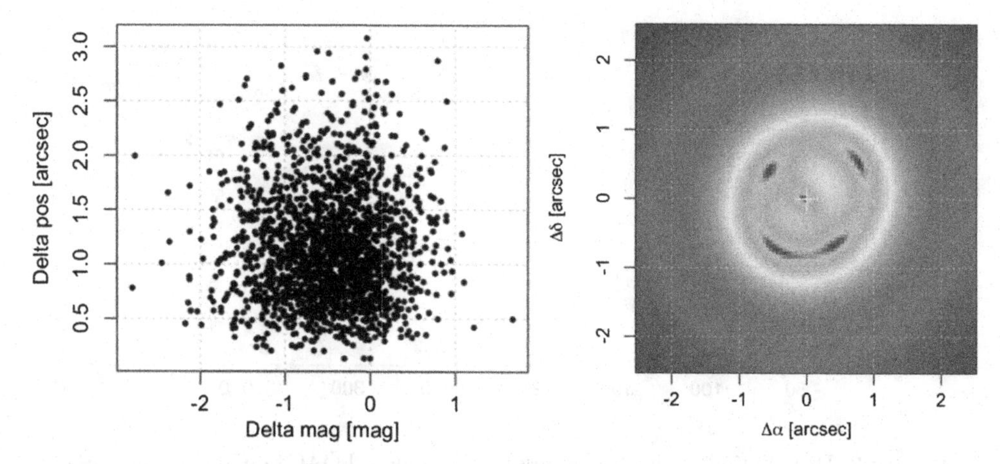

Figure 3. Left : known properties (differences of magnitude, differences of position) of images of lensed quasars (as given by model from Wertz and Surdej (2014)). **Right** : SELenA simulated "happy" lens.

5. Conclusion

We presented here three strategies to detect and recover gravitational lenses from the Gaia DR1. A preliminary list of good candidates has been established and is under evaluation for validation.

Acknowledgment

This work has made use of data from the European Space Agency (ESA) mission *Gaia* (https://www.cosmos.esa.int/gaia), processed by the *Gaia* Data Processing and Analysis Consortium (DPAC, https://www.cosmos.esa.int/web/gaia/dpac/consortium). Funding for the DPAC has been provided by national institutions, in particular the institutions participating in the *Gaia* Multilateral Agreement.

We are grateful to the French and Brazilian organisms COFECUB, FAPESP, CAPES and CNPq for financial support.

References

Adelman-McCarthy J. *et al.* 2017, *ApJS*, 175, 297-313

de Souza, R. E., Krone-Martins, A., dos Anjos, S., Ducourant, C., & Teixeira, R., 2014, *A&A*, 568A, 124D

Finet, F. & Surdej, J., 2016, *A&A*, 590, A42

Gaia Collaboration, Brown, A. G. A., Vallenari, A., Prusti, T., de Bruijne, J. H. J., Mignard, F. *et al.*, 2016a, *A&A*, 595A, 2G

Gaia Collaboration, Prusti, T. *et al.*, 2016b, *A&A*, 595, A1

Inada, Naohisa, Oguri, Masamune, Shin, Min-Su *et al.* 2012, *AJ*, 143, 119

Mignard, F., 2012, *MmSAI*, 83, 918

More, A. *et al.* 2016, *MNRAS*, 456, 1595M

J. A. Munoz, E. E. Falco, C. S. Kochanek, J. Lehar, B. A. McLeod, C. D. Impey, H.-W. Rix, C. Y. Peng arXiv:astro-ph/9902131, 1999

Perryman, M. A. C., de Boer, K. S., Gilmore, G., *et al.*, 2001, *A&A*, 369, 339

Souchay, J., Andrei, A. H., Barache, C., Kalewicz, T., Gattano, C., Coelho, B., Taris, F., Bouquillon, S., & Becker, O., 2015, *A&A*, 583, 75

Wertz, O. & Surdej, J., 2014, *MNRAS*, 442, 428

Astronomy and Astrophysics in the Gaia sky
Proceedings IAU Symposium No. 330, 2017
A. Recio-Blanco, P. de Laverny, A.G.A. Brown & T. Prusti, eds.

doi:10.1017/S1743921317005907

Local tests of gravitation with Gaia observations of Solar System Objects

Aurélien Hees[1], Christophe Le Poncin-Lafitte[2], Daniel Hestroffer[3] and Pedro David[3]

[1]Department of Physics and Astronomy, University of California, Los Angeles, CA 90095, USA
email: aheesQastro.ucla.edu

[2]SYRTE, Observatoire de Paris, PSL Research University, CNRS, Sorbonne Universités, UPMC Univ. Paris 06, LNE, 61 avenue de l'Observatoire, 75014 Paris, France

[3]IMCCE, Observatoire de Paris, PSL Research University, CNRS, Sorbonne Universités, UPMC Univ. Paris 06, Univ. Lille, 77 av. Denfert-Rochereau, 75014 Paris, France

Abstract. In this proceeding, we show how observations of Solar System Objects with Gaia can be used to test General Relativity and to constrain modified gravitational theories. The high number of Solar System objects observed and the variety of their orbital parameters associated with the impressive astrometric accuracy will allow us to perform local tests of General Relativity. In this communication, we present a preliminary sensitivity study of the Gaia observations on dynamical parameters such as the Sun quadrupolar moment and on various extensions to general relativity such as the parametrized post-Newtonian parameters, the fifth force formalism and a violation of Lorentz symmetry parametrized by the Standard-Model extension framework. We take into account the time sequences and the geometry of the observations that are particular to Gaia for its nominal mission (5 years) and for an extended mission (10 years).

Keywords. gravitation, ephemerides, minor planets, asteroids

1. Introduction

Although General Relativity (GR) is currently very well tested (see e.g. Will 2014), there exist strong motivations to pursue searches for modified gravitational theory like e.g. the development of a quantum theory of gravitation, the development of models of dark matter and dark energy, etc. Launched in December 2013, the ESA Gaia mission is scanning regularly the whole celestial sphere once every 6 months providing high precision astrometric data for a huge number ($\approx$ 1 billion) of celestial bodies. In addition to stars, it is also observing Solar System objects (SSOs), in particular asteroids. One can estimate that about 350,000 asteroids will be regularly observed. The high precision astrometry (at sub-mas level) will allow us to perform competitive tests of gravitation and to provide new constraints on alternative theories of gravitation. These constraints will be complementary to the ones existing currently since relying on different bodies, on different type of observations and therefore sensitive to other systematics. In this communication, we report preliminary results of a sensitivity study of Gaia SSOs observations to several modifications of the gravitational theory.

2. Methodology

In this work, we have considered SSOs from the ASTORB database. A match between their expected trajectories and the Gaia scanning law is performed to find the observation times for each SSO. Two scenarios are considered: (i) the 5 years nominal mission and

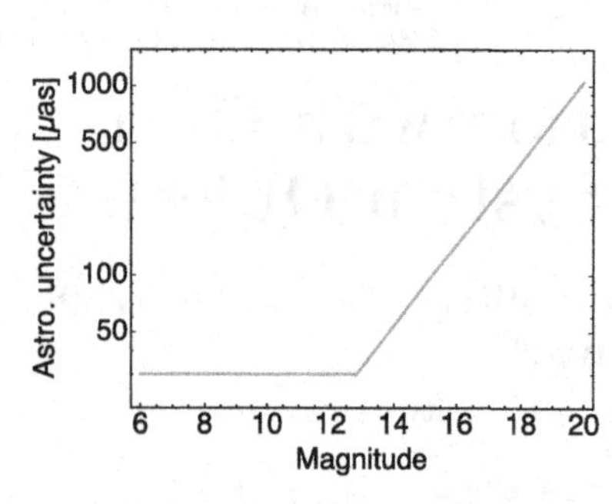

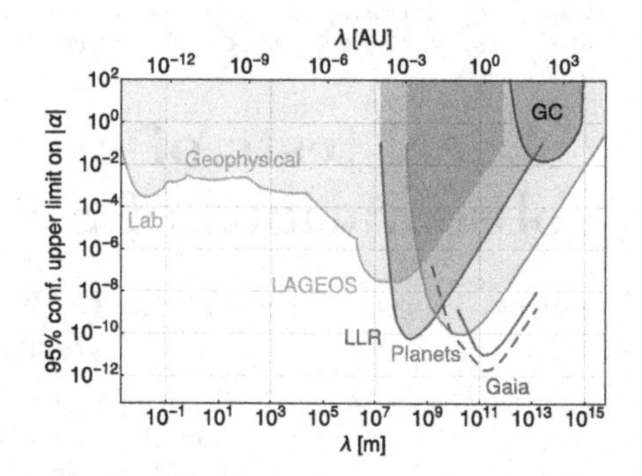

Figure 1. Left: astrometric uncertainty of SSOs observations as a function of the magnitude. Right: Existing constraints on the fifth force parameters. The solid Gaia curve corresponds to constraint reachable using 5 years of SSOs observations with Gaia while the dashed Gaia line corresponds to an extended mission.

(ii) a case where the nominal mission is extended by an additional 5 years. For the 5 years nominal mission, 342,449 SSOs are observed for a total of 20,450,775 observations while for a 10 years mission, this corresponds to 391,518 SSOs for a total of 41,004,470 observations. The astrometric observational uncertainty of each observation depends on the magnitude following a relation which is illustrated on the left panel of Fig. 1.

For each SSO, we integrate the standard post-Newtonian equations of motion in a heliocentric frame. The Sun oblateness J_2 is considered and the perturbations from all the planets and the Moon are modeled using the INPOP10e ephemerides (Fienga *et al.* 2013). On the other hand, in this preliminary analysis, the mutual interactions between the SSOs are neglected as well as non-gravitational forces. Simultaneously with the equations of motion, we integrate the variational equations to obtain the partial derivatives of the observables with respect to all the estimated parameters, i.e. six initial conditions for each SSO and global parameters like the Sun J_2 and the parameters characterizing the gravitational theory (for a detailed presentation of the method, see Hestroffer *et al.* (2010), Mouret (2011) and Hees *et al.* (2015)). Our sensitivity analysis is based on the Fisher information matrix (or covariance matrix) which gives an estimate of the uncertainty for each parameter as well as correlation coefficients. Therefore, the uncertainties presented in this communication correspond only to statistical uncertainties and our analysis does not include any hypothetical systematics.

3. Sensitivity study to various modifications of General Relativity

In the following, we consider modifications of the gravitational theory that do not impact significantly the light propagation and we focus only on the impact on the orbital dynamics of the SSOs. We report a sensitivity study performed by a global inversion that includes the 6 initial conditions for each of the SSO, the Sun J_2 and the parameters characterizing the deviations from GR.

3.1. *The parametrized post-Newtonian framework and the Nordtvedt effect*

The parametrized post-Newtonian (PPN) formalism is a phenomenological framework in which the space-time metric is parametrized by 10 dimensionless coefficients (see Will 2014 and references therein). The two most important PPN parameters are γ which describes the spatial space-time curvature and β which parametrizes the non-linearity in the time component of the space-time metric. In this analysis, we use $\gamma = 1$ since this parameter is better determined by other types of observations like the Shapiro time delay

Table 1. Statistical uncertainties reachable using Gaia observations to determine J_2, β and η parameters. In the left part of the table, these three parameters are supposed to be independent while on the right part of the table, η is supposed to be directly linked to β.

	J_2, β and η independent			$\eta = 4\beta - 4$	
	σ_{J_2}	σ_β	σ_η	σ_{J_2}	σ_β
5 years mission	4.4×10^{-8}	4×10^{-4}	3×10^{-4}	4.1×10^{-8}	8×10^{-5}
10 years mission	1.5×10^{-8}	9×10^{-5}	3×10^{-5}	1.3×10^{-8}	8×10^{-6}

(Bertotti *et al.* 2003) and by Gaia itself that will be able to constrain it at the level of 10^{-6} by observing light deflection (see e.g. Mignard & Klioner 2010). In addition, we also consider a violation of the Strong Equivalence Principle which appears in many (if not all) alternative gravitational theories. One effect produced by a violation of the SEP is that the trajectories of self-gravitating bodies depend on their gravitational self-energy Ω. It is characterized by a difference between the gravitational and the inertial mass usually parametrized by the Nordtvedt parameter η defined by $m_g = m_i + \eta\frac{\Omega}{c^2}$, where m_g is the gravitational mass and m_i is the inertial mass.

The left part of Table 1 shows the statistical uncertainties that can be reached using SSOs observations from Gaia assuming J_2, β and η to be independent. The uncertainties reachable in the case of the nominal mission are not as competitive as the ones obtained using planetary ephemerides (see e.g. Fienga *et al.* 2015). A 5 years extension of Gaia would result in an improvement of the estimate of J_2 by a factor 3, of β by a factor 5 and of η by one order of magnitude.

In addition, in metric gravitational theories, the Nordtvedt parameter is univocally related to the PPN parameters through $\eta = 4\beta - \gamma - 3$ (Will 2014). Assuming this relation improves significantly the estimations of the β PPN parameter, as can be seen from the right part of Table 1. An extended mission would give results as accurate as the ones from the planetary ephemerides.

3.2. *The fifth force formalism*

A fifth force is predicted by several theoretical scenarios motivated by the development of unification theories and models of dark matter. This framework considers deviations from Newtonian gravity in which the gravitational potential takes the form of a Yukawa potential characterized by two parameters: a length λ and a strength of interaction α. A summary of the constraints on these two parameters can be found for example in Fig. 2 from Hees *et al.* (2017). The right panel of Fig. 1 shows the current constraints on the fifth force parameters as well as the statistical uncertainties expected from Gaia for the nominal and extended missions. The expected results improve the current constraints above 1 astronomical unit. This result is only preliminary and the correlation with the Sun GM still needs to be assessed.

3.3. *A breaking of Lorentz symmetry*

A breaking of Lorentz symmetry is predicted in various unification theories, in a quantum theory of gravity, in non-commutative geometry, ... The Standard-Model Extension (SME) framework has been developed in order to systematically search for a breaking of Lorentz symmetry in all sectors of physics. In the gravitational sector, at the lowest order, a breaking of Lorentz symmetry is parametrized by a symmetric traceless tensor $\bar{s}^{\mu\nu}$ (see Bailey & Kostelecký 2006). These coefficients have already been constrained by various observations (for a review, see Hees *et al.* 2016 and references therein). Nevertheless, current observations did not manage to decorrelate satisfactorily all the SME coefficients. One strong advantage of SSOs observations with Gaia comes from the

Table 2. Statistical uncertainties reachable using Gaia observations to determine the SME $\bar{s}^{\mu\nu}$ coefficients considering a 5 years nominal mission and an extended mission of 10 years.

	$\bar{s}^{XX} - \bar{s}^{YY}$ $[10^{-12}]$	$\bar{s}^{XX} + \bar{s}^{YY} - 2\bar{s}^{ZZ}$ $[10^{-12}]$	$\bar{s}^{XY}$ $[10^{-12}]$	$\bar{s}^{XZ}$ $[10^{-12}]$	$\bar{s}^{YZ}$ $[10^{-12}]$	$\bar{s}^{TX}$ $[10^{-9}]$	$\bar{s}^{TY}$ $[10^{-9}]$	$\bar{s}^{TZ}$ $[10^{-9}]$
5 years mission	3.8	6.5	1.7	0.93	1.7	5.7	8.9	16.7
10 years mission	1.5	2.1	0.71	0.38	0.59	1.1	2.1	4.1

high number of objects observed and from the variety of their orbital parameters, which allows to decorrelate the SME coefficients. As a result, the estimations of the SME coefficients expected by Gaia, presented in Table 2, are improving the current best constraints by more than one order of magnitude. These results are highly promising.

4. Conclusion

In conclusion, Gaia offers a great opportunity to probe fundamental physics by measuring the deflection of light by the gravitational potential of the Sun (Mignard & Klioner 2010) and by using SSOs observations. The main advantage of Gaia SSOs observations to test GR comes from the wide variety of orbital parameters that can help to decorrelate different parameters. The sensitivity analysis presented in this proceeding relies only on a statistical analysis (no systematic effect has been considered so far). We show that in the PPN framework, an extended mission is expected to provide results competitive with the current best constraints on the PPN parameters while an improvement is expected in the fifth force framework. The more spectacular result is expected within the SME framework where at least an order of magnitude improvement is expected with respect to the current best constraints.

References

C. Will 2014, *Living Rev. Relativ.*, 17, 04

A. Fienga, H. Manche, J. Laskar, M. Gastineau, & A. Verma 2013, *arXiv:*1301.1510

D. Hestroffer, S. Mouret, F. Mignard, P. Tanga, & J. Berthier 2010, in: S. A. Klioner, P. K. Seidelmann & M. H. Soffel, (eds.) *Proceedings of the International Astronomical Union*, (Cambridge University Press), 5(S261), p. 325

S. Mouret 2011, *Phys. Rev. D*, 84, 122001

A. Hees, D. Hestroffer, C. Le Poncin-Lafitte, & P. David 2015, in: F. Martins, S. Boissier, V. Buat, L. Cambrésy, & P. Petit (eds.), *Proceedings of the Annual meeting of the French Society of Astronomy and Astrophysics*, p. 125

B. Bertotti, L. Iess, & P. Tortora 2003, *Nature*, 425, 374

F. Mignard & S. Klioner 2010, in: S. A. Klioner, P. K. Seidelmann & M. H. Soffel, (eds.) *Proceedings of the International Astronomical Union*, (Cambridge University Press), 5(S261), p. 306

A. Fienga, J. Laskar, P. Exertier, H. Manche, & M. Gasineau 2015, *Celest. Mech. Dyn. Astr.*, 123, 325

A. Hees, T. Do, A. M. Ghez, *et al.* 2017, *Phys. Rev. Letters*, 118, 211101

Q. Bailey & V. A. Kostelecký 2006, *Phys. Rev. D*, 74, 045001

A. Hees, Q. Bailey, A. Bourgoin, H. Pihan-La Bars, C. Guerlin, & C. Le Poncin-Lafitte, *Universe* 2, 30

Astronomy and Astrophysics in the Gaia sky
Proceedings IAU Symposium No. 330, 2017
A. Recio-Blanco, P. de Laverny, A.G.A. Brown
& T. Prusti, eds.

doi:10.1017/S1743921317005348

GaiaNIR – A future all-sky astrometry mission

David Hobbs[1] and Erik Høg[2]

[1]Lund Observatory, Box 43, 22100, Lund Sweden
email: david@astro.lu.se

[2]Niels Bohr Institute, Juliane Maries Vej 30, 2100 Copenhagen Ø, Denmark
email: ehoeg@hotmail.dk

Abstract. With the launch of Gaia in December 2013, Europe entered a new era of space astrometry following in the footsteps of the very successful Hipparcos mission. A weakness of Gaia is that it only operates at optical wavelengths. However, much of the Galactic centre and the spiral arm regions are obscured by interstellar extinction. An obvious improvement on Gaia is to include the Near-Infra-Red (NIR) which requires the use of new types of detectors. Additionally, to scan the entire sky and measure global absolute parallaxes the spacecraft must have a constant rotation resulting in a moving image that must be compensated for by, for example, operating the detectors in Time Delayed Integration (TDI) mode. If these technical issues can be solved a new Gaia-like mission separated by a 20 year interval would give; 1) NIR all-sky astrometry and photometry to penetrate the obscured regions and to observe intrinsically red objects with almost diffraction limited resolution; 2) improved proper motions with fourteen times smaller errors than from Gaia alone opening up new science cases, such as long period exoplanets and accurate halo measurements; 3) allow the slowly degrading accuracy of the Gaia reference frame, which will be the basis for future astronomical measurements, to be reset.

Keywords. Astrometry, Catalogs, Reference systems, Galaxy: kinematics and dynamics

---

1. Introduction

The current Gaia mission has only just begun to revolutionize our understanding of the Galaxy. The first Gaia data release gave 5 parameter astrometry for more than 2 million sources but this gives just a hint of what is about to come with the second release. In April 2018 we expect to release 5 parameter astrometry for more than 1 billion sources and subsequent releases will give increasingly accurate and comprehensive sets of astrophysical data. Gaia will eventually provide positions, absolute parallaxes and proper motions, to unprecedented accuracies (20–25 μas (yr^{-1}) at G=15), with the addition of all-sky homogeneous multi-colour photometry and spectroscopy. These unique capabilities go well beyond and are complementary to the science cases being addressed by ground based surveys (such as RAVE, SDSS, Pan-Starrs, APOGEE, LSST, etc).

The most obvious way to improve on Gaia's capabilities is to extend them to all-sky absolute NIR astrometry allowing the new mission to probe through the Galactic dust to observe the structure and kinematics of the star forming regions in the disk, the spiral arms and the bulge region to give model independent distances and proper motions in these obscured parts of the sky. A new mission launched with an interval of 20 years (around 2035-2040) would allow new measurements of objects already in the Gaia catalogue to be combined with older data giving improved proper motions with fourteen times smaller errors. Parallaxes would also be improved in such joint solutions by a factor of $\sqrt{2}$ assuming the two missions are of equal duration. After the publication of the final Gaia catalogue the positions of stars will be accurately known at the chosen reference epoch (currently 2017) and linked to the VLBI reference frame. However, this accurate positional information will slowly degrade due to the small uncertainties in the proper motions of the stars. Hence, it is necessary to repeat the measurements of Gaia after about 20 years to maintain the positional accuracy and the optical reference frame.

The accuracy of the mission should be at least that of Gaia using tried and trusted instrumentation, techniques, and lessons learned from Gaia. To achieve these goals we need to explore the feasibility and technological developments needed to manufacture space qualified optical and NIR (400–2000 nm) TDI sensors with passive cooling. To maintain ESA's leadership in all-sky space astrometry it is highly desirable to develop such detector technology within Europe. The most promising NIR sensors today seem to be hybrid HgCdTe-CMOS multiplexing sensors which can also support TDI mode. A more limited sensitivity from 900 to 1600 nm in a part of the field of view may be considered if CCDs are also used. In 2016 we successfully proposed such a technology study to ESA (Hobbs *et al.* (2016)) in a call for "New Science Ideas" to be investigated for technologies not yet sufficiently mature. It is hoped that these ideas may become candidates for future missions in the ESA Science Program.

2. NIR astrometry and photometry

Much of the Galactic centre and the spiral arm regions are obscured by interstellar extinction so a NIR mission is needed to astrometrically map these regions for the first time. Linking the individual motions of obscured objects with those from the earlier Gaia optical survey would give a much more complete picture of the dynamics of the Galaxy's interior. The complex nature of the Galaxy with its bulge, bar and spiral structure can excite stars to radially migrate and induce disk heating (Friedli *et al.* (1994), Sellwood and Binney (2002)) and accurate measurements of the 3-dimensional motion and properties of these obscured stars are needed to trace the dynamical history and evolution of our Galaxy. To study dark matter in the inner disk of the Galaxy we need NIR measurements, avoiding the need to use extinction maps, that may eventually resolve questions regarding the nature of dark matter by showing us whether the Galaxy has a cored or cusped dark matter halo (Governato *et al.* (2012)), whether there are thin, disc-like components of dark matter, and whether the spiral arms have their own dark matter components. Similarly, NIR is needed to study stars belonging to clusters which have similar age and metallicity, having formed from the same cloud of gas and dust, and can be used to probe the galactic disk structure and formation rate (Kuhn *et al.* (2015)).

Gaia is not good at detecting very faint red objects especially in extinction regions, examples include Red Dwarfs (RD), cool White Dwarfs (WD), Brown Dwarfs (BD) and free-floating planets. An NIR capable mission would allow such objects to be detected within a large volume but also crucially in the extinction (open clusters and star forming) regions that are of great interest and would shed light on the Initial Mass Function (IMF) of these regions. An NIR facility will allow to characterize the cool WD population much better than Gaia in the optical, which has limited capabilities (Carrasco *et al.* (2014)) both in terms of detection and parametrization. The study of WDs provides key information about the late stages of the star's life and comparison of the empirical and theoretical Luminosity Functions (LF) of WDs one can derive the age of the Galaxy and its star formation rate. Free-floating planets are interesting to study if they were ejected from planetary systems or formed in collapsing dust clouds in a similar manner to stars.

High-precision astrometry in the NIR will have an impact on a wide variety of stellar physics topics particularly in obscured regions in the Galactic plane and Galactic centre which are the birthplaces of most stellar clusters and associations. Gaia will greatly extend our current knowledge of clusters but they will mostly be located at high galactic latitudes and may not be fully representative of the conditions in the disk. The binary orbits of exotic objects, such as neutron stars and stellar-mass black holes, will benefit from improved mass estimates and may allow their equations-of-state to be accurately constrained for the first time. Extension to NIR will also enable extensive local tests of stellar standard candles and their period-luminosity relations since the key uncertainties of variable extinction and metallicity are significantly reduced at longer wavelengths. Finally, AGB stars are very bright in the NIR. Mira's are much more numerous than Cepheids and the period-luminosity relation of OH/IR stars are not well known. GaiaNIR can better characterize these objects as standard candles improving the distance scale.

3. Improved proper motions and parallaxes

Gaia is a fabulous mission for detecting and characterising nearby streams that cross the disk of the Milky Way but it will not be sufficient to discover and characterise most of the stream-like structures in the halo. A new mission could be combined with the older Gaia catalogue to give a much longer baseline, with very accurate proper motions (a factor of 14 better in the two components) and improved parallaxes needed to reach larger distances. Dynamical studies in the outer halo would be greatly enhanced resolving tangential motions in streams and local dwarf galaxies, with a potential accuracy of 2–3 km s^{-1} for samples out to $\sim$100 kpc. This will provide great insight into the gravitational potential in the outer reaches of the Milky Way and halo streams are sensitive probes. Finding "gaps" in the streams may reveal the influence of dark matter sub-haloes in the Milky Way's halo and allow us to determine the dark matter distribution at large radii, including any flattening of the potential, and the total mass of the Galaxy.

Hyper-Velocity Stars (HVSs) originate from gravitational interactions with massive black holes. Very accurate proper motion measurements are a key tool to study these objects. When combined with radial velocities the three dimensional space velocity is obtained. Unfortunately, known HVSs are distant and on largely radial trajectories. Some HVSs originate in the Galactic centre while others have an origin in the disk but an origin in the Magellanic Clouds or beyond is also possible. Accurate proper motions are needed to reconstruct their trajectories and distinguish between the different possible origins. Gnedin *et al.* (2005) have shown that precise proper motions of HVSs would give significant constraints on the structure (axis ratios and orientation of triaxial models) of the Galactic halo. Adding NIR would also allow us to probe more deeply into the Galactic centre and to detect small populations of HVSs closer to their ejection location.

Astrometrically resolving internal dynamics of nearby galaxies, such as M31, dwarf spheroid galaxies, globular clusters, the Large and Small Magellanic Clouds (LMC, SMC), sets requirements on the accuracy. For example, the LMC has a parallax of 20 μas and an accuracy of about 10% is needed, which is just within the reach of Gaia. Precise mapping of dark matter (sub-)structure throughout local group and beyond is possible with accurate proper motions. Gaia will not be able to directly measure internal motions of the nearby galaxies. However, by combining Gaia with a new Gaia-like mission opens up the new tantalising possibility of measuring their internal motions and thus astrometrically resolving the dynamics within the Local Group.

A new astrometric mission would allow the detection of planets with significantly longer periods than by Gaia alone. The longest detectable period using astrometry roughly coincides with the mission lifetime. This sets a maximum period $P \sim$5 years for the Gaia nominal mission, or $P \sim$10 years if an extension is granted. A new mission launched 20 years after Gaia would increase the detectable period to about 30 years making Saturn

($P = 29$ years) like planets detectable. Microlensing and imaging can detect some long period planets but only astrometry can properly characterise tens of thousands of planetary systems. This is required to fully understand planetary systems and their evolution, including the effect of migration of giant outer planets through the habitable zone and to find planetary systems like our own, assuming they exist! A similar limit holds for long-period binary stars which have much larger astrometric signatures, it may be possible to detect them even if their period is somewhat longer than the limit above, possibly to $\sim$40 years. Accurate astrometry over a longer time span would also benefit solar system measurements improving our understanding of asteroid families and the mechanisms that lead to the injection of asteroids on Earth-crossing trajectories.

4. The reference frame

The Gaia optical reference frame based on quasars will slowly degrade over time due to errors in its orientation and spin and due to small proper motion patterns which are not accounted for. Additionally, the catalogue accuracy will decay more rapidly due to errors in the measured proper motions. Dense and accurate reference grids are needed for forthcoming Extreme, Giant and Overwhelming telescopes but also for smaller instruments currently operating or being planned. The detection of the reference frame quasars solely from zero linear proper motion and parallax is possible, reducing the need for spectra. Quasars will have some apparent proper motions due to time-dependent source structure. These proper motions are probably not linear and this can be used to distinguish them.

An important aspect of reference frames is to link them, cross-matching with absolute coordinates, to reference frames at other wavelengths to produce reference grids for various surveys. This requires the maintenance of the accuracy of the Gaia optical reference frame at an appropriate density that is useful for new surveys and is a science objective in itself as it lies at the heart of fundamental astrometry. The Gaia celestial reference frame will immediately supercede any other optical reference frames and will be the standard optical reference frame for the astronomical community for many years. Nevertheless, a new Gaia-like mission will be necessary in the coming decades in order to maintain the optical realization of the celestial reference frame. This on its own is a strong science case but the addition of NIR astrometry will increase the density of this grid in obscured regions and provide a link to the ICRF in a new wavelength region.

5. Outlook

The proposal to ESA's call for "New Science Ideas" to develop NIR detectors with TDI mode (Hobbs *et al.* (2016)) was successful. Since then work has begun to clearly define the science requirements for such a mission and further studies are on-going to determine how best to implement the detector technology.

References

Hobbs, D. *et al.* 2016, *arXiv - http://adsabs.harvard.edu/abs/2016arXiv160907325H*
Friedli, D. *et al.* 1994, *Astrophysical Journal*, 430, L105-L108
&Sellwood, J. A. and Binney, J. J. 2002, *Monthly Notices of the RAS*, 336, 785-796
Governato, F. *et al.* 2012, *Monthly Notices of the RAS*, 422, 1231-1240
Kuhn, D. *et al.* 2015, *Astrophysical Journal*, 812, 131
Carrasco, J. M. *et al.* 2014, *Astronomy & Astrophysics*, 565, A11
Gnedin, O. Y. *et al.* 2005, *Astrophysical Journal*, 634, 344-350

Astronomy and Astrophysics in the Gaia sky
Proceedings IAU Symposium No. 330, 2017
A. Recio-Blanco, P. de Laverny, A.G.A. Brown
& T. Prusti, eds.

doi:10.1017/S1743921317006731

Gaia DR1 compared to VLBI positions

François Mignard[1] **and Sergei Klioner**[2]

[1]Laboratoire Lagrange, Université Nice Sophia-Antipolis, Observatoire de la Côte d'Azur, CNRS, CS 34229, F-06304 Nice Cedex, France
email: francois.mignard@oca.eu

[2]Lohrmann-Observatorium, Technische Universität Dresden, 01062 Dresden, Germany
email: Sergei.Klioner@tu-dresden.de

Abstract. Comparison of the Gaia DR1 auxiliary quasar solution to recent ground based VLBI solutions for ICRF2 sources.

Keywords. astrometry, reference systems, surveys, quasars

1. Introduction

In addition to observing above a billion galactic stars, Gaia catches in its net around one million distant QSOs that materialises a non-rotating reference frame directly accessible in the visible domain. The lack of global rotation rests on the assumption that distant quasars are globally at rest with respect to the CMB and can then be used to materialise a triad of inertial directions as prescribed in the ICRS principles. Regarding the orientation of this triad there is no compelling physical principle which will favour a particular choice, and the best solution is to ensure the metrological continuity by selecting the pole and the origin of right-ascension as close as possible to the current ICRF2. This alignment is achievable without loss of accuracy provided that the sources selected with radio position have an accuracy compatible with Gaia and that the radio-optical offset is smaller and random in direction. Therefore it is of utmost importance to carefully investigate the random and systematic differences between the best optical and radio astrometric catalogues of QSOs. While such a comparison has been done between the Gaia quasar solution and the ICRF2 in Mignard *et al.* 2016, we present here further analyses carried out on more recent radio solutions produced within the ongoing preparation of the ICRF3. Validating Gaia on a small subset of QSOs common to VLBI program brings confidence on the general Gaia astrometric solution for the larger set of several 10^5 found in the DR1, which materialises the optical frame.

2. Data used in these comparisons

2.1. *Gaia*

The sources are taken from 2191 ICRF2 sources published within the Gaia DR1 and available in the Gaia archive. The Gaia DR1 formal accuracy for this solution is given in Mignard *et al.* (2016). Unlike ICRF2 there is no difference of quality between the ICRF2 subsets of defining, VLBA calibrator or not. As the differences with Gaia are much smaller for the defining sources than for the calibrators, it was reasonable to conclude that the scatter seen in $(\Delta\alpha, \Delta\delta)$ reflected primarily the limited accuracy of the non-defining source in ICRF2 rather than Gaia itself.

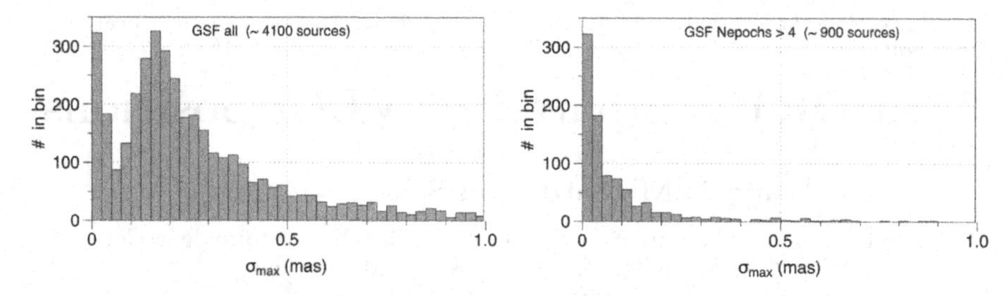

Figure 1. Formal positional accuracy of the GSFC VLBI solution. All sources with $\sigma < 1$mas (left), subset of sources observed over at least 5 epochs (right).

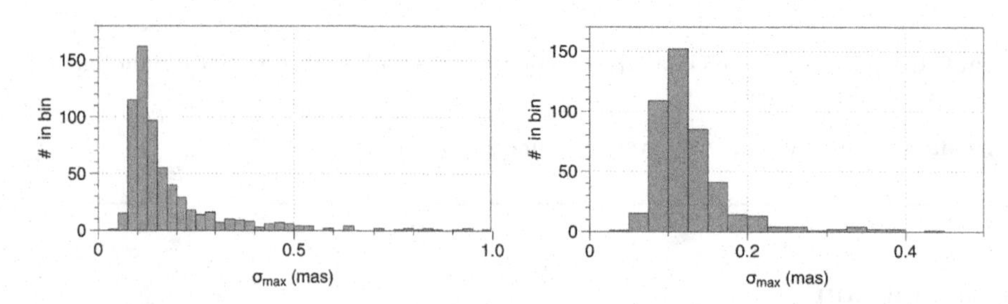

Figure 2. Formal positional accuracy of the X/Ka VLBI solution. All 673 sources (left), subset of 450 sources observed at least 25 times (right).

2.2. *GSFC*

For this new comparison we have used the most recent VLBI solution made available by the NASA/GSFC as `gsf2016a`. This catalogue includes all sources detected in X/S bands with astro/geo VLBI from 1979.7 through 2016.8 and comprises 4196 entries. Compared to ICRF2 many of the VLBA calibrators, but not all, have been observed at least at a second epoch and are of much better quality. The solution is an ICRF2-based solution, meaning the 295 ICRF2 defining sources were set to their ICRF2 positions and held to a no-net-rotation constraint. All ICRF2 sources are included and the accuracy is at the sub-mas level. The catalogue was cross-matched to Gaia observations and we found in the Gaia raw data that 3200 sources have been detected at least once by Gaia and 2700 have got a solution in the Gaia DR1 general solution. The formal error computed with the semi-major axis of the dispersion ellipse is shown in Fig. 1. The subset of well observed sources is nominally significantly better than Gaia DR1. The extended tail seen for the whole catalogue comes from the VLBA calibrators with only one epoch.

2.3. *X/Ka*

A new realisation of the Celestial Reference Frame has been produced recently in the X/Ka band (8.4/32 GHz) by Garcia-Miro *et al.* (2014) using baselines from the combined NASA and ESA Deep Space Networks for approximately 100 sessions each between July 2005 and September 2014. This results in a frame based on 654 sources (673 in the most recent update of the catalogue used in this analysis), with 525 common to ICRF2. Thanks to the addition of the ESA antenna located in Argentina in 2013, the catalogue includes 138 sources around the southern polar cap accessible for the first time. The formal positional error, computed as the semi-major axis of the error ellipse is plotted in Fig. 2 for the full catalogue (left) and the subset of the best observed sources (right). The median positional error is just above 100 μas and the tail at values larger than 500 μas is hardly visible, or even absent for the selected subset. If these uncertainties are

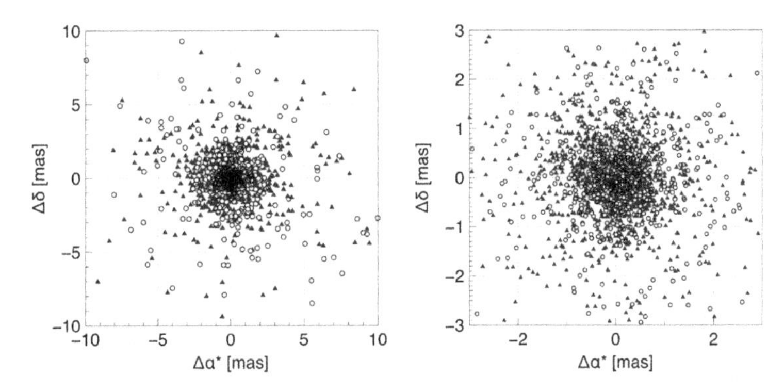

Figure 3. Positional difference between Gaia DR1 auxiliary solution and the GSF solution for the ≈ 2200 common sources within a window of ± 10 mas (left) and ± 3 mas (right). The black triangles stand for the VLBA calibrators and the open circles for the other sources, including the defining sources.

realistic, this catalogue constitutes the best reference one can use to compare with Gaia DR1 auxiliary solution and is fully independent of ICRF2 or GSF. In addition to the random errors, the authors mentioned evidence of systematic errors at the 100 μas level from tropospheric turbulence and limited calibration. The cross-match with the Gaia DR1 quasar solution ended up with 435 common sources with good Gaia solutions, most in magnitude range 18–20, but also about 80 with $G < 18$ mag.

3. Comparison to Gaia

The Gaia DR1 solution discussed here is based on the 2191 ICRF2 sources published separately in the DR1 from the Gaia auxiliary solution. Details can be found in Lindegren *et al.* (2016). In this analysis no new tie has been performed, but we checked that the remaining rotations between Gaia DR1 and the two comparison catalogues were negligible.

3.1. *GSF solution*

The main result is shown in Fig. 3, giving the scatter distribution of the positional differences between Gaia DR1 and the GSF solution within a window of 10 mas (left) and 3 mas (right). The ICRF source categories have been retained since, despite the new observations, there is still a quality difference between the defining source and the non calibration sources (open circles) and the larger set comprising the VLBA calibrators (black triangles). Generally the calibrators are more scattered throughout the diagram than the other sources. As already pointed out in Mignard *et al.* (2016), Gaia reveals here essentially the accuracy of the ICRF2 for this category of sources and this is in agreement with the quoted accuracy in ICRF2 or GSF. For Gaia there is no difference between the three classes unlike the VLBI data. Compared to the analysis done by Mignard *et al.* (2016) using the ICRF2 solution, the comparison with GSF displays a real improvement for the VLBA calibrators that have been observed at a second epoch, and no visible overall change for the other sources, since the Gaia uncertainty could be the larger source of scatter, in particular for the small set of defining sources.

3.2. *X/Ka solution*

The comparison in right ascension and declination is shown in Fig. 4 for a window of 3 mas and two selections made from Gaia formal accuracy: all common sources with Gaia

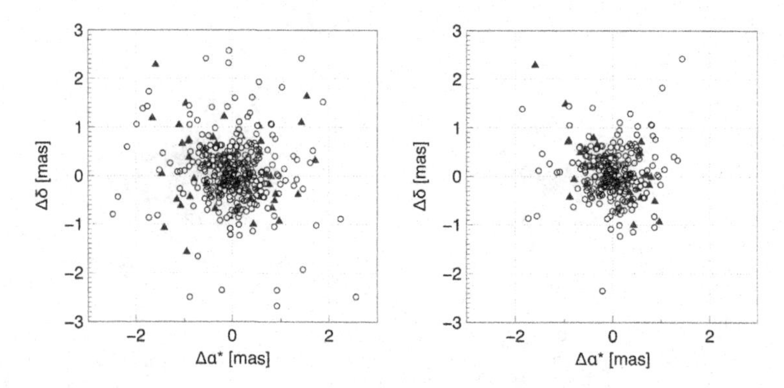

Figure 4. Positional difference between Gaia DR1 and the X/Ka solution for the common solutions within $\pm$3mas and Gaia uncertainty < 5 mas (left , ≈ 410 sources) and < 1 mas, (right, ≈ 290 sources). The black triangles stand for the ICRF2/VLBA calibrators and the open circles for the other sources.

accuracy < 5 mas (left) and < 1 mas (right). One has 411 common solutions in the first case and 291 in the second. We have kept the distinction between the usual categories of sources inherited from the ICRF2, although this is not relevant in this case with no difference expected in accuracy in Gaia and the X/Ka VLBI solution. The reference set from the X/Ka solution is expected to be better than Gaia DR1 and the scatter is normally mostly due to Gaia uncertainty. This is clearly seen when the data are filtered with the Gaia formal accuracy at 5 mas and 1 mas. Most of the large discrepancies seen in the left diagram have been eliminated when this filter is applied. This is a strong evidence that Gaia (and X/Ka) formal errors are probably realistic with the core of the Gaia solution really in the sub-mas range. An important feature of both plots is the very regular distribution around the centre, without any visible bias in either coordinate.

4. Conclusion

We have presented new analyses of the Gaia DR1 reference frame against two independent ground based VLBI solutions. Both confirm the quality of the Gaia data and the absence of systematic offset between the radio and optical frame.

Acknowledgments

We are grateful to C.S. Jacobs for having provided us with an early access to the X/Ka solution before its completion. This research has made use of NASA Goddard Space Flight Center's VLBI source position solution gsf2016a, prepared by David Gordon and he is gratefully acknowledged for his assistance. This work has made use of data from the European Space Agency (ESA) mission *Gaia*, processed by the *Gaia* Data Processing and Analysis Consortium, (DPAC, `https://www.cosmos.esa.int/gaia`).

References

Garcia-Miro, C, Sotuela, I, Jacobs, C. S., *et al.* 2014, *Proceedings of the 12th European VLBI Network Symposium and Users Meeting (EVN 2014)*, `http://pos.sissa.it/cgi-bin/reader/conf.cgi?confid=230`

Lindegren, L., Lammers, U., Bastian, U. *et al.* 2016, *Ast.Astroph*, 595, A4

Mignard, F., Klioner, S., Lindegren, L. *et al.* 2016, *Ast.Astroph*, 595, A5

Astronomy and Astrophysics in the Gaia sky
Proceedings IAU Symposium No. 330, 2017
A. Recio-Blanco, P. de Laverny, A.G.A. Brown
& T. Prusti, eds.

doi:10.1017/S1743921317005646

The LQAC-4, last update of the Large Quasar Astrometric Catalogue

J. Souchay[1], A. H. Andrei[2], C. Barache[1], F. Taris[1], C. Gattano[1] and B. Coelho[2]

[1]SYRTE, Observatoire de Paris
PSL research Univ., CNRS Sorbonne Univ. Paris 06, LNE
61, avenue de l'observatoire, 75014 Paris, France
email: Jean.Souchay@obspm.fr

[2]Observ. do Valongo, Universidade Federal do Rio de Janeiro
Ladeira Pedro Antonio 43, Saude, Rio de Janeiro-REJ, CEP 20080-90, Brazil
email: bcoelho@astro.ufrj.br

Abstract. Thanks to their point-like structure and to their lack of significant proper motion, quasars represent the ideal objects for modeling quasi-inertial directions in space. For that reason the present primary conventional reference frame, the ICRF-2, is constructed from the set of the celestial coordinates of a sample of extragalactic objects, whose the very large majority are quasars. Thus any newly discovered quasar must be considered as a potential future ideal astrometric marker. Therefore compiling all the recorded quasars at a given epoch looks as a useful task. This constitutes the aim of the LQAC (Large Quasar Astrometric Catalogue). We present here the contents of the future release of this catalogue quoted as the LQAC-4, insisting on the related strategy of compilation. Preliminary results concerning the cross-identification with the Gaia DR1 catalogue are emphasized

Keywords. quasars : general, catalogs, surveys

---

1. Introduction

Soon after the first quasar was discovered, in the early 60s, continuous efforts have been made to compile the whole set of known quasars with regular updates. As any other type of celestial object, the interest of gathering in a same catalogue all the quasars can be considered as a commendable effort for the astronomical and astrophysical community. First this kind of work constitutes a useful tool for any researcher wanting to know the positions of all the already recorded objects. Second it gives some information about some of the fundamental characteristics as the redshift, the apparent magnitudes at several wavelengths, the absolute magnitude computed from suitable algorithms etc... Third it serves as a basis for general studies as statistics with respect to the parameters above, as well to sky coverage. Fourth it can be used as a kind of reference catalogue to identify the quasars present in the recently published Gaia DR1 release.

For these reasons continuous efforts have been made for more than 25 years by the same authors (Véron-Cetty and Véron (1984), Véron-Cetty and Véron (2010)) to construct an all-sky compilation of all the quasars detected so far. Their last release in 2010, the 13 th one, contained 133 336 quasars together with an additional set of 34 231 active galaxies (AGNs).

Then such a compilation work was continued by a new group (Souchay *et al.* (2009)) in the form of what is named the Large Quasar Astrometric Catalogue (LQAC), containing 113666 objects. Two up-dated catalogues followed, the LQAC-2 (Souchay *et al.* (2012)) and the LQAC-3 (Souchay *et al.* (2015)) with respectively 187504 and 321957 objects.

These considerable consecutive increases are essentially due to the inclusion of successive releases of the Sloan Digital Sky Survey (SDSS), by far the main contributor.

In the following we discuss the construction of the fourth up-date of the LQAC, named as LQAC-4, which is scheduled to publication in the very near future.

2. The construction of the LQAC-4

The principle of compilation of the LQAC-4 (Gattano *et al.* (2017)) is identical to the LQAC-3 (Souchay *et al.* (2015)). It is built in a large part from a compilation of relatively large catalogues of quasars, four of them in the radio VLBI domain, five of them in the optical one. In the following we present the set of these nine original large catalogues involved. They are ranged in the following in an a priori decreasing order of accuracy of determination of the celestial coordinates of the objects. An alphabetic flag, from A to I is given according to this order.

<u>Four radio VLBI catalogues</u>

The four following radio catalogues have been elaborated from Very Long Baseline Interferometry (VLBI) observations. They are characterized by a relatively small number of objects but a very good astrometric accuracy.

• The ICRF-2 (flag "A")

The International Celestial Reference Frame, 2nd. release, called the ICRF-2 (Ma *et al.* (2009), Fey *et al.* (2015)) is the primary reference frame for all astronomers. It contains the positions, i.e. the celestial coordinates of 3414 compact radio-sources (mainly quasars but also some AGNs and BLAC). These coordinates are given with a noise floor estimated at 40 μas. They implicitly define the axes of the frame with a stability of 10 μas. Consequently the ICRF-2 is by far the most accurate catalogue of extragalactic objects in terms of celestial coordinates.

• The RFC/ VLBA (flag "B")

The Radio Fundamental Catalog (RFC) comes from a VLBI Astrometric global solution using all the available VLBI observations carried out in the frame of various surveys in different bandwidths, mainly at 8.6 GHz and 2.2 GHz. The leading observational program is the VCS (VLBA calibrator survey) consisting in several thousands of 24h continuous VLBI sessions playing a leading role in the astrometric performance. The catalogue is regularly updated with more sources and more observations per source. For the LQAC-4 we used the RFC2013 release already considered in the LQAC-3 and containing 7213 objects instead of the 5198 objects present in the LQAC-2.

• The VLA Catalogue (flag "C")

The Very Large Array (VLA) interferometer consists of 27 radio antennas of 25 m diameter dispatched in a Y shape at St. Augustin (New Mexico, USA). This gives the same resolution as one antenna of 36 km-across, with the sensitivity of a 130 m dish. The catalog used in the LQAC-4 is exactly the same as in the LQAC-3. It contains 1701 quasars with an astrometric accuracy of roughly $10mas$. (Claussen (2006).

• The JVAS Catalogue (flag "D")

The Jodrell Bank VLA Astrometric Survey (JVAS) catalog (Browne *et al.* (2009), Wilkinson *et al.* (1998)) contains 2118 compact radiosources uniformly distributed in the northern hemisphere with a peak flux density $> 50mJy$ at a resolution of 200 mas at 8.4 GHz. The rms accuracy of the position of sources ranges between 10 mas and $55mas$

Five optical catalogues

The five following big catalogues of quasars contain information at optical wavelenths. Notice that two of them, the FIRST and the Hewitt and Burbridge ones, include substantial additional information in the radio domain.

• The SDSS Catalogue (flag "E")

The Sloan Digital Sky Survey (SDSS) constitutes by far the largest catalogue of the LQAC-4 compilation. Thanks to its recent up-date, the DR12Q by Pâris *et al.* (2017), it gathers 384 834 elements. That constitutes a big increase with respect to the 262 535 SDSS quasars in the LQAC-3, and represents 86.7 % of the total number of objects in the LQAC-4. Nevertheless it covers roughly only one quarter of the sky, staring from observations with a dedicated 2.5 m telescope at Apache Point, New Mexico (USA). Images are obtained in five broad optical bands, designated as u, g, r, i, z. Astrometric calibration yields an accuracy of $45mas$ per coordinate when reduced using the USNO CCD Astrograph Catalogue (Pier *et al.*, 2003.

• The 2QZ Catalogue (flag "F")

The 2-degree Field (2dF) quasar redshift survey, quoted as 2QZ (Croom *et al.*, 2004) is the second largest contributor, with a number of 23660 objects. It is based on a preselection of quasar candidates starting from color criteria and a spectroscopic follow-up completed by a minimization technique fitting each spectrum to a number of stellar and extragalactic objects templates. The survey area includes 30 fields arranged in $75° \times 5°$ declination strips passing across the South and North Galactic caps.

• The 2SLAQ Catalogue (flag "G")

The 2SLAQ Survey, which stands for the 2dF-Sloan Digital Sky Survey luminous red galaxy (LRG) and QSO Survey (2dF-SDSS LRG and QSO) (da Angela *et al.* 2008), combines photometric and spectroscopic data respectively from the Sloan telescope and from the 2dF. It is an extension of the 2QZ survey above, at fainter magnitudes, and includes a total of 9058 QSO's.

• The FIRST Catalogue (flag "H")

The FIRST catalogue used in the LQAC-4 consists in a combination of three successive releases matching the radio NRAO VLA survey (Becker *et al.*, 2001) with an optical catalogue obtained by digitization of plates coming from the PSS (Palomar Sky Survey).

• Hewitt and Burbridge Catalogue (flag "H")

The HB catalogue (Hewitt and Burbridge, 1993) contains 6720 objects representing a majority of the quasars known up to the epoch it was completed, that is to say in 1992.

One complementary catalogue of quasar (Véron-Cetty and Véron (2010)), flag "M"

For the LQAC-4 the VV2010 catalogue (Véron-Cetty and Véron (2010)) was used to extract the 14509 quasars not included in the compilation obtained from the nine quasar catalogues described above. That constitutes a mere 3.26 % of the whole LQAC-4.

3. Cross-identification with Gaia DR1

Awaited results concern the cross-match with the DR1, (Gaia coll., 2016, Arenou *et al.*, 2017) with underlying fundamental questionings as : what are the quasars recognized by the Gaia release and how many objects are concerned ? Results will be detailed in the LQAC-4 paper scheduled soon. We give a partial result concerning the ratio of cross-matchs of individual catalogues of the LQAC-4 with respect to he DR1. Notice that

Catalog name	Flag	Nature Radio / Optical	Nbs of quasars in LQAC-4	Nb. of quasars in DR1	Perc.
ICRF2	A	R	3 414	2314	67.78
VLBA	B	R	7 213	4287	68.25
VLA	C	R	1 858	1228	60.71
JVAS	D	R	2 118	1373	64.82
SDSS	E	0	384 834	218 720	56.83
2QZ	F	O	23 660	18 075	76.39
LRG	G	O	9 058	2 474	27.31
FIRST	H	O	969	909	93.80
HB	I	O & R	6 720	5 991	89.15

Table 1. Number of quasars for each catalog of the LQAC-4 compilation with the corresponding number of quasars in common with the Gaia DR1 catalogue

the LQAC-4 contains at all 443725 quasars from which 249071 have a DR1 counterpart, within a 1" search radius. This represents a 56.13% completness ratio.

Acknowledgements

This work has made use of data from the European Space Agency (ESA) mission *Gaia* (`https://www.cosmos.esa.int/gaia`), processed by the *Gaia* Data Processing and Analysis Consortium (DPAC, `https://www.cosmos.esa.int/web/gaia/dpac/consortium`). Funding for the DPAC has been provided by national institutions, in particular the institutions participating in the *Gaia* Multilateral Agreement.

References

Arenou, F., Luri, X., Babusiaux, C. *et al.* 2017, *A & A* 599A, 50A
Becker, R. H., White, R. L., Gregg, M. D. *et al.* 2001, *ApJ. Suppl.*135,227
Browne, I. W. A., Patnaik, A. R., Wilkinson, P. N., & Wrobel, J. M. 1998, *MNRAS* 293, 257.
Claussen, M. 2006, *VLA Calibrator Manual*
Croom, S. M., Smith, R. J., Boyle, B. J. *et al.* 2004, *MNRAS* 349, 1397
da Angela, J., Shanks, T., Croom, S. M., *et al.* 2008, *MNRAS* 383, 565
Fey, A. L., Gordon, D., Jacobs, C. S., *et al.* 2015, *AJ* 150, 58.
Gaia Collaboration, Brown, A. G. A., Vallenari, A., Prusti, T. *et al.* 2016, *A & A*595A, 2G
Hewitt, A. & Burbridge, G. 1993 *ApJ.Suppl.* 87, No.2, 451
Ma, C., Arias, E. F., Bianco, G. *et al.* 2009, *IERS Technical Note* 35
Pâris, I., Petitjean, P., Nicholas, P. R. *et al.* 2017, *A & A* 597, A79
Pier, J. R., Munn, J. A., Hindsley, R. B. *et al.* 2003, *AJ* 125, 1559.
Souchay J., Andrei, A. H., Barache, C., *et al.* 2009, *A & A*494, 815.
Souchay, J., Andrei, A. H., Barache, C. *et al.* 2012, *A & A*537, A99.
Souchay, J., Andrei, A. H., Barache, C. *et al.* 2015, *A & A* 583, A75.
Véron-Cetty, M. P., & Véron, P., 1984, *A & A*, ESO Scientific Report, No. 1
Véron-Cetty, M. P., & Véron, P., 2010, *A & A* 518, A10
Wilkinson, P. N., Browne, I. W. A., Patnaik, A. R. *et al.* 1998, *MNRAS* 300, 790.

Astronomy and Astrophysics in the Gaia sky
Proceedings IAU Symposium No. 330, 2017
A. Recio-Blanco, P. de Laverny, A.G.A. Brown
& T. Prusti, eds.

doi:10.1017/S174392131700583X

The Differential Astrometric Reference Frame on short timescales in the Gaia Era

Ummi Abbas, Beatrice Bucciarelli, Mario G. Lattanzi, Mariateresa Crosta, Mario Gai, Richard Smart, Alessandro Sozzetti and Alberto Vecchiato

INAF - Osservatorio Astrofisico di Torino,
Via Osservatorio 20, Pino Torinese I-10025, Italy
email: abbas@oato.inaf.it

Abstract. We use methods of differential astrometry to construct a small field inertial reference frame stable at the micro-arcsecond level. Using Gaia measurements of field angles we look at the influence of the number of reference stars and the stars magnitude as well as astrometric systematics on the total error budget with the help of Gaia-like simulations around the Ecliptic Pole in a differential astrometric scenario. We find that the systematic errors are modeled and reliably estimated to the μas level even in fields with a modest number of 37 stars with G <13 mag over a 0.24 sq. degrees field of view for short timescales of the order of a day for a perfect instrument and with high-cadence observations. Accounting for large-scale calibrations by including the geometric instrument model over such short timescales requires fainter stars down to G=14 mag without diminishing the accuracy of the reference frame.

Keywords. astrometry, reference systems, methods: statistical

---

1. Modelling and Systematic Effects

In the presence of Earth's atmosphere, limitations to the differential astrometric precision are caused by effects such as refraction, turbulence, etc (Sozzetti, 2005). In its absence, for differential space-based measurements (based on reference objects that are all within a small field), we need to address effects such as: light aberration that is of the order of ~20 arcseconds to first order and a few mas to second order (Klioner 2003); gravitational deflection terms that lead to effects of several mas even at the Ecliptic Pole due to the monopole moment of the Sun (Crosta & Mignard, 2006); parallaxes and proper motions of stars that can be either removed apriori or accounted for in the model; and for changes in the geometric instrument model due to thermal variations and imperfections in the instrument that need to be efficiently calibrated (Lindegren *et al.*, 2012).

The simulation used was produced with AGISLab (Holl *et al.*, 2012) and takes advantage of the high-cadence observations of Gaia during the Ecliptic Pole Scanning Law to simulate the AL and AC field angles of stars taken from the IGSL catalogue (Smart & Nicastro 2014) that lie close to the North Ecliptic Pole (see Abbas *et al.* (2017) for details). In a nutshell, the observing times of the set of stars is restricted to within ± 15 seconds of the NEP t_{obs} for the same CCD column. Successive observations are separated by the time it takes the star to cross from one fiducial line to the next (approx. 4.42 secs). We then adopt the first configuration, i.e. t_{obs} of the NEP at the fiducial line of the first CCD column, on the first scan as the reference frame thereby obtaining the plate/CCD parameters that can transform coordinates on any other frame onto this reference.

The overlapping frames are solved using the Gaussfit software (Jefferys, 1988) through a differential procedure that involves determining the plate solution coefficients through a least squares adjustment and then applying the plate solution to obtain the corresponding coordinates of the target star on the frame. Gaussfit solves the set of equations through a least squares procedure that minimizes the sum of squares of the residuals constrained by the input errors alongwith appropriate constraints on the proper motions and calibration parameters. The distribution of residuals then informs us as to how well the model accounts for various physical or instrumental effects.

2. Results

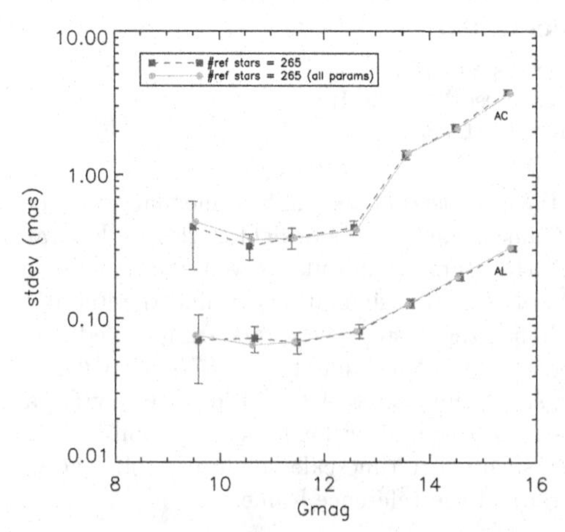

Figure 1. The standard deviations (in mas) from using a linear model to fit the overlapping frames of observations around the NEP in bins of the star's G-magnitude simulated with a perfect instrument and no proper motions (red dashed lines) superposed to the full linear model with all physical and instrumental effects included (green solid lines). The standard deviations shown in AL (lower) and AC (upper curve) have input errors that follow the standard CCD-level location estimation errors. The good agreement implies that a fully linear model is sufficient to describe the various physical and instrumental effects.

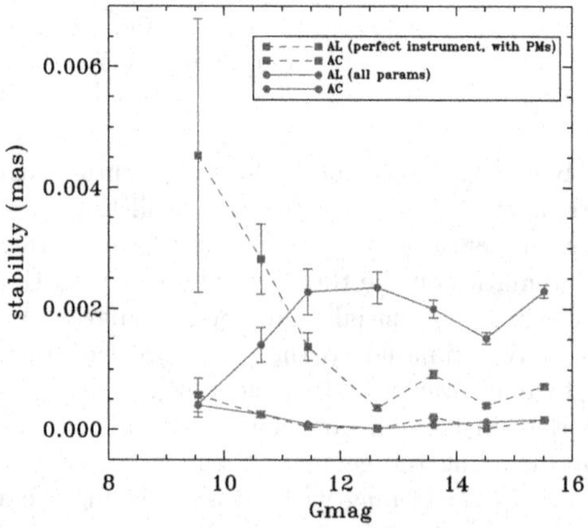

Figure 2. The absolute differences between the standard deviations of the estimated residuals and the input standard uncertainties of the Differential Astrometric Reference Frame in bins of the star's G-magnitude for different models in the AL and AC directions with poisson errors. The dashed lines are for a perfect instrument model (only physical effects included), whereas the solid lines are for the full linear model that involves the inclusion of 576 more unknowns and maintains the μas stability. Lower red lines and upper blue lines are for the AL and AC scan directions respectively. [Colour only in online version.]

References

Abbas, U., *et al.* 2017, *PASP*, 129, 4503
Crosta, M. & Mignard 2006, *Classical and Quantum Gravity*, 23, 4853
Holl, B., *et al.* 2012, *A&A*, 543, A15
Jefferys, W. H., Fitzpatrick, M. J., & McArthur, B. E. 1988, *Celestial Mechanics*, 41, 39
Klioner, S. A. 2003, *AJ*, 125, 1580
Lindegren, L., Lammers, U., Hobbs, D., *et al.* 2012, *A&A*, 538, A78
Smart, R. L. & Nicastro, L. 2014, *A&A*, 570, A87
Sozzetti, A. 2005, *PASP*, 117, 1021

Astronomy and Astrophysics in the Gaia sky
Proceedings IAU Symposium No. 330, 2017
A. Recio-Blanco, P. de Laverny, A.G.A. Brown
& T. Prusti, eds.

doi:10.1017/S174392131700607X

The PMA Catalogue as a realization of the extragalactic reference system in optical and near infrared wavelengths

Volodymyr S. Akhmetov, Peter N. Fedorov and Anna B. Velichko

Institute of Astronomy V.N.Karazin Kharkiv National University,
61022, 35 Sumska Str, Kharkiv, Ukraine
email: akhmetovvs@gmail.com

Abstract. We combined the data from the *Gaia* DR1 and Two-Micron All Sky Survey (2MASS) catalogues in order to derive the absolute proper motions more than 420 million stars distributed all over the sky in the stellar magnitude range 8 mag < G < 21 mag (Gaia magnitude). To eliminate the systematic zonal errors in position of 2MASS catalogue objects, the 2-dimensional median filter was used. The PMA system of proper motion has been obtained by direct link to 1.6 millions extragalactic sources. The short analysis of the absolute proper motion of the PMA stars Catalogue is presented in this work. From a comparison of this data with same stars from the TGAS, UCAC4 and PPMXL catalogues, the equatorial components of the mutual rotation vector of these coordinate systems are determined.

Keywords. astrometry, catalogue, surveys, reference systems.

---

1. Introduction

The Hipparcos Celestial Reference Frame (HCRF), according to IAU Resolution B1.2 of the XXIVth IAU GA, has been the optical realization of the International Celestial Reference System (ICRS, (Arias *et al.* 1995)). The Tycho-2 catalogue (Høg *et al.* 2000) that contains positions and proper motions of about 2.5 million stars, is the HCRF extension towards the large stellar magnitudes domain, approximately up to $V = 11.5$ mag. The PPMXL (Roeser *et al.* 2010), UCAC4 (Zacharias *et al.* 2013), SPM4 (Girard *et al.* 2011), 2MASS (Skrutskie *et al.* 2006) and others catalogues which extend the HCRF system towards the faint of the stellar magnitudes range, use the Hipparcos (Kovalevsky *et al.* 1997), (van Leeuwen, 2007) and Tycho-2 stars as the reference ones. In this work, we investigate the problem of mutual rotation of the Hipparcos/Tycho-2 system with respect to absolute proper motion of the PMA stars.

2. Comparison of the PMA with other catalogues data

In September 2016 the first *Gaia* data were released based on the first 14 months of regular in-orbit operations (Gaia Collaboration *et al.* 2016a). *Gaia* Data Release 1 (DR1) contains astrometric results for more than one billion stars brighter than magnitude 20.7. The PMA Catalogue (Akhmetov *et al.* 2017) has been derived from a combination of two catalogues - 2MASS and *Gaia* DR1 (Gaia Collaboration *et al.* 2016b). The difference of epochs of observations for these catalogues is approximately 15 years. In order to eliminate the distortions we used a two-dimensional median filter that provided corrections by eliminating systematic errors in the 2MASS positions and, reducing them to the *Gaia* DR1 system. The absolute calibration procedure (zero-pointing of the proper motions) was fulfilled with the use of about 1.6 million positions of extragalactic sources. To creation the sample of extragalactic sources, we intersected the sample of SSA galaxies

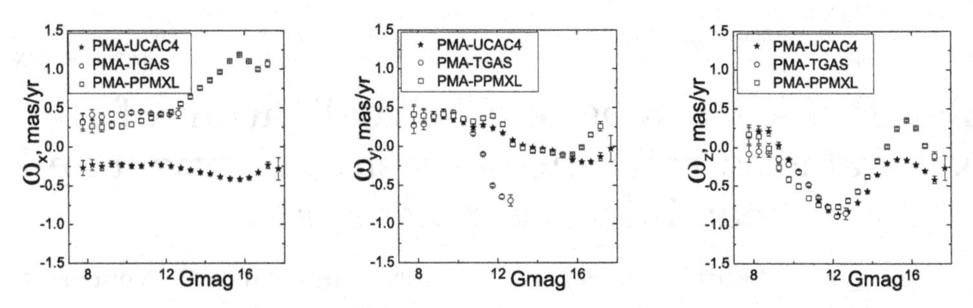

Figure 1. Components w_x, w_y and w_z of the mutual rotation vector of the Hipparcos/Tycho-2 system with respect to absolute proper motion of the PMA stars catalogue.

(Hambly *et al.* 2001a, Paper I), (Fedorov *et al.* 2014) with the WISE Catalogue (Wright *et al.* 2010). The colour diagram $(B - I)$ versus $(j_m - W1)$ has been used to separation stars and extragalactic sources. The mean formal error of the absolute calibration is less than 0.35 mas/yr.

To determine the value of the solid-body rotation of the Hipparcos/Tycho-2 system with respect to absolute proper motion of the PMA stars we use the well-known equations (Lindegren & Kovalevsky, 1995). The PMA-UCAC4, PMA-TGAS(Michalik *et al.* 2015) and PMA-PPMXL stellar proper motion difference have been used for solved by the least-squares method. The obtained components of the mutual rotation vector (Fig. 1) were made with the aim only of demonstrating their existence in the stellar proper motion these catalogues. For analysis and interpretation of the results of the comparison are need a separate and comprehensive investigation.

3. ACKNOWLEDGEMENTS

This work has made use of data from the European Space Agency (ESA) mission *Gaia* (`https://www.cosmos.esa.int/gaia`), processed by the *Gaia* Data Processing and Analysis Consortium (DPAC, `https://www.cosmos.esa.int/web/gaia/dpac/consortium`). Funding for the DPAC has been provided by national institutions, in particular the institutions participating in the *Gaia* Multilateral Agreement.

References

Arias E. F., Charlot P., Feissel M., & Lestrade J. F. 1995, *A&A* , 303, L604
Akhmetov V. S., Fedorov, P. N., Velichko A. B., & Shulga V. M. 2017, *MNRAS*, 469, 763
Brown, A. G. A., Vallenari, A., & Prusti, T. 2016, *A&A*, 595, id.A2, 23 pp.
Fedorov, P. N., Akhmetov, V. S., & Shulga, V. M. 2014, *MNRAS*, 440, 624
Girard, T. M. at al. AJ 142, 1538-3881
Høg, E. *et al.* 2000, *A&A*, 355, L27
Hambly, N. C. *et al.* 2001a, *MNRAS*, 326, 1279
Kovalevsky, J. *et al.* 1997, *A&A*, 323, L620
van Leeuwen, F. 2007, *A&A*, 494, L799,
Lindegren L.& Kovalevsky J. 1995, *A&A* 304, 189
Michalik, D., Lindegren, L., & Hobbs, D. 2015, *A&A*, 574, A115
Prusti, T., *et al.* 2016, *A&A* 595, id.A1, 36 pp.
Roeser, S., Demleitner, M., & Schilbach, E. 2010, *AJ*, 139, 2440
Skrutskie, M. F. *et al.* 2006, *ApJ*, 131, 1163
Wright, E. L., *et al.* 2010 *AJ*, 140, 1868
Zacharias, N., Finch, C. T., Girard, T. M., Henden, A., Bartlett, J. L., Monet, D. G., & Zacharias, M. I. 2013, *AJ*, 145, 14 pp.

Astronomy and Astrophysics in the Gaia sky
Proceedings IAU Symposium No. 330, 2017
A. Recio-Blanco, P. de Laverny, A.G.A. Brown
& T. Prusti, eds.

doi:10.1017/S1743921317005853

New Astronomical Reduction of Old Observations (the NAROO project)

Jean-Eudes Arlot[1], **Vincent Robert**[1,2], **Valery Lainey**[1], **Coralie Neiner**[3] **and Nicolas Thouvenin**[1]

[1]IMCCE, Observatoire de Paris,UMR 8028 CNRS, PSL,
77 avenue Denfert-Rochereau, F-75014 Paris, France
email: arlot@imcce.fr

[2]IPSA, Institut Polytechnique des Sciences Avancees,
63 boulevard de Brandebourg, F-94200 Ivry sur Seine, France
email: vincent.robert@imcce.fr

[3]LESIA, Observatoire de Paris, UMR 8109 CNRS, PSL,
email: coralie.neiner@obspm.fr

Abstract. The Gaia astrometric reference catalogue will provide star proper motions with an accuracy of one mas one century ago for stars of magnitude 14 or brighter. Our project is to re-reduced the old observations with the new catalogue allowing to have an astrometric accuracy only limited by the observational biases and not by reference stars. Then, we plan to get an accuracy of 50 mas where the old reductions were not better than 500 mas!

For our purpose, we will digitize old photographic plates with a sub-micrometric scanner. Tests were made using the UCAC catalogue showing that old photographic plates have an intrinsect accuracy of 30 to 60 mas.

Keywords. astrometry, photographic plates, photometry, spectrometry, solar system.

1. Introduction: the interest for old observations

Old observations are interesting for modelling transient events or periodic behaviors (observable through astrometry for dynamics or through photometry for variable objects). For that purpose, old data are used as published at the epoch they were made. However, the accuracy was not good since reference used for the reduction were not well-known. We propose to re-reduce these old observations with the Gaia reference star catalogue.

2. Solar system objects

First are the natural planetary satellites which are typical for our project. They are fast running objects so that many dynamical perturbations are easy to model provided that the sample of data is sufficiently large and sufficiently accurate. By increasing the astrometric accuracy of one order of magnitude, we will be able to quantify small effects such as the dissipation of energy due to tides in the natural satellite systems: internal structure and scenarios for the formation should be validated. We explain the thermal equilibrium of Io and the geysers on Enceladus by looking for an acceleration in the motion of the icy satellites. Concerning the Saturnian system, we look for old data helping to validate the scenarios of formation and evolution of the satellite system.

Second are the asteroids and comets. Digitizing and analyzing old Schmidt or astrographic plates may allow to make pre-discoveries of NEA/NEO or TNO. A large time interval of data of high accuracy may help to model non-gravitational effects of asteroids and comets.

3. Spectrometry

It will be possible to digitize the large collection of variable stars spectra (Be stars). The data base BeSS, available at LESIA, provides a catalogue of all known (around 2000) Be stars and owns (around 130 000) available spectra of these objects. The study of the long-term spectroscopic variability of Be stars is very important to understand their sporadic ejection of matter. At the present time, the data start only in 1990 and the analysis of plates from 1950 to 1980 will enlarge the time series of data and will allow to detect a period in the ejections.

4. The first results

We had collaborations with laboratories owning sub-micrometric scanners and photographic plates (USNO-Washington DC, ROB-Bruxelles, Bucharest Astronomical Institute, QMUL-London, OCA-Nice). We made tests by scanning Galilean, Saturnian and Martian plates with ROB-DAMIAN digitizer (Robert *et al.* (2016).

Our results show that the accuracy reachable on old plates may be from 30 to 60 mas at a time where the accuracy of the data extracted from the same plates were from 200 to 500 mas! Using the proper motions of the final Gaia catalogue could improve our first results and completely renew the astrometric reduction.

5. The NAROO project

For our purpose, we are installing in Meudon observatory a sub-micrometric scanner the accuracy of which being chosen to guaranty the astrometric accuracy from the plates. We plan to start scanning old plates in 2018.

Our goal is now:

- making inventories of plate archives with objects, field, quality of plates, dates, and selecting plates to be analyzed
- choosing criteria and parameters for digitization and preparing reduction softwares taking into account the increase in accuracy
- making a specific database for files of digitized plates made available to the scientific community

6. Conclusion

The new reduction of old data started a few years ago. Unfortunately, the available catalogues such as UCAC did not allow to go back farther thanl the 1970's. At the present time, reducing older observations do not bring any improvement compared to the data published at the time they were made. The arrival of the final Gaia reference catalogue will completely renew the reduction of old data mainly for astrometry but also in the field of photometry and spectrometry.

Reference

Robert, V., Pascu, D., Lainey, V., Arlot, J.-E., De Cuyper, J.-P., Dehant, V., & Thuillot, W. 2016, *Astron. Astrophys.*, 596, A37

Astronomy and Astrophysics in the Gaia sky
Proceedings IAU Symposium No. 330, 2017
A. Recio-Blanco, P. de Laverny, A.G.A. Brown
& T. Prusti, eds.

doi:10.1017/S1743921317005841

Using Gaia as an Astrometric Tool for Deep Ground-based Surveys

Dana I. Casetti-Dinescu[1], **Terrence M. Girard**[2] **and Michael Schriefer**[1]

[1]Southern Connecticut State University,
501 Crescent Street, New Haven, CT, USA
email: dana.casetti@gmail.com

[2]14 Dunn Road, Hamden, CT 06518, USA
email: terrence.girard@gmail.com

Abstract. Gaia DR1 positions are used to astrometrically calibrate three epochs' worth of Subaru SuprimeCam images in the fields of globular cluster NGC 2419 and the Sextans dwarf spheroidal galaxy. Distortion-correction "maps" are constructed from a combination of offset dithers and reference to Gaia DR1. These are used to derive absolute proper motions in the field of NGC 2419. Notably, we identify the photometrically-detected Monoceros structure in the foreground of NGC 2419 as a kinematically-cold population of stars, distinct from Galactic-field stars. This project demonstrates the feasibility of combining Gaia with deep, ground-based surveys, thus extending high-quality astrometry to magnitudes beyond the limits of Gaia.

Keywords. astrometry, Galaxy: kinematics and dynamics, galaxies: dwarf

---

1. Introduction

Deep and wide-field ground-based programs such as the Dark Energy Camera Surveys, or the upcoming Large Synoptical Survey Telescope (LSST) to name only two, will routinely produce catalogues with billions of objects. Such data will uncover sparse and distant stellar systems such as clusters, dwarf galaxies, and tidal streams and over-densities which are key to our understanding of the Galaxy and the Local Group in the current cosmological paradigm. Here, we focus on the astrometric potential of such surveys which will deliver the much needed proper motions of very faint systems. We develop a methodology to astrometricaly calibrate mosaic imagers on large telescopes, and apply it to Subaru SuprimeCam data.

2. Observational Data

SuprimeCam images covering three epochs were downloaded from the Subaru telescope archive (SMOKA) in the fields of globular cluster NGC 2419, and dwarf spheroidal galaxy (dSph) Sextans. The first two epochs consist of V-filter, 15 to 360-sec exposures. The 2012 data are in Strömgren b and y, with exposures of 18 to 360 sec. The camera underwent a major upgrade in 2008, including the installation of new detectors, therefore separate astrometric calibrations are required. Gaia DR1 positions were extracted from the web-based Gaia archive facility (Gaia Collaboration *et al.* 2016). An overview of the data is presented in Table 1. Having offset exposures is important and allows the construction of astrometric calibration maps of the field of view and each detector. The pattern of offsets and dithers is shown in Fig. 1.

3. Distortion-Correction Maps

Maps of position residuals are constructed by transforming detector coordinates of each SuprimeCam chip into Gaia DR1 positions using third-order polynomials. The ten

Table 1. Characteristics of the Observations

	NGC 2419	**Sextans**
Epochs / No. of exposures	**2002** / 207 **2005** / 48 **2012** / 22	 **2005** / 62 **2012** / 60
Gaia DR1	**2015**	**2015**

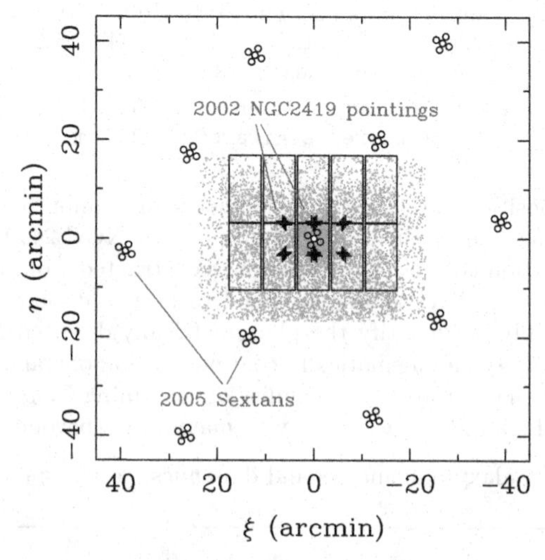

Figure 1. Dither pattern of SuprimeCam pointings in the field of NGC 2419 and Sextans. The dark frame illustrates a single exposure's field of view.

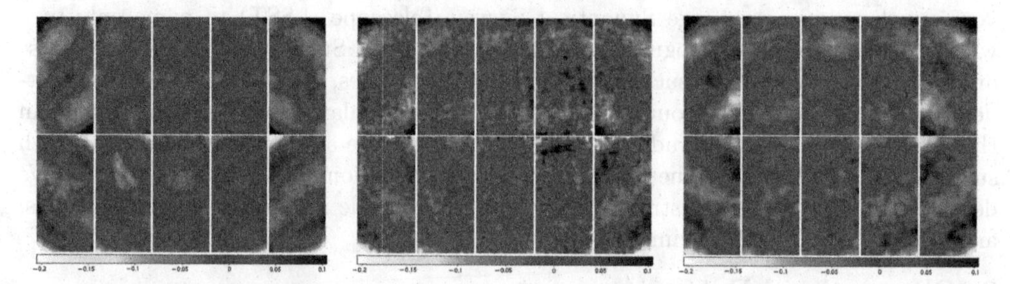

Figure 2. Distortion-correction maps, beyond third-order field terms, for SuprimeCam in 2002 (left), 2005 (middle) and 2012 (right). Corrections in the y-coordinate are shown, while those in x are generally similar. Units are pixels with a pixel scale of 0.2"/pix.

individual chips of SuprimeCam are apparent in Fig. 2, in which we display the y-coordinate maps. The 2012 data are close to the epoch of Gaia DR1 positions (2015), so these residuals can be used directly to construct a correction map (Fig. 2, right panel). Residuals from both the Sextans and NGC 2419 fields are stacked to produce this epoch's map. For the 2002 and 2005 data, the first step is a transformation into Gaia DR1, after which an average catalog at each epoch is constructed. This is followed by a transformation of each chip and frame into the appropriate average catalog; the subsequent residuals are used to construct these two epochs' maps. Averaging positions for the same object from multiple locations in the chip and field of view approximates a systematics-free average catalog. The more offsets and dithers we have, the better the average catalog.

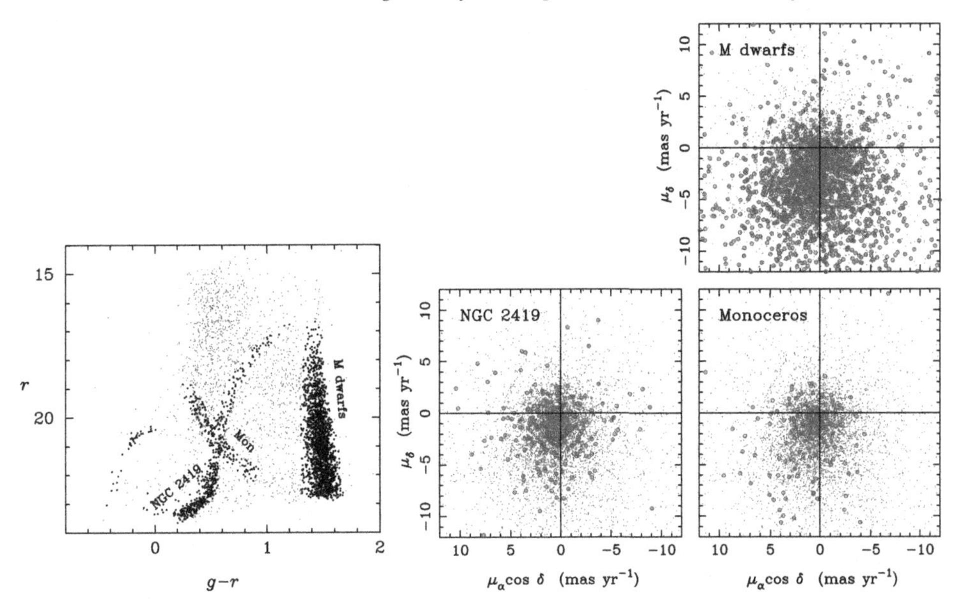

Figure 3. Left: Color-magnitude diagram in the field of NGC 2419 (beyond 5' from cluster center). NGC 2419 stars, Monoceros stars, and disk M dwarfs are highlighted and labeled. Right: Proper-motion diagrams of the three stellar populations highlighted in the left panel.

Among our data sets, those in the field of NGC 2419 for epoch 2002, and in the field of Sextans for epoch 2005, are appropriate for such mapping.

4. Preliminary Results

Preliminary proper motions are determined in the field of NGC 2419. Proper motions are on an inertial reference frame using background galaxies. The galaxies are selected using the 2002 and 2005 data classification (i.e., FWHM-based neural-network classification employed by the SExtractor code). In the foreground of NGC 2419 lies a diffuse and extended stellar system, originally discovered from SDSS data, that is referred to as the Monoceros ring, and is part of a more extended structure toward the galactic anticenter. Recently, a photometric study by Carballo-Bello *et al.* (2015) has shown a very distinctive main sequence of the Monoceros structure in the particular field of NGC 2419. Here, we use their photometry to select various stellar populations and examine their proper motions. In Fig. 3, we show the color-magnitude diagram (left) for this field, with various stellar populations highlighted, while the right panel shows the corresponding proper-motion distributions. Both cluster stars and Monoceros stars show a tight clump in proper motions, indicating kinematically cold populations, for which the scatter is dominated by our proper-motion errors. Conversely, nearby disk stars, as represented by the M dwarfs, show a large scatter that is due to the intrinsic velocity dispersion of these stars.

We acknowledge travel support from NSF grant #AST-1517824 to attend this IAU Symposium. This project was begun thanks to a grant from the NASA Connecticut Space Grant program. This work has made use of data from the European Space Agency (ESA) mission *Gaia*, processed by the *Gaia* Data Processing and Analysis Consortium (DPAC).

References

Carballo-Bello, J. A., Munõz, R. R., Carlin, J. L., Côté, P., Geha, M., Simon, J. D., & Djorgovski, S. G. *ApJ*, 805, 51

Gaia Collaboration *et al.* 2016, *A&A*, 595, 1

Astronomy and Astrophysics in the Gaia sky
Proceedings IAU Symposium No. 330, 2017
A. Recio-Blanco, P. de Laverny, A.G.A. Brown
& T. Prusti, eds.

doi:10.1017/S1743921317005282

Remarks of Gaia DR1 magnitude using ground-based optical monitoring of QSOs

Goran Damljanović[1], François Taris[2] and Alexandre Andrei[3]

[1]Astronomical Observatory, Volgina 7, 11060 Belgrade, Serbia
email: `gdamljanovic@aob.rs`

[2]Observatoire de Paris - SYRTE, CNRS/UPMC, 61 av. de l'Observatoire, 75014 Paris, France
email: `Francois.Taris@obspm.fr`

[3]Observatório Nacional/MCTI, Rua Gal. José Cristino 77, Rio de Janeiro, RJ CEP 20921-400, Brasil
email: `oat1@ov.ufrj.br`

Abstract. Since September 2016, the first release (DR1) of the Gaia catalogue was appeared. The optical Gaia positions of sources will be linked to the ICRF (VLBI radio positions of mostly quasars, QSOs). For high accurate link we need to investigate variations of optical flux of QSOs via their magnitude variations using data of ground-based telescopes. To do that, from 2013 we observed 47 QSOs and other sources; nine optical telescopes were used for that monitoring. To increase the total number of objects for the link, after a first set of 70 objects (Bourda *et al.* 2008), Bourda *et al.* (2011) established a second set of 47 objects. It is necessary to investigate the photometry and morphology of these objects. We collected ground-based data of QSOs (B, V and R mag) and compared with G mag of Gaia DR1; some results are presented here.

Keywords. Astrometry, reference systems, quasars: general.

1. Instruments and results

The installation of the 60 cm telescope at new site Astronomical Station Vidojevica - ASV (of Astronomical Observatory in Belgrade - AOB, Serbia) was in 2010, and of new 1.4 m ASV was in mid-2016. During 2013 we established the Serbian-Bulgarian mini-network of 6 telescopes. Also, we did with the 1.5 m telescope at the Leopold-Figl Observatorium für Astrophysics LFOA (Vienna Observatory, Universität Wien) after its reconstruction. And, we used the two TAROT telescopes (Taris *et al.* 2013, 2016) and robotic 0.8m Telescope Joan Oró - TJO (Observatori Astronòmic del Montsec, Spain).

The main information of the Serbian-Bulgarian telescopes and LFOA are:

1. ASV (AOB) Cassegrain $D/F(cm) = 60/600$ (longitude is 21.5 deg E, latitude is 43.1 deg N, altitude is 1140 m), CCD camera Apogee Alta U42, 2048x2048 pixels, 13.5x13.5 mkm pixel size, scale is 0.46 arcses, field of view (FoV) is 15.8x15.8 arcmin,

2. ASV (AOB) Ritchey-Chrétien 140/1142 (21.5 E, 43.1 N, 1150 m), Apogee Alta U42, 2048x2048 pixels, 13.5x13.5 mkm, scale is 0.24 arcsec, FoV is 8.3x8.3 arcmin,

3. Rozhen (NAO BAS) Ritchey-Chrétien 200/1577 (24.7 E, 41.7 N, 1730 m), VersArray 1300B, 1340x1300 pixels, 20x20 mkm, scale is 0.26 arcsec, FoV is 5.6x5.6 arcmin,

4. Rozhen (NAO BAS) Cassegrain 60/740 (24.7 E, 41.7 N, 1760 m), FLI PL09000, 3056x3056 pixels, 12x12 mkm, scale is 0.33 arcsec, FoV is 16.8x16.8 arcmin,

5. Belogradchik AO Cassegrain 60/740 (22.7 E, 43.6 N, 650 m), FLI PL09000, 3056x3056 pixels, 12x12 mkm, scale is 0.33 arcsec, FoV is 16.8x16.8 arcmin,

6. LFOA R.C. $D(cm) = 152$ (48.1 E, 15.9 N, 880m), SBIG ST-10 XME, 2184x1472 pixels, 6.8x6.8 mkm, scale is 0.15 arcsec, FoV is 3.8x5.6 arcmin.

The NAO BAS means National Astronomical Observatory of Bulgarian Academy of Sciences. Also, the 60 cm ASV was used with the CCD SBIG ST-10 XME: scale = $0.''23$, FoV is 8.4x5.6 arcmin. About TJO, see www.oadm.cat/en/home.htm. We collected about 7000 images. The Johnson-Cousins filters were available. The TAROT image reduction was described in papers by Taris *et al.* (2013, 2016). The standard bias, dark and flat-fielded corrections were done (also, hot/dead pixels were removed); this step was achieved using the PRISM commercial software (see www.prism-america.com). The next step, astrometric and photometric reduction, was done by the Gaia-GBOT Astrometric Reduction Pipeline (Bouquillon *et al.* 2014). The TJO magnitudes are relative ones, and they are calculated via a least square adjustment of the instrumental magnitudes of all known objects in the FoV. With other six telescopes, we observed targets in B,V and R bands (usually 3 CCD images per filter), and got the photometric results via comparison stars. The comparison stars were taken from the SDSS (or APASS) catalogue using suitable transformation (Chonis and Gaskell 2008). The calculated magnitude (B,V,R) is the average value (of 3 CCD images per filter) with st.error. Some our photometric results of 0049+003 are (using the 60 cm ASV telescope):

- Sep. 6th 2013, $JD = 2456542.48866$ for $B = 16.669 \pm 0.027$ mag, $JD = 2456542.47938$ for $V = 16.296 \pm 0.021$ mag, $JD = 2456542.49410$ for $R = 15.877 \pm 0.014$ mag,
- Sep. 7th 2013, $JD = 2456543.58255$, $B = 16.383 \pm 0.030$ mag, $JD = 2456543.57874$, $V = 16.280 \pm 0.030$ mag, $JD = 2456543.58638$, $R = 15.855 \pm 0.020$ mag.

The polynomial expression was used to get the G mag (from ground-based V and R ones), and to compare with G mag of the Gaia DR1: $G - V = -0.0120 - 0.3502(V - R) - 0.6105(V - R)^2$.

2. Conclusion

In the Gaia DR1 there is not epoch for each Gaia G-mag of QSOs (there is only the average G value) to compare with our ground-based results. The flux of QSOs is not constant with time. It is not clear what value of the Aperture Radius was used for the Gaia photometry reduction (which is very important for some QSOs as extended sources), etc. We hope, these values will be included into the next Gaia realize dataset.

Acknowledgements. This work has made use of data from the European Space Agency (ESA) mission *Gaia* (https://www.cosmos.esa.int/gaia), processed by the *Gaia* Data Processing and Analysis Consortium (DPAC, https://www.cosmos.esa.int/web/gaia/dpac/consortium). Funding for the DPAC has been provided by national institutions, in particular the institutions participating in the *Gaia* Multilateral Agreement. GD acknowledges observing grant of Institute of Astronomy and Rozhen NAO BAS. This work is part of the Project No 176011 (Dynamics and kinematics of celestial bodies and systems), supported by Ministry of Education, Science and Technological Development of R. Serbia.

References

Bourda, G. *et al.* 2008, *Astronomy and Astrophysics*, 490, 403
Bourda, G. *et al.* 2011, *Astronomy and Astrophysics*, 526, A102
Bouquillon, S. *et al.* 2014, *SPIE*, 9152
Chonis, T. & Gaskell, C. 2008, *Astronomical Journal*, 135, 264
Taris, F. *et al.* 2013, *Astronomy and Astrophysics*, 552, A98
Taris, F. *et al.* 2016, *Astronomy and Astrophysics*, 587, A112

Astronomy and Astrophysics in the Gaia sky
Proceedings IAU Symposium No. 330, 2017
A. Recio-Blanco, P. de Laverny, A.G.A. Brown
& T. Prusti, eds.

doi:10.1017/S1743921317005270

Outline of Infrared Space Astrometry missions:JASMINE

N. Gouda[1] and JASMINE working group

[1]National Astronomical Observatory of Japan,
2-21-1 Osawa, Mitaka, Tokyo, Japan, 181-8588
email: naoteru.gouda@nao.ac.jp

Abstract. Japanese group is promoting infrared space astrometry missions, JASMINE project series, in international collaboration with Gaia DPAC team. In this paper, the outline of Nano-JASMINE and Small-JASMINE missions is shown.

Keywords. infrared space astrometry missions, the Milky Way, the Galactic nuclear bulge, supermassive black hole

1. Introduction

JASMINE(Gouda(2011)) is an abbreviation for Japan Astrometry Satellite Mission for INfrared Exploration. We are now focusing on the development of two projects; those are Nano-JASMINE and Small-JASMINE whose missions are complementary to the Gaia mission.

2. Nano-JASMINE

The Nano-JASMINE micro-satellite project, with a primary mirror aperture of 5-cm class, is planned to produce scientific results based on the astrometric information of bright objects in the neighboring space as the first foray into space astrometry in Japan.The size and weight of the satellite are $(50\text{cm})^3$ and about 35 kg, respectively. Nano-JASMINE will operate in zw- band ($0.6 \sim 1.0\mu$m) to perform an all sky survey with a precision of 3 mas for positions, annual parallaxes and proper motions of stars brighter than zw=7.5 magnitude. The combination of the observational data from Nano-JASMINE and the Hipparcos Catalogue is expected to produce more precise data on proper motions (precision $\sim$ 0.1mas/yr) and annual parallaxes (precision $\sim$ 0.75mas). Assembly of the flight model that will be actually launched into space was completed in 2010. The original launch schedule was August 2011. However, the launch date has been delayed due to complex international situations and we are now looking for another opportunity for the launch. Steady progress has been also made in the development of algorithms and software required to determine astrometric information from raw observational data at the required level of precision with good international cooperation with the data analysis team (DPAC) for Gaia.

3. Small-JASMINE

An additional plan is underway to launch Small-JASMINE in around 2023. We have been aiming at the realization of the Small-JASMINE mission as a mission of the small science satellite program(JAXA Competitive M-class missions (Epsilon rocket missions)).

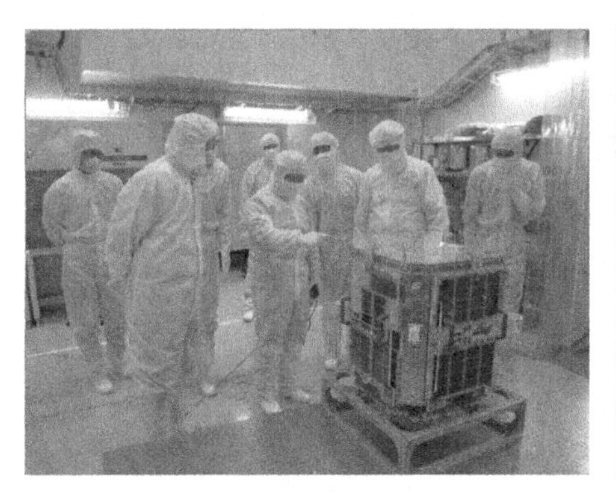

Figure 1. Flight model of the Nano-JASMINE satellite.

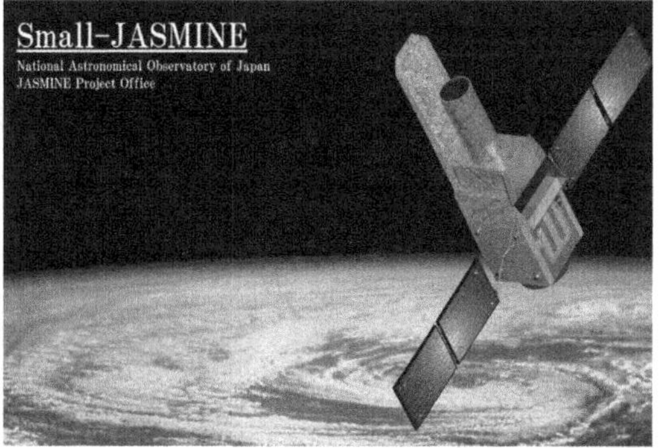

Figure 2. Artist'impression of the Small-JASMINE satellite.

The objective of the Small-JASMINE project is to perform infrared astrometric observations (Hw band: $1.1 \sim 1.7\mu$m) by the use of a three-mirror optical system telescope with a primary mirror aperture of 30 cm. The goal is to measure annual parallaxes with a precision of $\sim 20\mu$as and proper motions with a precision of $<\sim 50\mu$as/year for stars brighter than Hw=12.5 magnitude in the direction of an area of few square degrees around the Galactic center within the nuclear bulge and the direction of a number of specific astronomical objects of interest in order to create a catalogue of the positions and movements of stars within these regions. The project is unique in that unlike Gaia, observation will be performed in the near-infrared band, in which the effect of absorption by dust is weak, and the same astronomical object can be observed frequently.

The main scientific objective of Small-JASMINE is to clarify the dynamical structure of the Galactic nuclear bulge. In particular, our main goal is that Small-JASMINE will provide an understanding of the past evolution processes of the supermassive black hole through knowledge of the phase space distribution of stars in the Galactic nuclear bulge and also a prediction of the future activities of our Galactic center through knowledge of the gravitational potential in the nuclear bulge, and that this understanding can contribute to a better understanding of the co-evolution of the supermassive black holes and bulges in external galaxies. Next to this primary goal, Small-JASMINE will have many other scientific targets. Small-JASMINE can measure the same target every 100 minutes, so it is useful to resolve phenomena with short periods such as X-ray binaries, extrasolar planetary systems and gravitational lens effects. For example, the orbital elements of the star accompanying Cygnus X-1 can be resolved by Small-JASMINE.

In 2016, we, JASMINE working group, submitted a mission proposal for JAXA's ISAS (Institute of Space and Astronautical Science) call for JAXA Competitive M-class missions (the small science satellite program (Epsilon rocket missions)). We should have multiple-stage selection processes executed by JAXA. Small-JASMINE is now under the ISAS Mission Definition Review(MDR) as the first stage selection process (the review committee at ISAS just recently recommended Small-JASMINE to pass the MDR).

Reference

Gouda, N. 2011, *Scholarpedia*, 6(10):12021, http://www.scholarpedia.org/article/JASMINE

Astronomy and Astrophysics in the Gaia sky
Proceedings IAU Symposium No. 330, 2017
A. Recio-Blanco, P. de Laverny, A.G.A. Brown
& T. Prusti, eds.

doi:10.1017/S1743921317005336

GIER: A Danish computer from 1961 with a role in the modern revolution of astronomy - II

Erik Høg

Niels Bohr Institute, University of Copenhagen,
DK-2100, Copenhagen Ø, Denmark
email: ehoeg@hotmail.dk

Abstract. A Danish computer, GIER, from 1961 played a vital role in the development of a new method for astrometric measurement. This method, photon counting astrometry, ultimately led to two satellites with a significant role in the modern revolution of astronomy. A GIER was installed at the Hamburg Observatory in 1964 where it was used to implement the entirely new method for the measurement of stellar positions by means of a meridian circle, at that time the fundamental instrument of astrometry. An expedition to Perth in Western Australia with the instrument and the computer was a success. This method was also implemented in space in the first ever astrometric satellite Hipparcos launched by ESA in 1989. The Hipparcos results published in 1997 revolutionized astrometry with an impact in all branches of astronomy from the solar system and stellar structure to cosmic distances and the dynamics of the Milky Way. In turn, the results paved the way for a successor, the one million times more powerful Gaia astrometry satellite launched by ESA in 2013. Preparations for a Gaia successor in twenty years are making progress.

Keywords. history of astronomy, stars: distances, Galaxy: kinematics and dynamics, dark matter, solar system: general

---

The following on the poster is a short version of the article by Høg (2017, hereafter EH2017) where further references are given.

1. A new astrometric method

The meridian circle was the fundamental astronomical instrument for measuring large angles between stars and thereby establishing an accurate coordinate system on the sky. Naked eye observation of the star was used as it crossed the telescope field of view, but an entirely new method for measuring the star, *photon counting astrometry*, was proposed in 1960 at the Hamburg Observatory by the present author, see Høg (2011 or 2017). The method was implemented on the Hamburg meridian circle and proved to be a success on the Hamburg expedition to Perth in Western Australia during the five years 1967-72.

The method was also well received by astronomers elsewhere, especially in France where the grid with inclined slits required for the method was called "une grille de Høg" in those years. Most significant was that Pierre Lacroute (1906-1993), director of Strasbourg Observatory, adopted the method for his great vision of astrometry from space, Lacroute (1967, 1974 - see references in EH2017).

In 1975 ESA began a study of these ideas to which I was invited and soon proposed a mission design with many new features, see Høg (2011). The resulting Hipparcos satellite mission was launched in 1989 and the astrometric results obtained by photon counting astrometry were published in 1997. A new era for astrometry had begun where all branches of astrophysics benefitted as explained in the abstract.

Figure 1. GIER computer room of Perth Observatory 1971. GIER, in the teak cupboard in the background at right, was one of the first transistorized computers. Mrs Ilse Holst at the reader for 8-channel punched tape, one roll could contain 120 kilobytes. The control panel is seen just behind the paper punch. The tape from reader or punch poured into a large basket and could then be quickly rolled up. The operator would type commands to start a program on the typewriter and e.g. error messages from the computer were typed on endless paper. - Photo: Bernd Loibl.

2. Peter Naur and the Danish computer from 1961

Peter Naur (1928-2016) played important roles in this development as detailed in EH2017. He was my tutor at Copenhagen University from September 1953 and introduced me to astrometry, electronics and computing, subjects not common in astronomical department in those years when astrophysics had become fashionable.

Naur himself had been an astronomer when he was a school boy. He worked in astrophysics on stellar evolution and in astrometry, e.g., at the new meridian circle installed at Brorfelde in 1953. But he was far from satisfied with the observatory leadership and left astronomy entirely. In 1959 Naur joined Regnecentralen, a new public institution where a new computer, GIER, was being developed, one of the first transistorised ones. He was leading in the development of the ALGOL programming language. He won the 2005 Turing-award, "the Nobel-prize of computer science".

In 1964 a GIER computer with a powerful ALGOL60 compiler was acquired for the Perth expedition because it was ten times faster than the fastest other affordable computer, the IBM1620, and GIER was 20 percent cheaper. The speed was crucial for the expedition since it was just about possible to keep pace with the many observations produced at the meridian circle in Perth. Without GIER the pioneering scientific work would have failed - but the success gave confidence in the new astrometric method and I could continue to believe in my ideas.

References

Høg, E. 2017, *GIER: A Danish computer from 1961 with a role in the modern revolution of astronomy.* Nuncius Hamburgensis, Vol. 20, Ed. G. Wolfschmidt. 19 pp, arXiv.1704.05828

Høg, E. 2011, *Astrometry Lost and Regained. Baltic Astronomy*, 20, 221–230. http://esoads.eso.org/abs/2011BaltA..20..221H

Astronomy and Astrophysics in the Gaia sky
Proceedings IAU Symposium No. 330, 2017
A. Recio-Blanco, P. de Laverny, A.G.A. Brown
& T. Prusti, eds.

doi:10.1017/S174392131700552X

Astrometry with A-Track Using Gaia DR1 Catalogue

Yücel Kılıç, Orhan Erece and Murat Kaplan

Akdeniz University, Dumlupinar Blv., Campus, 07058, Antalya, Turkey
email: yucel.kilic@linux.org.tr

Abstract. In this work, we built all sky index files from Gaia DR1 catalogue for the high-precision astrometric field solution and the precise WCS coordinates of the moving objects. For this, we used build-astrometry-index program as a part of astrometry.net code suit. Additionally, we added astrometry.net's WCS solution tool to our previously developed software which is a fast and robust pipeline for detecting moving objects such as asteroids and comets in sequential FITS images, called A-Track. Moreover, MPC module was added to A-Track. This module is linked to an asteroid database to name the found objects and prepare the MPC file to report the results. After these innovations, we tested a new version of the A-Track code on photometrical data taken by the SI-1100 CCD with 1-meter telescope at TÜBİTAK National Observatory, Antalya. The pipeline can be used to analyse large data archives or daily sequential data. The code is hosted on GitHub under the GNU GPL v3 license.

Keywords. techniques: image processing, catalogs, minor planets, asteroids, astrometry.

---

1. A-Track

A-Track is a fast, open-source, cross-platform pipeline, for detecting the moving objects (asteroids and comets) in sequential telescope images in FITS format. The pipeline is coded in Python 3. The moving objects are detected using multiple image line detection algorithm, called MILD.

Once the candidate objects are found for each image (catalog file) by SExtactor Bertin & Arnouts (1996), MILD looks for points from different images that would form a line when plotted on the same graph. MILD investigates all of the possible 3-point combinations obtained from all of the candidate files (.cnd), each point coming from a different image. For each 3-point combination, it performs three checks (Atay, T. *et al.* (2016)).

2. Building GAIA Index Files for Astrometry.net

In order to use GAIA DR1 catalogue in astrometry.net code, we built new astrometry.net index files for high precision astrometric solution. To build such index files "hp-split" and "build-astrometry-index" commands were used respectively Lang, D. *et al.* (2016) for five (0 - 4, <24) different scales. However, GAIA DR1 catalogue as FITS table is approximately 350 GB. To compare the results with 2MASS catalogue and to reduce the size of GAIA DR1 some limitations similar to 2MASS were applied to GAIA DR1. Besides that to reach high accurate astrometric results, stars with the errors of RA and Dec greater than 1 milliarcsecond were excluded. After this elimination, we produced new index files (GAIA Index Files version 1) with 48 healpixes from homogeneously built GAIA DR1 catalogue ($\sim$ 178 million stars).

3. Using GAIA Index Files version 1 (GIFv1)

A newly developed module is added to the A-Track which obtains WCS coordinates of CCD frame by using GIFv3. With the obtained WCS coordinates, known asteroids passing through the region are listed with the query of MPC and SkyBoT and are compared with the moving objects found by A-Track. The objects found in the comparison are listed to the user in MPC format. If the moving object detected by A-Track has no match with the database, it is added to the MPC file as a new object.

4. Results and Comparisons for GIFv1 and 2MASS Index Files

The archived data taken from 1-meter telescope equipped with SI 1100 CCD was solved astrometrically using both 2MASS index files and GIFv1. Then all the asteroids in the frames were queried by the SkyBot database and the residuals of the coordinates were plotted (Fig. 1 and Fig. 2).

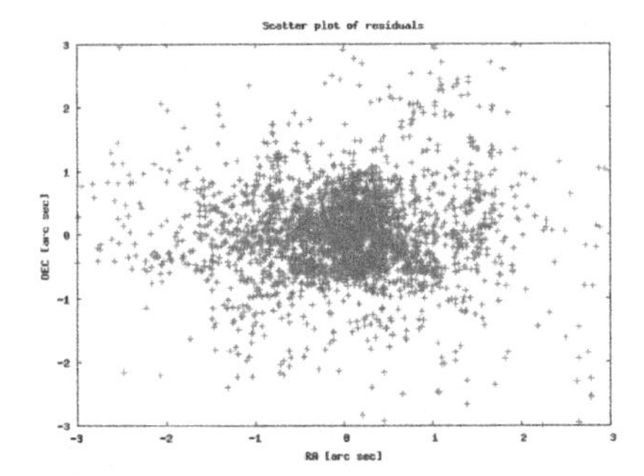

Figure 1. Scatter plot of residuals for 2MASS index files

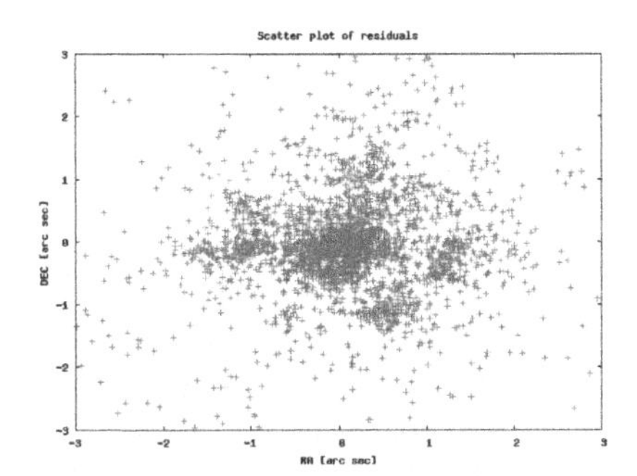

Figure 2. Scatter plot of residuals for GIFv1

5. Conclusions

As shown in Table 1a and Table 1b, the results obtained through GIFv1 are just as sensitive as 2MASS index files.

Residual (")	Detected Asteroids	Percentage (%)
<0.2	235	7.3
<0.5	1000	30.9
<1.0	2079	64.3
<2.0	2834	87.7
<2, <8	399	12.3
<8	3233	100

(a) Residuals for 2MASS index files

Residual (")	Detected Asteroids	Percentage (%)
<0.2	327	9.3
<0.5	1107	33.4
<1.0	1850	55.8
<2.0	2898	87.5
<2, <8	415	12.5
<8	3313	100

(b) Residuals for GIFv1

Table 1. Residuals

Acknowledgement

We acknowledge TÜBİTAK National Observatory for a partial support in using T100 telescope with project number 12BT100-327. This research has made use of IMCCE's SkyBoT VO tool. This work has made use of data from the European Space Agency (ESA) mission *Gaia*.

References

Bertin, E. & Arnouts, S. 1996, *A&AS*, 393-404, 117

Atay, T., Kaplan, M., Kilic, Y., & Karapinar, N. 2016, *Comput Phys. Commun.*, 524-530, 207

Lang, D., Hogg, D. W., Mierle, K., Blanton, M., & Roweis, S. 2010, *AJ*, 1782-1800, 139

Astronomy and Astrophysics in the Gaia sky
Proceedings IAU Symposium No. 330, 2017
A. Recio-Blanco, P. de Laverny, A.G.A. Brown
& T. Prusti, eds.

doi:10.1017/S1743921317005701

Astrometry for New Reductions: The ANR method

Vincent Robert[1,2] and Christophe Le Poncin-Lafitte[3]

[1]IMCCE, Observatoire de Paris, PSL Research University, CNRS, Sorbonne Universités, UPMC Univ. Paris 06, Univ. Lille 1,
77 avenue Denfert-Rochereau, F-75014, Paris, France
email: vincent.robert@obspm.fr

[2]IPSA, Institut Polytechnique des Sciences Avancées,
63 bis boulevard de Brandebourg, F-94200, Ivry-sur-Seine, France

[3]SYRTE, Observatoire de Paris, PSL Research University, CNRS, Sorbonne Universités, UPMC Univ. Paris 06, LNE,
61 avenue de l'Observatoire, F-75014, Paris, France
email: christophe.leponcin@obspm.fr

Abstract. Accurate positional measurements of planets and satellites are used to improve our knowledge of their orbits and dynamics, and to infer the accuracy of the planet and satellite ephemerides. With the arrival of the Gaia-DR1 reference star catalog and its complete release afterward, the methods for ground-based astrometry become outdated in terms of their formal accuracy compared to the catalog's which is used. Systematic and zonal errors of the reference stars are eliminated, and the astrometric process now dominates in the error budget.
We present a set of algorithms for computing the apparent directions of planets, satellites and stars on any date to micro-arcsecond precision. The expressions are consistent with the ICRS reference system, and define the transformation between theoretical reference data, and ground-based astrometric observables.

Keywords. Astrometry, Catalogs, Methods: miscellaneous

---

1. Introduction

Accurate orbit determination requires a large amount of observations over a large time span to allow for the best precision and extrapolation. Such kind of observations were processed a long time ago with inaccurate star catalogs and methods, compared to recent ones. No real efforts have been attempted to reanalyze these data a new time, considering the amount of time, mean and energy required. Using photographic plates of planetary satellites, we demonstrated that a new reduction of old observations can improve significantly the ephemerides (Arlot *et al.* 2012, Robert *et al.* 2011, Robert *et al.* 2015, Robert *et al.* 2016, Lainey *et al.* 2017). More important, we showed that ephemerides can sometimes be significantly biased while their extrapolation quickly diverges, due to several reasons and in particular the imprecision of past observations that are introduced in the adjustments.
In this framework and with support of the Gaia mission, the NAROO project has been initiated at Paris Observatory with the primary aim to reprocess the old astrometric observations with the best instrumental, algorithmic and numerical techniques. We attempt to answer the following question: how to ensure the Gaia accuracy over one century ?

2. The astrometry in the Gaia era

The "classical" astrometric method to compute high-precise apparent places of the objects consists in correcting reference positions for all known effects (Kaplan *et al.* 1989). In particular the proper motions with first-order models, the parallaxes with geometric transformations, the aberrations and light deflection with geometric transformations and relativistic terms, and the atmospheric refraction with semi-analytical solutions. The key point is that the "classical" accuracy of the overall astrometric process is about 20 mas, that is to say 60 km at the distance of Jupiter, or 120 km at the distance of Saturn.

The example of the proper motions. 1st-order models of propagation are enough accurate since the star catalog errors are dominating and $\dot{r}$ is not known:

$$\begin{cases} \Delta\alpha\cos\delta = \mu_\alpha(t - t_0) \\ \Delta\delta = \mu_\delta(t - t_0) \end{cases} \tag{2.1}$$

But since the Gaia catalog errors are up to 1000 times more accurate than other astrometric references, one should use 3rd-order models of propagation to ensure the catalog accuracy:

$$\begin{cases} \Delta\alpha\cos\delta = \mu_\alpha(t - t_0) - (\dot{r}\mu_\alpha - \tan\delta_0\mu_\alpha\mu_\delta)(t - t_0)^2 + \xi_\alpha(t - t_0)^3 \\ \Delta\delta = \mu_\delta(t - t_0) - (\dot{r}\mu_\delta + \frac{1}{2}\tan\delta_0\mu_\alpha^2)(t - t_0)^2 + \xi_\delta(t - t_0)^3 \end{cases} \tag{2.2}$$

The example of the calibration models. N-order conventional calibration models are used to map the transformation between the measured $(x; y)$ and tangential $(X; Y)$ frames. But since the Gaia external accuracy is up to 1000 times more accurate than other astrometric references, one should refine the developments at least with 1st-order coma and magnitude, 2nd-order tilt, 3rd-order distorsion, and temperature terms to traduce and measure all physical effects that could not be analyzed before:

$$\begin{cases} X = \rho_x\cos\theta_x x - \rho_y\varphi(T)\sin\theta_x y + \Delta_x + C_x x\Delta m + px^2 + qxy + Dx(x^2 + y^2) \\ Y = \rho_x\sin\theta_y x + \rho_y\varphi(T)\cos\theta_y y + \Delta_y + C_y y\Delta m + pxy + qy^2 + Dy(x^2 + y^2) \end{cases} \tag{2.3}$$

3. The ANR project

Our work aims at providing complete set of algorithms to compute the apparent place of the objects in the sky, observed from ground. While the most accurate star catalogs provided an overall accuracy about of 1 mas, the Gaia catalog improved the positioning accuracy with that of 1μas. In the same way and while the method of Kaplan *et al.* (1989) provided an overall accuracy about of 1 mas, we aim for that of 1 μas for ground observations.

References

Arlot, J.-E., Desmars, J., Lainey, V., & Robert, V. 2012, *PSS*, 73, 1

Kaplan, G. H., Hughes, J. A., Seidelmann, P. K., Smith, C. A., & Yallop, B. D. 1989, *AJ*, 97

Lainey, V., Jacobson, R. A., Tajeddine, R., Cooper, N. J., Murray, C., Robert, V., Tobie, G., Guillot, T., Mathis, S., Remus, F., Desmars, J., Arlot, J.-E., De Cuyper, J.-P., Dehant, V., Pascu, D., Thuillot, W., Le Poncin-Lafitte, C., & Zahn, J.-P. 2017, *Icarus*, 281

Robert, V., de Cuyper, J.-P., Arlot, J.-E., de Decker, G., Guibert, J., Lainey, V., Pascu, D., Winter, L., & Zacharias, N. 2011, *MNRAS*, 415, 1

Robert, V., Lainey, V., Pascu, D., Pasewaldt, A., Arlot, J.-E., De Cuyper, J.-P., Dehant, V., & Thuillot, W. 2015, *A&A*, 582, A36

Robert, V., Pascu, D., Lainey, V., Arlot, J.-E., De Cuyper, J.-P., Dehant, V., & Thuillot, W. 2016, *A&A*, 596, A37

Astronomy and Astrophysics in the Gaia sky
Proceedings IAU Symposium No. 330, 2017
A. Recio-Blanco, P. de Laverny, A.G.A. Brown
& T. Prusti, eds.

doi:10.1017/S1743921317006275

Optimisation of JWST operations with the help of Gaia

J. Sahlmann, E. G. Nelan, P. Chayer, B. McLean and M. Lallo

Space Telescope Science Institute, 3700 San Martin Drive, Baltimore, MD 21218, USA
email: jsahlmann@stsci.edu

Abstract. The James Webb Space Telescope (JWST) is scheduled for launch in 2018. To operate and observe efficiently, JWST will rely on various external astrometric and photometric catalogues, in particular the HST Guide Star Catalog (GSC), for instance to locate sources accurately on the sky. The incorporation of the Gaia astrometric catalog will improve the absolute astrometry of the GSC and is therefore relevant for JWST operations. We outline how the JWST Science and Operations Center hosted at the Space Telescope Science Institute (STScI) intends to use the Gaia survey results to improve upon operational aspects such as the guiding and the geometric focal plane characterisation of JWST.

Keywords. space vehicles: instruments, catalogs, astrometry

1. Context

JWST (Gardner *et al.* 2006), the space observatory succeeding the Hubble Space Telescope (HST) has a 6.5 m diameter primary mirror that feeds four science instruments (MIRI, NIRSpec, NIRISS, NIRCam) that cover the near-to-mid-infrared spectral range with powerful and versatile observational modes. Two cameras (FGS) are used for guiding operations. STScI is in the process of incorporating Gaia astrometric catalog data into the HST Guide Star Catalog (GSC) and the JWST calibration field catalog in order to improve their astrometric accuracy. JWST will use the GSC for the selection of guide stars and it will observe the calibration field for various purposes. The Gaia astrometric catalogs will therefore be the backbone for the astrometric calibrations of JWST, e.g. for monitoring the observatory's focal plane alignment and for determining the geometric distortion of the instruments.

2. Focal plane geometric calibration

Accurate focal plane calibration is necessary for efficient observatory operations, e.g. for target acquisition and for maintaining pointing stability. For some modes it is critical, because stringent requirements at milli-arcsecond level exist, e.g. multi-object slit spectroscopy and coronagraphy. In-orbit focal plane alignment will be established and monitored by observing the JWST calibration field with several instruments in parallel and locking the observed star fields onto the Gaia reference system to determine relative positions and orientations (Figure 1). Because most Gaia stars lie at the bright end of stars observable with JWST full-frame imaging, we will bridge that magnitude gap using the GSC itself, SDSS, Pan-STARRS, VISTA, and other surveys, and propagate the exquisite absolute and relative astrometric accuracy of Gaia's catalogs.

3. The JWST astrometric calibration field

A field in the Large Magellanic Cloud was chosen in 2005 (Rhoads 2006) because it has a high density of faint stars with small proper motions and is situated in JWST's

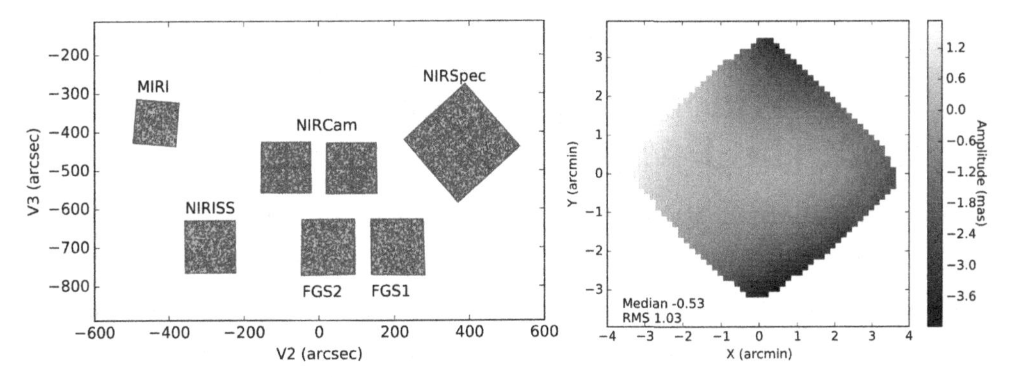

Figure 1. *Left*: JWST apertures in the observatory coordinate system. Dots indicate Gaia sources for a pointing in the JWST calibration field. *Right*: Offsets in Declination corresponding to higher-order distortions between the HST and Gaia catalogs of the JWST calibration field.

continuous viewing zone. It was observed with HST in 2006 and a catalog spanning 5'x5' with milli-arcsecond relative astrometry and deep optical photometry (V$\lesssim$25) was obtained (Anderson & Diaz 2011). This provides the basis for monitoring the astrometric calibration of the telescope and its instruments.

Gaia's first data release (Gaia Collaboration *et al.* 2016a,b) gave us the opportunity to verify the accuracy of the HST calibration field catalog. There are 2020 high-fidelity sources in common between the Gaia and HST catalogs. We mapped the two catalogs with a standard third-degree bivariate polynomial. Imposing Gaia as reference, we found offset, rotation, and scale terms that were within the known limitations of the HST absolute astrometric calibration. We also identified higher-order distortion terms at the level of ~1 mas RMS across the field, see Figure 1 (Sahlmann 2017). The residual dispersion of the mapping is ~2 mas, which roughly corresponds to the expected amplitude of the field's internal proper motion dispersion (~50 km/s at a distance of ~50 kpc) over the 9 years between the HST and Gaia epochs. We recently acquired a second HST epoch of the field to generate a catalog on the Gaia reference frame that includes the faint-star proper motions and thus will conserve its accuracy during the JWST mission.

Initial work is being done with Gaia's first data release but we anticipate that the second data release catalogs will be used during JWST commissioning and early operations, capitalising on the improved astrometry and availability of proper motions for faint stars.

Acknowledgements

This work has made use of data from the European Space Agency (ESA) mission *Gaia* (`https://www.cosmos.esa.int/gaia`), processed by the *Gaia* Data Processing and Analysis Consortium (DPAC, `https://www.cosmos.esa.int/web/gaia/dpac/consortium`).

References

Anderson, J. & Diaz, R. 2011, Validation of the Astrometry in the JWST Calibration Field, Technical Report JWST-STScI-002474, STScI

Gaia Collaboration, Brown, A. G. A., Vallenari, A., *et al.* 2016a, *A&A*, 595, A2

Gaia Collaboration, Prusti, T., de Bruijne, J. H. J., *et al.* 2016b, *A&A*, 595, A1

Gardner, J. P., Mather, J. C., Clampin, M., *et al.* 2006, *Space Science Reviews*, 123, 485

Rhoads, J. E. 2006, in The 2005 HST Calibration Workshop: Hubble After the Transition to Two-Gyro Mode, ed. A. M. Koekemoer, P. Goudfrooij, & L. L. Dressel, 360

Sahlmann, J. 2017, Astrometric accuracy of the JWST calibration field catalog examined with the first Gaia data release, Analysis Report JWST-STScI-005492, STScI

Astronomy and Astrophysics in the Gaia sky
Proceedings IAU Symposium No. 330, 2017
A. Recio-Blanco, P. de Laverny, A.G.A. Brown & T. Prusti, eds.

doi:10.1017/S1743921317006068

Kinematics of our Galaxy from the PMA and TGAS catalogues.

Anna B. Velichko, Volodymyr S. Akhmetov and Peter N. Fedorov

Institute of Astronomy, V. N. Karazin Kharkiv National University,
61022, 35 Sumska str., Kharkiv, Ukraine
email: Velichko.Anna.B@gmail.com, akhmetovvs@gmail.com, pnfedorov@gmail.com

Abstract. We derive and compare kinematic parameters of the Galaxy using the PMA and *Gaia* TGAS data. Two methods are used in calculations: evaluation of the Ogorodnikov-Milne model (OMM) parameters by the least square method (LSM) and a decomposition on a set of vector spherical harmonics (VSH). We trace dependencies on the distance of the derived parameters including the Oort constants A and B and the rotational velocity of the Galaxy V_{rot} at the Solar distance for the common sample of stars of mixed spectral composition of the PMA and TGAS catalogues. The distances were obtained from the TGAS parallaxes or from reduced proper motions for fainter stars.

The A, B and V_{rot} parameters derived from proper motions of both catalogues used show identical behaviour but the values are systematically shifted by about 0.5 mas/yr.

The Oort B parameter derived from the PMA sample of red giants shows gradual decrease with increasing the distance while the Oort A has a minimum at about 2 kpc and then gradually increases.

As for models chosen for calculations, first, we confirm conclusions of other authors about the existence of extra-model harmonics in the stellar velocity field. Secondly, not all parameters of the OMM are statistically significant, and the set of parameters depends on the stellar sample used.

Keywords. astrometry, stellar kinematics, Galactic structure

1. Introduction

The quality of kinematic investigations is mainly determined by accuracy and amount of the used catalogue data. The *Gaia* DR1 catalogue (Gaia collaboration *et al.* (2016a)) provides a huge number of astrometric data that open new opportunities in different domains of astrometry and stellar astronomy. It allowed to build a new catalogue of absolute proper motions called PMA (Akhmetov *et al.*(2017)). The PMA was derived from a combination of the *Gaia* DR1 and 2MASS (Skrutskie *et al.* (2006)) catalogues. It contains more than 420 million stars with positions, proper motions as well as photometry in G, J, H and K bands taken from both catalogues used.

Also the *Gaia* mission has produced a sub-set of about 2 million stars to be common with the Tycho-2 catalogue, *Tycho-Gaia astrometric solution* (TGAS, Michalik *et al.*(2015), Gaia collaboration *et al.* (2016), Lindegren *et al.* (2016)), containing stellar positions, proper motions and parallaxes. This data is a qualitatively new material for kinematic investigations of our Galaxy.

We use two different ways for deriving kinematic parameters. The first one involves some physical model containing a set of parameters with the known physical meaning. The values of this parameters are derived after applying the LSM to the set of the model equations. Several conventional models of various degrees of complexity are described in literature. To select the most suitable one is the hard challenge, because it is unknown a priori which of constituents are present in the observed stellar velocity field. Instead, there is an opportunity to peak out all statistical significant harmonics in the stellar velocity field and then choose the adequate model using decomposition on a set of orthonormal functions on the sphere. We use 2-dimensional (in the tangential plane) VSH first applied by Mignard & Morando (1990) to analyze systematic differences between the *Hipparcos* (Perryman *et al.* (1997)) and FK5 catalogues. Later this method was expanded by Vityazev & Shuksto (2005) to investigation of stellar kinematics. It was applied to analysis of stellar proper motions of the *Hipparcos* catalogue.

The goal of this work is to compare kinematic parameters derived from the PMA and TGAS catalogues as well as verify the completeness of the OMM.

2. Working equations

Within framework of the OMM (du Mont, 1977, Rybka, 2004) the stellar velocity field can be represented as follows:

$$\mathbf{V} = \mathbf{V}_0 + \boldsymbol{\Omega} \times \mathbf{r} + \mathbf{M}^+ \times \mathbf{r} \qquad (2.1)$$

It is assumed that the systematic constituent of stellar spatial motions $\mathbf{V}$ consists of translational motion of the Sun $\mathbf{V}_0 = (U, V, W)$ relative to the centroid used, the angular velocity of the rigid-body rotation $\boldsymbol{\Omega} = (\omega_x, \omega_y, \omega_z)$ of the stellar sample as well as the symmetric deformation tensor $\mathbf{M}^+ = (M_{12}^+, M_{13}^+, M_{23}^+, M_{11}^+, M_{22}^+, M_{33}^+)$. The first three parameters of the deformation tensor correspond to deformations of the stellar velocity field in $(x, y), (x, z)$ and (y, z) Galactic planes while the others - to its contraction - expansion along three principal Galactic axes.

In case of the VSH method the stellar velocity field can be represented as the sum of toroidal and spheroidal coefficients with toroidal and spheroidal harmonics:

$$\mathbf{U}(l,b) = \sum_{nkp} t_{nkp}\mathbf{T}_{nkp} + \sum_{nkp} s_{nkp}\mathbf{S}_{nkp} \qquad (2.2)$$

After projection of these two equations onto the Galactic coordinate system we obtain the following sets of equations to be solved by the LSM:

$$\begin{aligned}
K\mu_l \cos b = &\quad U/r \sin l \;-\; V/r \cos l \;-\; \omega_1 \sin b \cos l \;-\; \omega_2 \sin b \sin l \;+\; \omega_3 \cos b \;+ \\
+ &\quad M_{12}^+ \cos b \cos 2l - M_{13}^+ \sin b \sin l \;+\; M_{23}^+ \sin b \cos l \;- \\
- &\quad 0.5\, M_{11}^+ \cos b \sin 2l + 0.5\, M_{22}^+ \cos b \sin 2l \\
K\mu_b = &\quad U/r \cos l \sin b \;+\; V/r \sin l \sin b \;-W/r \cos b \;+\; \omega_1 \sin l - \omega_2 \cos l \;- \\
- &\quad 0.5\, M_{12}^+ \sin 2b \sin 2l - \; M_{13}^+ \cos 2b \cos l \;+\; M_{23}^+ \cos 2b \sin l \;- \\
- &\quad 0.5\, M_{11}^+ \sin 2b \cos^2 l - 0.5\, M_{22}^+ \sin 2b \sin^2 l + 0.5\, M_{33}^+ \sin 2b
\end{aligned} \qquad (2.3)$$

and

$$\begin{aligned}
K\mu_l \cos b = &\quad \sum_{nkp} t_{nkp}\mathbf{T}^l_{nkp}(l,b) + \sum_{nkp} s_{nkp}\mathbf{S}^l_{nkp}(l,b) \\
K\mu_b = &\quad \sum_{nkp} t_{nkp}\mathbf{T}^b_{nkp}(l,b) + \sum_{nkp} s_{nkp}\mathbf{S}^b_{nkp}(l,b)
\end{aligned} \qquad (2.4)$$

3. Kinematic parameters based on OMM

At first, we compare the kinematic parameters derived from the PMA and TGAS proper motions within the OMM. For this, the list of 2,048,407 common stars of the PMA

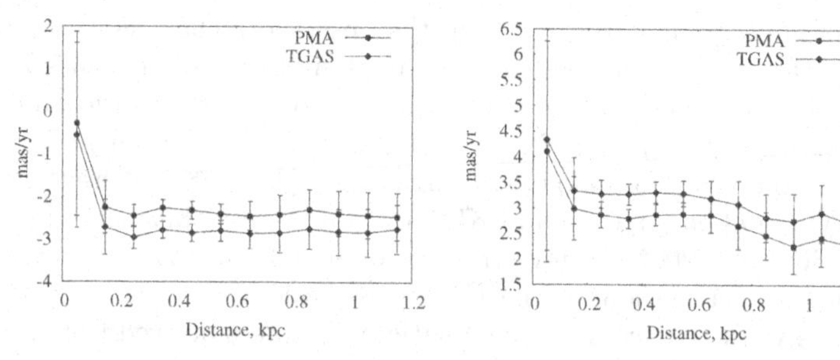

Figure 1. Kinematic parameters based on the OMM: Oort constant B (left panel), Oort constant A (right panel). The PMA data is designated by filled circles while the TGAS data - by filled diamonds.

and TGAS catalogues has been compiled. The distance to each star was determined from its parallax: $D = 1/\pi$. Stars with good parallaxes ($\sigma_\pi/\pi < 0.3$) and distances in the range from 0 kpc to 1.2 kpc were selected. The data were binned by distances with the 100 pc-sized step. The kinematic parameters were derived from the system of equations (2.3) by the standard LSM. The resulting behaviours of the Oort A and Oort B depending on the distance are shown in Fig. 1.

It can be seen from these figures that the Oort A and B constants derived from the stellar velocity field of the nearest stars with distances from 0 to 100 pc have huge root-mean-square (rms) errors. This suggests that the peculiar constituent dominates in proper motions of these stars. As the distance increases the accuracy of the parameters determination improves significantly. One more feature to be noted is the systematic differences between the PMA and TGAS data probably caused by systematic shifts between proper motions of these two catalogues in the Northern hemisphere (see Fig. 6 in Akhmetov *et al.* (2017)).

4. Applying VSH

According to the Vityazev's algorithm, applying the VSH method involves the following steps:

1) The coordinate grid regularization to make the coordinate distribution uniform throughout the sphere. For this, the Healpix (Gorsky *et al.* (2015)) pixelization scheme was used. The whole sphere was partitioned into 1200 pixels of equal area about 34.4 sq. deg. The proper motions were averaged over the pixels and referred to their centers.

2) The substraction of Solar motion effects from the stellar velocity field.

3) Detection of all statistical significant harmonics using the χ^2-test.

4) Calculation of toroidal and spheroidal decomposition coefficients t_j, s_j with all statistical significant harmonics from the system of equations (2.4).

5) Determining parameters of the OMM using the relations between the decomposition coefficients and the OMM parameters derived by Vityazev & Tsvetkov(2009).

As a result, it was found out that the stellar velocity field of the nearby stars with distances from 100 to 300 pc contains almost all constituents of the OMM. As the distance increases, only Solar motion components as well as Oort constants A and B remain in the proper motions of the TGAS. In addition to them the $M_{11}^{+} - M_{22}^{+}$ is significant for the PMA proper motions even for the farthest stars. Besides, several extra-model harmonics were detected. Among them the t_{310} and t_{211} have the greatest amplitudes. To find their physical meaning is a separate task.

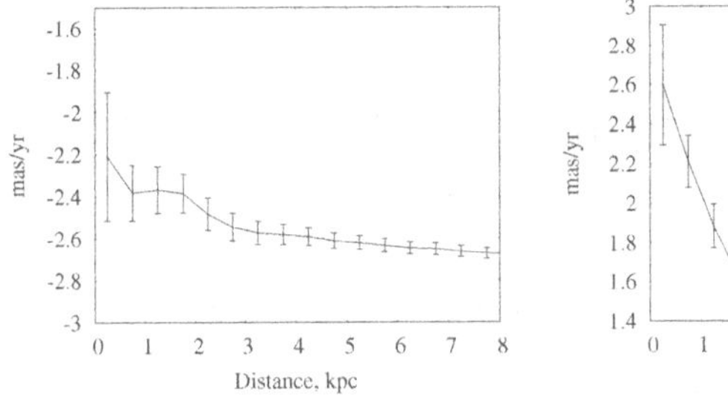

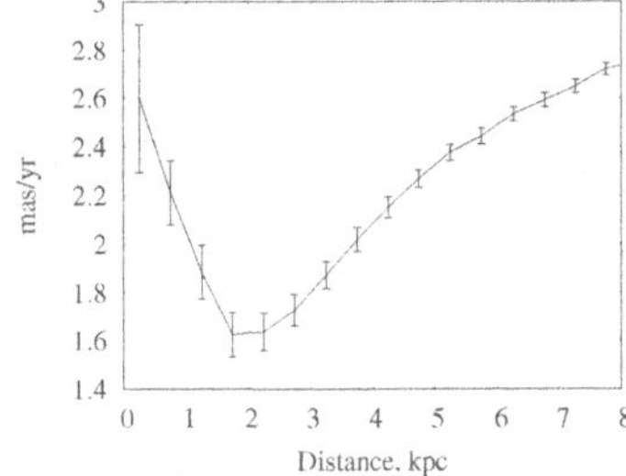

Figure 2. Kinematic parameters based on the VSH from red giants of the PMA: Oort constant B (left panel), Oort constant A (right panel).

The results just presented are related to bright range of magnitudes constituting only a small fraction of all PMA stars. To estimate distances to stars which do not have parallaxes we use reduced proper motions: $M'_{K_S} = K_S + 5 + 5lg(\mu)$. The sample of about 50 million red giants was compiled. For this sample the dependencies of the Oort constants A and B derived from the VSH method have been traced up to 8 kpc. It can be seen from Fig.2 that the Oort B gradually decreses with distance while the Oort A has a minimum at about 2 kpc and then increases. It would be interesting for theorists to explain this behaviour.

5. Acknowlegements

This work has made use of data from the European Space Agency (ESA) mission *Gaia* (`https://www.cosmos.esa.int/gaia`), processed by the *Gaia* Data Processing and Analysis Consortium (DPAC, `https://www.cosmos.esa.int/web/gaia/dpac/consortium`). Funding for the DPAC has been provided by national institutions, in particular the institutions participating in the *Gaia* Multilateral Agreement. It is a pleasure to acknowlege financial support from IAU.

References

Akhmetov, V. S., Fedorov, P. N., Velichko, A. B., & Shulga, V. M. 2017, *MNRAS*, 469, 763.

Clube, S. V. M. 1972, *MNRAS*, 159, 289

Gorsky, K. M., Hivon, E., Banday, A. J., Wandelt, B. D., Hansen, F. K., Reinecke, M., Bartelmann, M. 2015 *Astrophys. J.*, 622, 759

Mignard, F. & Morando, B. 1990, *Journées 1990. Systèmes de Référence Spatio-Temporels*, 151.

du Mont, B. 1977, *A& A*, 61, 127

Perryman, M. A. C., Lindegren, L., Kovalevsky, J., Høg, E., Bastian, U., Bernacca, P. L., Crézé, M., Donati, F., Grenon, M., Grewing, M., van Leeuwen, F., van der Marel, H., Mignard, F., Murray, C. A., Le Poole, R. S., Schrijver, H., Turon, C., Arenou, F., Froeschlé, M., & Petersen, C. S., 1997, *A& A*, 323, L49

Rybka, S. P. 2004, *Kinem. i Fiz. Neb. Tel* 20, 133

Skrutskie, M. F., Cutri, R. M., Stiening, R., Weinberg, M. D., Schneider, S., Carpenter, J. M., Beichman, C., Capps, R., Chester, T., Elias, J., Huchra, J., Liebert, J., Lonsdale, C., Monet, D. G., Price, S., Seitzer, P., Jarrett, T., Kirkpatrick, J. D., Gizis, J. E., Howard, E., Evans, T., Fowler, J., Fullmer, L., Hurt, R., Light, R., Kopan, E. L., Marsh, K. A., McCallon, H. L., Tam, R., Van Dyk, S., & Wheelock, S. 2006, *ApJ*, 131, 1163

Vityazev, V. V. & Tsvetkov, A. S. 2009, *Astron. Letters*, 35, 100.

Vityazev, V. V. & Shuksto, A. K. 2005, *Vestn. Spb. Gos. Univ. Ser. 1*, 1, 116.

Astronomy and Astrophysics in the Gaia sky
Proceedings IAU Symposium No. 330, 2017
A. Recio-Blanco, P. de Laverny, A.G.A. Brown
& T. Prusti, eds.

doi:10.1017/S1743921317005269

Nano-JASMINE and small-JASMINE data analysis

Yoshiyuki Yamada[1], Yuji Shirasaki[2] and Ryoichi Nishi[3]

[1]Department of Physics, Kyoto University, Kyoto 606-8502 Japan
email: yamada@scphys.kyoto-u.ac.jp / yamada@amesh.org

[2]Astronomical Data Center, National Astronomical Observatory of Japan, 181-xxxx
email: yuji.shirasaki@nao.ac.jp

[3]Department of Physics, Niigata University,
email:nishi@astro.sc.niigata-u.ac.jp

Abstract. Space astrometry missions Nano-JASMINE and small-JASMINE are planned in Japan. Data analysis tasks are performed under Gaia-JASMINE collaboration in long time. We expected to achieve 3 mas accuracy in Nano-JASMINE, and 20 micro arcsec in small-JASMINE of astrometric performance. Gaia DR1 publication and instruction is done from NAOJ and Niigata University.

Keywords. astrometry, infrared: galaxies

---

1. Introduction

We are planning a series of space astrometry satellite mission(Gouda (2011)). Nano-JASMINE is a global astrometry mission with 5 cm aperture aiming to 3 mas. Small-JASMINE is a differential astrometry mission of very narrow region with near infrared band aiming to 20 micro arcseconds.

2. Overview

2.1. *Nano-JASMINE data analysis*

Nano-JASMINE data analysis will be done by the collaboration with Gaia DPAC AGIS team. Nano-JASMINE is a scanning astrometric satellite with two beam. Its observational strategy is the same as that of Gaia. By simply replacing IDT (Initial Data Treatment) and PDB (Parameter Database), we can apply Gaia AGIS for Nano-JASMINE data analysis. Checking applicability of AGIS to Nano-JASMINE parameters has been done by Dr. Daniel Michalik with Lund and ESAC members. We confirmed that 2.5 mas parallax accuracy will be expectable.

We are now implementing IDT for Nano-JASMINE. Calculating PSF fitted center is the most important parts of IDT. Principal component approach of template PSFs are already implemented and checked. The algorithm assums that we have knowledge of PSF shape. We are now replacing running solution for calculating PSF fitted center position of the each stellar images.

2.2. *small-JASMINE data analysis*

As small JASMINE is not a scanning satellite, but it observes by step stare observation, the strategy is different from that of Hipparcos and Gaia. So Gaia software cannot be used for its analysis. The essential parts are 1)calculate PSF fitted center of the each stellar image, and 2)plate adjustment / overlapping.

By using many stellar images, we calculate effective PSF(Anderson & King (2000)). Effective means that it is not purely optical PSF, but convoluted with detector response. It is observable. For checking the accuracy of PSF fitted center, we show pixel phase error (Anderson & King (2000)) i.e. the difference between true center and estimated center in sub pixel level in Fig. 1. We can achieve 1/300 pixel (Peak) accuracy for constructing ePSF from simulated PSFs. We also check real observed data provided by HST group. By comparing positions of PSF fitted center by using ePSF provided by HST group and ePSF which we construct, the difference is less than 1/100 pixels.

For adjusting plates, we use bi-polynomial. The order is flexible. We check 4th or 5th order.

$$x(t) = \xi_0(t) + a_{10}(t)\xi + a_{01}(t)\eta + a_{20}(t)\xi^2 + a_{11}(t)\xi\eta + a_{02}(t)\eta^2 + \cdots \quad (2.1)$$

$$y(t) = \eta_0(t) + b_{10}(t)\xi + b_{01}(t)\eta + b_{20}(t)\xi^2 + b_{11}(t)\xi\eta + b_{02}(t)\eta^2 + \cdots \quad (2.2)$$

Accuracy of plate adjustment is evaluated by simulation and by using HST plates.

There may also be expected to exist pixel wise size in uniformity. For correcting this, we also consider the calibration operation. We observe globular cluster or some other dense region, and take picture by dithering telescope. In 50 min, assumption that all stars on the picture has the same position. By using this constraint, we can solve the geometry of the detector. It was done by Heidelberg group.

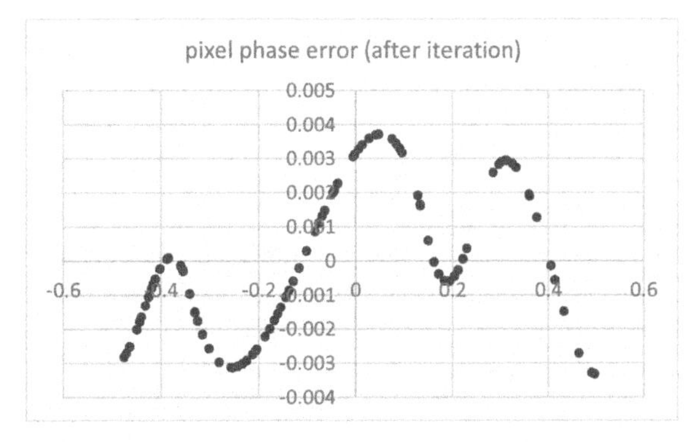

Figure 1. Pixel phase error of generated ePSF. The abscissa is the pixel phase of true center, and the ordinate is the difference between true and estimated center in the unit of pixel.

3. Gaia Data release from Japan

Within the collaboration of Gaia-JASMINE in GENIUS, NAOJ becomes one of the partner data center of Gaia data release. Characteristics of the database is very high performance system. `http://jvo.nao.ac.jp/portal/gaia.do`.

Also within the collaboration, Japanese explanation of Gaia data characteristics and instruction of the usage of DR1 data (`http://astro1.sc.niigata-u.ac.jp/~nishi/Gaia/GaiaDR1top.html`) which helps many Japanese professional and amateur astronomers.

This work has made use of data from the European Space Agency (ESA) mission *Gaia* (`https://www.cosmos.esa.int/gaia`), processed by the *Gaia* Data Processing and Analysis Consortium (DPAC, `https://www.cosmos.esa.int/web/gaia/dpac/consortium`). Funding for the DPAC has been provided by national institutions, in particular the institutions participating in the *Gaia* Multilateral Agreement.

References

Gouda, N. 2011, *Scholarpedia*, 6(10), 12021

Anderson, J. & King, I. R. 2000, *Publ. Astron. Soc. Pacific*, 112, issue 776, 1360

Astronomy and Astrophysics in the Gaia sky
Proceedings IAU Symposium No. 330, 2017
A. Recio-Blanco, P. de Laverny, A.G.A. Brown & T. Prusti, eds.

doi:10.1017/S1743921317005245

Light propagation in the Solar System for astrometry on sub-micro-arcsecond level

Sven Zschocke

Institute of Planetary Geodesy - Lohrmann Observatory, Dresden Technical University, Helmholtzstrasse 10, D-01069, Dresden, Germany
email: sven.zschocke@tu-dresden.de

Abstract. We report on recent advancement in the theory of light propagation in the Solar System aiming at sub-micro-arcsecond level of accuracy:

(1) A solution for the light ray in 1.5PN approximation has been obtained in the field of N arbitrarily moving bodies of arbitrary shape, inner structure, oscillations, and rotational motion.

(2) A solution for the light ray in 2PN approximation has been obtained in the field of one arbitrarily moving pointlike body.

Keywords. astrometry, relativity, gravitation

1. Introduction

In order to trace a light ray received by an observer back to the celestial light source, one has to determine the trajectory of the light ray by solving the geodesic equation, which in terms of the Barycentric Celestial Reference System $(ct, \boldsymbol{x})$ reads

$$\frac{d^2 x^i(t)}{c^2 dt^2} + \Gamma^i_{\mu\nu} \frac{dx^\mu(t)}{cdt} \frac{dx^\nu(t)}{cdt} = \Gamma^0_{\mu\nu} \frac{dx^\mu(t)}{cdt} \frac{dx^\nu(t)}{cdt} \frac{dx^i(t)}{cdt}, \tag{1.1}$$

where the Christoffel symbols $\Gamma^\alpha_{\mu\nu} = g^{\alpha\beta} \left(g_{\beta\mu,\nu} + g_{\beta\nu,\mu} - g_{\mu\nu,\beta}\right)/2$ are functions of the metric tensor. For a unique solution of (1.1) initial-boundary conditions must be imposed,

$$\boldsymbol{x}_0 = \boldsymbol{x}(t_0), \quad \boldsymbol{\sigma} = \lim_{t \to -\infty} \frac{\dot{\boldsymbol{x}}(t)}{c}, \tag{1.2}$$

where $\boldsymbol{x}_0$ is the position of the light-source and $\boldsymbol{\sigma}$ defines the unit tangent vector of the light ray at past null-infinity. The first and second integration of geodesic equation yields the coordinate velocity and trajectory of the light signal, given by

$$\dot{\boldsymbol{x}}(t) = \int_{-\infty}^{t} \ddot{\boldsymbol{x}}(t')\, dt' = c\,\boldsymbol{\sigma} + \Delta\dot{\boldsymbol{x}}(t), \tag{1.3}$$

$$\boldsymbol{x}(t) = \int_{t_0}^{t} \dot{\boldsymbol{x}}(t')\, dt' = \boldsymbol{x}_0 + c\,(t - t_0)\,\boldsymbol{\sigma} + \Delta\boldsymbol{x}(t), \tag{1.4}$$

where $\Delta\dot{\boldsymbol{x}}$ and $\Delta\boldsymbol{x}$ are small corrections to the unperturbed (i.e. straight) light ray.

2. Light trajectory in post-Newtonian expansion

Because the gravitational fields of the Solar system are weak and velocities of the bodies are slow, $v \ll c$, one may utilize the post-Newtonian (PN) expansion of the metric,

$$g_{\alpha\beta} = \eta_{\alpha\beta} + h^{(2)}_{\alpha\beta} + h^{(3)}_{\alpha\beta} + h^{(4)}_{\alpha\beta} + \mathcal{O}\left(c^{-5}\right), \tag{2.1}$$

where $h^{(n)}_{\alpha\beta} = \mathcal{O}\left(c^{-n}\right)$ are metric perturbations of the flat space-time $\eta_{\alpha\beta}$. The Damour-Soffel-Xu approach as well as the Brumberg-Kopeikin formalism provide expressions for the metric perturbations in terms of intrinsic multipoles, M_L^A and S_L^A, allowing for arbitrary shape, inner structure, oscillations, and rotational motion of the Solar System bodies. Inserting (2.1) into (1.1) results in a post-Newtonian expansion of the light ray,

$$\dot{\boldsymbol{x}}(t) = c\boldsymbol{\sigma} + \Delta\dot{\boldsymbol{x}}_{1\mathrm{PN}}(t) + \Delta\dot{\boldsymbol{x}}_{1.5\mathrm{PN}}(t) + \Delta\dot{\boldsymbol{x}}_{2\mathrm{PN}}(t) + \mathcal{O}\left(c^{-5}\right), \tag{2.2}$$

$$\boldsymbol{x}(t) = \boldsymbol{x}_0 + c\,(t - t_0)\,\boldsymbol{\sigma} + \Delta\boldsymbol{x}_{1\mathrm{PN}}(t) + \Delta\boldsymbol{x}_{1.5\mathrm{PN}}(t) + \Delta\boldsymbol{x}_{2\mathrm{PN}}(t) + \mathcal{O}\left(c^{-5}\right), \tag{2.3}$$

where $\Delta\boldsymbol{x}_{1\mathrm{PN}} = \mathcal{O}\left(c^{-2}\right)$, $\Delta\boldsymbol{x}_{1.5\mathrm{PN}} = \mathcal{O}\left(c^{-3}\right)$, and $\Delta\boldsymbol{x}_{2\mathrm{PN}} = \mathcal{O}\left(c^{-4}\right)$.

3. Light trajectory in 1PN and 1.5PN approximation

In the investigations of Zschocke (2015) and Zschocke (2016 a) a solution for $\Delta\dot{\boldsymbol{x}}_{1\mathrm{PN}}$, $\Delta\boldsymbol{x}_{1\mathrm{PN}}$ and $\Delta\dot{\boldsymbol{x}}_{1.5\mathrm{PN}}$, $\Delta\boldsymbol{x}_{1.5\mathrm{PN}}$, respectively, has been determined for the case of N arbitrarily moving bodies with full intrinsic multipole structure.

4. Light trajectory in 2PN approximation

A solution for $\Delta\dot{\boldsymbol{x}}_{2\mathrm{PN}}$, $\Delta\boldsymbol{x}_{2\mathrm{PN}}$ for one arbitrarily moving pointlike body has been given by Zschocke (2016 b) in terms of vectorial functions $\boldsymbol{A}_1, \boldsymbol{A}_2, \boldsymbol{A}_3$ and $\boldsymbol{B}_1, \boldsymbol{B}_2, \boldsymbol{B}_3$. The vectorial functions $\boldsymbol{A}_2$ and $\boldsymbol{B}_2$ contain tiny vectorial parameters, ϵ_1 and ϵ_2, whose expressions were not presented by Zschocke (2016 b) and will, therefore, be given here:

$$\begin{aligned}\epsilon_1 = & -\frac{v^2}{c^2}\frac{\boldsymbol{\sigma}\times(\boldsymbol{x}\times\boldsymbol{\sigma})}{x-\boldsymbol{\sigma}\cdot\boldsymbol{x}}\frac{1}{x} + 2\left(\frac{\boldsymbol{v}\cdot\boldsymbol{x}}{cx}\right)^2\frac{\boldsymbol{\sigma}\times(\boldsymbol{x}\times\boldsymbol{\sigma})}{x-\boldsymbol{\sigma}\cdot\boldsymbol{x}}\frac{1}{x} - 2\left(\frac{\boldsymbol{\sigma}\cdot\boldsymbol{v}}{c}\right)^2\frac{\boldsymbol{\sigma}\times(\boldsymbol{x}\times\boldsymbol{\sigma})}{x-\boldsymbol{\sigma}\cdot\boldsymbol{x}}\frac{1}{x}\\ & +4\left(\frac{\boldsymbol{\sigma}\cdot\boldsymbol{v}}{c}\right)\left(\frac{\boldsymbol{v}\cdot\boldsymbol{x}}{cx}\right)\frac{\boldsymbol{\sigma}\times(\boldsymbol{x}\times\boldsymbol{\sigma})}{x-\boldsymbol{\sigma}\cdot\boldsymbol{x}}\frac{1}{x} + 4\frac{\boldsymbol{v}}{c}\left(\frac{\boldsymbol{v}\cdot\boldsymbol{x}}{cx}\right)\frac{1}{x} - 4\frac{\boldsymbol{v}}{c}\left(\frac{\boldsymbol{\sigma}\cdot\boldsymbol{v}}{c}\right)\frac{1}{x}\\ & -\frac{v^2}{c^2}\frac{\boldsymbol{\sigma}}{x} + 2\left(\frac{\boldsymbol{v}\cdot\boldsymbol{x}}{cx}\right)^2\frac{\boldsymbol{\sigma}}{x} + 2\left(\frac{\boldsymbol{\sigma}\cdot\boldsymbol{v}}{c}\right)^2\frac{\boldsymbol{\sigma}}{x},\end{aligned} \tag{4.1}$$

$$\epsilon_2 = -\frac{v^2}{c^2}\frac{\boldsymbol{\sigma}\times(\boldsymbol{x}\times\boldsymbol{\sigma})}{x-\boldsymbol{\sigma}\cdot\boldsymbol{x}} + \frac{v^2}{c^2}\boldsymbol{\sigma}\ln\left(x-\boldsymbol{\sigma}\cdot\boldsymbol{x}\right). \tag{4.2}$$

The upper limit for their absolute values can be estimated as follows:

$$|\epsilon_1| \leqslant \frac{10}{|\boldsymbol{\sigma}\times\boldsymbol{x}|}\frac{v^2}{c^2} \quad \text{and} \quad |\epsilon_2| \leqslant \frac{v^2}{c^2}\sqrt{\frac{4\,x^2}{|\boldsymbol{\sigma}\times\boldsymbol{x}|^2} + \ln^2\left(x-\boldsymbol{\sigma}\cdot\boldsymbol{x}\right)}. \tag{4.3}$$

As noticed by Zschocke (2016 b), ϵ_1 and ϵ_2 are negligible for light deflection measurements on nano-arcsecond level and time delay measurements on pico-second level, respectively.

5. Outlook

For sub-micro-arcsecond astrometry the approach needs to be further developed. Especially, the following issues have to be treated: (i) retardation effects, (ii) light ray in the observers reference system, (iii) 2PN light ray in the field of arbitrarily shaped bodies.

References

Zschocke, S. 2015, *Phys. Rev. D*, 92, 063015
Zschocke, S. 2016 a, *Phys. Rev. D*, 93, 103010
Zschocke, S. 2016 b, *Phys. Rev. D*, 94, 124007

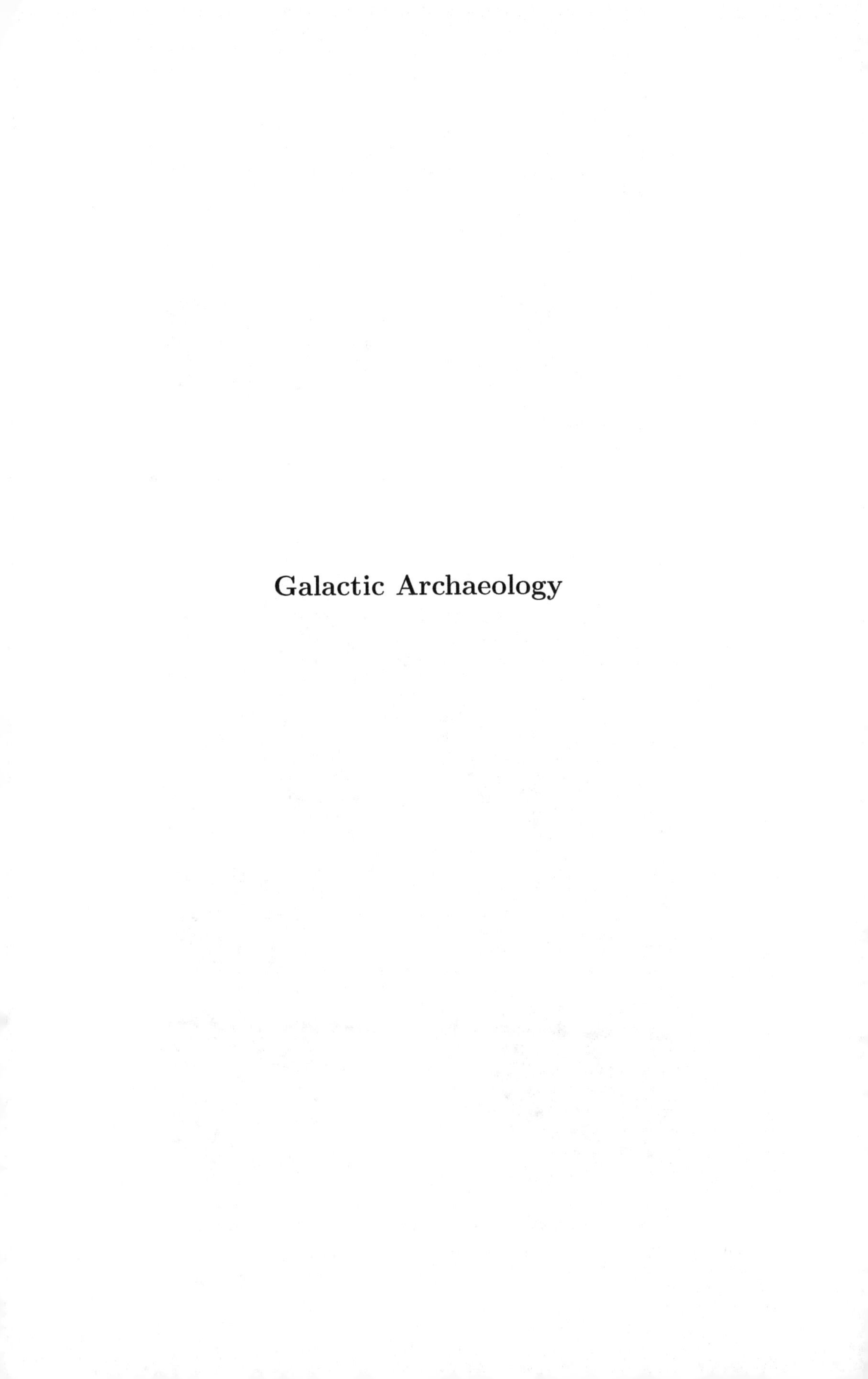

Galactic Archaeology

A scientific session in the Palais de la Méditerrannée.

The DR1 Gaia sky superimposed on the Verrière of the Negresco hotel, where the conference dinner was held (Courtesy from A. Robin and N. Lagarde).

Astronomy and Astrophysics in the Gaia sky
Proceedings IAU Symposium No. 330, 2017
A. Recio-Blanco, P. de Laverny, A.G.A. Brown
& T. Prusti, eds.

doi:10.1017/S1743921317007049

Self-consistent modelling of our Galaxy with Gaia data

James Binney

Rudolf Peierls Centre for Theoretical Physics, University of Oxford, 1 Keble Road, Oxford OX1 3NP, UK
email: binney@thphys.ox.ac.uk

Abstract. Galaxy models are fundamental to exploiting surveys of our Galaxy. There is now a significant body of work on axisymmetric models. A model can be defined by giving the DF of each major class of stars and of dark matter. Then the self-consistent gravitational potential is determined. Other modelling techniques are briefly considered before an overview of some early work on non-axisymmetric models.

Keywords. stellar dynamics, Galaxy: kinematics and dynamics, (cosmology:) dark matter.

---

1. The Standard Galaxy

In 2027 our understanding of the Galaxy will be encapsulated in a software tool that I'll call the Standard Galaxy (SG). The SG, like a Wiki page, will always be a work in progress. When a survey is planned, the SG will be used to simulate the survey's contents in light of its selection function (SF). When the survey is completed, the SG will be updated by maximising the likelihood of the new data with respect to the SG's parameters and the priors from earlier surveys. The SG will describe what's actually out there, which each survey sees only partially.

1.1. *What's in the SG?*

- The SG will model the distribution in $(\mathbf{x}, \mathbf{v})$ of many types of stars:
  - O, B, A, F, G K, M dwarfs, Cepheid variables, RR Lyrae stars, BHB stars, red clump stars, white dwarfs, neutron stars, ...
  - Most stellar types will be subdivided by age, [Fe/H] and [α/Fe] and some will be subdivided by other abundance ratios.
- The SG will also specify the distribution in $(\mathbf{x}, \mathbf{v})$ of dark matter (DM).
- The SG will include 3-dimensional models of HI, H_2, the density of H^+ and possibly other chemical species. It will include a model of the interstellar velocity field, $\mathbf{v}(\mathbf{x})$.

1.2. *The SG and Jeans' theorem*

DM plays an essential role in Galactic structure. But we cannot (yet) actually see it, and there is no guarantee that DM particles will have been detected by 2027. Indeed, even if current efforts to detect DM particles underground bear fruit, measurements of their density and velocity distribution on Earth will add only moderately to what we have already discovered about DM by modelling its contribution to the Galaxy's gravitational field.

The process by which we currently constrain DM by modelling its gravitational field is completely reliant on the assumption that the Galaxy is close to statistical equilibrium. If we drop the assumption that the Galaxy looked pretty much the same 200 Myr ago, and will look essentially the same 200 Myr hence, we can infer nothing about the Galaxy's gravitational field from the kinematics of stars and gas, because without this assumption of statistical equilibrium any current distribution of matter in phase space is consistent with any gravitational field.

From our reliance on statistical equilibrium to track $\sim 95\%$ of the Galaxy's mass, it follows that the foundations of the SG will rest on an equilibrium model of the Galaxy. In reality, the Galaxy is not in equilibrium, even in the rotating frame of the bar, because the stellar halo hosts tidal streams and the disc displays ephemeral spiral arms and a warp. These non-stationary phenomena play key roles in the Galaxy's evolution (e.g Aumer & Binney 2017) and promise to be valuable probes of the Galaxy's assembly history (e.g. Erkal *et al.* 2016), but they must be excluded from the basic model, which will inevitably be an equilibrium model. Only after its construction will it be decorated with spiral arms, warps and streams.

The natural way to construct an equilibrium model is to exploit Jeans' theorem: the distribution function (DF) of an equilibrium stellar system may be presumed to depend on $(\mathbf{x}, \mathbf{v})$ through the constants of stellar motion $I_1(\mathbf{x}, \mathbf{v}), I_2(\mathbf{x}, \mathbf{v}), \ldots$. Since any function of constants of motion is itself a constant of motion, we have infinite freedom in the choice of the constants of motion that we use as arguments of the DF $f(\mathbf{I})$. There is, however, only one rational choice: the action integrals J_i (e.g. Binney & McMillan 2016). Until recently the use of angle-action variables in galactic dynamics was very limited because we lacked effective ways to convert action-angle coordinates to and from conventional phase-space coordinates. Several numerical schemes for making these transformations are now available (Sanders & Binney 2016). The "Stäckel Fudge" (Binney 2012a) provides the most widely used mapping from $(\mathbf{x}, \mathbf{v})$ to $(\boldsymbol{\theta}, \mathbf{J})$ while Torus Mapping (Binney & McMillan 2016) provides the best inverse map. In particular cases the existing tools do have limitations, so further work is required in this area.

1.3. *How to assemble a model*

First for each component one wishes to consider, one chooses either a fixed density distribution $\rho(\mathbf{x})$ or a DF $f(\mathbf{J})$. The normalisation of either ρ of f is adjusted to ensure that the component's mass is reasonable. Then one makes a guess at the gravitational potential $\Phi(\mathbf{x})$ of the final model, and using this potential, the Stäckel Fudge and the DFs one evaluates the model's density by integrating over velocities at each point of a suitable spatial grid. Then one computes the resulting potential $\Phi(\mathbf{x})$ and uses this potential instead of the originally guessed potential to re-compute the density at the grid points. This cycle of computing the density using one potential and from it deriving a new potential converges after 4 to 5 iterations (Binney 2014). From this potential and the DFs one can predict *any* observable. The parameters in each component's DF are adjusted to optimise the fit between the model and observational data (Binney & Piffl 2015).

Given the long tradition of including energy $E = \frac{1}{2}v^2 + \Phi$ in the arguments of the DF, it's worth noting that model construction proceeds smoothly as just described only if E is excluded from the arguments of f. When E is included it's hard to converge on the correct potential, and harder still to ensure that each component has an observationally motivated mass.

1.4. *Choice of DFs*

It turns out that simple analytic functions $f(\mathbf{J})$ generate models that closely resemble familiar models that are defined by simple functional forms of the density $\rho(\mathbf{x})$. The isochrone sphere can be exactly generated from an analytic $f(\mathbf{J})$ (Binney 2014). Posti *et al.* (2015) showed that the Hernquist, Jafffe and NFW (Navarro, Fren & White 1997) spheres can be generated to high precision by simple analytic functions $f(\mathbf{J})$. Jeffreson *et al.* (2017) have given an analytic $f(\mathbf{J})$ which generates a good approximation to a Plummer model. By tweaking the given forms of $f(\mathbf{J})$, a spherical model can be endowed with either tangential or radial anisotropy, and it can be flattened or made prolate by the anisotropy (Binney 2014; Binney & Piffl 2015).

It is useful to break the DF into a parts $f_+(\mathbf{J})$ and $f_-(\mathbf{J})$ that are even and odd in J_ϕ, respectively. The part f_- does not contribute to the density but instead controls the model's rotation. Consequently, a model's rotation can be readily adjusted to match observational data after its density distribution has been perfected.

Binney & McMillan (2011) introduced the "quasi-isothermal" DF for discs:

$$f_+(\mathbf{J}) = \frac{\Sigma_0 \Omega}{\kappa^2} \exp(-R_\mathrm{c}/R_\mathrm{d}) \frac{\exp(-\kappa J_r/\sigma_r^2)}{\sigma_r^2 \kappa^{-1}} \frac{\exp(-\nu J_z/\sigma_z^2)}{\sigma_z^2 \nu^{-1}}$$

as the simplest DF that creates a plausible disc. Here Σ_0 and R_d are constants that, respectively, set the disc's mass and approximate scale length, while $R_\mathrm{c}, \Omega, \kappa, \nu, \sigma_r$ and σ_z are all functions of J_ϕ. Specifically, R_c, Ω, κ and ν should be the radius, angular velocity, in-plane and vertical epicycle frequencies of a circular orbit of angular momentum J_ϕ in some potential that is similar to that of the Galaxy. The velocity-dispersion parameters σ should be decreasing functions of J_ϕ and the normal hypothesis is

$$\sigma_i = \sigma_{i0} \exp\left[-\frac{R_\mathrm{c} - R_0}{R_{\sigma i}}\right]$$

where σ_{i0} is a constant that sets the ith velocity dispersion at the Sun and $R_{\sigma i}$ is a constant that determines how rapidly $\langle v_i^2 \rangle$ declines with distance from the Galactic centre. In a realistic Galactic potential the quasi-isothermal DF produces a disc with an approximately exponential surface density $\Sigma(R) \simeq \Sigma_0 \mathrm{e}^{-R/R_\mathrm{d}}$ and a vertical density profile that is sub-exponential in the sense that $|\mathrm{d}\log\rho/\mathrm{d}z|$ is a slowly increasing function of distance z from the plane. However, a vertical density profile in which $|\mathrm{d}\log\rho/\mathrm{d}z|$ decreases with z as is observed (Gilmore & Reid 1983; Juric *et al.* 2008) emerges naturally when one models the observed secular increase in $\langle v_z^2 \rangle$ with age τ by modelling each coeval population of stars by a quasi-isothermal with, for example

$$\sigma_{i0}(\tau) = \sigma_{i*} \left(\frac{\tau + \tau_1}{\tau_\mathrm{m} + \tau_1}\right)^\beta,$$

where σ_{i*} and τ_m are the velocity dispersion and age of the oldest disc stars, σ_1 determines the velocity dispersion of these stars at birth and $\beta \sim 0.5$ controls how the velocity dispersion increases with age.

To date f_- has been taken to be

$$f_-(\mathbf{J}) = \tanh(J_\phi/J_0) f_+(\mathbf{J}),$$

where J_0 is a constant. This ansatz eliminates counter-rotating stars at angular momenta significantly larger than J_0, which is assumed to be much smaller that $R_0 v_\mathrm{c}(R_0)$.

2. What's been done so far

Several papers have fitted models based on DFs $f(\mathbf{J})$ to data by evaluating moments of the DF in an assumed gravitational potential $\Phi(\mathbf{x})$. Adopting a plausible $\Phi(\mathbf{x})$ rather than solving for the self-consistent potential saves a great many CPU cycles because it is then necessary to compute moments only at locations for which we have data, rather than throughout the vast extent of the Galaxy (which extends out to > 100 kpc). Moreover, with $\Phi(\mathbf{x})$ assumed, we only need the moments of components for which we have data. In particular, we don't need to compute moments for the Galaxy's principal component, dark matter.

Binney (2010) fitted a disc DF to data from the Geneva-Copenhagen survey (GCS) (Nordström *et al.* 2004; Holmberg *et al.* 2007). The most significant finding of this paper was that the Sun's peculiar velocity $V_\odot$ needed to be revised upwards from $5.2\,\mathrm{km\,s^{-1}}$ to $\sim 11\,\mathrm{km\,s^{-1}}$. Schönrich *et al.* (2010) explained how the standard extraction of $V_\odot$ from the Hipparcos data had been undermined by the metallicity gradient in the disc. Binney (2012b) upgraded his earlier work by using an improved the quasi-isothermal DF (hereafter the "2012 DF") and the just introduced Stäckel Fudge rather than the adiabatic approximation to compute actions. Binney *et al.* (2014) showed that the 2012 DF had great success in predicting the data from the RAVE survey, which reaches distances in excess of 2 kpc whereas the GCS is essentially confined to distances < 0.1 kpc. The extent to which the 2012 DF captures the strong non-Gaussianity of the velocity distributions in v_ϕ and v_z is remarkable.

Sanders & Binney (2015) proposed an extended DF (EDF) in which [Fe/H] appears alongside $\mathbf{J}$ as an argument. Since stellar age already appeared in the 2012 DF as a nuisance parameter, with [Fe/H] added to the argument list it became possible to employ stellar isochrones to compute the probability that a star in the model would be included in a given survey. Hence Sanders & Binney (2015) were able to take properly into account survey SFs, which Binney (2012b) and Binney *et al.* (2014) had neglected to do. They showed that when the SF of the GCS is taken into account, significantly larger values of σ_{i*} are required because the GCS is strongly biased towards young stars, so the stars picked up in the survey have atypically small random velocities.

By fitting an EDF to a sample of SDSS K giants Das & Binney (2016) found evidence for two subpopulations. The EDF Das *et al.* (2016) fitted to BHB stars showed the older stars to be more tightly confined in action space than the younger stars. Binney & Wong (2017) explored DFs for the Galaxy's disc and halo globular clusters. This exercise revealed that a featureless DF $f(\mathbf{J})$ is liable to generate a density distribution $\rho(\mathbf{x})$ on which the gravitational potential has imprinted a feature associated with its transition around $r \sim 10$ kpc from disc- to halo-domination.

The moments yielded by a given DF depend on the adopted potential Φ. Consequently, Φ can be constrained by fitting to data the real-space and velocity-space distributions obtained by a particular pair (Φ, f). Bovy & Rix (2013) computed the likelihoods of 43 groups of stars over a 5-dimensional grid in (Φ, f) space. Each stellar group comprised the ~ 400 G dwarfs from the SEGUE survey that lie in a cell in the ([Fe/H],[α/Fe]) plane. The DF was constrained to be a quasi-isothermal and Φ was generated by a spherical bulge and dark halo plus double-exponential stellar and gas discs. The bulge was a fixed Hernquist model and the dark halo was a power-law model, so its free parameters were its logarithmic slope and its local density. The mass and scale lengths of the stellar disc were free parameters. The likelihood of each stellar group was computed over a 5-dimensional grid in (Φ, f) space. Unfortunately, the likelihood distributions in (Φ, f) space of the different populations were not mutually consistent. This is not surprising since the phase-space distribution of stars of a given chemical composition cannot be well modelled by a quasi-isothermal. In particular, stars with low [α/Fe] and low [Fe/H], being relatively young low-metallicity stars, have to be confined to an annulus centred beyond R_0.

Piffl *et al.* (2014) took a different approach to choosing a (Φ, f) pair, and they exploited the kinematics of $\sim 200\,000$ giant stars from the RAVE survey rather than $\sim 17\,000$

SEGUE stars. They computed a χ^2 statistic for the fit provided by (Φ, f) to both the RAVE kinematics and the density of stars $\rho(z)$ extracted from SDSS star counts by Juric *et al.* (2008) and two diagnostics of the Galaxy's circular-speed curve: terminal velocities measured from CO and HI data, and astrometry of maser sources. They homed in on the best (Φ, f) pair by first adopting a local dark-halo density $\rho_{\rm DM}$, finding the disc that then best reproduces the diagnostics of the circular speed and the kinematics of RAVE giants and then computing the resulting vertical density profile of the disc. If the initially assumed $\rho_{\rm DM}$ is small, a massive disc is required to match constraints on the rotation curve, and then given the RAVE kinematics the disc is too thin to match the SDSS $\rho(z)$. When $\rho_{\rm DM} \simeq (0.013 \pm 0.002)\,M_\odot\,{\rm pc}^{-3}$ all data are fitted quite nicely. The principal uncertainty is the shape of the dark halo: a more oblate dark halo requires a less massive disc to complement it. It turns out that the mass of the dark halo at $R < R_0$ is almost independent of the halo's axis ratio.

Piffl *et al.* (2015) first took the major step of specifying the dark halo by a DF $f(\mathbf{J})$ rather than a parametrised density distribution. Once this step is taken, it no longer makes sense to use a parametrised potential Φ and the model is completely specified through the self-consistency condition by the DFs of its constituent populations. For the disc, Piffl *et al.* adopted the DF fitted by Piffl *et al.* (2014), while for the dark halo they chose the DF that would in isolation self-consistently generate the NFW density profile determined by Piffl *et al.* (2014). When the self-consistent model implied by these two DFs was constructed, its circular-speed curve proved inconsistent with the data at $r \lesssim 3\,{\rm kpc}$ because the dark halo was pulled inwards and towards the plane by the gravitational field of the disc.

Binney & Piffl (2015) took the obvious next step: search the space $(f_{\rm disc}, f_{\rm DM})$ for a pair of DFs that self-consistently generate a model that is consistent with all the data assembled by Piffl *et al.* (2014). In this search the functional forms of the DFs were the same as those adopted by Piffl *et al.* (2015) but the parameters were free. A model that satisfied the observational constraints was found. It did however, have a remarkably large disc scale-length (3.7 kpc) and it did not produce as high an optical depth to microlensing bulge stars as microlensing surveys require (Sumi & Penny 2016). The reason for these short-comings was clear: in order to keep the circular speed at $R \sim 3\,{\rm kpc}$ within the observational upper limit, stars had been shifted outwards by increasing $R_{\rm d}$, leaving dark matter, which does not cause microlensing, the dominant mass density at small radii.

This finding establishes a tension between (i) the local density of DM that one deduces from $v_{\rm c}(R_0)$ and the kinematics and thickness of the disc, (ii) the assumption that DM has been adiabatically compressed by the gravitational field of slowly inserted baryons, and (iii) the density of stars at $R \lesssim 3\,{\rm kpc}$ required by the microlensing data. The clear next step is to drop the assumption that DM has responded adiabatically to the insertion of baryons.

The NFW density profile implies that $f_{\rm DM}$ diverges as $\mathbf{J} \to 0$, because the divergence of $\rho_{\rm DM}(r)$ as $r \to 0$ is accompanied by decreasing velocity dispersion. This prediction of simulations of cosmological clustering in the absence of baryons must be a consequence of the assumed extremely large density of DM at high redshift. The very high phase-space density of DM at the centre of an NFW halo will be drastically lowered when a passing clump of baryons imparts even a tiny velocity kick to DM particles. Once the phase-space density of DM has been reduced in this way, by Liouville's theorem it cannot be increased. Moreover, the efficient scattering of particles in a confined region of phase space will drive the phase-space density to a constant, which is the state of maximum entropy subject to constrained mean phase-space density. So Cole & Binney (2017) took the view that a probable $f_{\rm DM}(\mathbf{J})$ is one that tends to a constant as $\mathbf{J} \to 0$, where scattering by baryons has been efficient, and tends to the NFW $f(\mathbf{J})$ at high energies, where scattering has been unimportant. The transition between these two asymptotic regimes must be consistent with conservation of DM by scattering. Cole & Binney (2017) devised a form for $f_{\rm DM}(\mathbf{J})$ that is consistent with these principles and successfully searched the space $(f_{\rm disc}, f_{\rm DM})$ for models consistent with the P14 data plus the microlensing constraints. Their new $f_{\rm DM}(\mathbf{J})$ has an additional parameter h_0, which specifies the transition between the regime of approximately constant $f_{\rm DM}$ and the regime in which $f_{\rm DM} \sim f_{\rm NFW}$. In the absence of baryons their favoured $h_0 \sim 150\,{\rm kpc\,km\,s^{-1}}$ implies a DM core radius $\sim 3\,{\rm kpc}$, but the dark halo doesn't have a core in presence of the baryons.

Through the sequence of models of increasing sophistication just described, the local density of DM has changed very little from the value $\rho_{\rm DM} \simeq 0.013\,M_\odot\,{\rm pc}^{-3}$ determined by Piffl *et al.* (2014).

2.1. *Models not based on actions*

Most models of external galaxies are based either on the Jeans equations or on the technique introduced by Schwarzschild (1979) (Akin *et al.* 2016). The best current models of the Galactic bar (Portail *et al.* 2017) use the "made-to-measure" variant of Schwarzschild's technique (Syer & Tremaine 1996; de Lorenzi *et al.* 2007). Jeans models certainly lack the rigour and flexibility required to do justice to Galactic data. In Scharzschild and made-to-measure models the initial conditions of numerically integrated orbits play the role of constants of motion and weights assigned to orbits play the role of the value taken by the DF on an orbit. It is relatively simple to fit such models to observational data, and neither deviations from axisymmetry nor resonant trapping are problematic. The major differences with $f(\mathbf{J})$ modelling are (i) the DFs have vastly more parameters than a typical $f(\mathbf{J})$, and (ii) the orbit labels they use are complex and devoid of physical meaning.

I see several reasons why these models are unlikely to rise to the challenge posed by data in the era of Gaia. First, such a model is cumbersome because it is specified by millions of weights with low individual information content. Consequently, it's hard to compare models – two models with identical physical content will have no two parameters the same because each model will sample phase space in a different way. Moreover, the lack of a systematic scheme for labelling orbits that's invariant under changes in the gravitational potential means that it's hard to refine a model by subdividing components into sub-groups. For example, as data on the age distributions of giants accumulate, one will want to subdivide stars on the giant branch into age cohorts. Such refinements will invariably be associated with a change in the self-consistent potential and it's important to be able to identify orbits before and after such a change.

An additional issue with models of the Schwarzschild type is that it is difficult to include DM in the modelling process. One issue is that the quantity and spatial extent of DM necessitates a huge increase in the particle count. A more profound issue is that *no* observational data directly constrain DM – all constraints come indirectly through observational constraints on visible matter. It is not clear that current techniques can assign weights to DM particles.

It will never be possible to fit a standard N-body model to the exquisite data for our Galaxy, but generic N-body models have a huge role to play because they uniquely enable us to model from first principles evolutionary processes such as secular heating and radial migration (e.g. Aumer *et al.* 2016; Aumer & Binney 2017).

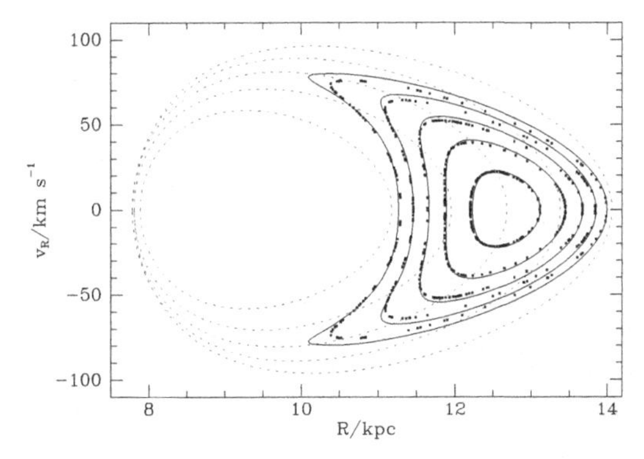

Figure 1. A surface of section $\phi = z = 0$ in a realistic barred Galaxy model. The full black curves are perturbatively constructed cross sections through tori trapped by the bar's OLR. The points are consequents on orbits started from a point on each of these curves. The dashed blue curves show cross sections through the underlying axisymmetric tori.

3. What next?

Not only is our Galaxy barred, but numerical simulations of the secular growth of a disc within a dark halo show that spontaneously generated, transient non-axisymmetries account very naturally for the structure of the thin disc (Aumer *et al.* 2016; Aumer & Binney 2017). Moreover, it now seems likely that the bar's corotation resonance lies as far out as $R \simeq 6\,\text{kpc}$, so the kinematics of solar-neighbourhood stars are significantly affected by the bar. Hence it is essential to regard axisymmetric models as spring-boards from which to progress to models that include both a bar and spiral structure.

As noted above, angle-action variables provide the natural arena for perturbation theory, and a major attraction of models built around analytic DFs $f(\mathbf{J})$ is the ease with which we can perturb them. The impact of a perturbing potential $\Phi_1(\boldsymbol{\theta}, \mathbf{J})$ on a model with DF $f(\mathbf{J})$ can be computed by linearising the Boltzmann equation: $f_0(\mathbf{J}) \to f_0(\mathbf{J}) + f_1(\boldsymbol{\theta}, \mathbf{J}, t)$. With $[a, b]$ denoting the Poisson bracket, we have

$$\frac{\partial f_1}{\partial t} + [f_0, \Phi_1] + [f_1, \Phi_0] = 0.$$

This equation is readily solved once one knows the Fourier expansion $\Phi_1 = \sum_{\mathbf{n}} \phi_{\mathbf{n}}(\mathbf{J})\mathrm{e}^{\mathrm{i}\mathbf{n}\cdot\boldsymbol{\theta}}$ of the perturbing potential. Monari *et al.* (2016) have used this approach to compute the response of thin-disc stars at the bar's outer Lindblad resonance (OLR), which Dehnen (2000) argued lies just beyond R_0.

Unless Φ_1 is extremely small, sufficiently near the resonance f_1 becomes larger than f_0. This is problematic because the sign of f_1 fluctuates, so once f_1 exceeds f_0 the total DF $f_0 + f_1$ is liable to be negative, which is unphysical.

The linearised Boltzmann equation breaks down when orbits become trapped by a resonance. Far from a resonance, the phase-space tori to which stars on regular orbits are confined are merely distorted by Φ_1. At a critical distance from the resonance, the tori abruptly rearrange themselves into a completely new pattern, and we say that the old tori/orbits have been trapped by the resonance. Angle-action variables enable us to reduce this phenomenon to a one-dimensional problem, closely analogous to the dynamics of a pendulum. Untrapped orbits correspond to a pendulum that rotates fast enough to pass top dead-centre and rotate always in the same sense, while trapped orbits correspond to a pendulum that swings to and fro as in a clock. Trapped orbits cannot be characterised by the same set of actions that characterise untrapped orbits, but must be assigned an entirely new "action of libration" in addition to two linear combinations of the original actions (e.g. Binney 2016). Consequently, for each family of resonantly trapped orbits we require a new DF $f(\mathbf{J})$ in addition to the familiar DF of the untrapped orbits.

Fig. 1 shows that when one uses the angle-action coordinates that torus mapping provides and the enhanced pendulum equation that Kaasalainen (1994) introduced, one can obtain remarkably accurate analytic models of trapped orbits. Specifically the figure shows a surface of section (R, v_R) at $\phi = 0, z = 0$ for motion in a realistic barred Galactic potential. The full curves are cross sections through tori trapped by the OLR of a realistic barred Galaxy model. These curves agree extremely well with the lines of dots, which are consequents of orbits integrated from an initial condition provided by one point on each curve. The blue dashed curves are cross sections through the axisymmetric tori that underpin the perturbative results.

References

Akin, Y., *et al.*, 2016, 456, 538
Aumer, M., Binney, J., & Schönrich, R., 2016, *MNRAS*, 459, 3326
Aumer, M. & Binney, J., 2017, *MNRAS* accepted (arXiv1705.09240)
Erkal, D., Belokurov, V, Bovy, J., & Sanders, J. L., 2016, *MNRAS*, 457, 3817
Arnold, V. I., 1978, *Mathematical methods of classical mechanics*, Berlin: Springer
Binney, J., 2010, *MNRAS*, 401, 2318
Binney, J., 2012a, *MNRAS*, 426, 1324
Binney, J., 2012b, *MNRAS*, 426, 1328
Binney, J., 2014, *MNRAS*, 440, 787
Binney, J., 2016, *MNRAS*, 462, 2792
Binney, J., Burnett, *et al.*, 2014, *MNRAS*, 439, 1231
Binney, J. & McMillan, P. J., 2011, *MNRAS*, 413, 1889
Binney, J. & McMillan, P. J., 2016, *MNRAS* 456, 1982
Binney, J. & Piffl, T., 2015, *MNRAS*, 454, 3653
Binney, J & Wong, L. K., 2017, *MNRAS*, 467, 2446
Das, P. & Binney, J., 2016, *MNRAS*, 460, 1725
Das, P., Williams, A., & Binney, J., 2016, *MNRAS*, 463, 3169
Bovy, J. & Rix, H.-W., 2013, *ApJ*, 779, 115
Cole, D. & Binney, J., 2017, *MNRAS*, 465, 798
Dehnen, W., 2000, *AJ*, 119, 800
de Lorenzi, F., Debattista, V. P., Gerhard, O., & Sambhus, N., 2007, *MNRAS*, 376, 71
Gilmore, G. & Reid, N., 1983, *MNRAS*, 202, 1025
Holmberg, J., Nordström, B. & Anderson, J., 2007, *A&A*, 475, 519
Jeffreson, S. M. R., 14 others 2017, *MNRAS*, submitted
Juric, M. *et al.*, 2008, *ApJ*, 673, 864
Kaasalainen, M., 1994, *MNRAS*, 268, 1041
Monari, G., Famaey, B., & Siebert, A., 2016, *MNRAS*, 457, 2569
Navarro, J., Frenk, C. S., & White, S. D. M., 1997, *ApJ*, 490, 493
Nordström, B. *et al.*, 2004, *A&A*, 418, 989
Piffl, T. *et al.*, 2014, *MNRAS*, 445, 3133
Piffl, T., Penoyre, Z., & Binney, J., 2015, *MNRAS*, 451, 639
Portail, M., Gerhard, O., Wegg, C & Ness, M., 2017, *MNRAS*, 465, 1621
Posti, L., Binney, J., Nipoti, C., & Ciotti, L., 2015, *MNRAS*, 447, 3060
Sanders, J. L. & Binney, J., 2015, *MNRAS*, 447, 2479
Sanders, J. L. & Binney, J., 2016, *MNRAS*, 457, 2107
Schönrich, R., Binney, J., & Dehnen, W., 2010, *MNRAS*, 403, 1829
Schwarzschil, M., 1979, *ApJ*, 232, 236
Sumi, T. & Penny, M. T., 2016, *ApJ*, 827, 139
Syer, D. & Tremaine, S., 1996, *MNRAS*, 282, 223

Astronomy and Astrophysics in the Gaia sky
Proceedings IAU Symposium No. 330, 2017
A. Recio-Blanco, P. de Laverny, A.G.A. Brown & T. Prusti, eds.

doi:10.1017/S1743921317005294

Stellar clusters in the Gaia era

Angela Bragaglia

INAF-Osservatorio Astronomico di Bologna
via Gobetti 93/3, 40129 Bologna, Italy
email: angela.bragaglia@oabo.inaf.it

Abstract. Stellar clusters are important for astrophysics in many ways, for instance as optimal tracers of the Galactic populations to which they belong or as one of the best test bench for stellar evolutionary models. Gaia DR1, with TGAS, is just skimming the wealth of exquisite information we are expecting from the more advanced catalogues, but already offers good opportunities and indicates the vast potentialities. Gaia results can be efficiently complemented by ground-based data, in particular by large spectroscopic and photometric surveys. Examples of some scientific results of the Gaia-ESO survey are presented, as a teaser for what will be possible once advanced Gaia releases and ground-based data will be combined.

Keywords. Space vehicles: instruments, catalogues, surveys, astrometry, spectroscopy, stars: abundances, stars: distances, stars: fundamental parameters, Galaxy: globular clusters, Galaxy: open clusters and associations

---

1. Introduction

We all know that Gaia will reach all-sky, exquisite precision in astrometric and photometric measurements, even if not in its first data release (DR1, September 2016), see e.g., Gaia Collaboration, Prusti, de Bruijne, *et al.* (2016), Gaia Collaboration, Brown, Vallenari, *et al.* (2016), Lindegren *et al.* (2016), Arenou *et al.* (2017), Evans *et al.* (2017), van Leeuwen *et al.* (2017). With the RVS, Gaia will also deliver radial velocities (RV) and chemistry for a more limited, but still huge, sample of stars, see e.g., Cropper *et al.* (2014), Recio-Blanco *et al.* (2016). Known Galactic clusters (see Fig. 1) comprise about 160 globulars (GC, Harris (1996) and web updates), and about 3000 open clusters (OC, e.g., Kharchenko *et al.* (2013)). At least for OCs, this is only the tip of the iceberg; if we extrapolate the solar vicinity to the whole disc, we may reach about 100000 clusters -and Gaia will discover many of them. As a role of thumbs we may say that, for a 15th mag star (for which also the RV will be available), the precision in parallax and proper motions will be better than 1% within 1 kpc, and 5% within 5 kpc. These limits will contain a good fraction of known OCs and also some GCs. Already in DR1/TGAS there are data for about 400 clusters (according to Vallenari in her presentation at the DR1 release event at ESA/Madrid). In future releases Gaia will deliver a dataset for both known and newly discovered clusters that will have an extraordinary impact on a large variety of topics, from cluster formation and eventual dissolution, to the use of clusters as test of stellar models and of the Galactic disc properties, etc. In the meanwhile, we have already results based on DR1 and on ground based surveys and a few examples are presented here.

2. Gaia DR1 and stellar clusters

For GCs, Gaia DR1 is clearly not ideal; for the vast majority of GCs, only positions and Gmag are available. However, this has not deterred its use, see Fig. 2 for results

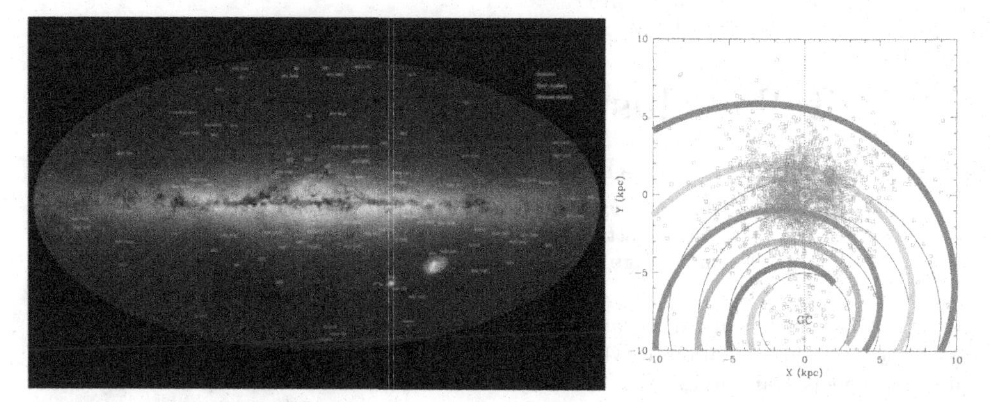

Figure 1. Left: First Gaia sky map, annotated; many of the named objects are stellar clusters [Credit: ESA/Gaia/DPAC. Acknowledgement: A. Moitinho & M. Barros (CENTRA University of Lisbon), F. Mignard (Observatoire de la Côte d'Azur), on behalf of DPAC)]. Right: Schematic view of the MW near the Sun position (here at 0,0), with the about 3000 clusters from Kharchenko *et al.* (2013) indicated by grey symbols.

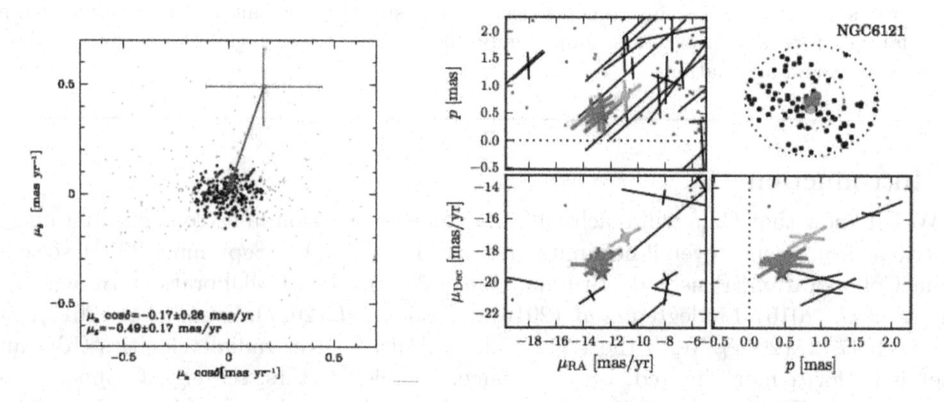

Figure 2. Left: vector point diagram for the stars in the HST catalogue for NGC 2419. Stars used to measure the average cluster PM are in black, likely non-members are in grey. The location of the background galaxy used to determine the absolute PM zero-point with its uncertainty is shown with a red symbol. Figure reproduced from Fig. 1 in Massari *et al.* (2017). Right: sky positions, parallaxes, and proper motions for the TGAS stars in NGC 6121 (coloured symbols: retained as members, black points: rejected; grey dots: Besançon model predictions). Figure reproduced from Fig. 2 in Watkins & van der Marel (2017), see the paper for details.

from two papers. Watkins & van der Marel (2017) tried to find GC stars in the TGAS catalogue, but after selecting stars around all MW GCs they had to exclude almost all of them because the stars were out of the clusters evolutionary sequences, their parallaxes and/or proper motions did not agree with literature, etc. After starting with more than 4000 stars in 142 GCs, they ended up with 20 good candidates in 5 GCs. Massari *et al.* (2017) made good use of the positions of stars in NGC 2419, a massive, metal-poor GC, combining them to first-epoch HST data. The very high precision of Gaia and HST positions permitted them to deduce the mean proper motion of this GC, so distant from the Sun (about 90 kpc); they also derived an orbit and suggested that NGC 2419 is associated to the disrupting Sagittarius dwarf spheroidal.

GCs are indeed difficult fields for Gaia, due to the crowding, both internal and external (i.e., the fore/background). Pancino *et al.* (2017) produced a set of simulated GCs with different combinations of concentration, distance, and background (disc, bulge, and

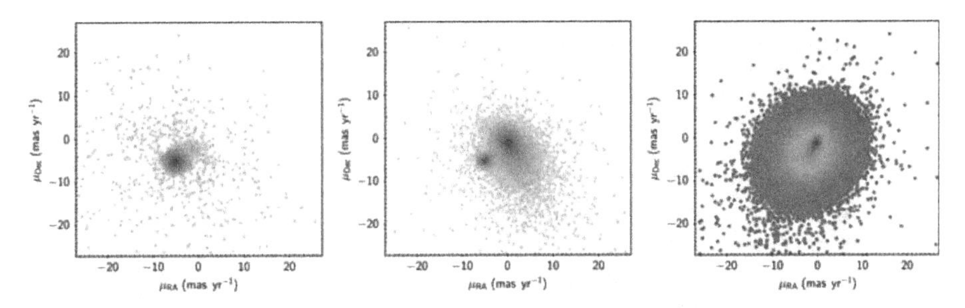

Figure 3. Proper motion diagrams for three simulated GCs (left, easy case: nearby, low concentration GC, halo-like background; centre, intermediate: more distant GC, disc background; right, difficult: even more distant GC, high concentration, bulge-like background). The GCs lie at (-5, -5) in all diagrams and are shown together with the background stars. Figure reproduced from Fig. 11 in Pancino *et al.* (2017), see the paper for details.

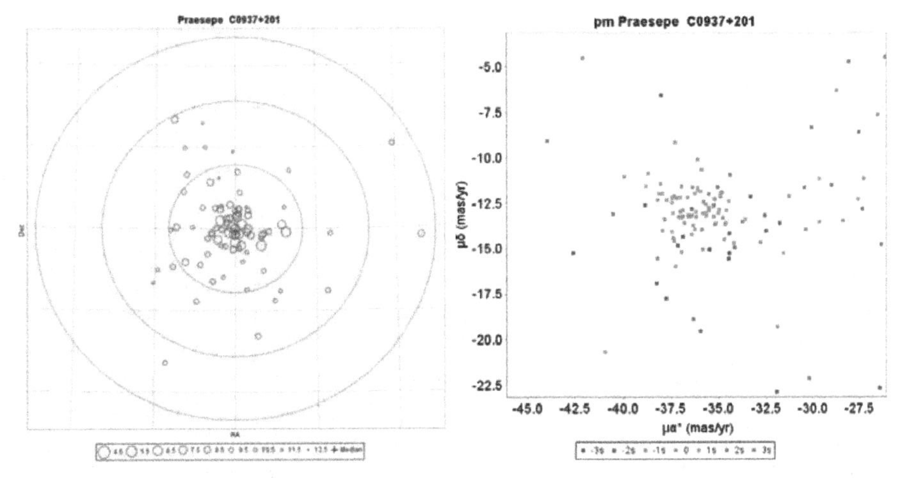

Figure 4. Left: map of the TGAS members of the Praesepe open cluster; the circles are ar 5, 10, and 15 pc from the cluster centre, while the grid is at 2 degrees intervals. Right: TGAS proper motions distribution for Praesepe, colour-coded according to difference from the cluster mean parallax. Figure adapted from Figs. D.7 and D.8 in Gaia Collaboration, van Leeuwen, Vallenari, *et al.* (2017), see the paper for details.

halo-like); Fig. 3 shows the proper motions distribution in three representative cases, one easy, one intermediate, and one difficult. They conclude that to reach the full potential of Gaia for globulars we need to wait until the end-of-mission, since the crowding problems will be alleviated only by the multiple scans. However, the future seems bright, the paper presents a long list of topics that will be possible to address. For instance, in the vast majority of GCs there will be $10^3 - 10^4$ clean stars and systemic proper motions and parallaxes will be determined to 1% or better for distances less than 15 kpc (i.e., 70% of all MW GCs; recall that an error of (less than) 1% in distance means an error less than 10% in age, absolute or relative for the cluster sub-populations).

Moving to open clusters, there is not much scientific exploitation of Gaia DR1 yet (but see Piatti (2017a), Piatti *et al.* (2017b) for the use of DR1 data to prove or disprove the nature of candidate clusters). However, the validation paper on OCs by Gaia Collaboration, van Leeuwen, Vallenari, *et al.* (2017), where 19 clusters closer than 500pc and already in the Hipparcos catalogue were examined, presents interesting results. They derive mean cluster parallaxes and proper motions taking into account the error

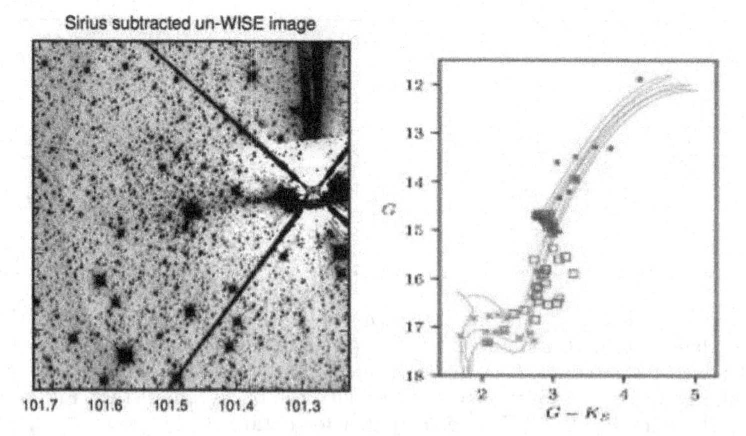

Figure 5. The first new cluster discovered in Gaia DR1 by Koposov *et al.* (2017) and confirmed as an old open cluster by Simpson *et al.* (2017). Left: $30' \times 30'$ image from the WISE survey showing Gaia 1, with the PSF of Sirius subtracted (reproduced from Fig. 3 of Koposov *et al.*). Right: Colour-magnitude diagram (using Gaia G and 2MASS Ks) showing only probable cluster members, confirmed by RV and metallicity, and isochrones; red filled symbols indicate HERMES targets, light blue open squares AAOmega targets (reproduced from Fig. 6 of Simpson *et al.*). See the original papers for details.

correlations within the astrometric solutions for individual stars, an estimate of the internal velocity dispersion, and the effects of the cluster depth. The conclusion is that, with the limitations of a first data release, we can derive membership based on proper motions and parallaxes and that we can study the whole extent of the clusters, not only their easily visible cores. In fact, cluster members were found to large distance, about 15pc from the centre (see Fig. 4 for an example on Praesepe). The paper also remarks on the narrowness of the cluster sequences in the colour-magnitude diagram (CMD), due to the very small error in distance; this is precious if we want to use cluster CMDs to test stellar models. Furthermore, it appears that the problem with the Pleiades parallax is solved: the TGAS value is in very good agreement with literature determinations, see e.g., van Leeuwen (2009) and Melis *et al.* (2014). A good fraction of these 19 OCs are also part of the Gaia-ESO survey (see Sect. 3) and analysis of TGAS plus Gaia-ESO information is under way (Randich *et al.*, in preparation).

Finally, I think that a special mention is due to the first new cluster discovered by Gaia, based only on positions. Koposov *et al.* (2017) essentially counted stars and detected about 260 overdensities, almost all corresponding to known clusters or dwarf galaxies. One of them, however, hidden behind Sirius, is a new stellar cluster and has been appropriately named Gaia 1. On the basis of Gaia and external photometric data, they proposed it to be a young and metal-rich GC. This discovery prompted immediate spectroscopic follow-up with the AAT: Simpson *et al.* (2017), using low and high-resolution spectra, identified about 40 members out of about 1000 observed stars and proposed instead that Gaia 1 is an old open cluster. Just recently, Mucciarelli *et al.* (2017) presented another follow-up, concentrated on the red clump stars of Gaia 1.

3. Complementing Gaia from the ground

A shortcoming of Gaia is its spectroscopic capability, with the limit for RVs about 5-6 mag brighter than for photometry and astrometry, with an accuracy well below that of transverse velocity and with an even brighter limit for abundance determination.

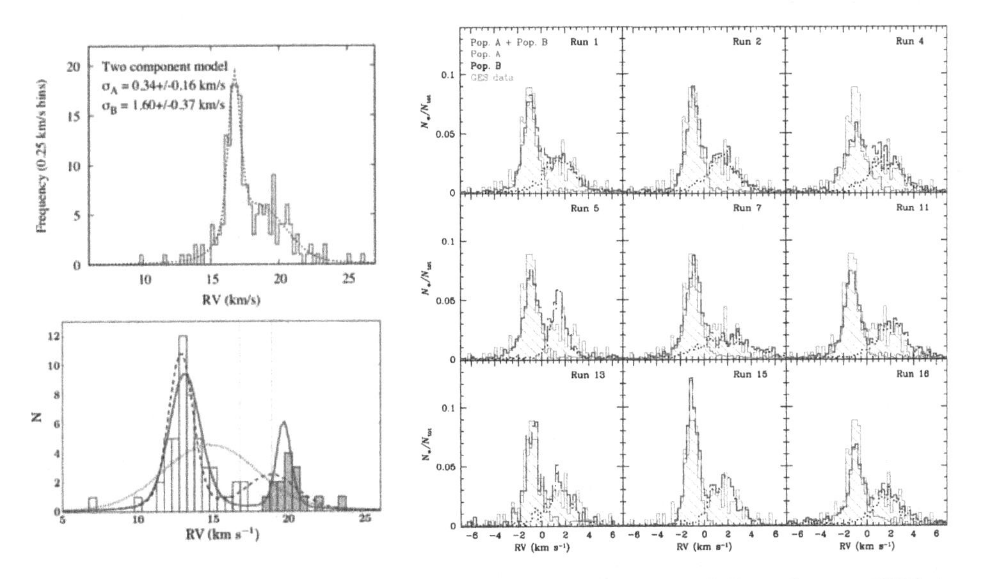

Figure 6. Gaia-ESO results on the Vela OB2 region. Left upper panel: Distribution of RVs in γ Vel, showing two kinematically distinct populations, from Jeffries *et al.* (2014); the two groups show different average RVs and velocity dispersions. Left lower panel: Examining the RVs in the field of NGC 2547, Sacco *et al.* (2015) discovered a secondary signature, consistent with one of the two populations in γ Vel. Mapelli *et al.* (2015) used N-body simulations of the two populations and suggested that γ Velorum formed from two sub-clusters, with slightly different age, one supervirial, the other in virial equilibrium. See the original papers for details.

Complementary data are required to obtain RVs accurate at better than 1 km/s to limits comparable to those of astrometry and photometry, and to measure metallicity and detailed abundance patterns.

To name only high-resolution (R=20000-45000) spectroscopic surveys, some are ongoing (APOGEE, GALAH, and Gaia-ESO), while WEAVE (at the 4m WHT), MOONS (at the 8m VLT), and 4MOST (on the 4m VISTA) are due to start in a short while. APOGEE, while not directly conceived to complement Gaia, is however very useful, because it works in the infrared and in crowded regions such as disc and bulge, where Gaia capabilities are more limited. Stellar clusters are not a dominant part of the survey, but some interesting results on them have been presented, see for instance Frinchaboy *et al.* (2013), Cunha *et al.* (2015) and Mészáros *et al.* (2015), Schiavon *et al.* (2017), Tang *et al.* (2017) for open and globular clusters, respectively. GALAH has a strong synergy with Gaia, since it observes only stars within the brightest 1% of Gaia targets, where the precision is at its best. So GALAH can combine the detailed chemistry and precise RVs from spectroscopy to the 5-dimensions of Gaia astrometry. Unfortunately, stellar clusters are not a main component of the GALAH survey, see Martell *et al.* (2017). WEAVE (see Dalton *et al.* (2014)) is the next high-resolution spectroscopic survey to start, at the 4m WHT on the Canary Islands. Apart from APOGEE, all other high-resolution surveys are in the Southern hemisphere, so WEAVE is particularly important; for instance, it will access the Galactic anticentre region, which is otherwise poorly covered. There is both a high-resolution mode (R=20000), reaching to about the limit of the Gaia RVS but with RVs of much better precision and with a full chemical characterisation, and a low-resolution more (R=5000), reaching down to the astrometric and photometric limit of Gaia. WEAVE has a dedicated survey for OCs, associations and star forming regions and also GCs will be observed, so it will be a good complement to Gaia for stellar clusters.

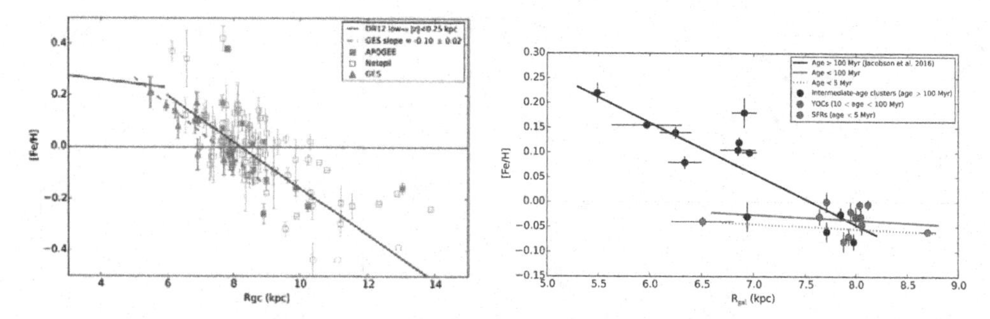

Figure 7. Left: The red filled triangles and dashed line show the radial metallicity gradient defined by the12 Gaia-ESO inner disk (Rgc$\leqslant$8 kpc), intermediate age open clusters (0.2-1.6 Gyr) in internal Data Release 4. The gradient is -0.10 ± 0.02 dex/kpc. For comparison, APOGEE clusters in Frinchaboy *et al.* (2013) and the compilation by Netopil *et al.* (2016) are also shown as grey filled and open squares, respectively. The solid blue line indicates the metallicity gradient in Hayden *et al.* (2014), based on APOGEE field giants. Right: Radial metallicity distribution of all the Gaia-ESO open clusters and star forming regions (iDR4). Different colours indicate star forming regions (red), young open clusters (blue, age 10-100 Myr), and older clusters (black, intermediate age). The lines indicate the gradient defined by the young clusters (the two red, flattish lines) and by the older clusters (the black, steeper line). Figure adapted from Fig. 2 in Jacobson *et al.* (2016) and Fig. 5 in Spina *et al.* (2017), respectively; see the original papers for details.

The large, public survey Gaia-ESO is on-going at the VLT using FLAMES and is obtaining moderate (R=20000) and high-resolution (R=45000) spectra of about 100000 stars of all Galactic components in pencil beams and of stellar clusters, see Gilmore *et al.* (2012) and Randich *et al.* (2013) for a presentation of the survey and its motivations and first results. Gaia-ESO has been designed to obtain spectra near to the Gaia astrometric limits for a fraction of the Gaia stars, thus adding fundamental information of RVs, metallicity, detailed chemistry, and astrophysical parameters. In combination with Gaia and with theoretical models, it will be possible for instance to derive precise ages of the observed stars and put robust constraints to the history of formation and evolution of our Galaxy.

In particular, Gaia-ESO is observing a large and significant sample of OCs and star forming regions, covering the whole range of open cluster properties (age, mass, metallicity, Galactic position). The goal is to understand how clusters form, evolve, and eventually dissolve, to study the chemo-dynamical evolution of the Galactic disc, and to use clusters as powerful tests of stellar evolution models. The full exploitation of the Gaia-ESO data to reach all these goals has to await for Gaia DR2 and later, but the survey has significant scientific value and strong legacy also per se. Some interesting results have already been presented and a few examples are shown here.

Gaia-ESO is observing stars in OCs from the pre-main sequence to evolved giants and is obtaining RVs with a precision of about 0.25 km/s. This permits to study the internal kinematics of clusters. Fig. 6 shows the case of γ Velorum, in the Vela OB2 star forming region. Jeffries *et al.* (2014) were able to resolve the velocity structure of its low-mass population and found two components, differing in mean RV, velocity dispersion, and age. Further evidence was presented by Sacco *et al.* (2015) and N-body simulations were used to propose a model for the formation of this structure, see Mapelli *et al.* (2015). OCs are privileged tracers of the disc, useful to understand how the disc formed and reached the metallicity distribution we observe. Gaia-ESO has extended the study of the radial metallicity distribution both to the inner disc region and to very young objects,

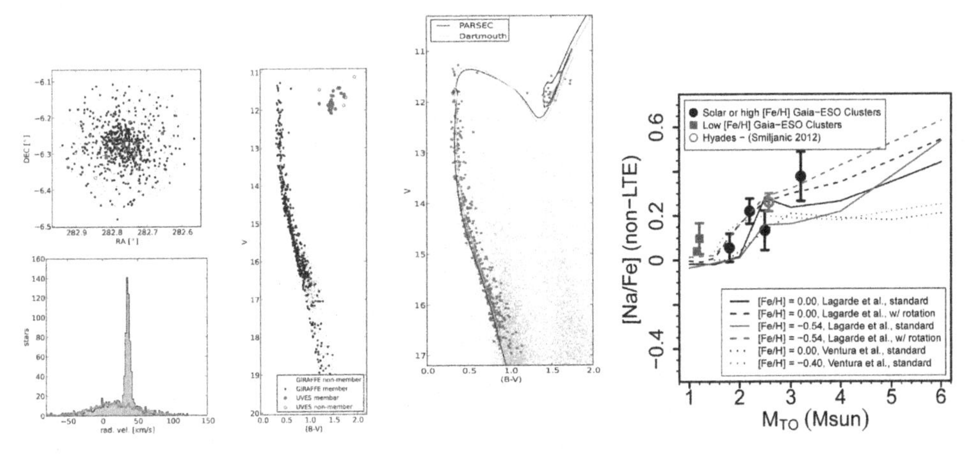

Figure 8. The three leftmost panels show the position of targets in the field of M11/NGC 6705, the distribution of the RVs with the cluster peak clearly standing out from the field stars, and the CMD for all observed targets, taken from the Gaia-ESO paper by Cantat-Gaudin *et al.* (2014). The middle panel shows the cluster CMD and two possible isochrones from different stellar models, also taken from the same paper. The rightmost panel shows the mean Na abundance for Gaia-ESO clusters (giant stars only) versus turn-off mass (i.e., cluster age), compared to lines indicating different models of extra-mixing; the figure is adapted from Smiljanic *et al.* (2016).

see Jacobson *et al.* (2016), Spina *et al.* (2017) and Fig. 7). Gaia-ESO results can be used to test stellar evolutionary models (see Fig. 8): CMDs can be cleaned from field interlopers using RVs and precise metallicity and chemical mixture information limit the degeneracies in model-to-data fit. Furthermore, the homogeneous determination of abundance ratios for many clusters of different properties is useful to constrain details of the models, such as mixing mechanisms. For all these topics, precise distances and efficient definition of cluster members from Gaia will be of course very useful to obtain more robust results.

To conclude, Gaia and complementary surveys and projects from the ground have already produced interesting results. Importantly, Gaia DR1 in not only useful for many more programs and topics, but it represents also a good test bench to get us ready for DR2 and following releases.

AB warmly thanks the organizers for the invitation to a very lively and informative Symposium and the Bologna Observatory for funding her.
This work has made use of data from the European Space Agency (ESA) mission *Gaia* (https://www.cosmos.esa.int/gaia), processed by the *Gaia* Data Processing and Analysis Consortium (DPAC, (https://www.cosmos.esa.int/web/gaia/dpac/consortium). Funding for the DPAC has been provided by national institutions, in particular the institutions participating in the *Gaia* Multilateral Agreement. This work used data products from observations made with ESO Telescopes at the La Silla Paranal Observatory under programme ID 188.B-3002 and following (Gaia-ESO Survey). These data products have been processed by the Cambridge Astronomy Survey Unit (CASU) at the Institute of Astronomy, University of Cambridge, and by the FLAMES/UVES reduction team at INAF/Osservatorio Astrofisico di Arcetri. This work made use of Vizier and SIMBAD, operated at CDS, Strasbourg, France, of arXiv, and of NASA's Astrophysical Data System.

References

Arenou, F., Luri, X., Babusiaux, C., *et al.* 2017, *A&A*, 599, A50
Cantat-Gaudin, T., Vallenari, A., Zaggia, S., *et al.* 2014, *A&A*, 569, A17
Cropper, M., Katz, D., Sartoretti, P., *et al.* 2014, *EAS Publications Series*, 67, 69
Cunha, K., Smith, V. V., Johnson, J. A., *et al.* 2015, *ApJ*, 798, L41
Evans, D. W., Riello, M., De Angeli, F., *et al.* 2017, *A&A*, 600, A51
Dalton, G., Trager, S., Abrams, D. C., *et al.* 2014, *SPIE*, 9147, 91470L
Frinchaboy, P. M., Thompson, B., Jackson, K. M., *et al.* 2013, *ApJ*, 777, L1
Gaia Collaboration, Brown, A. G. A., Vallenari, A., *et al.* 2016, *A&A*, 595, A2
Gaia Collaboration, Prusti, T., de Bruijne, J. H. J., *et al.* 2016, *A&A*, 595, A1
Gaia Collaboration, van Leeuwen, F., Vallenari, A., *et al.* 2017, *A&A*, 601, A19
Gilmore, G., Randich, S., Asplund, M., *et al.* 2012, *The Messenger*, 147, 25
Harris, W. E. 1996, *AJ*, 112, 1487
Hayden, M. R., Holtzman, J. A., Bovy, J., *et al.* 2014, *AJ*, 147, 116
Jacobson, H. R., Friel, E. D., Jílková, L., *et al.* 2016, *A&A*, 591, A37
Jeffries, R. D., Jackson, R. J., Cottaar, M., *et al.* 2014, *A&A*, 563, A94
Kharchenko, N. V., Piskunov, A. E., Schilbach, E., Röser, S., & Scholz, R.-D. 2013, *A&A*, 558, A53
Koposov, S. E., Belokurov, V., & Torrealba, G. 2017, *arXiv:1702.01122*
Lindegren, L., Lammers, U., Bastian, U., *et al.* 2016, *A&A*, 595, A4
Mapelli, M., Vallenari, A., Jeffries, R. D., *et al.* 2015, *A&A*, 578, A35
Martell, S. L., Sharma, S., Buder, S., *et al.* 2017, *MNRAS*, 465, 3203
Massari, D., Posti, L., Helmi, A., Fiorentino, G., & Tolstoy, E. 2017, *A&A*, 598, L9
Melis, C., Reid, M. J., Mioduszewski, A. J., Stauffer, J. R., & Bower, G. C. 2014, *Science*, 345, 1029
Mészáros, S., Martell, S. L., Shetrone, M., *et al.* 2015, *AJ*, 149, 153
Mucciarelli, A., Monaco, L., Bonifacio, P., & Saviane, I. 2017, *arXiv:1706.01504*
Netopil, M., Paunzen, E., Heiter, U., & Soubiran, C. 2016, *A&A*, 585, A150
Pancino, E., Bellazzini, M., Giuffrida, G., & Marinoni, S. 2017, *MNRAS*, 467, 412
Piatti, A. E. 2017, *MNRAS*, 466, 4960
Piatti, A. E., Dias, W. S., & Sampedro, L. M. 2017, MNRAS, 466, 392
Randich, S. & Gilmore, G., Gaia-ESO Consortium 2013, *The Messenger*, 154, 47
Recio-Blanco, A., de Laverny, P., Allende Prieto, C., *et al.* 2016, *A&A*, 585, A93
Sacco, G. G., Jeffries, R. D., Randich, S., *et al.* 2015, *A&A*, 574, L7
Schiavon, R. P., Johnson, J. A., Frinchaboy, P. M., *et al.* 2017, *MNRAS*, 466, 1010
Simpson, J. D., De Silva, G. M., Martell, S. L., *et al.* 2017, *arXiv:1703.03823*
Smiljanic, R., Romano, D., Bragaglia, A., *et al.* 2016, *A&A*, 589, A115
Spina, L., Randich, S., Magrini, L., *et al.* 2017, *A&A*, 601, A70
Tang, B., Cohen, R. E., Geisler, D., *et al.* 2017, *MNRAS*, 465, 19
van Leeuwen, F. 2009, *A&A*, 497, 209
van Leeuwen, F., Evans, D. W., De Angeli, F., *et al.* 2017, *A&A*, 599, A32
Watkins, L. L. & van der Marel, R. P. 2017, *ApJ*, 839, 89

Astronomy and Astrophysics in the Gaia sky
Proceedings IAU Symposium No. 330, 2017
A. Recio-Blanco, P. de Laverny, A.G.A. Brown & T. Prusti, eds.

doi:10.1017/S174392131700638X

Galaxy simulations in the Gaia era

Ivan Minchev

Leibniz-Institut für Astrophysik Potsdam (AIP), An der Sternwarte 16, D-14482, Potsdam, Germany
email: iminchev@aip.de

Abstract. We live in an age where an enormous amount of astrometric, photometric, asteroseismic, and spectroscopic data of Milky Way stars are being acquired, many orders of magnitude larger than about a decade ago. Thanks to the Gaia astrometric mission and followup ground-based spectroscopic surveys in the next 5-10 years about 10-20 Million stars will have accurate 6D kinematics and chemical composition measurements. KEPLER-2, PLATO, and TESS will provide asteroseismic ages for a good fraction of those. In this article we outline some outstanding problems concerning the formation and evolution of the Milky Way and argue that, due to the complexity of physical processes involved in the formation of disk galaxies, numerical simulations in the cosmological context are needed for the interpretation of Milky Way observations. We also discuss in some detail the formation of the Milky Way thick disk, chemodynamical models, and the effects of radial migration.

Keywords. stellar dynamics, Galaxy: kinematics and dynamics, (cosmology:) dark matter, history and philosophy of astronomy, space vehicles.

1. Introduction

The goal of Galactic Archeology (Freeman & Bland-Hawthorn 2002) is to dissect the Milky Way into its various components (discs, bulge, bar and halo) and thus to disentangle the various processes that contributed to their formation and evolution. The importance of this topic is manifested in the number of Galactic surveys dedicated to obtaining spectroscopic information for a large number of stars, e.g., RAVE (Steinmetz *et al.* 2006), SEGUE (Yanny *et al.* 2009), APOGEE (Majewski *et al.* 2010), HERMES (Freeman 2010), Gaia-ESO (Gilmore *et al.* 2012), and LAMOST (Zhao *et al.* 2006). This effort will soon be complemented by more than a billion stars observed by the Gaia space mission (Perryman *et al.* 2001). Millions of these will have accurate proper motions and parallaxes, which together with existing spectroscopic data, and especially with the advent of the dedicated Gaia follow-up ground-based surveys WEAVE (Dalton *et al.* 2012) and 4MOST (de Jong *et al.* 2012), will enable Galactic Archaeology as never before.

While in axisymmetric discs energy and angular momentum are conserved quantities and are, thus, integrals of motion (Binney & Tremaine 2008), this is not true for the more realistic case of potentials including perturbations from a central bar and/or spiral arms. In the case of one periodic perturbation there is still a conserved quantity in the reference-frame rotating with the pattern – the Jacobi integral $J = E - L\Omega_p$, where E is the energy of the particle, L is its angular momentum, and Ω_p is the pattern angular velocity. This is no longer the case, however, when a second perturbation with a different patterns speed is included.

It has now been well established that the Milky Way disc contains both a bar (as in more than 50% of external disc galaxies) and spiral structure moving at different pattern speeds, making it difficult to solve such a dynamical system analytically. Instead, different types of numerical methods are usually employed, from simple test-particle integrations, to preassembled N-body and SPH systems, to unconstrained, fully cosmological simulations of galaxy formation. All of these techniques have their strengths and weaknesses. Test particles are computationally cheap, allow for full control over the simulation parameters (such as spiral and bar amplitude, shape, orientation and pattern speed) but lack self-gravity. N-body simulations offer self-consistency but bar and especially spiral structure parameters are not easy to derive and not well controlled. Finally, in addition to being very computationally intensive, the outcomes of hydrodynamical cosmological simulations are even less predictable, with merging satellites and infalling gas making it yet harder to disentangle the disc dynamics; these are, however, much closer to reality in their complexity and a necessary ultimate step in the interpretation of observational data.

1.1. *Some of the questions we would like to answer*

- What are the spiral structure parameters?
- Bar parameters?
- Bulge structure?
- Disk chemo-kinematical structure as a fn of radius and distance from plane?

The above four questions are concerned with the current disk state and need to be answered before we can proceed to the following ones:

- How did the Milky Way thick disk form?
- How and when did the bulge/bar form?
- How much radial mixing happened in the disk (fn of time and radius)?
- Did the disk from from the inside-out?

For the latter set of questions a disk formation model covering the entire Milky Way history is needed. The complex dynamics of stars in the Galaxy demands the use of N-body simulations. Only then can we properly account for the perturbative effect of spiral arms, central bar, and minor mergers resulting from infalling satellites.

2. The need for simulations in the cosmological context

Here we present some of the complexity of physical processes encountered in the formation and evolution of galactic discs.

2.1. *Resonances in galactic discs*

Galactic discs rotate differentially with nearly flat rotation curves, i.e., constant circular velocity as a function of galactic radius. In contrast, density waves, such as a central bar and spiral structure, rotate as solid bodies. Therefore stars at different radii would experience different forcing due to the non-axisymmetric structure. Of particular interest are locations in the disc where the stars are in resonance with the perturber. The corotation resonance (CR), where stars move with the pattern, occurs when the angular rotation rate of stars equals that of the perturber. The Lindblad resonances (LRs) occur when the frequency at which a star feels the force due to a perturber coincides with the star's epicyclic frequency, κ. As one moves inward or outward from the CR circle, the relative frequency at which a star encounters the perturber increases. There are two values of r for which this frequency is the same as the radial epicyclic frequency. This is where the inner and outer Lindblad resonances (ILR and OLR) are located. Quantitatively, LRs occur when the pattern speed $\Omega_p = \Omega \pm \kappa/m$, where m is the multiplicity of the pattern†. The negative sign corresponds to the ILR and the positive to the OLR. While

† $m = 2$ for a bar or a two-armed spiral structure and $m = 4$ for a four-armed spiral.

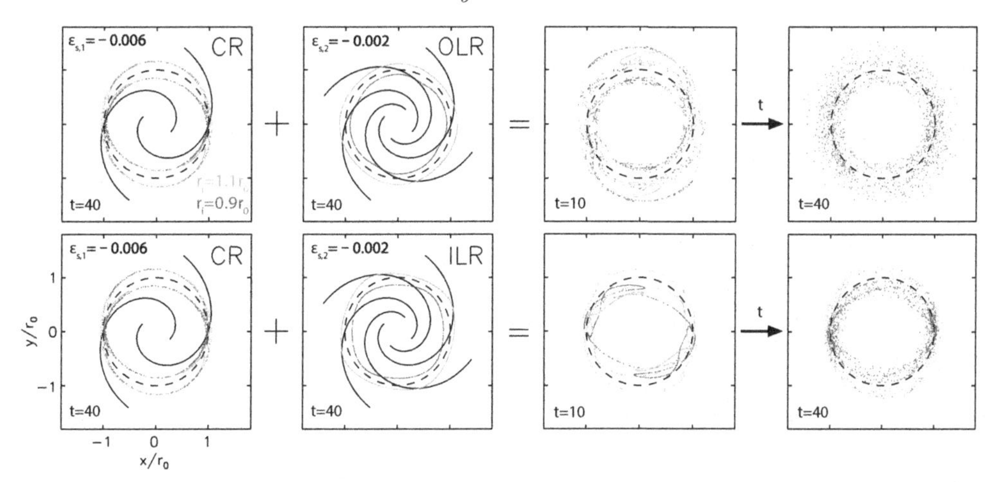

Figure 1. The effect on stellar orbits of two spiral perturbation with different pattern speeds acting together. **First column:** The red/green particles initially start on rings just inside/outside the CR (dashed black circle) of a two-armed spiral wave. Distances are in units of the solar radius, r_0. The top and bottom plots are identical. **Second column:** Same initial conditions, but near the 4:1 OLR (top) or 4:1 ILR (bottom) of a four-armed wave. **Third column:** Same initial conditions, but particles are perturbed by both spiral waves shown at t=10 rotation periods, or about 2.5 Gyr. Note that a secondary wave of only 1/3 the strength of the first one is enough to disrupt the horseshoe orbits near the CR. **Fourth column:** Same setup as in the third column but at t=40, or about 10 Gyr. Simulations from Minchev & Quillen (2006).

Bertil Lindblad defined these for the case of an $m = 2$ pattern (thus strictly speaking the ILR/OLR are the 2:1 resonances), for an $m = 4$ pattern the ILR/OLR must be the 4:1 resonances.

2.2. *Multiple patterns in galactic discs*

According to the theoretical work by Tagger *et al.* (1987) and Sygnet *et al.* (1988), two patterns can couple non-linearly as they overlap over a radial range, which coincides both with the CR of the inner one and the ILR of the outer one. This coincidence of resonances results in efficient exchange of energy and angular momentum between the two patterns.

Figure 1 shows the effect of multiple patterns on a ring of stars, i.e., initially uniformly distributed in azimuth and at the same galactic radius. One can see that the orderly behavior under the influence of a single perturber becomes stochastic when two spiral waves moving at different patterns speeds are considered.

3. Effects of radial migration on disk thickening

Several works have previously suggested that radial migration can give rise to thick disc formation by bringing out high-velocity-dispersion stellar populations from the inner disc and the bulge. Such a scenario was used, for example, in the analytical model of Schönrich & Binney (2009b), where the authors claimed to explain the Milky Way thick- and thin-disc characteristics (both chemical and kinematical) without the need of mergers or any discrete heating processes. Similarly, the increase of disc thickness with time found in the simulation by Roškar *et al.* (2008) has been attributed to migration in the work by Loebman *et al.* (2011).

Migration induces disc flaring in isolated discs: A first effort to demonstrate how exactly radial migration affects disc thickening in dynamical models was done by Minchev *et al.* (2012a). It was shown that stellar samples arriving from the inner disc have slightly higher velocity dispersions, which will result in them being deposited at higher distances above the galactic midplane. However, the opposite effect arrises from samples arriving from the outer disc (with lower velocity dispersions). Therefore, the *overall* migration effect on the disc thickening is minimal throughout most of the disc extend, except in the very inner/outer parts of the disc, where only inward/outward migrators are deposited. This naturally results in disc flaring, as shown in Fig. 7 by Minchev *et al.* (2012a). We explained this as the conservation of vertical action as opposed to conservation of the vertical energy assumed before. Several independent groups, using different simulation techniques and setups, have confirmed that migration does not thicken the disc (Martig *et al.* 2014; Vera-Ciro *et al.* 2014; Grand *et al.* 2016).

Migration suppresses disc flaring when infalling satellites are present: It is well known from both observations and cosmological simulations that minor mergers take place in the formation of galactic discs. Such interactions will have the effect of heating more the disc outskirts at any given time of the disc growth (more so at high redshift), because of the low mass density there.

Interestingly, and to complicate matters, when this more realistic scenario is considered, migration has the opposite effect on the disc vertical profile compared to the effect of an isolated galaxy – the role of outward and inward migrators is reversed in that they now cool and heat the disc, respectively. More on this can be found in Minchev *et al.* (2014a) and the review by Minchev (2017).

4. Formation of galactic thick discs by the flaring of mono-age populations

Stellar disc density decomposition into thinner and thicker components in external edge-on galaxies find that thicker disc components have larger scale-lengths than the thin discs (e.g., Yoachim & Dalcanton 2006; Pohlen *et al.* 2007). While this is consistent with results for the Milky Way when similar morphologically (or structural) definition for the thick disc is used (e.g., Robin *et al.* 1996; Ojha 2001), it is in contradiction with the more centrally concentrated older or $[\alpha/Fe]$-enhanced stellar populations (e.g., Cheng *et al.* 2012; Bovy *et al.* 2012). This apparent discrepancy may be related to the different definition of thick discs - morphological decomposition or separation in chemistry.

Additionally, while no flaring is observed in external edge-on discs (van der Kruit & Searle 1982; Comerón *et al.* 2011), numerical simulations suggest that flaring cannot be avoided due to a range of different dynamical effects. The largest source is most likely satellite-disc interactions (e.g., Villalobos & Helmi 2008), which have been found to increase an initially constant scale-height by up to a factor of ~ 10 in 3-4 disc scale-lengths. Other sources of disc flaring include misaligned gas infall (Scannapieco *et al.* 2009) and reorientation of the disc rotation axis.

Minchev *et al.* (2015) studied the formation of thick discs using two suites of simulations of galactic disc formation, one of which we present here. This is a full cosmological zoom-in hydro simulation, using initial conditions from one of the Aquarius Project haloes (Springel *et al.* 2008; Scannapieco *et al.* 2009). Further details about this simulation can be found in Aumer *et al.* (2013), their model Aq-D-5.

In the left panel of Fig. 2 we plot the scale-height variation with galactocentric radius, r, in the region 1-5 disc scale-lengths, h_d. Both the radius and scale-height, h_z, are in units of h_d. It can be seen that significant flaring is present, which increases for older coeval populations.

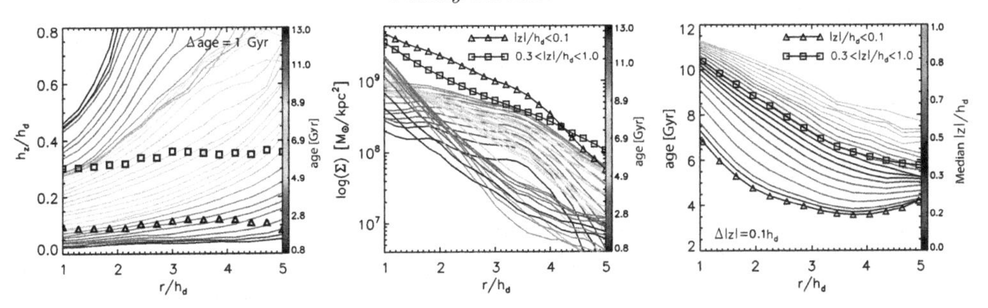

Figure 2. Left: Variation of disc scale-height, h_z, with galactic radius for a cosmological disc formation simulation. Color lines show mono-age populations, as indicated. Overlapping bins of width Δage $= 1$ Gyr are used. Overlaid also are the thin (triangles) and thick (squares) discs obtained by fitting a sum of two exponentials to stars of all ages. No significant flaring is found for the thin and thick discs. **Middle:** Disc surface density radial profiles of mono-age populations. Older discs are more centrally concentrated, which explains why flaring diminishes in the total population. Also shown are the surface density profiles of stars close to (triangles) and high above (squares) the disc midplane. The thicker disc component extends farther out than the thin one, consistent with observations of external galaxies. **Right:** Variation of mean age with radius for samples at different distance from the disc midplane, as indicated. Slices in $|z|$ have thickness $\Delta|z| = 0.1h_d$. Overlaid are also the age radial profiles of stars close to (triangles) and high above (squares) the disc midplane. Age gradients are predicted for both the (morphologically defined) thin and thick discs. Adapted from Minchev *et al.* (2015).

In contrast to the flaring found for all but the youngest mono-age populations, the thin and thick disc decomposition of the total stellar population including all ages, results in no apparent flaring. This is shown by the triangle and square symbols overlaid in the left panel of Fig. 2.

What is the reason for the flaring of mono-age discs? In numerical simulations flaring is expected to result from a number of mechanisms related to galactic evolution in a cosmological context (e.g., Kazantzidis *et al.* 2008; Villalobos & Helmi 2008). Even in the absence of environmental effects, flaring is unavoidable due to secular evolution alone (radial migration caused by spirals and/or a central bar, Minchev *et al.* 2012a). It should be stressed here that, while migration flares discs in the lack of external perturbations, during satellite-disc interactions it works *against* disc flaring (Minchev *et al.* 2014a). Yet, this is not sufficient to completely suppress the flaring induced by orbiting satellites, as evident from the left panel of Fig. 2. This suggests that external effects are much more important for the disc flaring in this simulation. Because the mass and intensity of orbiting satellites generally decreases with decreasing redshift, so does the flaring induced. It can be expected that at a certain time secular evolution takes over the effect of external perturbations.†

What is the reason for the lack of flaring in the total disc population? In an inside-out formation scenario, the outer disc edge, where flaring is induced, moves progressively from smaller to larger radii because of the continuous formation of new stars in disc subpopulations of increasing scale-length. At the same time the frequency and masses of perturbing satellites decreases. Because of the inside-out disc growth, which results in more centrally concentrated older samples (see Fig. 2, middle panel), the younger the stellar population, the further out it dominates in terms of stellar mass. The geometrically defined thick disc, therefore, results from the imbedded flares of different coeval populations, as seen in the left panel of Fig. 2.

† ? suggested that the time at which internal evolution takes over can also be inferred from the shape of the [α/Fe]-velocity dispersion relation of narrow metallicity samples.

The right panel of Fig. 2 shows that a geometrical thick disc is expected to have a negative age gradient. For this particular simulation the mean age decreases from ~ 10.5 to ~ 6 Gyr in four disc scale-lengths. Such an age drop of mean stellar age at high distances, $|z|$, from the disc midplane explains the inversion in $[\alpha/\mathrm{Fe}]$ gradients with increasing mean z found by Anders *et al.* (2014) in APOGEE data.

The recent work by Bovy *et al.* (2016) estimated the variation of disk scale-height for mono-abundance populations (MAPs) of APOGEE red clump giants and found that the high-$[\alpha/\mathrm{Fe}]$ subpopulations did not show flaring, thus disagreeing with the scenario presented in this section. It was then shown by Minchev *et al.* (2017) that flaring in MAPs can be lost if they are not mono-age populations as has been thus far assumed. This is because, due to the inside-out disk formation, a MAP displays a negative age gradient (except for the lowest $[\alpha/\mathrm{Fe}]$ MAPs). Flaring in the oldest APOGEE red clump stars was subsequently confirmed by Mackereth *et al.* (2017), who used the same data and methods as Bovy *et al.* and ages estimated by Martig *et al.* (2016). This example emphasizes the importance of good ages estimates for disentangling the structure and formation of the Milky Way.

5. Chemo-dynamical modeling of the Milky Way

So far we have focused on the dynamics of the Milky Way disc, which tells us mostly about its current state. To be able to go back in time and infer the Milky Way evolutionary history, however, we need to include both stellar chemical and age information.

A major consideration in a disc chemo-dynamical model is taking into account the effect of radial migration, i.e., the fact that stars end up away from their birth places. Below we briefly summarize models which include radial migration.

• Semi-analytical models tuned to fit the local metallicity distribution, velocity dispersion, and chemical gradients, etc., today (e.g., Schönrich & Binney 2009a; Kubryk *et al.* 2015) or Extended distribution functions (Sanders & Binney 2015):
– Easy to vary parameters
– Provide good description of the disc chemo-kinematic state today
– Typically not concerned with the Milky Way past history
– Time and spatial variations of migration efficiency due to dynamics resulting from non-axisymmetric disc structure is hard to take into account.

• Fully self-consistent cosmological simulations (e.g., Kawata and Gibson 2003; Scannapieco *et al.* 2005; Kobayashi and Nakasato 2011; Brook *et al.* 2012):
– Dynamics self-consistent in a cosmological context
– Can learn about disc formation and evolution
– Not much control over final chemo-kinematic state
– Problems with SFH and chemical enrichment due to unknown subgrid physics
– Much larger computational times needed if chemical enrichment included.

• Hybrid technique using simulation in a cosmological context + a classical (semi-analytical) chemical evolution model (Minchev *et al.* 2013:
– Avoids problems with SFH and chemical enrichment in fully self-consistent models
– Can learn about disc formation and evolution
– Not easy to get Milky Way-like final states.

6. Conclusions

In this work we argued that numerical simulations of disk formation in the cosmological context are invaluable for the interpretation of the massive new data sets expected in the

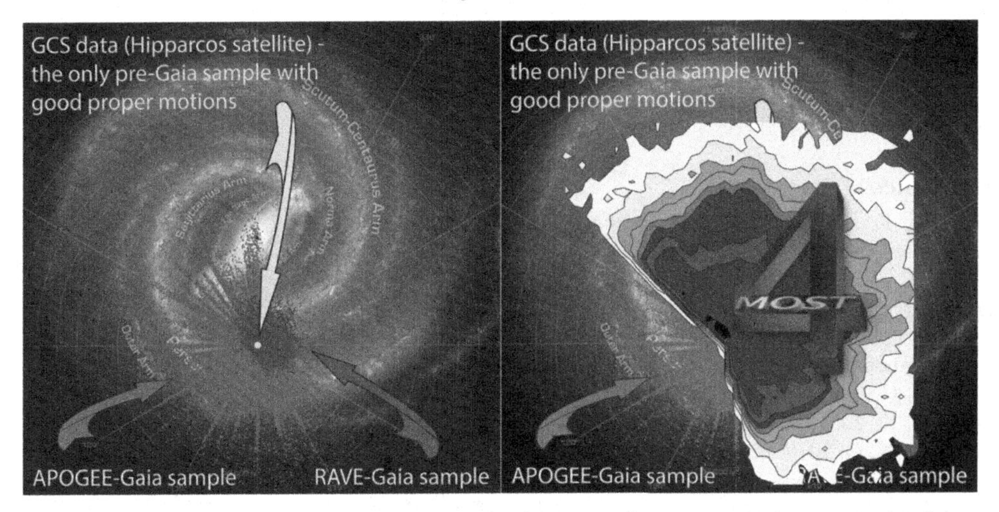

Figure 3. Left: Illustrating the vast increase of stars with precise 6D kinematical information and chemical abundances after Gaia's first two data releases. Distribution of RAVE-Gaia and APOGEE-Gaia stellar samples overlaid on top of R. Hurt's map of the Milky Way (SSC–Caltech). The Galactic disc rotation is in the clockwise direction. **Right**: The further increase expected from the ESO funded 4MOST spectroscopic survey, starting operation in 2022, that will provide radial velocities and chemistry for $10 - 15 \times 10^6$ Gaia stars.

very near future. Here are some examples of outstanding problems that we can hope to be able to answer in the next 5-10 years:

- What is the pattern speed and length of the Galactic bar: fast bar, about 3 kpc long (Dehnen 2000, Fux *et al.* 2001, Minchev *et al.* 2007, 2010, Antoja *et al.* 2008, Monari *et al.* 2017) or Slow bar, about 5 kpc long (Wegg *et al.* 2015, Perez-Villegas 2017)?
- What is the nature of the Hercules stream: results from the bar's Outer Lindblad Resonance or from the bar's Corotation?
- What causes the vertical wave patterns seen in the Milky Way disk: caused by bar/spirals or by the tidal effect of the Sgr dwarf galaxy?
- Are the local vertical disk asymmetries and Monoceros Ring part of the same global structure?
- How did the bulge form: inward stellar migration gives rise to different stellar populations in the bulge (Di Matteo 2016) or bulge chemistry distinct from that of the thick disk (Johnson *et al.* 2014)?
- What is the shape of the age-velocity relation: increase of stellar velocity dispersion with age as a power law or a step up exists at 8-10 Gyr due to the heating effect of the last massive merger?

The field of Galactic Archaeology will soon be transformed by the Gaia end-of-mission astrometry and followup ground-based spectroscopic surveys such as RAVE, APOGEE, LAMOST, GALAH, WEAVE, and 4MOST, which in the next 5-10 years will deliver 10-20 Million stars with accurate 6D kinematics and chemical abundance measurements. Finally, we would like to emphasize the importance of good age estimates needed to break degeneracies and refine chemo-dynamical models, which we expect to get for a good fraction of the data from the asteroseismic missions KEPLER-2, PLATO, and TESS.

7. Acknowledgements

IM acknowledges support by the Deutsche Forschungsgemeinschaft under the grant MI 2009/1-1.

References

Freeman, K. & Bland-Hawthorn, J. 2002, ARAA, 40, 487
Steinmetz, M., Zwitter, T., & Siebert, *et. al.* 2006, *AJ*, 132, 1645
Yanny, B., Rockosi, C., Newberg, H. J., *et al.* 2009, *AJ*, 137, 4377
Di Matteo, P. 2016, *PASA*, 33, e027
Majewski, S. R., Wilson, J. C., Hearty, F., Schiavon, R. R., & Skrutskie, M. F. 2010, in IAU Symposium, Vol. 265, IAU Symposium, ed. K. Cunha, M. Spite, & B. Barbuy, 480–481
Gilmore, G., Randich, S., Asplund, M., *et al.* 2012, *The Messenger*, 147, 25
Ojha, D. K. 2001, *MNRAS*, 322, 426
Freeman, K. C. 2010, in Galaxies and their Masks, ed. D. L. Block, K. C. Freeman, & I. Puerari, 319
Zhao, G., Chen, Y.-Q., Shi, J.-R., *et al.* 2006, CJAA, 6, 265
Perryman, M. A. C., de Boer, K. S., Gilmore, G., *et al.* 2001, *A&A*, 369, 339
Dalton, G., Trager, S. C., Abrams, D. C., *et al.* 2012, in Procspie, Vol. 8446, Ground-based and Airborne Instrumentation for Astronomy IV, 84460P
de Jong, R. S., Bellido-Tirado, O., Chiappini, C., *et al.* 2012, 8446, 84460T
Binney, J. & Tremaine, S. 2008, Galactic Dynamics: Second Edition, ed. J. Binney & S. Tremaine (Princeton University Press)
Minchev, I. & Quillen, A. C. 2006, *MNRAS*, 368, 623
Sygnet, J. F., Tagger, M., Athanassoula, E., & Pellat, R. 1988, *MNRAS*, 232, 733
Tagger, M., Sygnet, J. F., Athanassoula, E., & Pellat, R. 1987, *ApJL*, 318, L43
Schönrich, R. & Binney, J. 2009b, *MNRAS*, 399, 1145
Roškar, R., Debattista, V. P., Quinn, T. R., Stinson, G. S., & Wadsley, J. 2008, *ApJL*, 684, L79
Loebman, S. R., Roškar, R., Debattista, V. P., *et al.* 2011, *ApJ*, 737, 8
Minchev, I., Famaey, B., Quillen, A. C., *et al.* 2012a, *A&A*, 548, A127
Robin, A. C., Haywood, M., Creze, M., Ojha, D. K., & Bienayme, O. 1996, *A&A*, 305, 125
Martig, M., Minchev, I., & Flynn, C. 2014, *MNRAS*, 443, 2452
Vera-Ciro, C., D'Onghia, E., Navarro, J., & Abadi, M. 2014, *ApJ*, 794, 173
Grand, R. J. J., Springel, V., Gómez, F. A., *et al.* 2016, *MNRAS*, 459, 199
Minchev, I., Chiappini, C., & Martig, M. 2014a, *A&A*, 572, A92
Minchev, I. 2017, arXiv:1701.07034
Minchev, I., Steinmetz, M., Chiappini, C., *et al.* 2017, *ApJ*, 834, 27
Yoachim, P. & Dalcanton, J. J. 2006, *AJ*, 131, 226
Pohlen, M., Zaroubi, S., Peletier, R. F., & Dettmar, R.-J. 2007, *MNRAS*, 378, 594
Cheng, J. Y., Rockosi, C. M., Morrison, H. L., *et al.* 2012, *ApJ*, 752, 51
Bovy, J., Rix, H.-W., Liu, C., *et al.* 2012, *ApJ*, 753, 148
van der Kruit, P. C. & Searle, L. 1982, *A&A*, 110, 61
Comerón, S., Elmegreen, B. G., Knapen, J. H., *et al.* 2011, *ApJL*, 741, 28
Villalobos, Á. & Helmi, A. 2008, *MNRAS*, 391, 1806
Scannapieco, C., White, S. D. M., Springel, V., & Tissera, P. B. 2009, *MNRAS*, 396, 696
Minchev, I., Martig, M., Streich, D., *et al.* 2015, *ApJL*, 804, L9
Springel, V., Wang, J., Vogelsberger, M., *et al.* 2008, *MNRAS*, 391, 1685
Aumer, M., White, S. D. M., Naab, T., & Scannapieco, C. 2013, *MNRAS*, 434, 3142
Kazantzidis, S., Bullock, J. S., Zentner *et al.* 2008, *ApJ*, 688, 254
Anders, F., Chiappini, C., Santiago, B. X., *et al.* 2014, *A&A*, 564, A115
Bovy, J., Rix, H.-W., Schlafly, E. F., *et al.* 2016, *ApJ*, 823, 30
Mackereth, J. T., Bovy, J., Schiavon, R. P., *et al.* 2017, arXiv:1706.00018
Martig, M., Fouesneau, M., Rix, H.-W., *et al.* 2016, *MNRAS*, 456, 3655
Sanders, J. L. & Binney, J. 2015, *MNRAS*, 449, 3479

Schönrich, R. & Binney, J. 2009a, *MNRAS*, 396, 203
Kubryk, M., Prantzos, N., & Athanassoula, E. 2015, *A&A*, 580, A126
Minchev, I., Chiappini, C., & Martig, M. 2013, *A&A*, 558, A9

Astronomy and Astrophysics in the Gaia sky
Proceedings IAU Symposium No. 330, 2017
A. Recio-Blanco, P. de Laverny, A.G.A. Brown & T. Prusti, eds.

doi:10.1017/S1743921317005786

Galactic Surveys in the Gaia Era

Rosemary F. G. Wyse

Johns Hopkins University, Department of Physics and Astronomy,
Baltimore, MD 21210, USA
email: wyse@jhu.edu

Abstract. The final astrometric data from the Gaia mission will transform our view of the stellar content of the Galaxy, particularly when complemented with spectroscopic surveys providing stellar parameters, line-of-sight kinematics and elemental abundances. Analyses with Gaia DR1 are already demonstrating the insight gained and the promise of what is to come with future Gaia releases. I present a brief overview of results and puzzles from recent Galactic Archaeology surveys for context, focusing on the Galactic discs.

Keywords. Galaxy: formation, evolution, structure

---

1. The Fossil Record: Galactic Archaeology

Galactic Archaeology, or Near-Field Cosmology, is possible due to the long life-times of low-mass stars and the fact that the kinematics and chemical abundances of such stars contain information about conditions at their birth. Nearby stars older than 10Gyr probe redshifts greater than 2 and are found throughout the Galaxy and its retinue of satellites. Photospheric elemental abundances are largely unchanged throughout the lifetime of a star (setting aside mass transfer in close binaries, and the very low level of pollution through accretion of ambient material through which the star passes) and hence reflect those of the interstellar gas from which the star formed. Certain orbital properties, such as energy and angular momentum, are approximate adiabatic invariants in realistic potentials and structure in chemical and kinematic phase space persists after coordinate space structure is erased by mixing (due to the finite velocity dispersion of the structure). The multivariate distribution functions of different Galactic components overlap, so that large samples of stars with well-understood selection functions are required. The study of resolved stars of all ages within one galaxy (limited at the present to members of the Local Group) as a means to decipher how galaxies evolve is complementary to the direct study of the integrated light of galaxies at high redshift: evolution of a few galaxies compared to snapshots of different galaxies at different times.

The detailed study of stellar populations in the Milky Way and satellite galaxies (also M 31) is of particular importance to efforts to identify signatures of dark sector physics. Different candidate dark matter particles make very different predictions for structure formation on, and below, the scales of large galaxies, while on large scalles, where gravity dominates the physics, there is little divergence (see the review by Ostriker & Steinhardt, 2003). These small scales are just those on which tensions are found between the predictions of ΛCDM models without fine-tuning and observations (e.g. Weinberg *et al.* 2015). Much current effort is directed both at investigations of alternative dark matter candidates and investigations of more complex/improved baryonic physics, particularly stellar feedback, to modify the distribution of dark matter. Baryonic physics is imprinted on the fossil stellar populations and more sophisticated models within the ΛCDM framework are being developed to test against the increasingly detailed and comprehensive observational data.

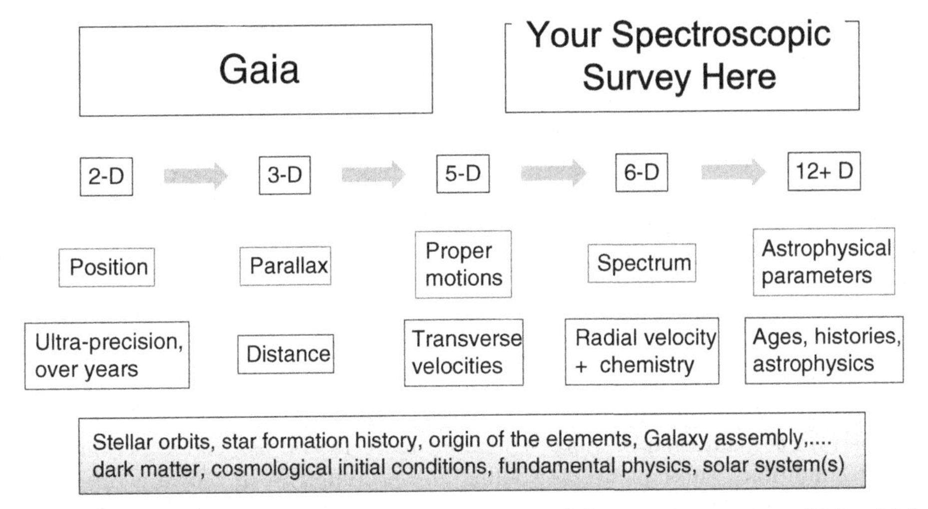

Figure 1. Illustration of the increase in dimensionality of the parameter space within which resolved stellar populations may be studied with the combination of astrometric and spectroscopic data. The observational quantities obtained from each type of data are in the boxes outlined in red, with the derived astrophysical quantities in the blue-outlined boxes. The 12+ dimensions of 'Astrophysical parameters' refers to the spectroscopic gravity, effective temperature and many individual elemental abundances that may be estimated from the spectra. The 'big-picture' physics questions that may then be addressed are in the blue-shaded box below. Adapted from Gilmore *et al.* (2012), their Fig. 2.

As I will discuss below, using the Galactic discs as examples, the distribution of stars in $n-$dimensional space is a complex function of how the Galaxy formed and evolved. The power of the combination of Gaia astrometric data with spectroscopic data is illustrated in Fig. 1, modified from the overview of the Gaia-ESO survey (Gilmore *et al.*, 2012).

There are several recent, on-going and planned stellar surveys using highly multiplexed spectrographs, each with its own niche defined by the selection function and characteristics of the spectra (such as resolution, wavelength coverage). One hopes that consistent conclusions will be reached from analyses of the different datasets. Those targeting brighter stars, such as the RAVE survey (see Kunder *et al.*, 2017 plus Andrea Kunder's contribution to this volume), the GALAH survey (see Martell *et al.*, 2017), the APOGEE survey (see Allende-Prieto's contribution to this volume) and the LAMOST survey (see Martin Smith's contribution to this volume) have significant overlap with the sample of the Tycho Gaia Astrometric Solution of Gaia DR1 and the range of talks and posters in this conference give a flavour of the wealth of science questions that can be addressed with these data, and point to the future analyses that will be enabled with the full Gaia dataset.

On-going and planned ground-based multi-band imaging surveys such as PanSTARRS1 (Kaiser *et al.* 2010), the Dark Energy Survey (Abbott *et al.* 2016) and LSST (Abell *et al.* 2012), together with space-based surveys from current and future missions, further extend the dimensionality of the dataset. Of course, the Gaia photometry and positions by themselves contain new information (for example see Deason's contribution to this volume for a clever use of photometry flags in Gaia DR1 to identify variable stars). This is truly an exciting time for Galactic astrophysics.

2. The Milky Way Discs: What can we learn from (old) disc stars?

Thin stellar discs are fragile and can be disturbed by external influences such as companion galaxies and mergers, in addition to internal gravitational perturbations such as spiral arms, bars and Giant Molecular Clouds. Stellar systems are collisionless and thus cannot 'cool' once heated, unlike gas whuch can radiate away energy in excited internal degrees of freedom. The vertical structure of the thin disc - with 'structure' intended to refer to all aspects, including number density, age distribution, chemical abundance distribution and kinematics - encodes the history of heating and minor merging/satellite accretion, relative rates of dissipational settling and star formation, adiabatic compression and further heating. The radial structure reflects the relative rates of star formation and gas flows, as a function of location, and contains imprints of the angular momentum distribution and re-arrangement. The thick disc probes the earliest phase of disc star formation and the nature, and rate, of the transition between the thick and thin discs constrain the duration of more significant mergers and associated heating and/or turbulent conditions, subsequent gas cooling and accretion to (re-)form a thin disc. Interactions with gravitational perturbations, particularly satellites, do not only heat and thicken stellar discs, they can also excite warps and 'bending' and 'breathing' modes in the thin disc (for example Widrow *et al.*, 2014). Some radial spreading of the disc is created as angular momentum is transferred from a satellite to disc stars; this change in mean orbital radius occurs together with transfer of random energy into the orbit i.e. associated heating (e.g. Bird *et al.* 2012). Induced radial displacement of stars can also occur without a change in the orbital circularity, due to interactions between stars and a perturbation (such as a transient spiral arm) at the corotation resonance of the perturbation (Sellwood & Binney 2002); this maintains the thinness of discs while stars move within them, and is commonly referred to as Radial Migration (even though other types of interaction can also cause a star's mean orbital radius to change). The outer regions of thin discs can be built-up by outward radial migration, and the quantification of this contribution to the stellar population at large radius, compared to *in situ* star formation, is of obvious importance.

2.1. *Vertical Structure: The Local Milky Way Thick Disc*

The thick disc was initially defined geometrically, through fits to star counts at the South Galactic Pole; two separate exponentially declining density laws were required (Gilmore & Reid 1983), with the thick disc having a significantly larger scale height than the thin disc. This geometric thick disc was subsequently shown (by many authors, see reviews by Gilmore, Wyse & Kuijken, 1989; Majewski, 1993) to have kinematics intermediate between those of the thin disc and stellar halo: the mean orbital rotation velocity about the Galactic centre lagging that of the thin disc by ~ 50 km/s, and the vertical velocity dispersion being ~ 40 km/s, consistent with the estimated scale height of ~ 1 kpc (there is a degeneracy between scale-height and local normalization in the fits to the star counts, but the scale-height has to be consistent with the kinematics and inferred mass surface density). These kinematics are too hot to have resulted from heating due to present-day internal disc perturbations (e.g. spiral arms, GMCs). A discontinuous trend in age-velocity dispersion from thin to thick disc suggested an exceptional heating event to form the thick disc. Early (multi-object) spectroscopic studies determined a mean metallicity for the thick disc of ~ -0.5 dex, with 'alpha-enhanced' elemental abundances ($[\alpha/\mathrm{Fe}] > 0$). The redder turn-off colour compared to the stellar halo (see Fig. 2, with this metallicity, implies that most thick disc stars are similar in age to the stellar halo, $\sim 10-12$ Gyr, thus forming at redshifts greater than 2. Since there are stars of

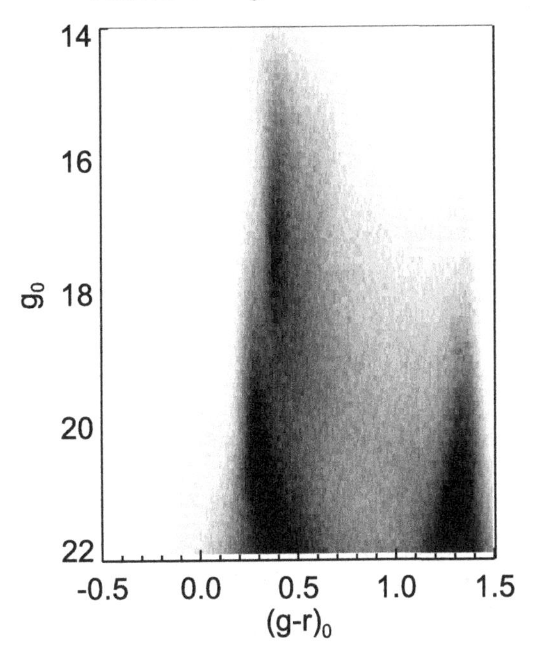

Figure 2. Taken from Jayaraman *et al.* (2013), showing star counts at intermediate latitudes across the SDSS Equatorial stripe. Each of the halo and thick disc show a well-defined - and distinct - blue edge, marking the main sequence turnoffs.

significantly younger ages in the local thin disc, which would be scattered into the thick disc should there have been a significant merger, or indeed any strong heating event, subsequent to their formation, the dominant old age in the (geometric) thick disc limits any such event to have occurred only at early epochs. The derived stellar mass of the thick disc is a significant fraction, in the range of 20% to 50%, of the thin disc mass, i.e. greater than $10^{10}M_{\odot}$. The narrow range of ages of the bulk of stars in the thick disc implies a high past star-formation rate.

2.2. *Chemically Defined Thick and Thin Discs*

More precise elemental abundance measurements for larger samples of stars have revealed that local (thin and thick) disc stars separate into two sequences in the plane of $[\alpha/\mathrm{Fe}]$ against [Fe/H]. Differences in the patterns of elemental abundances largely reflects differences in star formation history (e.g. Gilmore & Wyse 1991), and the two sequences indeed are in line with the inferred star formation histories of the thick and thin discs, albeit that the extent of the overlap in iron abundance is large. Further, the mean properties of chemically defined 'high-alpha' and 'low-alpha' discs compare well with those of the geometrically defined discs: the 'high-alpha' sequence is on average more metal-poor, consists of older stars, has 'hot' kinematics, and is taken to represent the thick disc, while the 'low-alpha' sequence is more metal-rich, contains young to old stars, has 'cold' kinematics and is taken to represent the thin disc (e.g. Bensby *et al.* 2014; Martig *et al.* 2016a). Indeed, the more metal-poor stars in the 'high-alpha' sequence (which actually have enhanced values of $[\alpha/\mathrm{Fe}]$) have old ages and heights above the plane ~ 1 kpc (Martig *et al.* 2016b) just as do stars in the geometrically defined thick disc. The two sequences merge at the metal-rich end(s), and there is real ambiguity about the assignment - and origins of - the metal-rich high-alpha stars, which also tend to be younger

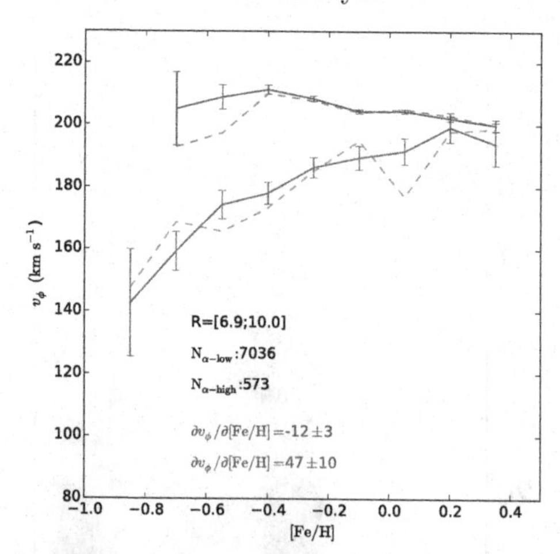

Figure 3. Taken from Kordopatis *et al.* (2017), showing the different trends of azimuthal streaming (rotation) velocity (v_ϕ) with iron abundance for stars in each of the two elemental abundance sequences (blue indicates the 'low-alpha' or thin disc sequence, while red indicates the 'high-alpha' or thick disc sequence). The data are from the APOGEE survey, in lines-of-sight selected to probe this velocity component without the use of tangential velocities, which are more uncertain prior to the improved proper motions and distances that Gaia will provide.

(e.g. Haywood *et al.* 2013, Chiappini *et al.* 2015). The younger stars at higher heights in the outer disc (Martig *et al.* 2016b) probably reflect flaring of the thin disc.

The distinctiveness of the kinematics of the two elemental abundance sequences is seen clearly in the very different trends of rotational velocity with iron abundance, illustrated in Fig. 3. Again we see the sequences merging at high iron abundances. The trend of increasing rotational velocity for decreasing iron abundance in the 'low-alpha' (thin disc) sequence is understandable as a consequence of the combination of a negative radial metallicity gradient and epicyclic motions, conserving orbital angular momentum - a star observed close to its perGalacticon, on an inward epicyclic excursion, will have a higher orbital rotation velocity and have a lower metallicity than the disc where it is observed, and *vice versa* for a star observed close to its maximum outward epicyclic excursion. The opposite trend for the 'high-alpha' (thick disc) sequence must reflect how the thick disc formed and evolved (see e.g. Schönrich & McMillan, 2017).

2.3. *The (Geometric) Thick Disc and the Earliest Phase of Disc Evolution*

Stars in the (geometric) thick disc are old, and formed at lookback time of $\sim$ 10 Gyr, corresponding to redshift $\sim$ 2: the epoch of peak cosmic star formation. The high stellar velocity dispersion and inferred short duration of star formation implies that the thick disc likely formed prior to equilibrium/virialization of the Milky Way's dark halo, during the epoch of active assembly/mergers, leading to highly turbulent conditions in the ISM (cf. Jones & Wyse 1983, Gilmore 1984, Brook *et al.* 2012, Bird *et al.* 2013, Ma *et al.* 2017). The derived stellar mass ($M_* > 10^{10} M_\odot$) and star-formation rate (several $M_\odot$/yr) are similar to those derived for star-forming discs observed at redshift $\sim$ 2. These high-redshift galaxies show clumpy, turbulent ionized gas discs, in organized rotational motion with amplitude of $\sim 100 - 200$ km/s, and a high internal velocity dispersion, $\sim 50 -$ 100 km/s (e.g. Wisnioski *et al.* 2015). Lower redshift star-forming gas discs show higher values of rotational velocity to random motions, and higher specific angular momentum (Swinbank *et al.* 2017). How this translates into the evolution of an individual disc consisting of stars and gas is model-dependent. The fossil record from discs in nearby galaxies provides the best guide.

2.3.1. *The Old Age of Thick Disc Stars Limits Recent Merger/Heating Events*

Mergers heat the thin stellar disc and input stars formed up to that epoch into the (geometric) thick disc (and perhaps the stellar halo). There are stars of a broad range of ages in the thin disc, reflecting continuous star formation since early times. The dominant old age of stars in the (geometric) thick disc implies that there has been no significant merger since the redshift at which the look-back time equals this old age of thick disc stars. An age of $\sim$ 10 Gyr means a quiescent merger history since redshift $\sim$ 2 (Wyse 2001). Such a quiet merger history is consistent with there being no evidence of a significant dark or stellar accreted disc (e.g. Ruchti *et al.*, 2015). Minor mergers/interactions, for example that ongoing with the Sagittarius dwarf spheroidal, affect primarily the outer thin disc, inducing outer spiral structure, warping and/or flaring (e.g. Purcell *et al.* 2011).

The underlying assumption of this dating technique gains support from the recent analysis by Ma *et al.* (2017) of the age structure of the thin/thick discs formed in their simulation of the formation of a Milky Way-mass disc galaxy in ΛCDM. Their model galaxy incurred its last significant merger at redshift $\sim$ 0.7, a lookback time of $\sim$ 6Gyr. Fig. 4 of Ma *et al.* shows that across the radial extent of the disc the mean age of stars at heights 1-2 kpc above the mid-plane is $\sim 4 - 7$ Gyr, with mean metallicity in the range -0.3 dex to the solar value. This mean age, for what would be identified as the geometric thick disc, indeed reflects the timing of the last significant merger $\sim$ 6Gyr ago (plus time for the merger/heating to complete), and the high mean enrichment is also consistent with a later heating event in the simulation than implied for the Milky Way.

Accurate and precise (old) stellar ages are crucial in this estimation of the timing of the last significant merger and the parallax data from Gaia will be extrenely important (see Tayar *et al.* 2017 for a discussion of the biases that must be dealt with in the estimation of ages for red giant stars based on astrosiesmology-based gravities). Distances are also of course fundamental to the determination of the vertical stucture. See Mackereth *et al.* 2017 for a recent pre-Gaia analysis of the age-metallicity structure of the discs from APOGEE data. The quiescent merger history inferred for the Milky Way is not typical in ΛCDM and we also need more detailed predictions of the merger history of typical Milky Way-mass disc galaxies, as a function of the orbit and mass ratio of the satellite/subhalo, from galaxy formation models, to interpret the derived observational limits.

3. Radial Structure: the Outer Disc

3.1. *The Ringing Disc*

A wealth of structure in the outer stellar disc has been revealed by the imaging data of SDSS (e.g. Belokurov *et al.*, 2006)) and Pan-STARRS (e.g. Slater *et al.*, 2014 and Bernard *et al.*, 2016). Systematic variations in star counts above and below the nominal mid-plane were identified by Xu *et al.* (2015) and interpreted as rings and radial waves in the disc. The 'Monoceros Ring', which had been speculated to be a remmant of an accreted satellite (e.g. Peñarrubia *et al.*, 2005) is clearly more simply an apparent overdensity due to structure within the thin disc. Oscillatory kinematic features in thin disc stars have also been identified in several kinematic surveys (e.g. Widrow *et al.*, 2012 and Williams *et al.*, 2013). These plausibly reflect breathing/bending modes of the perturbed thin disc (Widrow & Bonner 2015), excited due to interaction by either internal perturbations, such as spiral structure and/or the bar (Debattista, 2014) or external perturbations, perhaps the Sagittarius dwarf spheroidal (Gómez *et al.*, 2013; see also Widrow *et al.*,

2014). These oscillations, combined with flaring of the thin disc, may extend to high enough distances from the plane that even apparent 'halo' substructure may actually consist of the perturbed thin disc (Li *et al.*, 2017).

There are several mechanisms by which a sub-population of thin disc stars can be scattered/heated into the halo, such as through binary interactions. This could be the explanation behind the discovery of rare metal-rich high-velocity stars (e.g. Hawkins *et al.*, 2015). As is always the case for the analysis of objects with extreme values of the parameters characterizing their parent population (kinematics and chemistry in this case), the accuracy and precision of those values are critical.

The main lesson is that overdensities in star counts and kinematic substructure are not necessarily tidal debris from satellites. Again, a more comprehensive understanding will be available with improved distances, three-dimensional velocities and ages from the Gaia data.

3.2. *Internal, Secular Evolution to Re-arrange Discs?*

As noted above, radial migration (Sellwood & Binney, 2002) can move thin-disc stars across distances that are of order the disc scale-length, during the lifetime of the disc, without associated kinematic heating (maintaining orbital circularity and the thinness of the disc). This mechanism is more effective for stars on closer-to-circular orbits, less so for populations of higher velocity dispersion/lower angular momentum orbits (e.g. Solway, Sellwood & Schönrich, 2012; Vera-Ciro *et al.*, 2014; Daniel & Wyse, 2017). The efficiency of radial migration also obviously depends on the parameters of the perturbation(s) driving it, such as amplitude, duty cycle, pattern speed and wave number (e.g. Daniel & Wyse, 2015; Debattista, Roskar & Loebman, 2017). The chemical evolution of the disc can be strongly affected (Schönrich & Binney, 2009). The existence of a significant population of thin-disc stars in the solar neighbourhood with super-solar metallicities is consistent with outward radial migration bringing stars from the inner, higher-metallicity, regions of the disc to the outer disc (e.g. Kordopatis *et al.*, 2015). The global importance of radial migration in the evolution of discs is uncertain, with simulations showing both a minimal influence (Bird *et al.*, 2013), and a very important role (Minchev *et al.*, 2012). It is clearly important to isolate the important phyical effects causing these different conclusions, and test the predictions of the models in detail.

4. Concluding remarks

'More data are needed': accurate and precise positions, distances, space motions, ages, and chemical abundances are all critical to the characterisation of the present stellar populations of the Galaxy and to the eventual understanding of the roles of different physical processes in its evolution. Happily this is what the combination of astrometric data from Gaia and data from large spectroscopic surveys promises to deliver. When analysing these data, we need to be clear to what we are referring when using the terms halo/bulge/thin disc/thick disc, since entities defined through different parameters (e.g. spatial distribution, chemistry) may have different histories.

There are truly exciting times ahead: wonderful observational data for stars plus improved simulations of galaxy formation in cosmological context(s), complemented by increasingly detailed data for high-redshift discs in formation.

Acknowledgements: I thank the organisers for inviting me, and the American Astronomical Society for the award of an NSF-funded International Travel Grant.

References

Abbott, T. *et al.* (Dark Energy Survey Team) 2016, *MNRAS*, 460, 1270
Abell, P. A. *et al.* (LSST) 2012, arXiv:0912.0201
Bensby, T., Feltzing, S., & Oey, M. S. 2014, *A & A*, 562, A71
Bernard, E. J., Ferguson, A. M. N., Schlafly, E. F., *et al.* 2016, *mnras*, 463, 1759
Bird, J. C., Kazantzidis, S., & Weinberg, D. H. 2012, *MNRAS*, 420, 913
Bird, J. C., Kazantzidis, S., Weinberg, D. H., *et al.* 2013, *ApJ*, 773, 43
Brook, C. B., Stinson, G. S., Gibson, B. K., *et al.* 2012, *MNRAS*, 426, 690
Chiappini, C., Anders, F., Rodrigues, T. S., *et al.* 2015, *A & A*, 576, L12
Daniel, K. J. & Wyse, R. F. G. 2015, *MNRAS*, 447, 3576
Daniel, K. J. & Wyse, R. F. G. 2017, *MNRAS*, submitted
Debattista, V. P. 2014, *MNRAS*, 443, L1
Debattista, V. P., Roskar, R., & Loebman, S. R. 2017, in: Eds. J. H. Knapen, J. C. Lee and A. Gil de Paz, (eds.) 'Outskirts of Galaxies', *ASSL*, (Berlin: Springer), in press (arXiv:1706.01996)
Gilmore, G. & Wyse, R. F. G. 1991, *ApJL*, 367, L55
Gilmore, G., Wyse, R. F. G., & Kuijken, K. 1989, *ARAA*, 27, 555
Gilmore, G., *et al.* 2012, *ESO Messenger*, 147, 25
Gómez, F. A., Minchev, I., O'Shea, B. W., *et al.* 2013, *MNRAS*, 429, 159
Hawkins, K., Kordopatis, G., Gilmore, G., *et al.* 2015, *MNRAS*, 447, 2046
Haywood, M., Di Matteo, P., Lehnert, M. D., Katz, D., & Gómez, A. 2013, *A & A*, 560, A109
Jones, B. J. T. & Wyse, R. F. G. 1983, *A & A*, 120, 165
Kaiser, N., *et al.* (PanSTARRS 2010, *SPIE*, 7733, id 77330E
Kordopatis, G., Wyse, R. F. G., Gilmore, G., *et al.* 2015, *A & A*, 582, A122
Kordopatis, G., Binney, J., Gilmore, G., *et al.* 2015, *MNRAS*, 447, 3526
Kunder, A., *et al.* 2017, *AJ*, 153, 75
Li, T. S., Sheffield, A. A., Johnston, K. V., *et al.* 2017, arXiv:1703.05384
Ma, X., Hopkins, P. F., Wetzel, A. R., *et al.* 2017, *MNRAS*, 467, 2430
Mackereth, J. T., Bovy, J., Schiavon, R. P., *et al.* 2017, arXiv:1706.00018
Majewski, S. R. 1993, *ARAA*, 31, 575
Martell, S., *et al.* 2017, *MNRAS*, 465, 3203
Martig, M., Fouesneau, M., Rix, H.-W., *et al.* 2016a, *MNRAS*, 456, 3655
Martig, M., Minchev, I., Ness, M., Fouesneau, M., & Rix, H.-W. 2016b, *ApJ*, 831, 139
Minchev, I., Famaey, B., Quillen, A. C., *et al.* 2012, *A & A*, 548, A126
Ostriker, J. P. & Steinhardt, P. 2003, *Science*, 300, 1909
Peñarrubia, J., Martínez-Delgado, D., Rix, H. W., *et al.* 2005, *ApJ*, 626, 128
Purcell, C. W., Bullock, J. S., Tollerud, E., Rocha, M., & Chakrabarti, S. 2011, *Nature*, 477, 301
Schönrich, R. & Binney, J. 2009, *MNRAS*, 396, 203
Schönrich, R. & McMillan, P. J. 2017, *MNRAS*, 467, 1154
Sellwood, J. A. & Binney, J. J. 2002, *MNRAS*, 336, 785
Slater, C. T., Bell, E. F., Schlafly, E. F., *et al.* 2014, *ApJ*, 791, 9
Solway, M., Sellwood, J. A., & Schönrich, R. 2012, *MNRAS*, 422, 1363
Swinbank, A. M., Harrison, C. M., Trayford, J., *et al.* 2017, *MNRAS*, 467, 3140
Tayar, J., Somers, G., Pinsonneault, M. H., *et al.* 2017, *ApJ*, 840, 17
Vera-Ciro, C., D'Onghia, E., Navarro, J., & Abadi, M. 2014, *ApJ*, 794, 173
Weinberg, D. H., Bullock, J. S., Governato, F., *et al.* 2015, *PNAS*, 112, 12249
Widrow, L. M., Gardner, S., Yanny, B., Dodelson, S., & Chen, H.-Y. 2012, *ApJL*, 750, L41
Widrow, L. M., Barber, J., Chequers, M. H., & Cheng, E. 2014, *MNRAS*, 440, 1971
Widrow, L. M. & Bonner, G. 2015, *MNRAS*, 450, 266
Williams, M. E. K., Steinmetz, M., Binney, J., *et al.* 2013, *MNRAS*, 436, 101
Wyse, R. F. G. 2001, in: J. G. Funes & E. M. Corsini (eds.) 'Galaxy Discs and Disc Galaxies', *Astronomical Society of the Pacific Conference Series*, 230, (San Francisco: ASP), p71
Xu, Y., Newberg, H. J., Carlin, J. L., *et al.* 2015, *ApJ*, 801, 105

Astronomy and Astrophysics in the Gaia sky
Proceedings IAU Symposium No. 330, 2017
A. Recio-Blanco, P. de Laverny, A.G.A. Brown & T. Prusti, eds.

doi:10.1017/S1743921317005579

Close stellar encounters with the Sun from the first Gaia Data Release

Coryn A. L. Bailer-Jones

Max Planck Institute for Astronomy, Königstuhl 17, 69117 Heidelberg, Germany
email: calj@mpia.de

Abstract. I present preliminary results of searching for close stellar encounters with the Sun using the first Gaia data release. Gl 710 is found to be a much closer encounter than found in pre-TGAS studies. More detailed results will be reported in an upcoming publication (Bailer-Jones 2017, in preparation).

Keywords. comets:general – Oort cloud – stars: general, kinematics and dynamics – methods: numerical, statistical – surveys: Gaia

---

1. Introduction

Ever since we've been able to measure three-dimensional stellar space velocities, people have wanted to identify stars which have – or which will – come close to the solar system. One motivation is to study the effect of encounters on the Oort cloud, as it is understood that the gravitational perturbations of passing stars is partially responsible for generating the long-period comets. Many searches for close encounters have been published over the past decades, may using Hipparcos (e.g. García-Sánchez *et al.* 1999, 2001; Dybczyński 2006; Bailer-Jones 2015; Dybczyński & Berski 2015; Bobylev 2010a,b; Mamajek *et al.* 2015; Berski & Dybczyński 2016). Here I report on preliminary results from an on-going study of encounters found using the Gaia data. A more detailed and updated report will appear in a forthcoming publication (Bailer-Jones 2017, in preparation).

2. Method

TGAS astrometry (Lindegren *et al.* 2016) from Gaia Data Release 1 (Gaia Collaboration *et al.* 2016) was cross-matched with various radial velocity (RV) catalogues which claim measurement precisions better of around $2\,\mathrm{km\,s^{-1}}$ or better (in particular RAVE-DR5, Kunder *et al.* 2017). This yielded a set of about 300 000 objects with complete 6D phase space information (3D position and 3D velocity). Note that a given star can appear in more than one RV catalogue, so the number of (what I call) objects is larger than the number of (unique) stars. Using the linear motion approximation (LMA; i.e. motion neglecting gravity; see Bailer-Jones 2015), I first computed the time and distance at which each object encounters the Sun (perihelion). 709 objects have perihelia less than 10 pc. The distribution of their data and the corresponding standard uncertainties are shown in Figure 1.

I then integrated the orbits of these 709 objects in a Galactic potential in order to compute more accurate perihelia (same model as in Bailer-Jones 2015). By resampling the data for each object and integrating the orbits for all of the resulting surrogates, we properly characterize the (generally asymmetric) distribution in the perihelion distance, time, and speed. I summarize each distribution using the median and the 90% equal-tailed confidence interval (CI). When doing this resampling it is vital to take into account

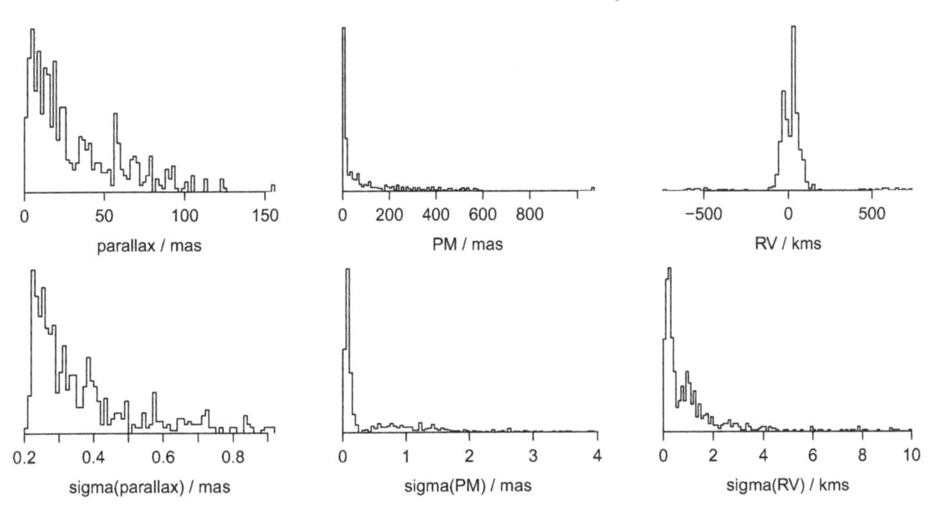

Figure 1. Distribution of the parallax, proper motion, and radial velocity, plus their standard uncertainties, for the 709 objects.

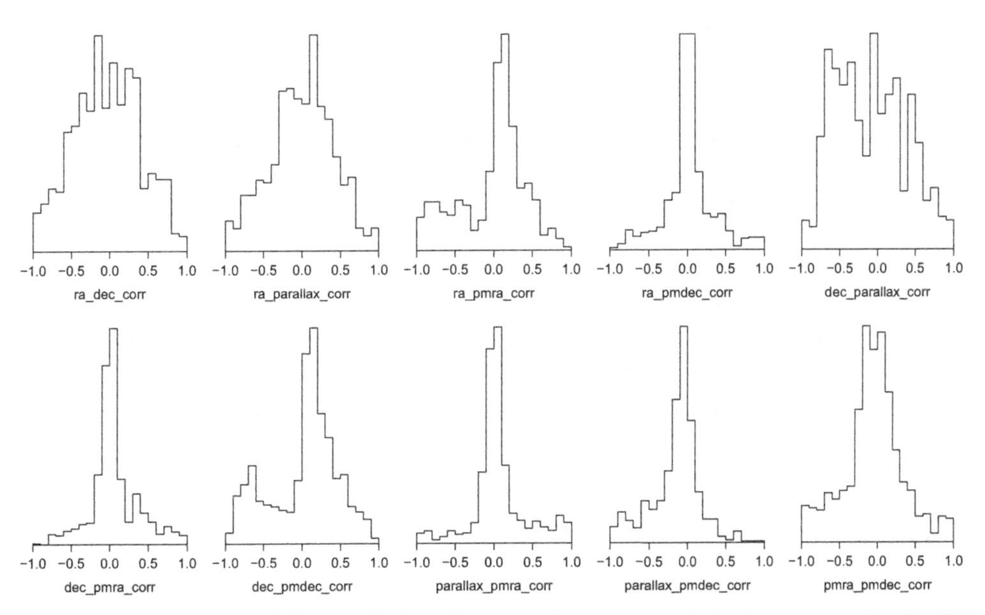

Figure 2. Distribution of the correlations between the astrometric measurements for the 709 objects.

the correlations between the different astrometric parameters for individual objects (as reported in the TGAS catalogue), as these can be very high (see Figure 2).

This procedure does not, of course, find all encounters in my input catalogue which come within 10 pc, as the initial LMA-selection it not guaranteed to include all objects which, when properly integrated, would come within 10 pc. Yet the LMA turns out to be a good approximation for most stars (the potential deflects paths little over timescales of a few Myr), so this approach is adequate for identifying encounters which come much closer than 10 pc.

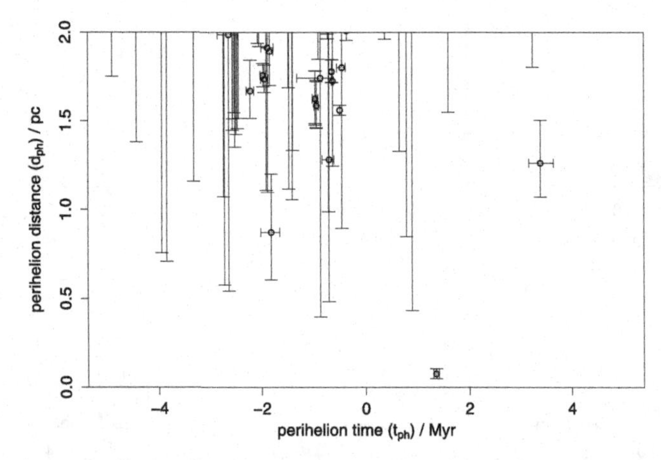

Figure 3. The closest stellar encounters to the Sun found by orbital integration in this study. The open circles show the median of the distribution and the (asymmetric) error bars show the 90% equal-tailed confidence intervals. This shows objects, not unique stars, so includes some duplicates. The closest encountering star is Gl 710.

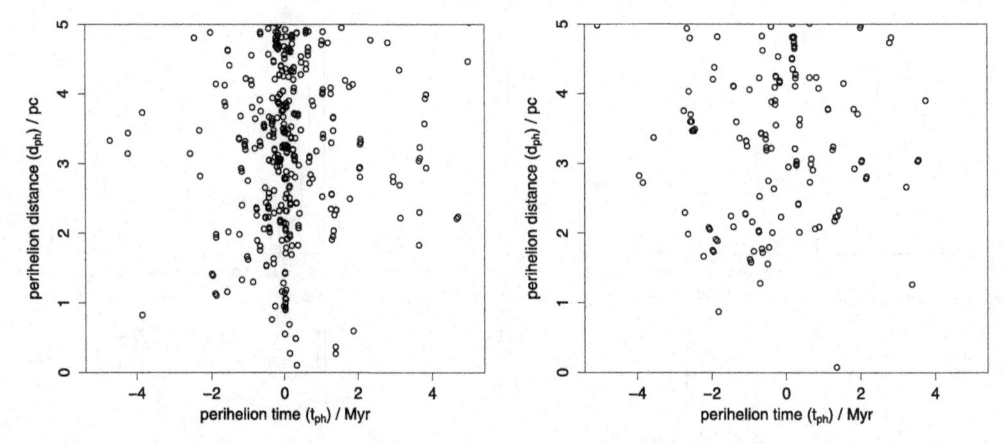

Figure 4. Comparison between object encounters found in this study from TGAS (left) with those found in Bailer-Jones (2015) based on Hipparcos data.

3. Results

The perihelion times and distances for the closest encounters found by the orbital integration are shown in Figure 3. This includes some duplicate stars. The closest encounter is Gl 710. This has been known for some time to be a close encounter: most previous studies have found it. Based on Hipparcos-2 data and a radial velocity of $-13.8 \pm 0.3\,\mathrm{km\,s^{-1}}$, I found a median encounter distance of 0.26 pc (90% CI 0.10–0.44 pc) (from Bailer-Jones (2015) and its online supplement). TGAS, however, assigns this star a much smaller proper motion: $0.50 \pm 0.16\,\mathrm{mas\,yr^{-1}}$ as opposed to $1.8 \pm 1.2\,\mathrm{mas\,yr^{-1}}$ in Hipparcos-2 (the parallax is the same to within 2%). Using the same radial velocity, my orbital integration now gives a median perihelion distance of 0.08 pc (90% CI 0.05–0.10 pc), equivalently 16 000 AU (90% CI 10 000–21 000 AU).

Berski & Dybczyński (2016) – who, unsurprisingly, found a very similar perihelion based on the same data and similar method – went on to explore the impact this stars would have on the Oort cloud. Although this close approach would take Gl 710 well within the Oort cloud – and its relative velocity at encounter is low ($14\,\mathrm{km\,s^{-1}}$) – it is a low mass (around $0.6\,M_\odot$) K7 dwarf, so its perturbing influence could be less than more distant but more massive encounters.

In my Hipparcos-based study (Bailer-Jones 2015) I found many more close encounters, as can be seen in Figure 4. As discussed in that paper, some of these encounters had dubious astrometric or radial velocity data, and I stressed that one should not blindly use Table 3 as a reliable list. Unfortunately, many of these interesting cases are not in TGAS, primarily because they are too bright (the TGAS bright limit is about $G = 6$ mag) or because there have not yet been enough observations to permit a reliable astrometric solution. This lack of bright stars probably explains why, in the present TGAS study, we see far fewer objects which are currently near perihelion (i.e. with t_{ph} near to zero in the left panel of Figure 4) compared to the Hipparcos-based study.

It must also be emphasised that the decrease in the number of encounters at larger absolute perihelion times is in part an observational selection effect due to the limiting magnitude of the survey (Hipparcos or TGAS). More distant stars will, in general, have encounters further in the past or present, and these stars are currently fainter and so less likely to be observed. However, some of this decrease in density at large times is, somewhat counter-intuitively, a consequence of the intrinsic spatial and velocity distribution of stars. Such effects are discussed in more detail in the upcoming publication, which also gives an incompleteness-corrected estimate of the current encounter rate.

4. Conclusions

A wide-ranging search for close stellar encounters with the solar system can only be done using large, precise, deep astrometric surveys. Gaia is a huge step forward. Although the first data release is limited, it has already allowed us to extend the list of close encounters and improve the precision for known encounters. TGAS is nonetheless limited not only by its depth, but also its complex selection function and bright limit. All of these issues will be addressed by future Gaia data releases. We are, however, already moving into a regime where our studies are limited not by the availability of astrometry, but by the availability of sufficiently accurate radial velocities.

Acknowledgements

This work has made use of data from the European Space Agency (ESA) mission Gaia (`https://www.cosmos.esa.int/gaia`), processed by the Gaia Data Processing and Analysis Consortium (DPAC). Funding for the DPAC has been provided by national institutions, in particular the institutions participating in the Gaia Multilateral Agreement. Funding for RAVE (`http://www.rave-survey.org`) has been provided by institutions of the RAVE participants and by their national funding agencies.

References

Bailer-Jones C. A. L., Mar. 2015, *A&A*, 575, A35
Berski F., Dybczyński P. A., Nov. 2016, *A&A*, 595, L10
Bobylev V.V., Mar. 2010a, *Astronomy Letters*, 36, 220
Bobylev V.V., Nov. 2010b, *Astronomy Letters*, 36, 816
Dybczyński P. A., Apr. 2006, *A&A*, 449, 1233
Dybczyński P. A., Berski F., May 2015, *MNRAS*, 449, 2459
Gaia Collaboration, Brown A. G. A., Vallenari A., *et al.*, Nov. 2016, *A&A*, 595, A2
García-Sánchez J., Preston R. A., Jones D. L., *et al.*, Feb. 1999, *AJ*, 117, 1042
García-Sánchez J., Weissman P. R., Preston R. A., *et al.*, Nov. 2001, *A&A*, 379, 634
Kunder A., Kordopatis G., Steinmetz M., *et al.*, Feb. 2017, *AJ*, 153, 75
Lindegren L., Lammers U., Bastian U., *et al.*, Nov. 2016, *A&A*, 595, A4
Mamajek E. E., Barenfeld S. A., Ivanov V. D., *et al.*, Feb. 2015, *ApJ*, 800, L17

Astronomy and Astrophysics in the Gaia sky
Proceedings IAU Symposium No. 330, 2017
A. Recio-Blanco, P. de Laverny, A.G.A. Brown & T. Prusti, eds.

doi:10.1017/S1743921317006159

Gaia DR1 completeness within 250 pc & star formation history of the Solar neighbourhood

Edouard J. Bernard

Université Côte dAzur, OCA, CNRS, Lagrange, France
email: ebernard@oca.eu

Abstract. We took advantage of the *Gaia* DR1 to combine TGAS parallaxes with *Tycho-2* and APASS photometry to calculate the star formation history (SFH) of the solar neighbourhood within 250 pc using the colour-magnitude diagram fitting technique. We present the determination of the completeness within this volume, and compare the resulting SFH with that calculated from the *Hipparcos* catalogue within 80 pc of the Sun. We also show how this technique will be applied out to ∼5 kpc thanks to the next *Gaia* data releases, which will allow us to quantify the SFH of the thin disc, thick disc and halo *in situ*, rather than extrapolating based on the stars from these components that are today in the solar neighbourhood.

Keywords. Hertzsprung-Russell diagram, Galaxy: disk, Galaxy: evolution, Galaxy: formation, solar neighbourhood

---

1. Introduction

Disc galaxies dominate the stellar mass density in the Universe, yet the details of their formation and evolution are still poorly understood. Even in the Milky Way, for which we have access to a tremendous amount of information, the onset of star formation and the evolution of the star formation rate (SFR) of the Galactic components are all but unknown. This seriously hinders the interpretation of the available observations and the quantitative comparisons with galaxy formation models. However, the details of the formation of stellar systems are encoded in the distribution of the stars in deep colour-magnitude diagrams (CMDs). Their star formation history (SFH), that is, the evolution of both the SFR and the metallicity from the earliest epoch to the present time, can thus be recovered using the robust CMD-fitting technique. This technique requires the precise knowledge of the intrinsic luminosity of each star, and therefore its distance. In the coming years, *Gaia* will deliver distances and proper motions for over a billion stars out to ∼10 kpc, thus covering all the structural components of our Galaxy. For the first time, this opens the possibility to map the spatial and temporal variations of the SFH back to the earliest epochs. We illustrate this potential by exploiting the Gaia DR1 and TGAS parallaxes for over 2 million stars in the solar neighbourhood (Gaia Collaboration *et al.* 2016, Lindegren *et al.* 2016), and calculating the SFH of the Milky Way disc within 250 pc of the Sun.

2. The solar neighbourhood CMD: photometry and completeness

While TGAS provides accurate parallaxes and G-band magnitudes for over 2 million stars, no colour information is available. On the other hand, the *Tycho-2* catalogue does include B_T, V_T for all TGAS stars, but the photometric quality quickly degrades at fainter magnitudes. We thus cross-matched TGAS with the *Tycho-2*, *Hipparcos*, and APASS (Henden *et al.* 2012) DR9 catalogues, after homogenizing their photometry to

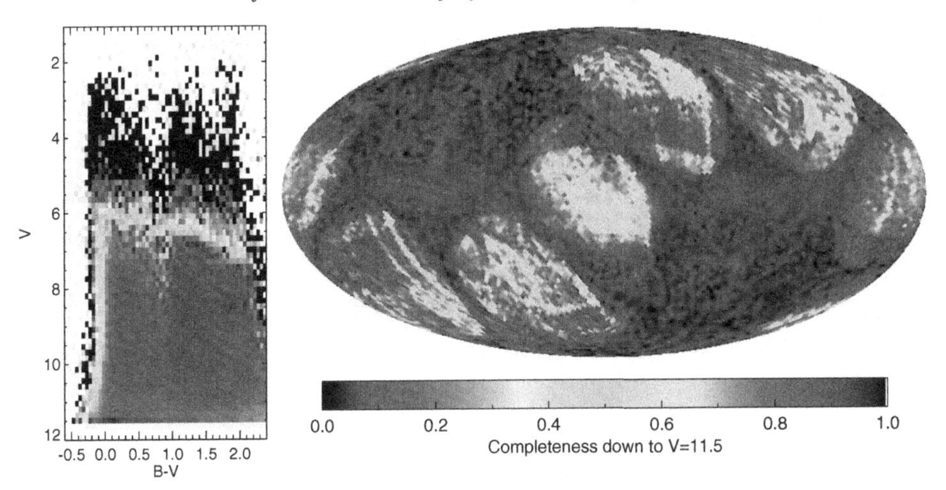

Figure 1. TGAS completeness down to V=11.5 relative to *Tycho-2* in colour-magnitude space (left) and in galactic coordinates (right), before the completeness corrections.

the Johnson B and V filters. For stars appearing in more than one catalogue, a weighted mean magnitude was calculated for each filter.

A further step before the CMD-fitting can be applied is a robust quantification of the completeness as a function of colour and magnitude. According to Høg *et al.* (2000), the *Tycho-2* completeness is over 90% down to $V\sim 11.5$, but decreases quickly at fainter magnitudes, so we only kept stars brighter than this limit. Even though TGAS is based on *Tycho-2*, about 20% of the stars from the latter catalogue are missing in TGAS, which means that we have no robust parallaxes for these stars. This is illustrated in Figure 1, which presents the completeness of TGAS versus *Tycho-2* as a function of both colour–magnitude (left) and spatial coordinates (right). The left panel shows that a significant fraction of the stars brighter than $V\sim 6.5$ or bluer than $B-V{\sim}0$ are missing in TGAS; however, most of these stars have *Hipparcos* parallaxes, which we combined with those from TGAS to erase completeness variations as a function of colour and magnitude. The right panels shows that completeness is below ~60% in about half of the sky due to the lower number of *Gaia* transits. Unfortunately, the completeness function in these regions is too complex to correct; we therefore simply excised 57% of the sky coverage where completeness down to $V\sim 11.5$ was <90%.

Finally, to obtain an accurate SFH back to the earliest epoch of star formation, a CMD reaching the oldest main-sequence turn-off (oMSTO, at M_V=4.5) is required. Given the completeness limits described above, the volume in which the SFH can be calculated is therefore limited to a distance modulus of $(V-M_V)$=7, corresponding to 250 pc. The resulting CMD, shown in the left panel of Figure 2, contains ~148,000 stars and is mostly complete down to the oMSTO within 250 pc.

3. SFH calculation and results

The preliminary SFH has been calculated using the technique of synthetic CMD-fitting following the methodology presented in Bernard *et al.* (2012, 2015a, 2015b). The synthetic CMD from which we extracted the simple stellar populations' CMDs is based on the BaSTI stellar evolution library (Pietrinferni *et al.* 2004). It contains 2×10^7 stars and was generated with a constant SFR over wide ranges of age and metallicity: 0 to 15 Gyr old and $0.0001 \leqslant Z \leqslant 0.03$ (i.e. $-2.3 \leqslant$ [Fe/H] $\leqslant 0.26$, assuming $Z_{\odot} = 0.0198$;

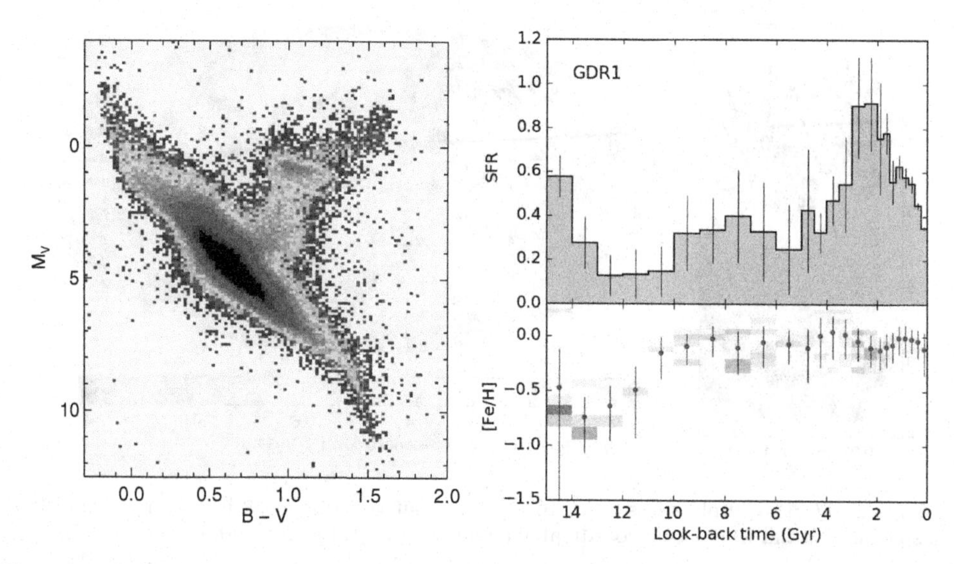

Figure 2. Left: CMD for the solar neighbourhood within 250 pc, which is ∼94% complete down to M_V=4.5. Right: Resulting SFH, showing the evolution of the SFR (top) and metallicity (bottom) as a function of time.

Grevesse & Noels 1993). We adopted a Kroupa (2002) initial mass function, and assumed a fraction of unresolved binary systems in TGAS of 10% with mass ratios between 0 and 1. Further tests with different fractions of binaries, a wider range of metallicity, and different prescriptions for the simulated photometric and parallactic uncertainties are necessary to better understand the possible systematic uncertainties.

While the full photometric uncertainties due to various observational effects are typically estimated using artificial stars tests on the original images (e.g. Gallart *et al.* 1999), this approach would be intractable for large photometric surveys, not to mention impossible in the case of *Gaia* for which (most of) the images are not sent back to Earth. Instead, we relied on the distributions of photometric errors as a function of colour and magnitude provided in the *Tycho-2*, *Hipparcos*, and APASS catalogues to simulate the uncertainties in the synthetic CMD.

The resulting SFH is presented in the right panel of Figure 2: the top and bottom plots show the evolution of the SFR and metallicity, respectively, as a function of time. Note that since the SFH was reconstructed based on the stars that are located *today* within the solar neighbourhood, the possible effects of secular evolution in discs (e.g. radial migrations, disc heating; e.g. Sellwood & Binney 2002, van der Kruit & Freeman 2011) have to be taken into account when interpreting the plots shown in Figure 2.

The SFH shows a roughly constant SFR for the first 10 Gyr or so, with a slight enhancement in the past 4 Gyr. This is in excellent agreement with the SFH calculated from *Hipparcos* data within a smaller volume (∼80 pc; Vergely *et al.* 2002, Cignoni *et al.* 2006), and with the known excess of stars younger than ∼4 Gyr old in the solar neighbourhood (e.g. Edvardsson *et al.* 1993, Anders *et al.* 2017). The age-metallicity relation (AMR), shown in the bottom-right panel of Figure 2, is mostly flat for the past 10 Gyr. Only the oldest stars show a lower mean metallicity ([Fe/H]∼ −0.7), which may correspond to the thick disc population. This is fully consistent with the independent results from other groups using different methods (e.g. Casagrande *et al.* 2011, Haywood *et al.* 2013, Bergemann *et al.* 2014).

4. Conclusions and future prospects

Taking advantage of the *Gaia* DR1 providing parallaxes for over 2 million stars in the solar neighbourhood, we produced a deep CMD that is mostly complete down to the magnitude of the oMSTO within 250 pc from the Sun. We applied the CMD-fitting technique to reconstruct the SFH of the Milky Way local thin disc. The preliminary results are fully consistent with those obtained previously using the *Hipparcos* data, despite the difficulty of dealing with the complex completeness function and photometric uncertainties from different catalogues. This is very promising for the prospects of using the same technique to upcoming *Gaia* data releases. With parallaxes and homogeneous photometry in 3 bands (G, BP, RP) for $>10^9$ stars, and not limited by the poorly understood completeness function of an input catalogue like *Tycho-2* was, it will allow us to extend this analysis out to about 5 kpc, and therefore to quantify the SFH of the thin disc, thick disc and halo *in situ* – rather than extrapolating based on the stars from these components that are today in the solar neighbourhood – and its spatial variations.

Acknowledgements

EJB acknowledges support from the CNES postdoctoral fellowship program. This work has made use of data from the European Space Agency (ESA) mission *Gaia* (https://www.cosmos.esa.int/gaia), processed by the *Gaia* Data Processing and Analysis Consortium (DPAC, https://www.cosmos.esa.int/web/gaia/dpac/consortium). Funding for the DPAC has been provided by national institutions, in particular the institutions participating in the *Gaia* Multilateral Agreement.

References

Anders, F., Chiappini, C., Rodrigues, T. S., *et al.* 2017, *A&A*, 597, A30
Bergemann, M., Ruchti, G. R., Serenelli, A., *et al.* 2014, *A&A*, 565, A89
Bernard, E. J., Ferguson, A. M. N., Barker, M. K., *et al.* 2012, *MNRAS*, 420, 2625
Bernard, E. J., Ferguson, A. M. N., Chapman, S. C., *et al.* 2015b, *MNRAS*, 453, L113
Bernard, E. J., Ferguson, A. M. N., Richardson, J. C., *et al.* 2015a, *MNRAS*, 446, 2789
Casagrande, L., Schönrich, R., Asplund, M., *et al.* 2011, *A&A*, 530, A138
Cignoni, M., Degl'Innocenti, S., Prada Moroni, P. G., & Shore, S. N. 2006, *A&A*, 459, 783
Edvardsson, B., Andersen, J., Gustafsson, B., *et al.* 1993, *A&A*, 275, 101
Gaia Collaboration, Brown, A. G. A., Vallenari, A., *et al.* 2016, *A&A*, 595, A2
Gallart, C., Freedman, W. L., Aparicio, A., Bertelli, G., & Chiosi, C. 1999, *AJ*, 118, 2245
Grevesse N., Noels A., 1993, in Prantzos N., Vangioni-Flam E., Cassé M., eds, Origin and Evolution of the Elements. Cambridge Univ. Press, Cambridge, p. 15
Haywood, M., Di Matteo, P., Lehnert, M. D., Katz, D., & Gómez, A. 2013, *A&A*, 560, A109
Henden, A. A., Levine, S. E., Terrell, D., Smith, T. C., & Welch, D. 2012, Journal of the American Association of Variable Star Observers (JAAVSO), 40, 430
Høg, E., Fabricius, C., Makarov, V. V., *et al.* 2000, *A&A*, 355, L27
Kroupa P., 2002, *Sci*, 295, 82
Lindegren, L., Lammers, U., Bastian, U., *et al.* 2016, *A&A*, 595, A4
Pietrinferni A., Cassisi S., Salaris M., Castelli F., 2004, *ApJ*, 612, 168
Sellwood, J. A. & Binney, J. J. 2002, *MNRAS*, 336, 785
van der Kruit, P. C. & Freeman, K. C. 2011, *ARA&A*, 49, 301
Vergely, J.-L., Köppen, J., Egret, D., & Bienaymé, O. 2002, *A&A*, 390, 917

Astronomy and Astrophysics in the Gaia sky
Proceedings IAU Symposium No. 330, 2017
A. Recio-Blanco, P. de Laverny, A.G.A. Brown & T. Prusti, eds.

doi:10.1017/S1743921317005932

Self-consistent Modelling of the Milky Way using Gaia data

David R. Cole[1] **and James Binney**[2]

[1]Rudolf Peierls Centre for Theoretical Physics, Keble Road, Oxford, OX1 3NP, United Kingdom
email: david.cole@physics.ox.ac.uk

[2]Rudolf Peierls Centre for Theoretical Physics, Keble Road, Oxford, OX1 3NP, United Kingdom
email: binney@physics.ox.ac.uk

Abstract. Angle/action based distribution function (DF) models can be optimised based on how well they reproduce observations thus revealing the current matter distribution in the Milky Way. Gaia data combined with data from other surveys, e.g. the RAVE/TGAS sample, and its full selection function will greatly improve their accuracy.

Keywords. Galaxy: disk, Galaxy: fundamental parameters, Galaxy: halo, solar neighborhood, dark matter

---

1. Introduction

Our knowledge of how baryons were accreted by galaxies such as the Milky Way is limited however but unprecedented amounts of data are becoming available from large scale surveys. It is vital that we use these data to improve our understanding of the local group and an excellent starting point is to find the current distribution of matter in the Galaxy. We cannot observe directly the distribution of dark matter so we must use the observable components as tracers to build dynamical models from which we can discover the distribution of dark matter. If we assume the Galaxy is in statistical equilibrium, we can exploit Jeans theorem (Jeans 1916) and presume that the distribution function (DF) $f(x, v)$ is a function of integrals of motion $I(x, v)$ only. There is an infinite choice of integrals because any function of the integrals is also an integral but the best choices are the actions J_i, which can be uniquely complemented by canonically conjugate variables to make a complete set (θ,J) for phase-space coordinates.

2. Modelling Process

The Milky Way can be built up from its components; the dark halo, thin and thick discs, stellar halo, bulge and gas disc. Each of these components is modelled either by a distribution function, $f(\mathbf{J})$ (first three components above), or a fixed potential. By making a sensible estimate for the parameters of these components we can make an estimate for the total potential. We then use the Stäckel Fudge (Binney 2012,2014) to find the actions and by integrating over velocity:

$$\rho(x) = \int d^3v f(\mathbf{J}(\mathbf{x}, \mathbf{v})) \tag{2.1}$$

we estimate the density. We then solve Poisson's equation to find the new potential and iterate this procedure until after a few iterations the total potential converges.

2.1. *Disc DF*

The DF of the discs in our models is superposition of the "quasi-isothermal" components introduced by Binney & McMillan (2011). It has the form

$$f(J_r, J_z, J_\phi) = f_{\sigma_r}(J_r, J_\phi) f_{\sigma_z}(J_z, J_\phi), \tag{2.2}$$

where f_{σ_r} and f_{σ_z} are

$$f_{\sigma_r}(J_r, J_\phi) \equiv \frac{\Omega\Sigma}{\pi\sigma_r^2\kappa}[1+\tanh(J_\phi/L_0)]\mathrm{e}^{-\kappa J_r/\sigma_r^2}, f_{\sigma_z}(J_z, J_\phi) \equiv \frac{\nu}{2\pi\sigma_z^2}\mathrm{e}^{-\nu J_z/\sigma_z^2}. \tag{2.3}$$

The thick disc is represented by a single quasi-isothermal DF, while the thin disc's DF is built up with a quasi-isothermal for each coeval cohort of stars. The velocity-dispersion parameters depend on J_ϕ and the age of the cohort. The star-formation rate in the thin disc decreases exponentially with time, with characteristic time scale $t_0 = 8\,\mathrm{Gyr}$. In addition a parameter F_{thk} controls the fraction of mass contributed by the thick disc.

2.2. *Dark halo DF*

Our dark halo DF is based on the form introduced by Posti *et al.* (2015) which in isolation self-consistently generates a density distribution which has a NFW (Navarro, Frenk & White 1997) profile (equation 2.4). This DF is a function of $h(\mathbf{J})$ which is an almost linear function (a homogeneous function of degree unity) of the actions J_i. The haloes generated are isotropic centrally and mildly radial when $r > r_s$. The anisotropy can be changed by varying the linear function of the J_i. These haloes closely resemble the haloes formed in dark-matter-only simulations. Specifically

$$f_{\mathrm{P}}(\mathbf{J}) = \frac{N}{J_0^3}\frac{(1+J_0/h)^{5/3}}{(1+h/J_0)^{2.9}} \tag{2.4}$$

The scale action J_0 encodes the scale radius around which the slope of the radial density profile shifts from -1 at small radii to -3 far out. From equation (2.4) it follows that $f_P(\mathbf{J}) \sim |\mathbf{J}|^{-5/3}$ as $|\mathbf{J}| \to 0$.

2.3. *Observational constraints*

We use several sets of observations to constrain the parameters of the DFs at various stages of the modelling. The observations include the astrometry of H_2O and SiO masing stars (Reid *et al.* 2014), the distribution of radio-frequency lines of HI and CO emission in the longitude-velocity plane, the stellar parameters and distance estimates in the fourth RAVE data release (Kordopatis *et al.* 2013) (see Section 4) and finally the vertical density profile from SDSS. We assume that the population from which the RAVE sample is drawn is identical to that studied by Jurić *et al.* (2008).

3. Results

Piffl *et al.* (2014) used RAVE data and SDSS Juric (2008) data to constrain the mass of DM within solar radius, R_0. The dark halo was included as a potential not a DF. Binney & Piffl (2015) used the DF in equation 2.4 for the dark halo in their self-consistent model of the Milky Way. In this model the halo was in self-consistent equilibrium with the other components so it had been adiabatically compressed by the baryons from its original NFW form. Hence it mimicked a scenario where baryons accumulated quiescently in the Galaxy's dark halo. An NFW halo becomes more centrally concentrated, and with so much dark matter at low radii (Fig. 1) the matching disc contains too few stars to satisfy the microlensing data (Popowski *et al.* 2005). This implies that the infinite phase-space

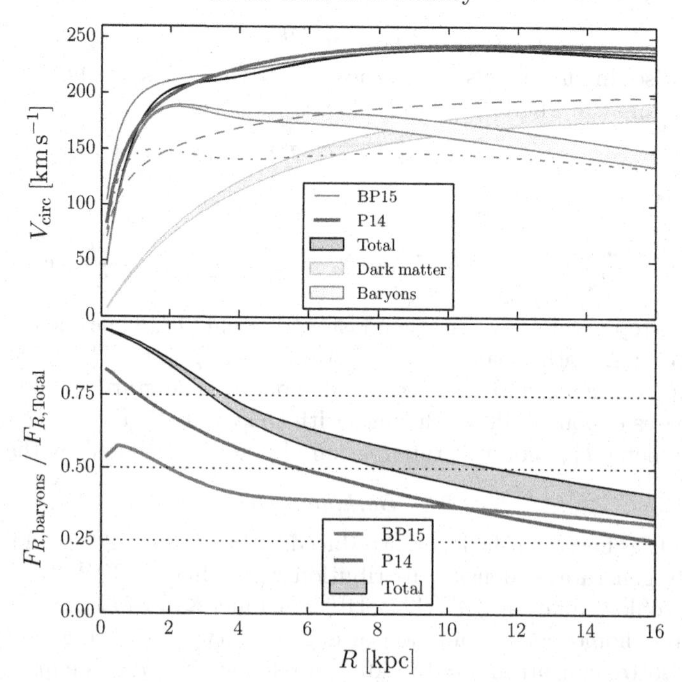

Figure 1. Upper panel: $v_c(R)$ for one of the Galaxy models of Cole & Binney (2017) (dark grey line) compared to the models of Binney & Piffl (2015) (BP15 black line) and Piffl *et al.* 2014 (P14 light grey line). The range of dark halo and baryonic contributions for Cole & Binney (2017) are shown as filled areas (baryonic in foreground). The dashed line is the dark halo contribution to the rotation curve of BP15 and the dash-dotted line is its baryonic component. Lower panel: the ratio of radial forces from the baryonic component to the total mass distribution for the same models shown above with Cole & Binney (2017) shown as a filled area.

density of particles at $J = 0$ characteristic of an NFW DF does not survive the accretion of baryons. The baryons cannot have accumulated entirely adiabatically but the most tightly bound dark matter particles were upscattered. This scattering of DM particles by baryons reduces their phase-space density and had greatest impact near $J = 0$.

In order to model this scattering process Cole & Binney (2017) modified the NFW DF by setting $f(\mathbf{J}) = g(h) f_{NFW}(h)$ with $g \sim h^{5/3}$ for small h and $g \sim 1$ at large h. This shifts particles from very low J to higher J. The functional form for g is

$$g(h) = \left[\frac{h_0^2}{h^2} - \beta\frac{h_0}{h} + 1\right]^{-5/6}, \tag{3.1}$$

where h_0 is an arbitrary constant with the dimensions of action that sets the scale of the almost constant-density core of the DF and β is determined by requiring the total mass of dark matter is conserved. The free parameters in $f_{\rm DM}$ are J_0, which sets the NFW scale radius, h_0, which sets the size of the dark halo's core, and the normalisation N.

Cole & Binney (2017) found (see Fig. 1) that with this modified dark halo DF their models of the Milky Way have a similarly good fit to the observations but the central regions are now dominated by baryons. The dark matter fraction is reduced and consistent with the results of surveys of microlensing events. In summary they found the local DM density $\rho_{DM} \gtrsim 0.012\ M_\odot\ \mathrm{pc}^{-3}$, stellar disc scale radius $R_d \sim 2.9$ kpc and stellar disc mass $\mathrm{M}_d \gtrsim 4 \times 10^{10}\ M_\odot$.

4. Using RAVE/TGAS

Our current modelling uses RAVE data in 8 spatial bins at $R_0 \pm 1$ kpc and at 0, 0.3, 0.6, 1.0, 1.5 kpc in $|z|$ to compute velocity distributions predicted by the DF at the mean positions. In order to take advantage of Gaia DR1 (Gaia Collaboration 2016) we are developing a method using the RAVE/TGAS observations (Kunder *et al.* 2017). In order to do this we will develop a full selection function $S(s)$ for this sample. Deriving an a priori calculation of $S(s)$ needs a full chemodynamical model of the MW disc $S(s, \tau, [Fe/H])$ such as used in Schönrich & Bergemann 2014 and in addition we need to model the exact distribution of stars in age and metallicity in the solar neighbourhood. Schönrich and Aumer 2017 did this using the RAVE selection function of Wojno *et al.* (2017) and found that at fixed metallicity $S(s, \tau)$ falls off exponentially with scale 0.12 kpc at $s > 0.2$ kpc.

The selection function for TGAS is biased towards younger stars which are more likely to be seen so the kinematics will appear cooler than they really are. Our models need to take this into account. They already have age but not metallicity and so we can add metallicity by use of a suitable metallicity DF. Then we can compute the likelihoods of our model based on the resulting selection function.

5. Action based modelling software library

Our modelling is currently being rewritten using AGAMA (Action-based Galaxy Modelling Architecture) which is a library of low-level programs containing interfaces and generic routines required to create the functions described here. The main sets of functions include gravitational potential and density interfaces, action/angle interface, interface for creating gravitationally self-consistent multicomponent galaxy models etc. The code can be downloaded from https://github.com/GalacticDynamics-Oxford/Agama

References

Binney, J. 2012, *MNRAS*, 426, 1324
Binney, J. 2014, *MNRAS*, 440, 787
Binney J. & McMillan P. 2011, *MNRAS*, 413, 1889
Binney J. & Piffl T. 2015, *MNRAS*, 454, 3653
Cole D. & Binney J. 2017, *MNRAS*, 465, 798
Gaia Collaboration, Brown A. G. A., Vallenari A., Prusti T., de Bruijne J., Mignard F., Drimmel R. & co-authors 2016 *A&A*, 595, A2
Jeans J. 1916, *MNRAS*, 76, 552
Kordopatis G., Gilmore G., Steinmetz M., Boeche C., Seabroke G. M., Siebert A., ZwitterT., Binney J. & co-authors 2013, *AJ*, 146, 134
Navarro J. F., Frenk C. S. & White S. D. M. 1997, *ApJ*, 490, 493
Piffl T., Binney J., McMillan P. J., Steinmetz M., Helmi A., Wyse R. F. G., Bienaymé O., Bland-Hawthorn J., Freeman K. & co-authors 2014, *MNRAS*, 445, 3133
Piffl T., Penoyre Z. & Binney J. 2015, *MNRAS*, 451, 639
Schönrich R. & Bergemann M. 2014, *MNRAS*, 443, 698
Popowski P., Griest K., Thomas C. L., Cook K. H., Bennett D. P., Becker A. C., Alves D. R., Minniti D., Drake A. J. & co-authors 2005, *ApJ*, 631, 879
Posti L., Binney J., Nipoti C. & Ciotti L. 2015, *MNRAS*, 447, 3060
Reid M. J., Menten, Brunthaler, Zheng, Dame, Xu, Wu, Zhang, Sanna & co-authors 2014, *ApJ*, 783, 130
Schönrich R. & Bergemann M. 2014, *MNRAS*, 443, 698
Schönrich R. & Aumer M. 2017, *ArXiv e-prints* 1704.01333
Wojno, J., Kordopatis, G., Piffl, T., Binney, J., Steinmetz, M., Matijevič, G., Bland-Hawthorn, J., Sharma & co-autors 2017, *MNRAS*, 468, 3368

Astronomy and Astrophysics in the Gaia sky
Proceedings IAU Symposium No. 330, 2017
A. Recio-Blanco, P. de Laverny, A.G.A. Brown & T. Prusti, eds.

doi:10.1017/S1743921317006081

Abundance ratios & ages of stellar populations in HARPS-GTO sample

E. Delgado Mena[1], M. Tsantaki[2], V. Zh. Adibekyan[1], S. G. Sousa[(1,3)], N. C. Santos[(1,3)], J. I. González Hernández[(4,5)] and G. Israelian[(4,5)]

[1]Instituto de Astrofísica e Ciências do Espaço,
Universidade do Porto, CAUP, Rua das Estrelas, PT4150-762 Porto, Portugal.
email: elisa.delgado@astro.up.pt

[2]Instituto de Radioastronomía y Astrofísica, IRyA, UNAM,
Campus Morelia, A.P. 3-72, C.P. 58089, Michoacán, Mexico

[3]Departamento de Física e Astronomía, Faculdade de Ciências, U. Porto, Portugal

[4]Instituto de Astrofísica de Canarias, C/ Via Lactea, s/n, 38205, La Laguna, Tenerife, Spain

[5]Departamento de Astrofísica, Universidad de La Laguna, 38206 La Laguna, Tenerife, Spain

Abstract. In this work we present chemical abundances of heavy elements (Z>28) for a homogeneous sample of 1059 stars from HARPS planet search program. We also derive ages using parallaxes from Hipparcos and Gaia DR1 to compare the results. We study the [X/Fe] ratios for different populations and compare them with models of Galactic chemical evolution. We find that thick disk stars are chemically disjunt for Zn adn Eu. Moreover, the high-alpha metal-rich population presents an interesting behaviour, with clear overabundances of Cu and Zn and lower abundances of Y and Ba with respect to thin disk stars. Several abundance ratios present a significant correlation with age for chemically separated thin disk stars (regardless of their metallicity) but thick disk stars do not present that behaviour. Moreover, at supersolar metallicities the trends with age tend to be weaker for several elements.

Keywords. stars: abundances - stars: fundamental parameters - Galaxy: evolution - Galaxy: disk - solar neighborhood

---

1. Introduction

In the era of large spectroscopic surveys such as APOGEE, Gaia-ESO Survey or RAVE, among others, the contribution of smaller samples with high-resolution and high quality spectra is of great importance to understand the Galactic Chemical Evolution (GCE). In this work we have derived abundances for Cu, Zn, Sr, Y, Zr, Ba, Ce, Nd and Eu (Delgado Mena *et al.* 2017) for 1111 stars within the volume-limited HARPS-GTO planet search sample in order to complement our previous works for light elements (Delgado Mena *et al.* 2014, Delgado Mena *et al.* 2015, Suárez Andrés *et al.* 2016, Bertrán de Lis *et al.* 2015), α- elements and Fe-peak elements (Adibekyan *et al.* 2012). The main purpose of this work is to evaluate the GCE evolution of those heavier elements and the dependence on stellar ages of different abundance ratios.

2. Stellar ages

We derive the masses, radii and ages with the PARAM v1.3 tool† using the PARSEC isochrones (Bressan *et al.* 2012) with our values for Teff and [Fe/H], the V magnitudes from the main Hipparcos catalogue (Perryman *et al.* 1997)) and the parallaxes from

† http://stev.oapd.inaf.it/cgi-bin/param

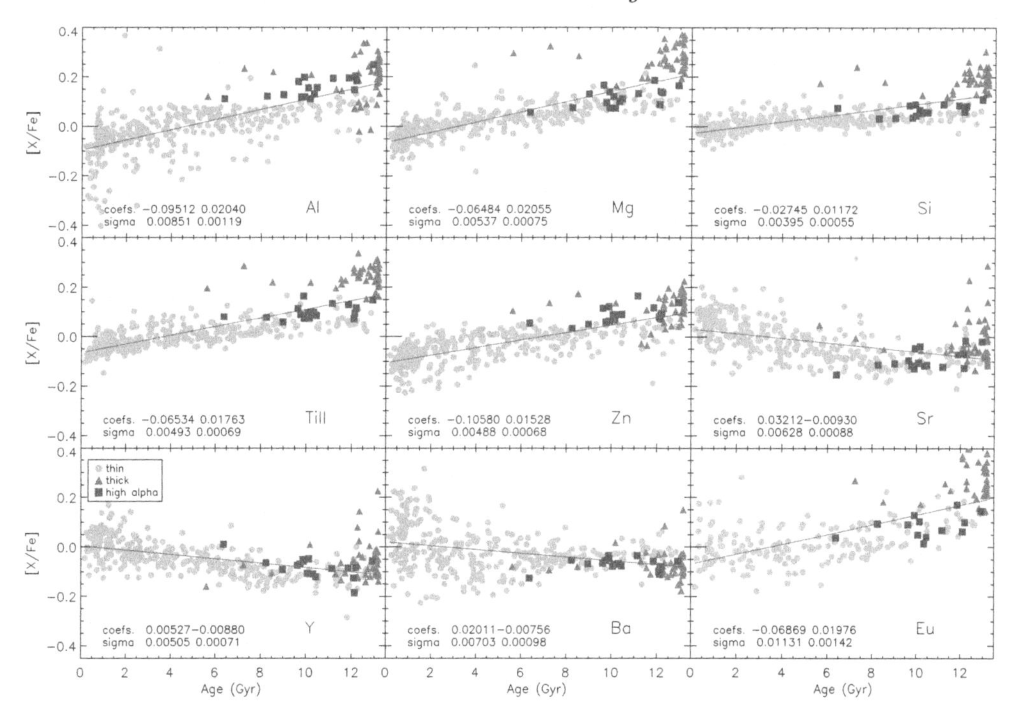

Figure 1. General [X/Fe] ratios as a function of age for the reduced sample with reliable ages. Thin disk stars, thick disk stars and *hαmr* are depicted with dots. triangles and squares, respectively. The linear fit to all the stars just serves as eye guiding.

the Hipparcos new reduction (van Leeuwen 2007) or from the first release (DR1) of Gaia (Lindegren *et al.* 2016). We note that we added a systematic error of 0.3 mas to the formal error of the Gaia DR1 parallaxes as recommended by the Gaia collaboration. Meanwhile Hipparcos provides parallaxes for 1051 out of the 1059 stars within our sample, only 923 stars have parallaxes in GAIA DR1. Moreover, there are significant differences in many cases, leading to non-negligible differences in age. In order to have a sample with ages as reliable as possible we decided to select the Hipparcos ages with a difference less than 1 Gyr with respect to the ages derived with GAIA parallaxes and with an error in age lower than 2 Gyr. This final sample is composed by 377 stars belonging to the thin disk, thick disk and high-α metal-rich stars (hereafter *hαmr*, a population with high α abundances at [Fe/H] > -0.2 dex discovered by Adibekyan *et al.* 2011).

3. Abundance ratios vs age

In Fig. 1 we can see how several elements depend on age. By combining elements that increase and decrease with age, respectively, it is possible to have steeper and more constrained trends. For example, [Mg/Fe] shows a tight increasing trend with age, meanwhile [Eu/Fe], [Zn/Fe] and [Al/Fe] also show this dependency though with more dispersion. This trend is expected since these elements are mainly formed by massive stars which started to contribute to the Galaxy chemical enrichment earlier than the lower mass stars responsible for Fe production. On the other hand, the light-*s* process elements Y and Sr show the most clear decreasing trends with age. These elements are formed by low-mass AGB stars so we can expect them to increase with time (for younger stars) due to the increasing and delayed contribution of low-mass stars as the Galaxy evolves. In Figs. 2 and 3 we show different combinations of previously mentioned elements at different

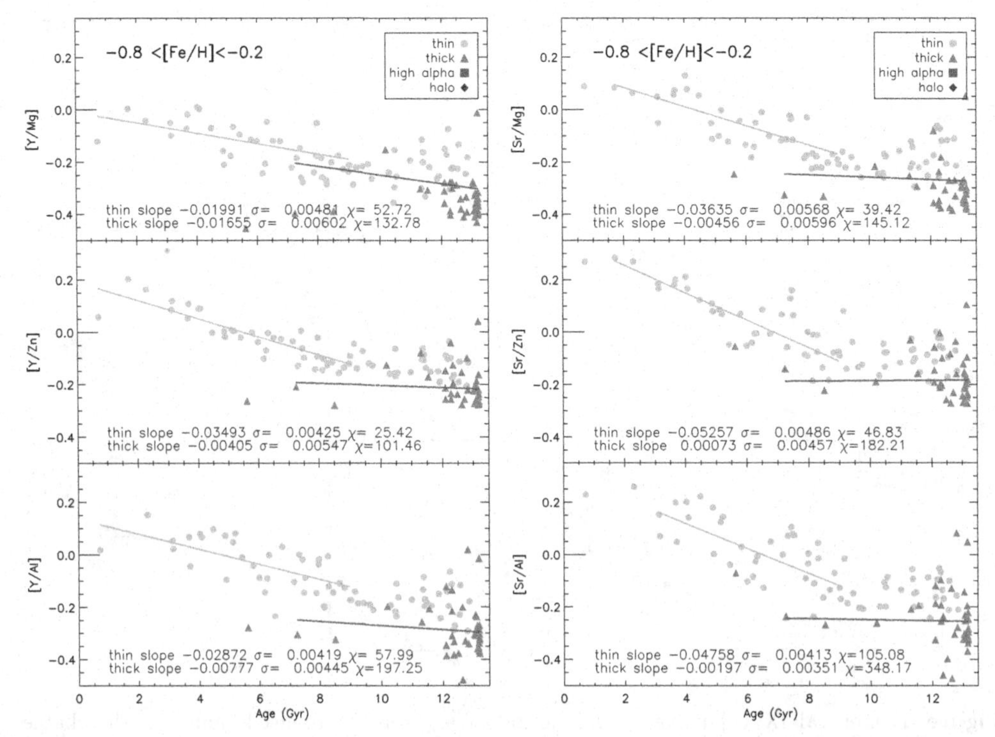

Figure 2. [Y/Mg],[Y/Zn],[Y/Al] and [Sr/Mg],[Sr/Zn],[Sr/Al] for -0.8<[Fe/H]<-0.2. Symbols as in Fig 1.

metallicity regions. Previous works have explored and confirmed the tight correlation of these abundance ratios with age (e.g. da Silva *et al.* 2012), Nissen 2015, Spina *et al.* 2016) but only using solar twins or solar analogues. However, Feltzing *et al.* (2017) noted that [Y/Mg] clock is not valid at [Fe/H] < -0.5 dex. In our sample we find that still at [Fe/H] < -0.5 dex the different abundance ratios show a correlation with age (steeper for [Sr/X] than for [Y/X]) but this is only valid for thin disk stars. We note however that our sample of thick disk stars with reliable ages is quite small. It is also clear that the trends become flat at ages $\gtrsim 8$ Gyr. On the other hand, at higher metallicities, in the bin -0.2 < [Fe/H] < 0.2 dex, the abundance ratios of Y and Sr (with respect to Mg, Zn and Al) present similar slopes. Nevertheless, we remark that meanwhile [Sr/Fe] presents a constant correlation with age at different metallicities, [Y/Fe] becomes flatter as [Fe/H] increases. Moreover, we can observe that thin disk stars present mostly no dependence on age for ages $\gtrsim 8$ Gyr but *hαmr* stars show a continuous dependence in the full age range for [Y/X] ratios. The improvement of parallaxes from GAIA DR2 will help to determine more precise ages for our stars increasing the sample size and allowing us to better understand the behaviour of the abundance-age trends for different populations in the Galaxy.

Acknowledgements

E.D.M., V.Zh.A., N.C.S. and S.G.S. acknowledge the support from Fundação para a Ciência e a Tecnologia (FCT) through national funds and from FEDER through COMPETE2020 by the following grants UID/FIS/04434/2013 & POCI-01-0145-FEDER-007672, PTDC/FIS-AST/7073/2014 & POCI-01-0145-FEDER-016880 and PTDC/FIS-AST/1526/2014 & POCI-01-0145-FEDER-016886. E.D.M., V.Zh.A., N.C.S. and S.G.S.

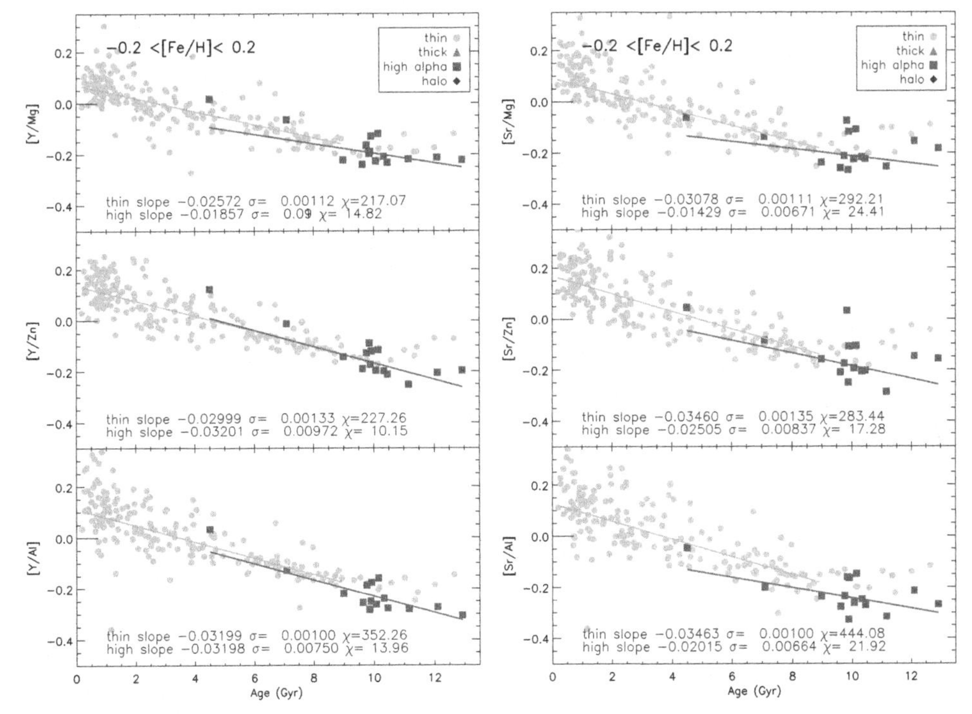

Figure 3. [Y/Mg],[Y/Zn],[Y/Al] and [Sr/Mg],[Sr/Zn],[Sr/Al] for -0.2<[Fe/H]<0.2. Symbols as in Fig 1.

also acknowledge the support from FCT through Investigador FCT contracts IF/00849/2015, IF/00650/2015, IF/00169/2012/CP0150/CT0002 and IF/00028/2014/CP1215/CT0002 funded by FCT (Portugal) and POPH/FSE (EC). This research has made use of the SIMBAD database operated at CDS, Strasbourg (France).

This work has made use of data from the European Space Agency (ESA) mission *Gaia* (https://www.cosmos.esa.int/gaia), processed by the *Gaia* Data Processing and Analysis Consortium (DPAC, https://www.cosmos.esa.int/web/gaia/dpac/consortium). Funding for the DPAC has been provided by national institutions, in particular the institutions participating in the *Gaia* Multilateral Agreement.

References

Adibekyan, V. Z., Santos, N. C., Sousa, S. G., & Israelian, G. 2011, *A&A*, 535, L11
Adibekyan, V. Z., Sousa, S. G., Santos, N. C., *et al.* 2012, *A&A*, 545, A32
Bertran de Lis, S., Delgado Mena, E., Adibekyan, V. Z., *et al.* 2015, *A&A*, 576, A89
Bressan, A., Marigo, P., Girardi, L., *et al.* 2012, *MNRAS*, 427, 127
da Silva, R., Porto de Mello, G. F., Milone, A. C., *et al.* 2012, *A&A*, 542, A84
Delgado Mena, E., Bertrán de Lis, S., Adibekyan, V. Z., *et al.* 2015, *A&A*, 576, A69
Delgado Mena, E., Israelian, G., González Hernández, J. I., *et al.* 2014, *A&A*, 562, A92
Delgado Mena, E., Tsantaki, M., Adibekyan, V. Z., *et al.* 2017, *arXiv:1705.04349*
Feltzing, S., Howes, L. M., & McMillan, P. J., Stonkutė, E. 2017, *MNRAS*, 465, L109
Lindegren, L., Lammers, U., Bastian, U., *et al.* 2016, *A&A*, 595, A4
Nissen, P. E. 2015, *A&A*, 579, A52
Perryman, M. A. C., Lindegren, L., Kovalevsky, J., *et al.* 1997, *A&A*, 323, L49
Spina, L., Meléndez, J., Karakas, A. I., *et al.* 2016, *A&A*, 593, A125
Suárez-Andrés, L., Israelian, G., González Hernández, J. I., *et al.* 2016, *A&A*, 591, A69
van Leeuwen, F. 2007, *A&A*, 474, 653

Astronomy and Astrophysics in the Gaia sky
Proceedings IAU Symposium No. 330, 2017
A. Recio-Blanco, P. de Laverny, A.G.A. Brown
& T. Prusti, eds.

doi:10.1017/S1743921317005725

The kinematics and surface mass density in the solar neighbourhood using TGASxRAVE

Jorrit H. J. Hagen and Amina Helmi

Kapteyn Astronomical Institute, University of Groningen,
Landleven 12, 9747 AD, Groningen, The Netherlands
email: hagen@astro.rug.nl

Abstract. We investigate the kinematics of stars in the Solar neighbourhood by combining radial velocities from the latest data release of the RAVE survey with the Tycho-Gaia Astrometric Solution presented in Gaia Data Release 1. We use moments of the velocity distribution to characterise the kinematics over a radial distance range of 6-10 kpc and up to 1 kpc away from the plane. Our ultimate goal is to use these to put new constraints on the (local) distribution of mass using the Jeans Equations.

Keywords. (Galaxy:) solar neighborhood, Galaxy: disk, Galaxy: kinematics and dynamics, (cosmology:) dark matter

---

1. Introduction

With the launch of the Gaia satellite in December 2013 a wealth of new data is becoming available on the motions and positions of stars in the Milky Way and its satellite galaxies (Gaia Collaboration *et al.* 2016a,b). For example, the Tycho Gaia Astrometric Solution (TGAS) set provides significantly improved proper motions and parallaxes of nearby stars making it possible to derive new kinematic maps of the Solar neighbourhood by combining this data with spectroscopic surveys.

The improved kinematic maps of the Solar neighbourhood can be used for example, to obtain new, more precise estimates of the local dark matter density (Read 2014; McKee *et al.* 2015), as well as to establish the presence of asymmetries associated to spiral arms, the Galactic bar or even bending waves in the Galactic disk (e.g. Williams *et al.* 2013).

2. Data

We derive new kinematic maps of the vicinity of the Sun by combining TGAS with the fifth Data Release of the Radial Velocity Experiment (RAVE DR5 Kunder *et al.* 2017). RAVE is a southern hemisphere magnitude-limited ($9 < I < 12$) spectroscopic survey, which contains radial velocities, astrophysical parameters such as surface gravities, temperatures and metallicities, as well as a spectro-photometric parallaxes for $\sim$450000 stars.

There are $\sim$210000 stars in common between RAVE and TGAS, which thus have full phase-space information. From this sample we select the best relative parallax error measurement for each star from either TGAS (provided that the TGAS parallax is positive) or the RAVE survey, the latter being chosen only when the signal to noise ratio exceeds 20 and when the ALGO_CONV flag is not equal to 1.

After selecting the best parallax measurement for each star, we only keep those stars that have radial velocity measurements with a maximum uncertainty of 8 km/s, a

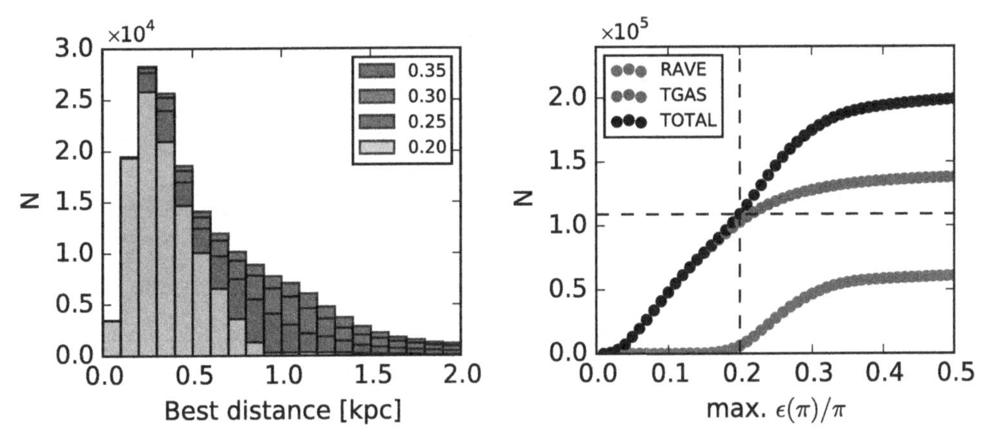

Figure 1. Left: Distance distribution of the stars with best distances determined from either RAVE or TGAS. We show the influence of varying the maximum relative parallax error on the final sample of stars. Right: The number of stars available as function of the maximum allowed relative parallax error. We also show the contribution of stars with TGAS and RAVE best parallaxes.

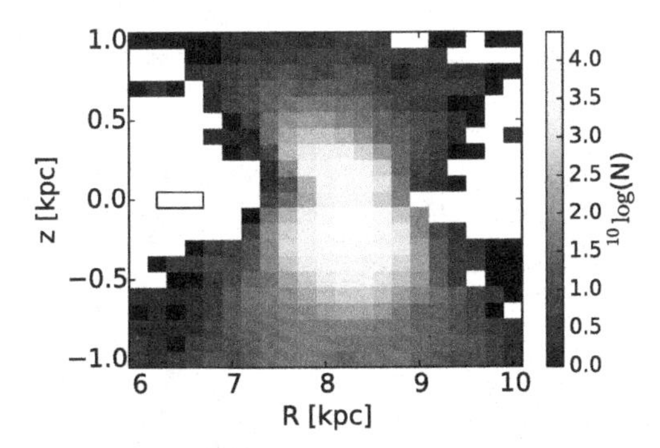

Figure 2. The spatial distribution of stars in our sample, where the colour coding indicates the number of stars found in each bin.

`CorrelationCoeff` > 10, and a maximum relative parallax error of 20%. Our final sample now contains 108916 stars and the majority of stars, 102738 stars, have best parallax measurements from TGAS.

In the left panel of Fig. 1 we show a histogram of the distances probed by varying the maximum allowed relative parallax error. Distances are obtained by taking the inverse of the parallax. In the right panel of Fig. 1 we show how varying the maximum allowed relative parallax error affects the number of stars in the sample, as well as the contribution from each of the surveys to the final sample. The spatial distribution of stars in our sample is shown in Fig. 2, where the colour coding indicates the number of stars in each bin.

When converting the observables to Galactocentric cylindrical coordinates and velocities, we assume $R_{\odot} = 8.3$ kpc (Schönrich 2012), $z_{\odot} = 0.014$ kpc (Binney *et al.* 1997), and $(U, V, W)_{\odot} = (11.1, 12.24, 7.25)$ km/s (Schönrich *et al.* 2010) for the peculiar velocity of the Sun. For the circular velocity at the Solar radius we adopt a value of 228.5 km/s.

3. Analysis and results

We sample the distribution of stars from $R = 6$ kpc to $R = 10$ kpc in bins of 0.25 kpc width and from $z = -1$ kpc to $z = 1$ kpc in bins of 0.05 kpc height. For

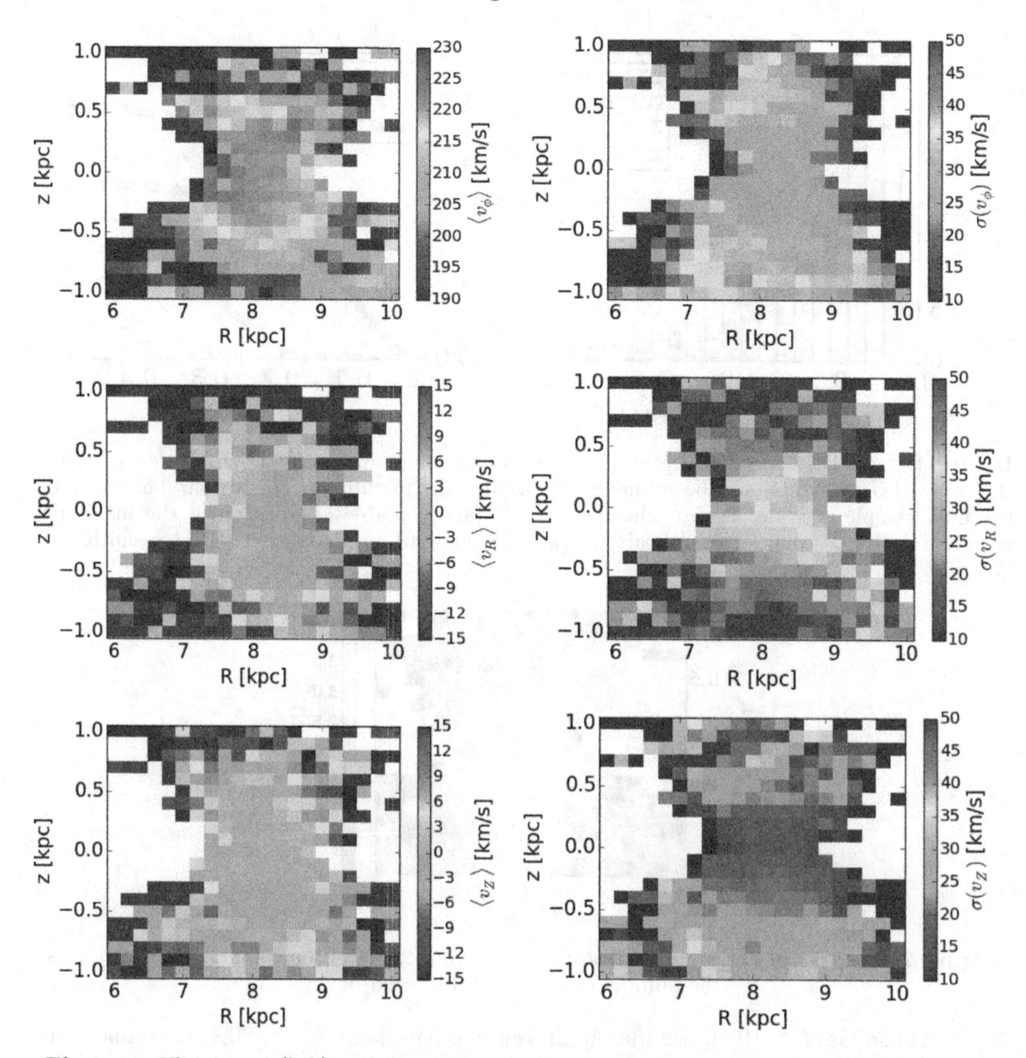

Figure 3. The means (left) and dispersions (right) of the cylindrical azimuthal, radial and vertical velocities in the meridional plane.

each bin we compute the first two velocity moments. The resulting kinematic maps are shown in Fig. 3. The asymmetric drift is clearly seen, as the mean rotational velocity decreases with increasing velocity dispersion. In these maps we do not see clear evidence for a wave-like behaviour in the vertical velocity component as reported in e.g. Williams *et al.* (2013).

In future work we plan to apply the Jeans equations to these data to compute the vertical gravitational force K_z. A key ingredient is the variation of the z-velocity dispersion in the vertical direction, which is shown in Fig. 4 for stars in our sample that are within 0.5 kpc in R from the Solar radius. In this case, we have folded the data with respect to the $z = 0$ plane. This figure shows that the velocity dispersion increases with distance from the Galactic plane as expected. The somewhat bumpy behaviour (e.g. at ~ 0.7 kpc) might be caused by the impact of distance errors at large galactic heights (McMillan, private communication).

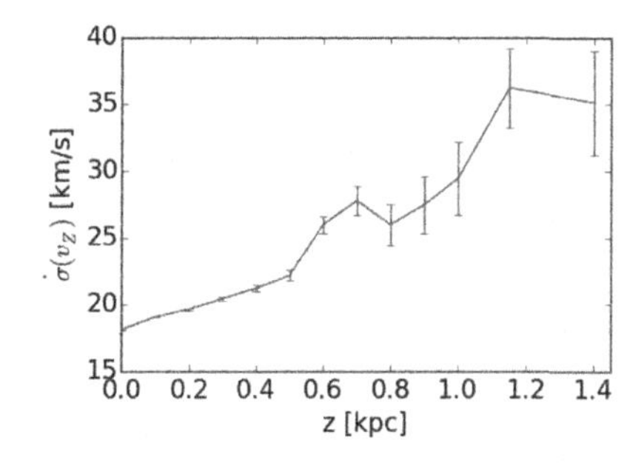

Figure 4. The vertical velocity dispersion of stars in our sample located within 0.5 kpc in R from the Solar circle.

It has been shown that to a good approximation $|K_z| \approx 2\pi G\Sigma(R,z)$, or that the relation at least gives a lower limit to the surface density $\Sigma(R,z)$ (Kuijken & Gilmore 1989; Bovy & Tremaine 2012). Comparing this quantity to the surface mass density that one would obtain by adding up the stellar contribution we can, if mass is missing, put new constraints on the local dark matter density.

Acknowledgements

We are grateful to Maarten Breddels and Jovan Veljanoski for many useful discussions and support when handling the datasets. This work has made use of data from the European Space Agency (ESA) mission *Gaia* (`https://www.cosmos.esa.int/gaia`), processed by the *Gaia* Data Processing and Analysis Consortium (DPAC, `https://www.cosmos.esa.int/web/gaia/dpac/consortium`). Funding for the DPAC has been provided by national institutions, in particular the institutions participating in the *Gaia* Multilateral Agreement.

References

Binney, J., Gerhard, O., & Spergel, D. 1997, *MNRAS*, 288, 365
Bovy, J. & Tremaine, S. 2012, *ApJ*, 756, 89
Gaia Collaboration, Brown, A. G. A., Vallenari, A., *et al.* 2016a, *A&A*, 595, A2
Gaia Collaboration, Prusti, T., de Bruijne, J. H. J., *et al.* 2016b, *A&A*, 595, A1
Kuijken, K. & Gilmore, G. 1989, *MNRAS*, 239, 571
Kunder, A., Kordopatis, G., Steinmetz, M., *et al.* 2017, *AJ*, 153, 75
McKee, C. F., Parravano, A., & Hollenbach, D. J. 2015, *ApJ*, 814, 13
Read, J. I. 2014, *Journal of Physics G Nuclear Physics*, 41, 063101
Schönrich, R. 2012, *MNRAS*, 427, 274
Schönrich, R., Binney, J., & Dehnen, W. 2010, *MNRAS*, 403, 1829
Williams, M. E. K., Steinmetz, M., Binney, J., *et al.* 2013, *MNRAS*, 436, 101

Astronomy and Astrophysics in the Gaia sky
Proceedings IAU Symposium No. 330, 2017
A. Recio-Blanco, P. de Laverny, A.G.A. Brown & T. Prusti, eds.

doi:10.1017/S174392131700610X

Dynamical effects of the spiral arms on the velocity distribution of disc stars

Kohei Hattori[1], Naoteru Gouda[2], Taihei Yano[2], Nobuyuki Sakai[2] and Hiromichi Tagawa[2]

[1]University of Michigan, 1085 S. University Ave, Ann Arbor, MI 48109, USA
email: khattori@umich.edu

[2]National Astronomical Observatory of Japan, 2-21-1, Osawa, Mitaka, Tokyo 181-8588, Japan

Abstract. Nearby disc stars in Gaia DR1 (TGAS) and RAVE DR5 show a bimodal velocity distribution in the metal-rich region (characterized by the Hercules stream) and mono-modal velocity distribution in the metal-poor region. We investigate the origin of this [Fe/H] dependence of the local velocity distribution by using 2D test particle simulations. We found that this [Fe/H] dependence can be well reproduced if we assume fast rotating bar models with $\Omega_{\rm bar} \simeq 52\,{\rm km\ s^{-1}\ kpc^{-1}}$. A possible explanation for this result is that the metal-rich, relatively young stars are more likely to be affected by bar's outer Lindblad resonance due to their relatively cold kinematics. We also found that slowly rotating bar models with $\Omega_{\rm bar} \simeq 39\,{\rm km\ s^{-1}\ kpc^{-1}}$ can not reproduce the observed data. Interestingly, when we additionally consider spiral arms, some models can reproduce the observed velocity distribution even when the bar is slowly rotating.

Keywords. Galaxy: disk, Galaxy: kinematics and dynamics, Galaxy: structure

---

1. Introduction: Does Hercules stream arise from slow-bar + spirals?

Since the discovery of the bimodal velocity distribution of the local disc stars, many authors have tried to explain the origin of the secondary peak, or the Hercules stream. The pioneering work by Dehnen (2000) demonstrated that the under-dense region between the Local Standard of Rest mode (LSR mode; first peak) and the Hercules stream (secondary peak) can arise from the bar's outer Lindblad resonance (OLR). Based on this idea, Dehnen (1999) estimated the bar's pattern speed to be $\Omega_{\rm b} = (53 \pm 3)\,{\rm km\ s^{-1}\ kpc^{-1}}$. Later studies refined this estimation by using other samples of nearby stars and obtained similar values of $\Omega_{\rm b}$ (Minchev *et al.* 2007; Antoja *et al.* 2014; Monari *et al.* 2017).

Unfortunately, this *fast bar* model is inconsistent with recent claims of long Galactic bar (half-length of 5 kpc; Wegg *et al.* 2015). Portail *et al.* (2017) argued that slowly rotating bar with $\Omega_{\rm b} = (39 \pm 3.5)\,{\rm km\ s^{-1}\ kpc^{-1}}$ is required to sustain the long bar.

This $\sim 30\%$ difference in $\Omega_{\rm b}$ between the *fast bar* and *slow bar* models is more serious than it sounds. For example, if $\Omega_{\rm b} \simeq 39\,{\rm km\ s^{-1}\ kpc^{-1}}$ (slow bar), OLR is unimportant in the Solar neighborhood, so the bimodal structure cannot be reproduced. Here we consider bar + spiral models to investigate the origin of the Hercules stream.

2. Data: TGAS+RAVE

Here we explain the observed data with which our models are compared.

First, we cross match the sample stars in Tycho Gaia Astrometric Solutions (TGAS) from the Gaia DR1 (Lindegren *et al.* 2016) and RAVE DR5 (Kunder *et al.* 2017). Then we use the 5D astrometric data from TGAS and the line-of-sight velocity and [Fe/H] from RAVE to derive the velocity distribution of stars within 200 pc from the Sun. Our sample is defined by the following criteria: (1) positive parallax ($\varpi > 0$); (2) distance cut

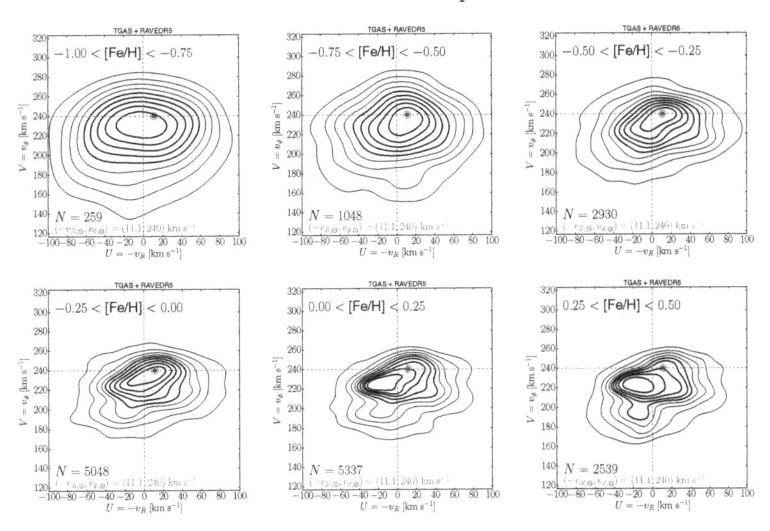

Figure 1. Distribution of nearby stars in the (U, V)-space at various [Fe/H] regions revealed by Gaia (TGAS) and RAVE data. Here we assume Solar motion of $(U_\odot, V_\odot) = (11.1, 240)\,\mathrm{km\ s^{-1}}$.

$(1/\varpi < 200\,\mathrm{pc})$; (3) small fractional error in parallax $(\delta\varpi/\varpi < 0.2)$; (4) small line-of-sight velocity error $(0 < \delta v_{\mathrm{los}}/(\mathrm{km\ s^{-1}}) < 5)$; (5) metallicity cut $(-1 \leqslant [\mathrm{Fe/H}] \leqslant 0.5)$; and (6) large S/N ratio in RAVE spectra (SNR> 40).

Figure 1 shows the [Fe/H] dependence of the velocity distribution. In the most metal-poor regions $(-1 \leqslant [\mathrm{Fe/H}] \leqslant -0.75$ and $-0.75 \leqslant [\mathrm{Fe/H}] \leqslant -0.5)$, we see a mono-modal distribution, which is associated with a large velocity dispersion (hot kinematics). As the [Fe/H] increases, the velocity dispersion becomes smaller (colder kinematics). At the most metal-rich regions $(0 \leqslant [\mathrm{Fe/H}] \leqslant 0.25$ and $0.25 \leqslant [\mathrm{Fe/H}] \leqslant 0.5)$, we see a clear secondary peak at $(U, V) = (-10, 195)\,\mathrm{km\ s^{-1}}$ (see also Pérez-Villegas *et al.* 2017).

3. Model

Here we describe our 2D test particle simulations. We assume that the stellar disc is formed at $t = -12\,\mathrm{Gyr}$ and that the star formation rate is constant as a function of time, t. Initially, the stellar disc is axisymmetric, and the non-axisymmetric components are adiabatically introduced at $t = T_{\mathrm{form}}$. At $t < T_{\mathrm{form}}$, we assume that the stellar disc is evolved due to internal heating. At $T_{\mathrm{form}} < t$, we follow the stellar orbits. Also, we use a simple model to assign the metallicity of stars based on the initial radius and the age.

The Galactic potential is modeled by the following components:

Axisymmetric component. For this component, we assume a singular isothermal potential of $\Phi_0(R) = v_0^2 \ln(R/R_0)$ with $(R_0, v_0) = (8\,\mathrm{kpc}, 220\,\mathrm{km\ s^{-1}})$.

Bar component. We assume a quadratic bar model introduced by Dehnen (2000), with the bar strength of $\alpha = 0.015$ (for bar-only models) or $\alpha = 0.01$ (for bar+spiral models). The current bar's angle with respect to the Sun-Galactic center line is set to be 25°.

Spiral component. We assume rigidly-rotating, two- or four-armed $(m = 2, 4)$ logarithmic spirals. The maximum amplitude of the spiral potential at $R = R_0$ is $(20\,\mathrm{km\ s^{-1}})^2$ and $(15\,\mathrm{km\ s^{-1}})^2$ for $m = 2$ and 4, respectively. The amplitude is varied as a function of R, in a manner that is motivated by Steiman-Cameron *et al.* (2010). We consider steady spirals without t-dependence of the amplitude, as well as transient spirals where the amplitude oscillates with a period of 200 Myr. In all the models, one of the spirals at $R = R_0$ is currently located at 45° ahead of the Sun.

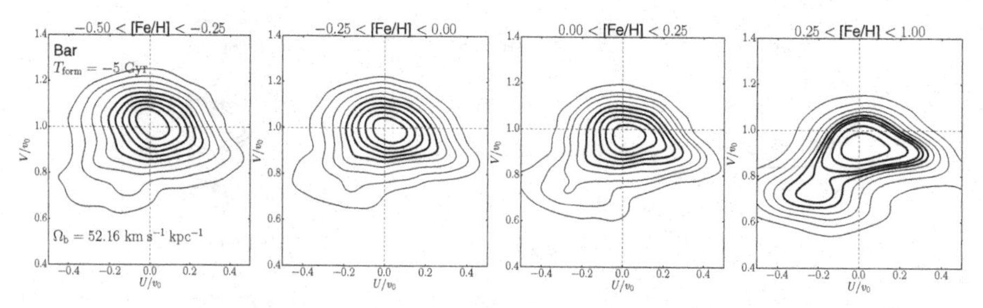

Figure 2. [Fe/H] dependence of the (U, V)-diagram in the fast-bar model ($T_{\rm form} = -5$ Gyr).

4. Results

Fast-bar model (no spirals). In the fast-bar model ($\Omega_{\rm b} = 52.16\,{\rm km\ s^{-1}\ kpc^{-1}}$), a prominent bimodal structure is seen at the metal-rich region but is not seen in metal-poor region (Fig 2), almost independent of $T_{\rm form}$. The value of V that separates the LSR and Hercules-like modes corresponds to the bar's OLR. These results are consistent with the Gaia+RAVE data (Fig 1). A possible explanation for these results is as follows: Disc stars with high (low) [Fe/H] tend to be young (old) and have cold (hot) kinematics, so they are more (less) likely to be affected by OLR. However, further investigation is required for a better understanding of these results.

Slow-bar model (no spirals). In the slow-bar model ($\Omega_{\rm b} = 39.12\,{\rm km\ s^{-1}\ kpc^{-1}}$), in contrast, we do not see bimodal structure in any [Fe/H] range.

Slow-bar + steady spirals. When $m = 4$ and $\Omega_{\rm s} = 20\,{\rm km\ s^{-1}\ kpc^{-1}}$, the [Fe/H] dependence of the (U, V)-diagram is similar to the observed one for certain values of $T_{\rm form}$ (Fig 3, top row); but if we slightly change $T_{\rm form}$, we see bimodal structure not only in metal-rich region but also in metal-poor region (Fig 3, bottom row), unlike Gaia data. It turned out that when bar and spirals have different pattern speeds, most of the disc orbits become chaotic. It seems that the [Fe/H] dependence of the bimodality behaves in a unpredictable manner due to the chaotic nature of disc orbits. Interestingly, the value of V that separates the two modes is the spiral's 4:1 inner Lindblad resonance (ILR), almost independent of $T_{\rm form}$. Therefore, the observed location of the Hercules stream may indicate the spiral's nature ($\Omega_{\rm s}$ that determines ILR in this model), while the [Fe/H] dependence of the bimodality is uninformative. Therefore, in this case we should not over-interpret the [Fe/H] dependence of the Hercules stream.

Slow-bar + transient spirals. When $m = 2$ and $\Omega_{\rm s} = 25\,{\rm km\ s^{-1}\ kpc^{-1}}$, the [Fe/H] dependence of the (U, V)-diagram is similar to the observed one for certain values of $T_{\rm form}$ (Fig 4); but it is not the case for other values of $T_{\rm form}$. Again, it seems this unpredictable [Fe/H] dependence of the bimodal structure is due to the chaoticity of disc orbits. Interestingly, the value of V of the secondary peak approximately satisfy a condition of (stellar radial period) $\simeq$ (spiral's half-period) ($= 100$ Myr in our case), almost independent of $T_{\rm form}$. Therefore, the observed location of the Hercules stream may indicate the spiral's properties (lifetime of the spirals in this model).

5. Conclusions

We show that fast-bar model can reproduce the observed [Fe/H] dependence of the local velocity distribution, while slow-bar model fails to explain it.

Also, when spiral arms are additionally considered, some models can explain the observed data even when the bar is slowly rotating. In such cases, the location of the bimodal structure may indicate some properties of spiral arms, such as $\Omega_{\rm s}$ or the lifetime

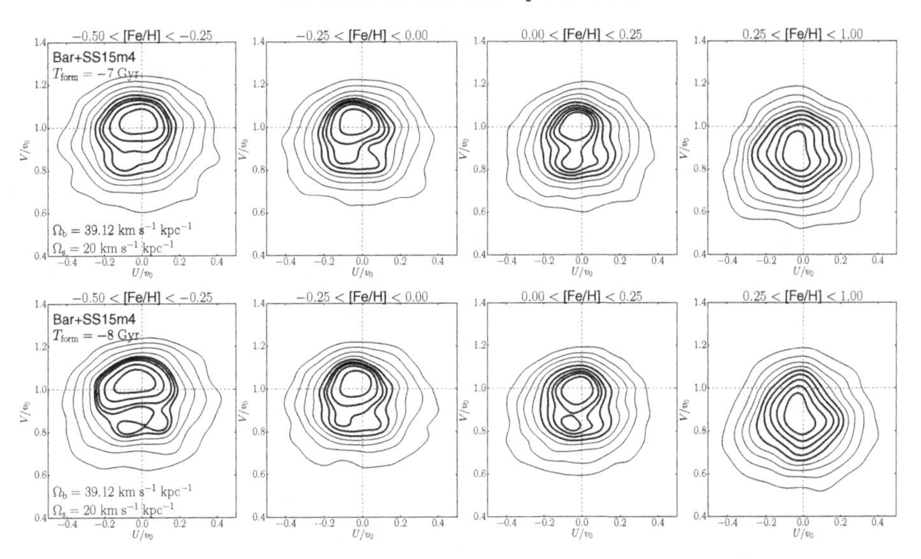

Figure 3. Slow-bar+steady spiral model ($m = 4$). ($T_{\rm form} = -7$ Gyr (top), -8 Gyr (bottom)).

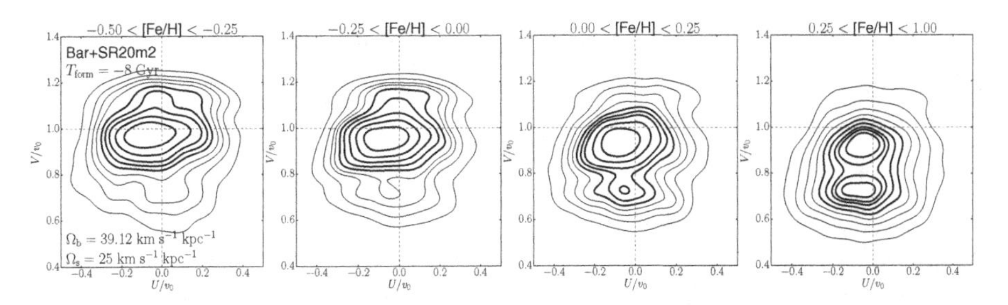

Figure 4. Slow-bar+transient spiral model ($m = 2$, $T_{\rm form} = -8$ Gyr).

of the spiral arms. However, in such cases, due to the chaotic nature of disc orbits, we should not over-interpret the [Fe/H] dependence of the bimodality.

Acknowledgments. This work has made use of data from the European Space Agency (ESA) mission *Gaia* (https://www.cosmos.esa.int/gaia), processed by the *Gaia* Data Processing and Analysis Consortium (DPAC, https://www.cosmos.esa.int/web/gaia/dpac/consortium). Funding for the DPAC has been provided by national institutions, in particular the institutions participating in the *Gaia* Multilateral Agreement. KH was supported by a grant from the Hayakawa Satio Fund awarded by the Astronomical Society of Japan.

References

Antoja, T., Helmi, A., Dehnen, W., *et al.* 2014, A&A, 563, A60
Dehnen, W. 1999, ApJ, 524, L35
Dehnen, W. 2000, AJ, 119, 800
Kunder, A., Kordopatis, G., Steinmetz, M., *et al.* 2017, AJ, 153, 75
Lindegren, L., Lammers, U., Bastian, U., *et al.* 2016, A&A, 595, A4
Minchev, I., Nordhaus, J., & Quillen, A. C. 2007, ApJ, 664, L31
Monari, G., Kawata, D., Hunt, J. A. S., & Famaey, B. 2017, MNRAS, 466, L113
Pérez-Villegas, A., Portail, M., Wegg, C., & Gerhard, O. 2017, ApJ, 840, L2
Portail, M., Gerhard, O., Wegg, C., & Ness, M. 2017, MNRAS, 465, 1621
Steiman-Cameron, T. Y., Wolfire, M., & Hollenbach, D. 2010, ApJ, 722, 1460
Wegg, C., Gerhard, O., & Portail, M. 2015, MNRAS, 450, 4050

Astronomy and Astrophysics in the Gaia sky
Proceedings IAU Symposium No. 330, 2017
A. Recio-Blanco, P. de Laverny, A.G.A. Brown
& T. Prusti, eds.

doi:10.1017/S1743921317006299

The evolution history of the extended solar neighbourhood

Andreas Just, Kseniia Sysoliatina and Ioanna Koutsouridou

Zentrum für Astronomie der Universität Heidelberg, Astronomisches Rechen-Institut,
Mönchhofstr. 12-14, 69120 Heidelberg
email: just@ari.uni-heidelberg.de

Abstract. Our detailed analytic local disc model (JJ-model) quantifies the interrelation between kinematic properties (e.g. velocity dispersions and asymmetric drift), spatial parameters (scale-lengths and vertical density profiles), and properties of stellar sub-populations (age and abundance distributions). We discuss a radial extension of the disc evolution model representing an inside-out growth of the thin disc with constant thickness. Based on metallicity distributions of APOGEE red clump stars we derive the AMR as function of galactrocentric distance and show that mono-abundance as well as mono-age populations are flaring. The predictions of the JJ-model are consistent with the TGAS-RAVE data, which provide a significant improvement of the kinematic data and unbiased distances for more than 250,000 stars.

Keywords. Galaxy: disk, Galaxy: kinematics and dynamics, Galaxy: evolution, Galaxy: solar neighbourhood, Galaxy: abundances

---

1. Dynamical disc model

The most elaborate model of the present day Milky Way is the Besançon Galaxy Model (BGM) including extinction, the bar, spiral arms and the warp (Czekaj, Robin, Figueras, *et al.* 2014). Nevertheless there are still open issues to be solved concerning the degeneracy of the SFR and IMF and a consistent chemical abundance model. Our alternative local disc model (JJ-model) based on the kinematics of main sequence stars (Just & Jahreiß 2010), the stellar content in the solar neighbourhood (Rybizki & Just 2015) and Sloan Digital Sky Survey (SDSS) star counts to the north Galactic pole (Just *et al.* 2011) has a significantly higher accuracy compared to the old BGM (Gao *et al.* 2013). In the JJ-model the local SFR, IMF, AVR (age–velocity dispersion relation) are determined self-consistently and it includes a simple chemical enrichment model. With the TGAS data as part of the first Gaia data release DR1 (Gaia Collaboration, Brown, & Vallenari, *et al.* 2016) we have independent and unbiased distances and significantly imroved proper motions for more than a million stars. This allows a sharp test of our local model (see Sect. 3).

In order to extend the JJ-model over the full radial range of the disc, we have used in a first step the Jeans equation for the asymmetric drift to connect local dynamics with the radial scale-lengths of mono-abundance populations (Golubov *et al.* 2013). Based on RAVE (RAdial Velocity Experiment) data we found an increasing scale-length with decreasing metallicity, which is consistent with a negative overall metallicity gradient of the disc. On the other hand Milky Way-like galaxies show a radial colour gradient of the disc to be bluer and younger in the outer part. Combining both observations immediately shows that the chemical enrichment in the inner disc must be faster/larger compared to the outer disc.

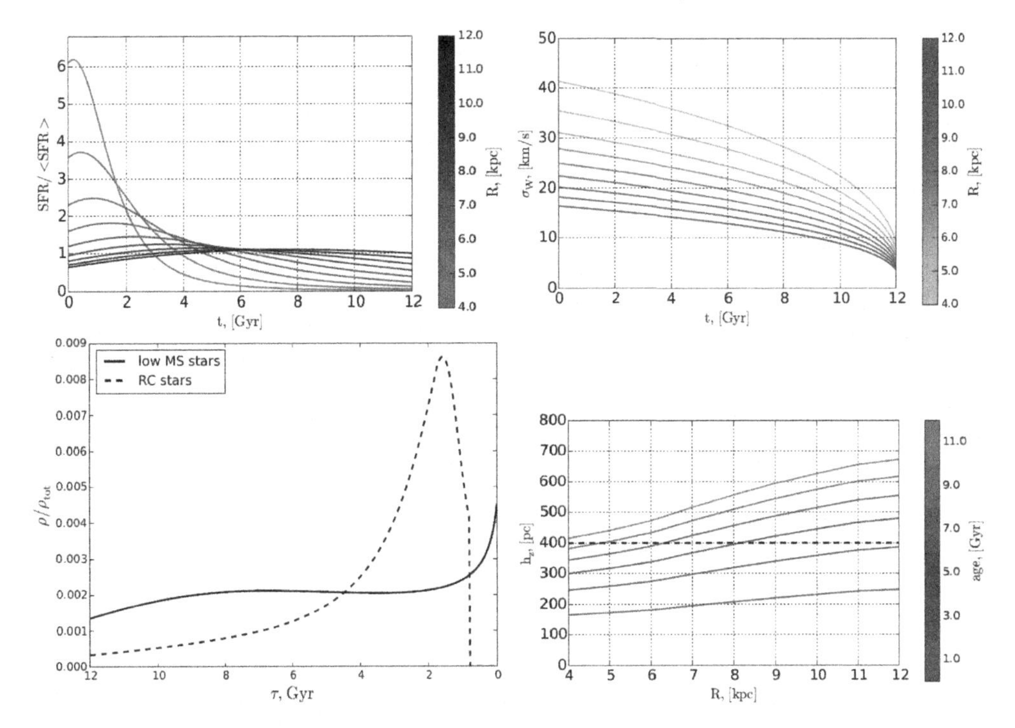

Figure 1. Top left: SFR at different galactrocentric distances R (increasing with declining maximum). Top right: Corresponding AVR. Bottom left: Age distributions of lower MS (full line) and of RC (dashed line) stars in the solar neighbourhood. Bottom right: Thickness profiles of mono-age populations (larger for older ages). The horizontal dashed line shows the thickness of the whole disc.

2. Inner structure of simple discs

Since most observed stellar samples, like main sequence (MS) stars, red clump (RC) stars or mono-abundance populations, are composed by different generations, it is crucial to derive self-consistent age distributions (see bottom left panel of Fig. 1 for lower MS and for RC stars in the solar neighbourhood at $(R, z) = (8\,\text{kpc}, 0\,\text{kpc})$) from the model, which vary strongly with galactrocentric distance R and vertical distance z, in order to derive number densities, velocity or abundance distributions. We are modelling an inside-out growth of the disc by varying the SFR in the range $4\,\text{kpc} < R < 12\,\text{kpc}$ in two models (model A with a strong variation of the SFR and model B with a weaker variation). Here we focus on model A, where the SFR has a strong peak at early times in the inner disc (top left panel of Fig. 1). The SFR is scaled to an exponential disc with a scalelength of 2.5 kpc. The AVR (top right panel of Fig. 1) is scaled in order to achieve a constant thickness $h = 400\,\text{pc}$ of the disc independent of R. Mono-age populations show an exponential radial density profile with increasing scalelength for decreasing age. Another generic consequence of the age shift from predominantly old to younger stars with increasing R combined with the larger thickness of older stars at each R is a flaring of all mono-age populations measured by the (half-)thickness h_z (lower right panel of Fig. 1).

We derive an empirical AMR by identifying the cumulative age distibutions of RC stars at each radius R with the observed cumulative metallicity distributions taken from APOGEE DR12. We find a fast enrichment to supersolar abundances in the inner disc and a much slower early enrichment in the outer disc reaching a lower present day metallicity

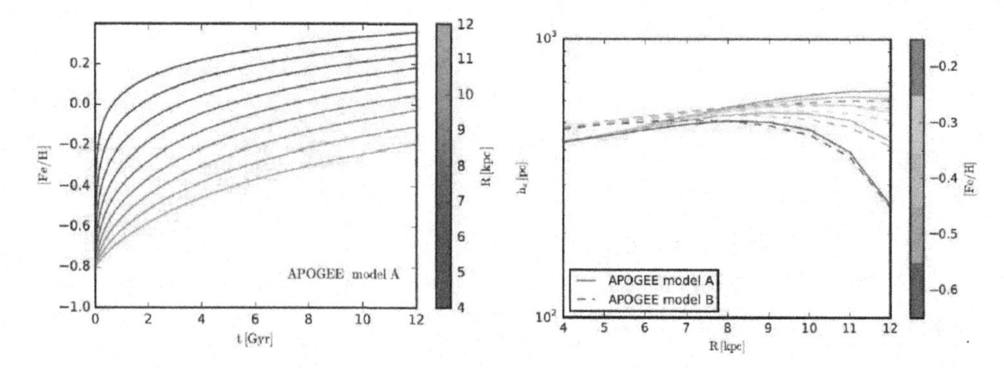

Figure 2. Left: AMR at different galactrocentric distances R (increasing from top to bottom line) derived from the APOGEE metallicity distribution and the RC age distribution of the model. Right: Radial thickness profiles of different metallicity bins.

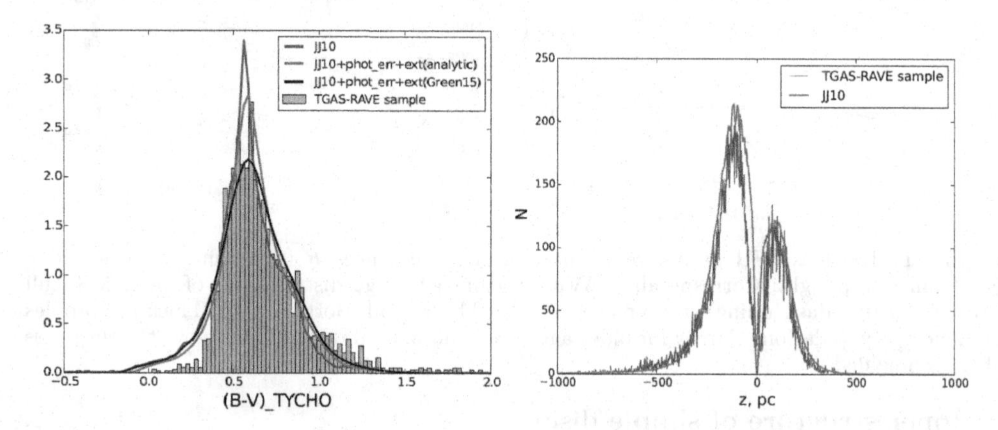

Figure 3. Left: Impact of reddening on the colour distribution (full lines) compared to the TGAS-RAVE sample (histogram). Right: Comparison of star counts in the local cylinder of model and data.

(see left panel of Fig. 2). The combination of the SFR and the AMR leads to the surface density profiles of mono-abundance populations. We find approximately exponential profiles with a scalelength range from 15 kpc to 4.5 kpc with increasing metallicity. Taking into account the self-consistent vertical density profiles, we find also for mono-abundance populations a flaring structure, but with an outer turndown due to the depletion of higher metallicity stars in the outer disc (see right panel of Fig. 2).

3. TGAS-RAVE data

For a detailed test of the local JJ-model we start with the ~250,000 stars in common of TGAS and RAVE providing improved 6-D phase space information. We select stars in the solar cylinder with radius 300 pc and relative parallax errors smaller than 30% and choose a subsample of 48,000 stars, which are most probably thin disc stars ([Fe/H]<-0.6 and [Mg/Fe]<0.2 dex). We apply the Tycho and RAVE selection functions as well as extinction and reddening to our model predictions. The left panel of Fig. 3 demonstrates the impact of the extinction model on the colour distribution of all stars at $|z| = 100$ pc and shows that the 3-D model based on Pan-STARRS and 2MASS photometry of Green *et al.*(2015) fits very well. No additional TGAS selection function is applied yet. The

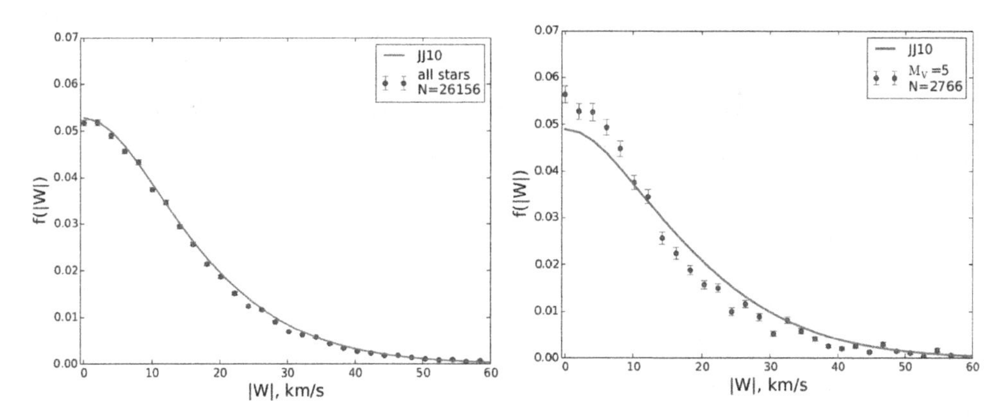

Figure 4. Left: Normalized distribution function of the vertical velocity component $|W|$ of the full sample (points) compared to the model (full line). Right: Similar distribution function for MS stars with $M_V = 5$ mag.

right panel of Fig. 3 shows that the vertical density profile including the asymmetry due to the RAVE selection is well reproduced by the model.

The large sample size and the high quality of the kinematic data allow a detailed test of the vertical component of the velocity distribution functions $f(|W|)$. The left panel of Fig. 4 shows that $f(|W|)$ of our full TGAS-RAVE sample is reproduced extremly well. On the other hand the right panel of Fig. 4 shows that the MS stars with $M_V = 5$ mag seem to have a slightly cooler core compared to our model.

We will use the potential of these new high quality data to improve the JJ-model by combining the number counts and kinematics of slices in z and sub-populations in the colour magnitude diagram.

Acknowledgements

This work was supported by Sonderforschungsbereich SFB 881 "The Milky Way System" (sub-project A6) of the German Research Foundation (DFG). This work has made use of data from the European Space Agency (ESA) mission Gaia, processed by the Gaia Data Processing and Analysis Consortium (DPAC). Funding for the DPAC has been provided by national institutions, in particular the institutions participating in the Gaia Multilateral Agreement.

References

Czekaj, M. A., Robin, A. C., Figueras, F., Luri, X., & Haywood, M. 2014, *A&A*, 564, A102

Gaia Collaboration, Brown, A. G. A., Vallenari, A., Prusti, T., de Bruijne, J. H. J., Mignard, F., *et al.* 2016, *A&A*, 595, A2

Gao, S., Just, A., & Grebel, E. K. 2013, *A&A*, 549, A20

Green, G. M., Schlafly, E. F., Finkbeiner, D. P., *et al.* 2015, *ApJ*, 810, 25

Golubov. O., Just, A., Bienaymé, O., *et al.* 2013, *A&A*, 557, A92

Just, A., Gao, S. & Vidrih, S. 2011, *MNRAS*, 411, 2586

Just, A. & Jahreiß, H. 2010, *MNRAS*, 402, 461

Koutsouridou, I. 2017, *Master thesis*, Ruprecht Karls Universität, Heidelberg

Kunder, A., Kordopatis, G., Steinmetz, M., *et al.* 2017, *ApJ*, 153, 75

Rybizki, J. & Just, A. 2015 *MNRAS*, 447, 3880

Astronomy and Astrophysics in the Gaia sky
Proceedings IAU Symposium No. 330, 2017
A. Recio-Blanco, P. de Laverny, A.G.A. Brown
& T. Prusti, eds.

doi:10.1017/S174392131700597X

Metallicity distribution functions using Gaia-DR1 data

Georges Kordopatis[1] **and RAVE collaboration**

[1]Université Côte d'Azur, Observatoire de la Côte d'Azur, CNRS, Laboratoire Lagrange, France
email: gkordo@oca.eu

Abstract. The metallicity distribution function (MDF) of the stellar components of the Milky Way hold valuable information regarding the processes that have taken place in the evolution of our Galaxy. In this proceeding, we investigate updates concerning the MDF now that the Tycho-Gaia Astrometric Solution (TGAS) catalogue has been released and that trigonometric distances are available. In particular, vertical changes and skewness of the MDF are investigated, together with the properties of the metal-rich stars in the sample, at different positions in the Galaxy.

Keywords. Milky Way, stellar content, evolution, metallicity

1. Introduction

The advent of large spectroscopic surveys has enabled the possibility to obtain precise abundance measurements of the photospheres for several hundreds of thousands stars, thought to reflect the chemical composition of the interstellar medium in which they were born. As a consequence, information such as the mean, dispersion and skewness of the metallicity distribution functions (MDF) at different regions of the Milky Way hold valuable information regarding the star formation history of those regions, about the presence of stars either not being born where they are being observed, and/or born with migrated or extra-galactic gas, and hence help us to disentangle the evolution of our Galaxy (Bland-Hawthorn & Gerhard 2016).

Up to recently, the study of the MDF with precise stellar locations (via parallax measurements) for FGK stars could be done only up to a few 100 pc from the Sun, i.e. within the Hipparcos volume. To go beyond this limit, so-called spectroscopic distances projecting on isochrones the stellar atmospheric parameters were used (e.g.: Pont & Eyer 2004; Kordopatis *et al.* 2011). Since Septembre 2016, the volume for which we have parallaxes has increased by more than an order of magnitude, thanks to the Gaia space mission (Gaia Collaboration *et al.* 2016a). The first Gaia data-release (Gaia Collaboration *et al.* 2016b), and in particular its sub-catalogue TGAS (Tycho-Gaia Astrometric Solution, Lindegren *et al.* 2016) published 2 million parallaxes and proper motions for stars brighter than the 11th magnitude, allowing to have a better view on both the tails of the MDFs and the way chemistry is linked to the three-dimensional kinematics of stars. In this proceeding, we show updates on that topic. Section 2 describes the datasets that have been combined, Sect. 3 shows some preliminary results, and Sect. 4 presents the perspectives.

2. Dataset used: the joint RAVE-TGAS catalogue

Our sample consists of $\sim 2 \cdot 10^5$ stars with measured atmospheric parameters, metallicities and radial velocities, coming from the RAVE fifth data release (Kunder *et al.*

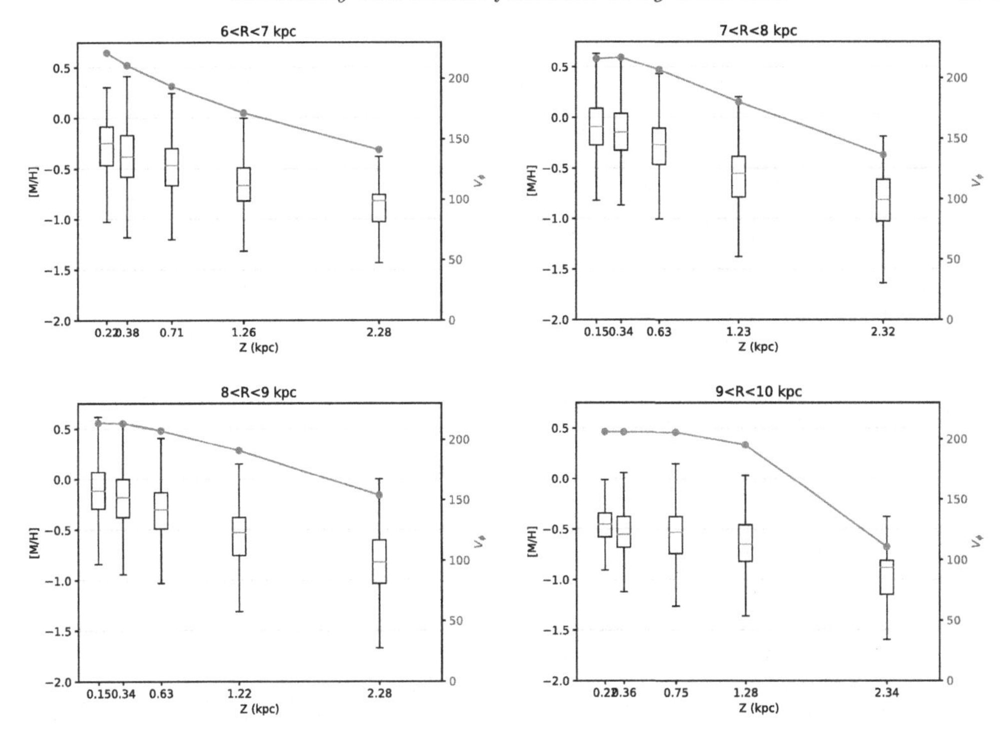

Figure 1. Box plots representing the median metallicity (orange line) as a function of absolute distance from the Galactic plane, Z, for different radial ranges, R, from the Galactic center. The actual boxes enclose the metallicity value of the first and third quartile, whereas the sizes of the bar span 99 per cent of the distribution. The median V_ϕ value, in km s^{-1}, for each spatial bin is also represented in blue (and right-hand-side y-axis).

2017), combined with UCAC5 proper motions (Zacharias *et al.* 2017) and improved RAVE spectroscopic distances (McMillan & RAVE 2017, this volume). The reason for not using the TGAS proper motions or parallaxes is that both UCAC5 and the new RAVE distances build on Gaia's measurements to improve the transverse velocities and distances. In particular, the new spectroscopic distances take into account not only the stellar atmospheric parameters but also Gaia's parallax in order to infer the most likely distance of the observed star (McMillan *et al.*, in prep.). This leads to more reliable distances, in particular for the stars for which the TGAS fractional uncertainty on the parallax is greater than 20 per cent (mostly corresponding to giants in our sample).

3. Results

Figure 1 shows box-plots representing the metallicity distributions as a function of height above the Galactic plane, at different Galactic radii (different panels). The $Z-$bins closest to the plane are suffering from selection biases and completeness issues that differ from one $R-$bin to the other and hence should not be used for the analysis, as they are not representative of the underlying population.

Excluding those bins, we measure the following vertical metalicity gradients, $\partial[\mathrm{M/H}]/\partial|Z|$, for the four 1 kpc-wide radial bins: $-0.27, -0.34, -0.33$ and -0.19 dex kpc^{-1}, going from 6 kpc to 10 kpc, respectively. These values are compatible, within the errors (of the order of 0.07 dex kpc^{-1}), with the ones of Schlesinger *et al.* (2014) of -0.243 ± 0.05 dex kpc^{-1}, obtained using SEGUE G dwarfs. We note, however, that our measurements should be

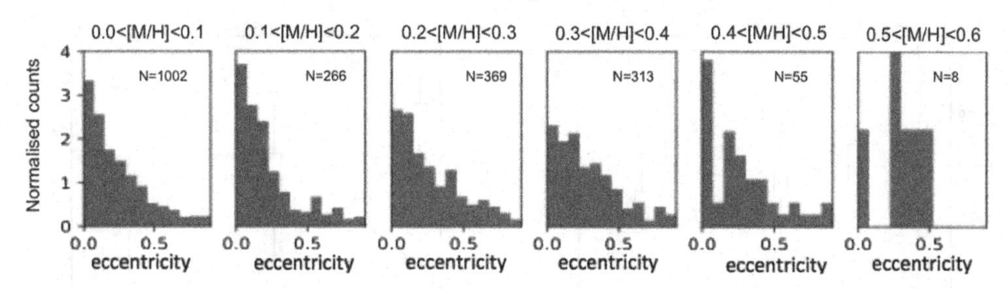

Figure 2. Normalised distribution of stellar eccentricities for the super metal-rich targets located between $7.5 \leqslant R \leqslant 8.5$ kpc and $|Z| < 0.25$ kpc.

considered more reliable compared to previous studies, thanks to the improved distances that are being used here.

Regarding the skewness of the the MDFs, no particular changes are noticed as a function of R; this is illustrated with the lack of change of the relative position of the median line within a box from one $R-$bin to the other, at a fixed Z. This is not in agreement with the inversion of the MDF's skewness highlighted in Hayden *et al.* (2015) using APOGEE data. Nevertheless, a more thorough analysis, taking into account the completeness of our sample (see Sect. 4), needs to be performed in order to confirm this statement.

We have investigated the impact on the MDFs of the new datasets available by computing similar plots using pure spectroscopic distances and UCAC4 proper motions (in practice, here, we have used the RAVE DR4 catalogue, Kordopatis *et al.* 2013). We have not found any significant change on the mean of the MDFs. That said, the updated tails of the MDFs make more sense using the Gaia data: many super metal-rich stars ([M/H] > 0, noted SMR hereafter) found previously to occupy large distances from the Galactic plane are found to be closer to the disc, whereas metal-poor stars are also shifted to higher Z. We note, however, that few SMR stars can still be detected, up to 1 kpc from the Galactic plane, confirming previous studies (see, for example, Kordopatis *et al.* 2015). The majority of these SMR stars, even though they have updated eccentricities, are found to be in majority on circular orbits ($e < 0.2$, see Fig. 2), therefore having radially migrated from the inner parts of the disc through mechanisms involving co-rotation resonances with the spiral arms (e.g. Sellwood & Binney 2002).

Finally, it should be highlighted that we find a remarkably flat trend for the mean V_ϕ as a function of Z for our outermost radial bin ($9 < R < 10$ kpc). Up to almost 750 pc from the plane, we find stars to have $< V_\phi >\approx 220\,\mathrm{km\,s^{-1}}$, i.e. associated to a stellar population dominated by the thin disc. This seems to confirm the thin disc flaring and the absence of the thick disc at the outer disc, as already suggested by other studies (Bensby *et al.* 2011; Minchev *et al.* 2014; Bovy *et al.* 2016; Kordopatis *et al.* 2017), but now clearly indicated with the simple kinematics.

4. Perspectives

In this preliminary work, we have combined the stars from the north and the south Galactic cap, and have not corrected for the selection function (noted SF, hereafter) of the sample.

On the one hand, the impact of the RAVE SF should not be dramatic, especially at intermediate and high Galactic latitudes. Indeed, Wojno *et al.* (2017) has shown that RAVE's SF does not bias the kinematics nor the metallicity distributions, for typical FGK stars under the condition that the stars located close to the Galactic plane (outside the

RAVE footprint) are removed. On the other hand, the SF of TGAS (Gaia Collaboration *et al.* 2016b), is more complicated and involves amongst other the scanning law of the satellite (only stars with five astrometric transits are included) and the proper motions of the stars (high proper motion stars are removed). In Kordopatis *et al.* (in prep.), we combine the Wojno *et al.* (2017) and Bovy (2017) tools to correct for the selection function and further investigate whether metallicity asymmetries exist relative to the Galactic plane.

Acknowledgements

This work has made use of data from the European Space Agency (ESA) mission *Gaia* (`https://www.cosmos.esa.int/gaia`), processed by the *Gaia* Data Processing and Analysis Consortium (DPAC, `https://www.cosmos.esa.int/web/gaia/dpac/consortium`). Funding for the DPAC has been provided by national institutions, in particular the institutions participating in the *Gaia* Multilateral Agreement. Funding for RAVE has been provided by: the Australian Astronomical Observatory; the Leibniz-Institut für Astrophysik Potsdam (AIP); the Australian National University; the Australian Research Council; the French National Research Agency; the German Research Foundation (SPP 1177 and SFB 881); the European Research Council (ERC-StG 240271 Galactica); the Instituto Nazionale di Astrofisica at Padova; The Johns Hopkins University; the National Science Foundation of the USA (AST-0908326); the W. M. Keck foundation; the Macquarie University; the Netherlands Research School for Astronomy; the Natural Sciences and Engineering Research Council of Canada; the Slovenian Research Agency; the Swiss National Science Foundation; the Science & Technology Facilities Council of the UK; Opticon; Strasbourg Observatory; and the Universities of Groningen, Heidelberg and Sydney. The research leading to these results has received funding from the European Research Council under the European Union's Seventh Framework Programme (FP7/2007-2013)/ERC grant agreement no. 321067. The RAVE web site is at `http://www.rave-survey.org`.

References

Bensby, T., *et al.*, 2011, *ApJ*, 735, 46
Bland-Hawthorn, J. & Gerhard, O., 2016, *ARA&A*, 54, 561
Bovy, J., *et al.*, 2016, *ApJ*, 823, 30
Bovy, J., 2017, *MNRAS*, submitted, arXiv:1704.05063
Gaia Collaboration, Brown, A. G. A., Vallenari, A., *et al.* 2016, *A&A*, 595, A2
Gaia Collaboration, Prusti, T., de Bruijne, J. H. J., *et al.* 2016, *A&A*, 595, A1
Hayden, M., *et al.*, 2015, *ApJ*, 808, 132
Kordopatis G., *et al.*, 2011, *A&A*, 535, A107
Kordopatis G., *et al.*, 2013, *AJ*, 146, 134
Kordopatis G., *et al.*, 2015, *MNRAS*, 447, 3526
Kordopatis, G., *et al.* 2017, *MNRAS*, 467, 469
Kunder, A., Kordopatis, G., Steinmetz, M., *et al.* 2017, *AJ*, 153, 75
Lindegren L., *et al.*, 2016, *A&A*, 595, A4
Minchev, I., Chiappini, C., & Martig, M. 2014, *A&A*, 572, 92
Pont F., Eyer L., 2004, *MNRAS*, 351, 487
Sellwood, J. A. & Binney, J. J., 2002, *MNRAS*, 336, 785
Schlesinger, K., *et al.*, 2014, *ApJ*, 791, 112
Wojno, J., Kordopatis, G., Piffl, T., *et al.* 2017, *MNRAS*, 468, 3368
Zacharias N., Finch C., Frouard J., 2017, *AJ*, 153, 166

Astronomy and Astrophysics in the Gaia sky
Proceedings IAU Symposium No. 330, 2017
A. Recio-Blanco, P. de Laverny, A.G.A. Brown & T. Prusti, eds.

doi:10.1017/S1743921317006160

RAVE-Gaia and the impact on Galactic archeology

Andrea Kunder[1,2]

[1]Leibniz-Institut für Astrophysik Potsdam (AIP), An der Sternwarte 16, D-14482 Potsdam, Germany
email: amkunder@gmail.com

[2]Saint Martin's University, 5000 Abbey Way SE, Lacey, WA 98503, USA

Abstract. The new data release (DR5) of the RAdial Velocity Experiment (RAVE) includes radial velocities of 520,781 spectra of 457,588 individual stars, of which 215,590 individual stars are released in the Tycho-*Gaia* astrometric solution (TGAS) in *Gaia* DR1. Therefore, RAVE contains the largest TGAS overlap of the recent and ongoing Milky Way spectroscopic surveys. Most of the RAVE stars also contain stellar parameters (effective temperature, surface gravity, overall metallicity), as well as individual abundances for Mg, Al, Si, Ca, Ti, Fe, and Ni. Combining RAVE with TGAS brings the uncertainties in space velocities down by a factor of 2 for stars in the RAVE volume – 10 km s^{-1} uncertainties in space velocities are now able to be derived for the majority (70%) of the RAVE-TGAS sample, providing a powerful platform for chemo-dynamic analyses of the Milky Way. Here we discuss the RAVE-TGAS impact on Galactic archaeology as well as how the *Gaia* parallaxes can be used to break degeneracies within the RAVE spectral regime for an even better return in the derivation of stellar parameters and abundances.

Keywords. astronomical data bases: RAVE, surveys, stars: kinematics, stars: abundances, stars: Hertzsprung-Russell diagram, Galaxy: kinematics and dynamics, Galaxy: stellar content, Galaxy: structure

---

1. Introduction

Our Milky Way galaxy contains stars that are distinctly closer and brighter to us than stars in neighbouring galaxies, so the level of detail with which the stellar populations in our Galaxy can be seen provide important information regarding the formation and evolution of large spiral galaxies. The motions of stars combined with their chemical abundances in particular place powerful constraints on the formation of spiral galaxies such as the Milky Way (e.g., Minchev, Chiappini & Martig 2013). Today, the astrometric satellite *Gaia* is providing its first measurements (Data Release 1, Gaia Collaboration *et al.* 2016), and the Tycho-*Gaia* Astrometric Solution (TGAS, Lindegren *et al.* 2016) contains positions, parallaxes, and proper motions for about 2 million of the brightest stars in common with the Hipparcos and Tycho-2 catalogues. With typical accuracies of $\sim$1 mas yr^{-1} and 0.3 mas in proper motion and parallax, respectively, this is comparable to the precision of Hipparcos, but on a sample that is more than an order of magnitude larger.

In TGAS, exquisite astrometry is given in the positions and proper motions of stars. Combined with external spectroscopy, the measure of stellar atmospheric parameters, individual chemical abundances and radial velocities allow a full definition of the motion of stars in the Galaxy. Among existing spectroscopic surveys, the Radial Velocity Experiment (RAVE, Steinmetz *et al.* 2006, Zwitter *et al.* 2008, Siebert *et al.* 2011, Kordopatis *et al.* 2013, Kunder *et al.* 2017) has the largest overlap with TGAS (>200,000) so is a

Table 1. Overlap of large spectroscopic surveys with *Gaia*-TGAS.

Survey	Number TGAS stars
RAVE DR5	215,600
LAMOST DR2	124,300
GALAH DR1	8,500
APOGEE DR13	21,700

particularly attractive database for astronomers seeking to simultaneously use chemical and dynamical information to complement the available *Gaia* astrometry.

2. RAVE Overview

RAVE is a magnitude-limited survey of stars randomly selected in the $9 < I < 12$ magnitude range, obtained from spectra with a resolution of R$\sim$7 500 covering the CaT regime. It currently containts the largest spectroscopic sample of stars in the Milky Way which overlaps with the *Gaia*-TGAS proper motions and parallaxes (Table 1).

Radial velocities are available for all RAVE stars, where the typical signal-to-noise (SNR) ratio of a RAVE star is 40 and the typical uncertainty in radial velocity is $< 2\,\mathrm{km\,s^{-1}}$. For a subsample of RAVE stars, stellar parameters are also provided. These temperatures, T_{eff}, gravities, $\log g$, and metallicities, [M/H], are obtained using the DR4 stellar parameter pipeline, which is built on the algorithms of MATISSE and DEGAS, with an updated calibration that improves the accuracy of especially the $\log g$ values of stars. The uncertainties vary with stellar population and SNR, but for the most reliable stellar parameters, the uncertainties in T_{eff}, $\log g$, and [M/H] are approximately 250 K, 0.4 dex and 0.2 dex, respectively. RAVE stars with the most reliable stellar parameters are those which have `Algo_Conv`=0 (meaning the stellar parameter algorithm converged), SNR > 40, and `c1`=n, `c2`=n and `c3`=n. (which means the star has a spectrum that is "normal"). Error spectra computed for each observed spectrum is used to assess the uncertainties in the radial velocities and stellar parameters.

The elemental abundances of Al, Si, Ti, Fe, Mg and Ni are derived for $\sim$2/3 of the RAVE stars, which have uncertainties of $\sim$ 0.2 dex, although their accuracy varies with SNR and, for some elements, also of the stellar population. Distances, ages, masses and the interstellar extinctions are computed using an upgraded method of what is presented in Binney *et al.*(2014).

RAVE DR5 further provides temperatures from the Infrared Flux Method, which are available for > 95% of all RAVE stars. For a sub-sample of stars that can be calibrated asteroseismically ($\sim$ 45% of the RAVE sample), an asteroseismically calibrated $\log g$, as detailed in Valentini *et al.*(2017) is provided. Stellar parameters of the RAVE stars are also found using the data-driven approach of *The Cannon* (Casey *et al.* 2017), for which T_{eff}, surface gravity $\log g$ and [Fe/H], as well as chemical abundances of giants of up to seven elements (O, Mg, Al, Si, Ca, Fe, Ni) is presented.

All of the above described information is publicly available, and can be downloaded via the RAVE Web site `http://www.rave-survey.org` or the Vizier database.

3. Reverse Pipeline

It is well-known that stellar spectra with a resolution R <10 000 suffer from spectral degeneracies at the Calcium triplet wavelength range. Specifically, at the RAVE wavelength and resolution, – see for example Figure 1 in Matijeviĉ *et al.* (2017). Parameter

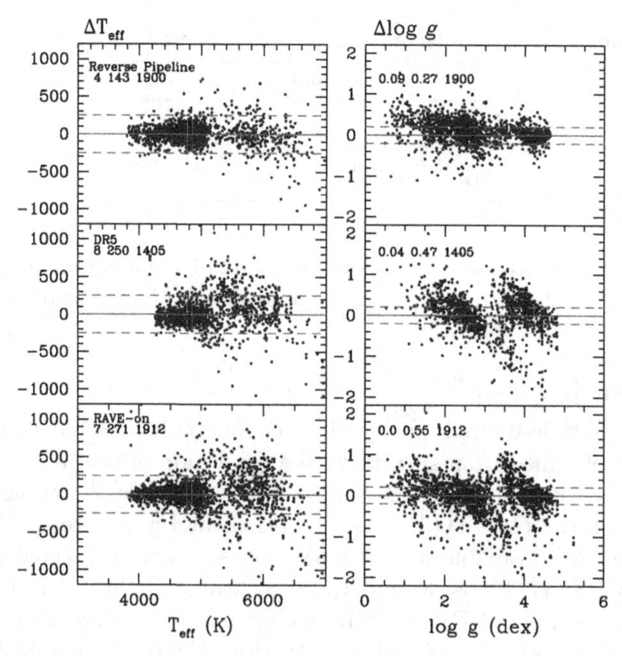

Figure 1. The reverse distance pipeline (top), DR5 main (middle), RAVE-on (bottom) temperatures and gravities compared to RAVE stars that overlap with high-resolution studies. The top left corner indicates the bias, dispersion and number of stars for each comparison. For the DR5 comparison, only stars with AlgoConv = 0 are shown.

degeneracy is usually less severe when the available information about the parameters increases: e.g., with a wider spectral range, higher spectral resolution, etc. The TGAS parallaxes can provide powerful extra information to break degeneracies, thereby constraining stellar parameters.

The RAVE distance pipeline (as described in Binney *et al.* 2014 and Kunder *et al.* 2017) takes as its input $T_{\rm eff}$, $\log g$, [M/H], J, H and K magnitudes and with this information combined with stellar isochrones, descriptions of the posterior probabilities of different properties of the stars (e.g., mass, age, line-of-sight extinction, distance) are generated. It has been modified to now also take the TGAS parallaxes, as well as AllWISE $W1$ and $W2$ magnitudes as an input (McMillan *et al.* 2017, in prep). Using the same prior as in Binney *et al.* (2014), a new $\log g$, $T_{\rm eff}$ and [M/H] is found, and for the first time, descriptions of the posterior probabilities for $T_{\rm eff}$, $\log g$ and [M/H] are obtained. We therefore refer to this as the 'reverse pipeline', because rather than just taking the stellar parameters as input, they are an end product. In fact, these are an inevitable byproducts of the distance pipeline, produced because each "model star" is compared to the data which has an associated $T_{\rm eff}$, $\log g$ and [M/H] as the likelihood is calculated.

Figure 1 shows how the reverse distance pipeline (top), DR5 main (middle), RAVE-on (Casey *et al.* 2017) (bottom) temperatures and gravities compare to RAVE stars that fortuitously overlap with high-resolution studies (e.g., Gaia-ESO, globular and open clusters, GALAH and field star surveys – see DR5 paper for details). The reverse pipeline yields temperatures and gravities that agree better to external, high-resolution studies of RAVE stars than both DR5 and RAVE-on. Note that the reverse pipeline temperatures and gravities are only available for TGAS stars.

The largest discrepancies between the DR5 and reverse pipeline main temperatures and gravities occur at the giant/dwarf interface. This is expected, as this is where the

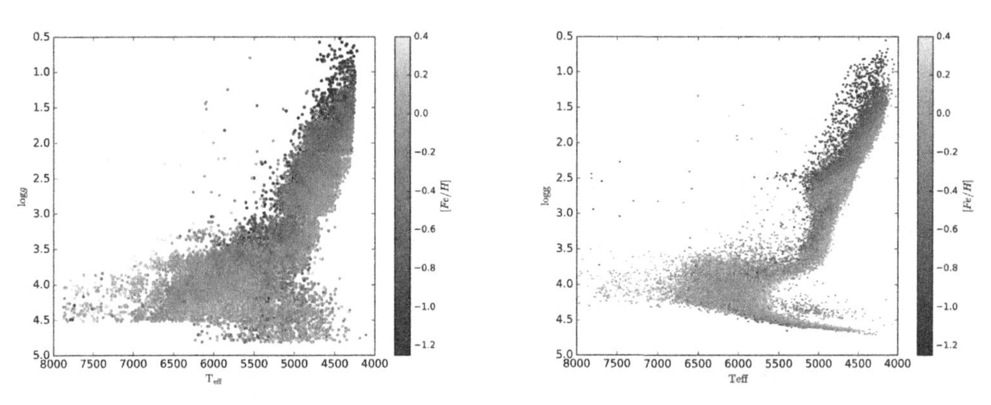

Figure 2. $T_{\rm eff}$-log g diagram for calibrated DR5 parameters (left) and reverse pipeline parameters (right).

degeneracies mentioned above are the most severe. Our preliminary tests show no signs of bias in the reverse pipeline stellar parameters as a function of parallax or TGAS parallax uncertainty. We are carrying out extensive tests to check if any and what kinds of subtle biases may exist when applying the reverse pipeline.

Figure 2 (right) shows the on the Hertzsprung Russel diagram using the DR5 and reverse distance pipeline temperatures and gravities. Of particular interest is the narrow main sequence, and a sequence of stars above the main-sequence separated by a clear gap. The majority of these stars are double-lined spectroscopic binaries (SB2s) which (in the absence of eclipses) are not variable and where the orbital period is short enough to permit any astrometric signature. Hence, they do not fall into photometrically or astrometrically peculiar classes, and are included in *Gaia* DR1.

4. Conclusions

RAVE is continuing to yield exciting results using the data products *Gaia* DR1. The reverse pipeline, described above, is allowing more accurate stellar parameters to be obtained, which can then be fed into a new elemental abundance pipeline designed specifically for the RAVE spectra (Guiglion *et al.* 2016, Guiglion *et al.* 2017, in prep). Jofre *et al.* (2017, submitted) has expanded the number of RAVE stars with TGAS parallax uncertainties less than 20% by applying the twin method to RAVE. McMillan *et al.*(2017, in prep) is using the TGAS parallaxes to find more precise distance estimates for all the RAVE stars, which also has the effect of an improvement in age uncertainties. Therefore, an exploration of the correlation between ages, metallicities, and velocities of stars in the solar neighborhood can be carried out (Wojno *et al.* 2017, in prep). Last but not least, 200 light curves of RAVE stars in the K2-Campaign 6 have been analysed, which will be used as calibration data for log g (Valentini *et al.* 2017, in prep).

This work makes use of data from the European Space Agency (ESA) mission *Gaia* (`https://www.cosmos.esa.int/gaia`), processed by the *Gaia* Data Processing and Analysis Consortium (DPAC, `https://www.cosmos.esa.int/web/gaia/dpac/consortium`). Funding for the DPAC has been provided by national institutions, in particular the institutions participating in the *Gaia* Multilateral Agreement. Funding for RAVE has been provided by: the Australian Astronomical Observatory; the Leibniz-Institut fuer Astrophysik Potsdam (AIP); the Australian National University; the Australian Research Council; the French National Research Agency; the German Research Foundation (SPP

1177 and SFB 881); the European Research Council (ERC-StG 240271 Galactica); the Istituto Nazionale di Astrofisica at Padova; The Johns Hopkins University; the National Science Foundation of the USA (AST-0908326); the W. M. Keck foundation; the Macquarie University; the Netherlands Research School for Astronomy; the Natural Sciences and Engineering Research Council of Canada; the Slovenian Research Agency; the Swiss National Science Foundation; the Science & Technology Facilities Council of the UK; Opticon; Strasbourg Observatory; and the Universities of Groningen, Heidelberg and Sydney.

References

Binney, J., Burnett, B., Kordopatis, G., *et al.* 2014a, *MNRAS*, 437, 351
Casey, A. R., Hawkins, K., Hogg, D. *et al.* 2017, *ApJ*, 840, 59
Guiglion, G., de Laverny, P., Recio-Blanco, A. *et al.* 2016, *A&A*, 595, 18
Kordopatis, G., Gilmore, G., Steinmetz, M., *et al.* 2013, *AJ*, 146, 134
Kunder, A. M., Kordopatis, G., Steinmetz, M. *et al.* 2017, *AJ*, 153, 75
Lindegren, L., *et al.* 2016, *A&A*, 595, 4
Matijevič, G., Zwitter, T., Bienaymé, O., *et al.* 2012, *ApJS*, 200, 14
Matijevič, G., Chiappini, C., Grebel, E. K. *et al.* 2017, arXiv:1704.05695
Minchev, I., Chiappini, C., & Martig, M. 2013, *A&A*, 558, 9
Siebert, A., Williams, M. E. K., Siviero, A., *et al.* 2011, *AJ*, 141, 187
Steinmetz, M., Zwitter, T., Siebert, A., *et al.* 2006, *AJ*, 132, 1645 (DR1)
Valentini, M., Chiappini, C., Davies, G.R. *et al.* *A&A*, 600, 66
Zwitter, T., Siebert, A., Munari, U., *et al.* 2008, *AJ*, 136, 421

Astronomy and Astrophysics in the Gaia sky
Proceedings IAU Symposium No. 330, 2017
A. Recio-Blanco, P. de Laverny, A.G.A. Brown & T. Prusti, eds.

doi:10.1017/S1743921317005671

Hypervelocity star candidates in *Gaia* DR1/TGAS

T. Marchetti[1], E. M. Rossi[1], G. Kordopatis[2], A. G. A. Brown[1], A. Rimoldi[1], E. Starkenburg[3], K. Youakim[3] and R. Ashley[4]

[1]Leiden Observatory, Leiden University, PO Box 9513 2300 RA Leiden, the Netherlands
email: marchetti@strw.leidenuniv.nl

[2]Université Côte d'Azur, Observatoire de la Côte d'Azur, CNRS, Laboratoire Lagrange, France

[3]Leibniz-Institut fur Astrophysik Potsdam (AIP), An der Sternwarte 16, 14482 Potsdam, Germany

[4]Department of Physics, University of Warwick, Gibbet Hill Road, Coventry, CV4 7AL, UK

Abstract. Hypervelocity stars (HVSs) are characterized by a total velocity in excess of the Galactic escape speed, and with trajectories consistent with coming from the Galactic Centre. We apply a novel data mining routine, an artificial neural network, to discover HVSs in the TGAS subset of the first data release of the *Gaia* satellite, using only the astrometry of the stars. We find 80 stars with a predicted probability $> 90\%$ of being HVSs, and we retrieved radial velocities for 47 of those. We discover 14 objects with a total velocity in the Galactic rest frame > 400 km s^{-1}, and 5 of these have a probability $> 50\%$ of being unbound from the Milky Way. Tracing back orbits in different Galactic potentials, we discover 1 HVS candidate, 5 bound HVS candidates, and 5 runaway star candidates with remarkably high velocities, between 400 and 780 km s^{-1}. We wait for future *Gaia* releases to confirm the goodness of our sample and to increase the number of HVS candidates.

Keywords. Astrometry, The Galaxy: kinematics and dynamics, The Galaxy: stellar content.

---

1. Introduction

Brown *et al.* (2005) discovered the first hypervelocity star (HVS): a star in the outer halo of the Milky Way with a heliocentric radial velocity of ~ 850 km s^{-1}, higher than the escape speed from the Galaxy. To explain such an extreme velocity, this star was suggested to be ejected according to the *Hills mechanism*, involving the interaction between a binary system and the massive black hole (MBH) in the Galactic Centre (GC) (Hills 1988). Observationally, what defines a HVS is that (i) its velocity is higher than the escape speed from the Galaxy at its position, and (ii) its trajectory is consistent with coming from the GC (Brown 2015). Following the first detection, a dedicated survey detected a total of 21 late B-type HVS candidates in the outer halo of the Milky Way (Brown *et al.* 2014).

HVSs have been proposed to study both the stellar population of the GC and the Galactic potential, since they can be used as dynamical tracers for the (dark) matter distribution (e.g. Gnedin *et al.* 2005, Yu & Madau 2007). However results have been hampered by the quality and quantity of the current sample (Rossi *et al.* 2017).

The first release of the *Gaia* satellite (DR1, Gaia Collaboration *et al.* 2016a,b), and in particular the TGAS subset (Lindegren *et al.* 2016), gives us the first opportunity to increase the current sample of HVSs. The challenge is in the disparity between the expected number of HVSs and that of other bound stars in our Galaxy. For all the details on the search for HVSs in Gaia DR1, see Marchetti *et al.* (2017). To estimate how many

HVSs we expect to find in the *Gaia* catalogue, we create mock populations using different assumptions for the unknown mass function, star formation rate, and binary population in the GC (Marchetti *et al.* in preparation). For the end of the mission catalogue, we anticipate from few hundreds to a thousand HVSs with relative error on total proper motion below 1%, and thousands with relative error on total proper motion below 10%. A few tens of objects will have relative error on parallax below 10%. Around 100 HVSs will be bright enough to have a radial velocity measurement. Repeating the same analysis considering the completeness limit of TGAS, we expect to find at maximum a few HVSs in this first release.

2. Data Mining Routine and Application to *Gaia* DR1/TGAS

In order not to bias our search towards particular spectral types, and to make as few assumption as possible on the stellar population in the GC, we decide to look for HVSs using only the information provided by TGAS: the 5 parameters astrometric solution. This consist of the projected position of the source on the sky, the parallax, and the two proper motions. We decide for a machine learning algorithm: an artificial neural network, trained on *mock* populations of HVSs and "normal" bound stars, the great majority of objects that *Gaia* will observe. We use the same mock populations mentioned in §1 for modelling HVSs, and we take normal stars from the *Gaia* Universe Model Snapshot (GUMS, Robin *et al.* 2012). This combined catalogue of the five parameters for the two populations is then used as a labelled training set for the neural network, which learns the general function mapping inputs (the five parameters) into the output (predicted probability of being a HVS).

The blind application of the neural network to the complete TGAS catalogue (2057050 sources) results in a total of 22263 stars with predicted probability $> 50\%$ of being HVSs, approximately 1% of the full catalogue. If we further require relative error on parallaxes to be lower than 1, we are left with 8175 stars, $\sim 0.4\%$ of the initial number of sources. We include uncertainties using Monte Carlo (MC) simulations: we draw 1000 realizations of the astrometry of each source, and we study the distribution of the output D of the neural network. We define our best candidates as those with $\bar{D} - \sigma_D > 0.9$, where $\bar{D}$ and σ_D are, respectively, the mean and the standard deviation of the probability distribution of the parameter D. This cut results in a total of 80 HVS candidates, selected without any prior assumption on the stellar type or on the photometry.

3. Results

In order to confirm or reject the nature of our candidates as HVSs, we need a measure of their total velocity. We successfully applied for director's discretionary time at the Isaac Newton Telescope (INT) in La Palma, Canary Island, where, on the night of the 5th of October 2016, we followed up spectroscopically 22 candidates. Furthermore, we cross-match our final sample with several spectroscopic surveys of the Milky Way. We are then able to recover a radial velocity measurement for 47 out of 80 stars. To derive distances from TGAS parallaxes, we implement a Bayesian approach similar to Astraatmadja & Bailer-Jones (2016), but considering also covariances between astrometric parameters. In addition, we determine spectroscopic distances for 22 stars. In the following we will present results assuming both distance estimates.

In Figure 1 we plot the total velocity in the Galactic rest frame $v_{\rm GC}$ as a function of Galactocentric distance for the 47 candidates with a reliable radial velocity measurement. Error bars are dominated by distance uncertainties. The left (right) panel corresponds

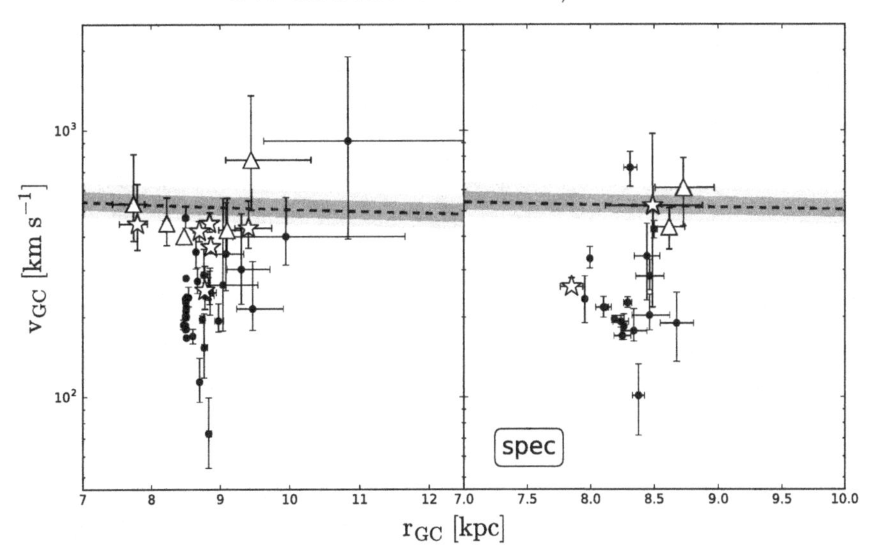

Figure 1. Total Galactic rest-frame velocity as a function of galactocentric distance for HVS candidates with a radial velocity information, using parallax-inferred distances (left panel) and, when available, spectroscopic distances (right panel). Stars (triangles) mark HVS/BHVS (runaway star) candidates, as in Table 1. The dashed line corresponds to the median escape velocity from the Milky Way (Williams *et al.* 2017), with the 68% (94%) credible interval shown as a dark (light) band.

to velocities computed adopting parallax-inferred (spectroscopic) distances. We overplot with a dashed line the median escape speed from the Milky Way from Williams *et al.* (2017), with corresponding 68% (94%) credible intervals shown as a dark (light) region. This plot shows how our algorithm succeeded in finding high velocity stars: 45 stars have median $v_{GC} > 150$ km s^{-1}, and 14 of these have median $v_{GC} > 400$ km s^{-1}. A total of 11 (3) stars are consistent with being unbound from the Galaxy in the left (right) panel.

To assess whether our candidates come from the GC, we integrate their orbits back in time in different Galactic potentials using the python package *galpy* (Bovy 2015). We draw 10^5 realizations of the initial condition of each star, and we check the consistency of the GC origin hypothesis for our candidates by recording the closest disc crossing (Galactic latitude = 0) to the GC. We find 8 (3) stars to have orbits consistent with coming from the GC using parallax-inferred (spectroscopic) distances. We define the probability of each star of being unbound from the Milky Way, P^u, as the fraction of MC realizations resulting in a total velocity higher than the escape speed at that position.

We classify stars as HVSs if (i) their velocity is > 350 km s^{-1} using at least one distance estimate, and (ii) if they are consistent with coming from the GC within 2σ when tracing back their obits. The Hills mechanism predicts as well a population of *bound* hypervelocity stars (BHVSs): stars with a velocity high enough to escape from the gravitational field of the MBH, but not sufficient to be unbound from the Galaxy (e.g. Bromley *et al.* 2006). Stars are then sub-classified as HVSs (BHVSs) if $P^u > 50\%$ ($P^u < 50\%$). We classify as *runaway stars* those objects with a velocity > 350 km s^{-1} and with orbits *not* consistent with coming from the GC. The 11 candidates whose ejection location interpretation is independent from the adopted distance are presented in Table 1. According to the adopted classification, we discover 1 HVS candidate, 5 BHVS candidates, and 5 runaway star candidates. Interestingly, 2 runaway stars have probabilities $> 50\%$ of being unbound from our Galaxy. One possible acceleration mechanism for these stars involves an asymmetric supernova explosion in a binary system (e.g. Tauris 2015).

Table 1. Derived kinematic properties for HVS, BHVS and runaway star candidates.

Tycho 2 ID	d (pc)	d_{spec} (pc)	v_{GC} (km s^{-1})	v_{GCspec} (km s^{-1})	P^{u}	P^{u}_{spec}
HVS / BHVS candidates						
2298-66-1	431^{+78}_{-55}	754 ± 569	248^{+58}_{-38}	519^{+451}_{-307}	0.1%	50.3%
8422-875-1[1]	1010^{+400}_{-218}	208 ± 124	446^{+186}_{-89}	259^{+21}_{-7}	29.1%	0.0%
2456-2178-1	976^{+358}_{-207}		430^{+117}_{-68}		22.7%	
2348-333-1	407^{+51}_{-40}		448^{+44}_{-32}		7.6%	
49-1326-1	304^{+38}_{-30}		419^{+38}_{-35}		1.2%	
5890-971-1	550^{+93}_{-72}		366^{+29}_{-20}		0.2%	
Runaway star candidates						
7111-718-1	1967^{+1413}_{-683}	1552 ± 430	776^{+576}_{-274}	611^{+176}_{-172}	82.2%	70.7%
8374-757-1	832^{+338}_{-179}		532^{+284}_{-147}		50.4%	
1071-404-1	439^{+91}_{-64}		449^{+113}_{-78}		23.7%	
4515-1197-1	881^{+292}_{-175}	902 ± 170	423^{+137}_{-76}	433^{+78}_{-76}	23.5%	15.6%
9404-1260-1	$67.0^{+1.0}_{-0.9}$		402^{+4}_{-4}		0.0%	

[1] The parallax-inferred distance d is more likely to be correct: this object is a RR Lyrae star, and this estimate consistent with the value obtained using a PLZ relation.
Notes: The subscript "spec" refers to quantities computed using the spectroscopic distance. Results are quoted in terms of the median of the distribution with uncertainties derived from the 16th and 84th percentiles.

With the advent of future *Gaia* releases we will be able to confirm the nature of our candidates, narrowing down their ejection location. Moreover, we will be able to use radial velocity as an extra feature to achieve a more precise classifier, dramatically increasing the number of expected HVSs.

This work has made use of data from the European Space Agency (ESA) mission *Gaia* (https://www.cosmos.esa.int/gaia), processed by the *Gaia* Data Processing and Analysis Consortium (DPAC, https://www.cosmos.esa.int/web/gaia/dpac/consortium). Funding for the DPAC has been provided by national institutions, in particular the institutions participating in the *Gaia* Multilateral Agreement.

References

Astraatmadja, T. L. & Bailer-Jones, C. A. L. 2016, *ApJ*, 832, 137
Bovy, J. 2015, *ApJS*, 216, 29
Bromley, B. C., *et al.* 2006, *ApJ*, 653, 1194
Brown, W. R., Geller, M. J., Kenyon, S. J., & Kurtz, M. J. 2005, *ApJ*, 622, L33
Brown, W. R., Geller, M. J., & Kenyon, S. J. 2014, *ApJ*, 787, 89
Brown, W. R. 2015, *ARA&A*, 53, 15
Gaia Collaboration *et al.* 2016a, *A&A*, 595, A1
Gaia Collaboration *et al.* 2016b, *A&A*, 595, A2
Gnedin, O. Y., Gould, A., Miralda-Escudé, J., & Zentner, A. R. 2005, *ApJ*, 634, 344
Hills, J. G. 1988, *Nature*, 331, 687
Lindegren, L., *et al.* 2016, *A&A*, 595, A4
Marchetti, T. *et al.* 2017, *MNRAS*, stx1304
Robin, A. C., *et al.* 2012, *A&A*, 543, A100
Rossi, E. M., Marchetti, T., Cacciato, M., Kuiack, M., & Sari, R. 2017, *MNRAS*, 467, 1844
Tauris, T. M. 2015, *MNRAS*, 448, L6
Williams, A. A., Belokurov, V., Casey, A. R., & Evans, N. W. 2017, *MNRAS*, 468, 2359
Yu, Q. & Madau, P. 2007, *MNRAS*, 379, 1293

Astronomy and Astrophysics in the Gaia sky
Proceedings IAU Symposium No. 330, 2017
A. Recio-Blanco, P. de Laverny, A.G.A. Brown
& T. Prusti, eds.

doi:10.1017/S1743921317006032

Search for Galactic warp signal in Gaia DR1 proper motions

E. Poggio[1,2], R. Drimmel[2], R. L. Smart[2,3], A. Spagna[2] and M. G. Lattanzi[2]

[1]Università di Torino, Dipartimento di Fisica, via P. Giuria 1, 10125 Torino, Italy
email: poggio@oato.inaf.it

[2]Osservatorio Astrofisico di Torino, Istituto Nazionale di Astrofisica (INAF), Strada Osservatorio 20, 10025 Pino Torinese, Italy

[3]School of Physics, Astronomy and Mathematics, University of Hertfordshire, College Lane, Hatfield AL10 9AB, UK

Abstract. The nature and origin of the Galactic warp represent one of the open questions posed by Galactic evolution. Thanks to Gaia high precision absolute astrometry, steps towards the understanding of the warp's dynamical nature can be made. Indeed, proper motions for long-lived stable warp are expected to show measurable trends in the component vertical to the galactic plane. Within this context, we search for the kinematic warp signal in the first *Gaia* data release (DR1). By analyzing distant spectroscopically-identified OB stars in the Hipparcos subset in *Gaia* DR1, we find that the kinematic trends cannot be explained by a simple model of a long-lived warp. We therefore discuss possible scenarios for the interpretation of the obtained results. We also present current work in progress to select a larger sample of OB star candidates from the *Tycho-Gaia* Astrometric Solution (TGAS) subsample in DR1, and delineate the points that we will be addressing in the near future.

Keywords. Warp, Milky Way, Kinematics, Galactic evolution

---

1. Introduction

The warp of the Milky Way is a well known feature of the outer disk, whose presence has been detected in the gas, dust and stars. However, its dynamical nature - as well as the formation mechanism - continues to remain an unsolved mystery. In the case of a long-lived static warp, systematic vertical motions would result in measurable trends in the stellar proper motions (Smart & Lattanzi 1996), which become most significant toward the Galactic anti-center. We select OB stars since they can be seen to large distances and are expected to trace the gaseous disk, in which the warp was originally detected. Given the unprecedented astrometric precision, we first search for the warp kinematic signal the Hipparcos subset in the first *Gaia* data release (DR1, Gaia Collaboration 2016) (see Poggio *et al.* 2017, hereafter Paper I). Section 2 gives a brief overview of the obtained results and discusses possible interpretations. Section 3 is dedicated to the selection of OB stars candidates in the *Tycho-Gaia* Astrometric Solution (TGAS) sample in DR1.

2. Hipparcos subset in Gaia DR1

Data selection. From the Hipparcos catalog (van Leeuwen 2008), we spectroscopically select the OB3 stars with apparent magnitude $m_V < 8.5$ and parallax $\varpi < 2$ mas, in order to remove local structures. The resulting selection contains 989 stars, among which 758 are present in the Hipparcos subsample in *Gaia* DR1. In the following, we present and discuss the results obtained with the smaller sample having *Gaia* superior astrometry.

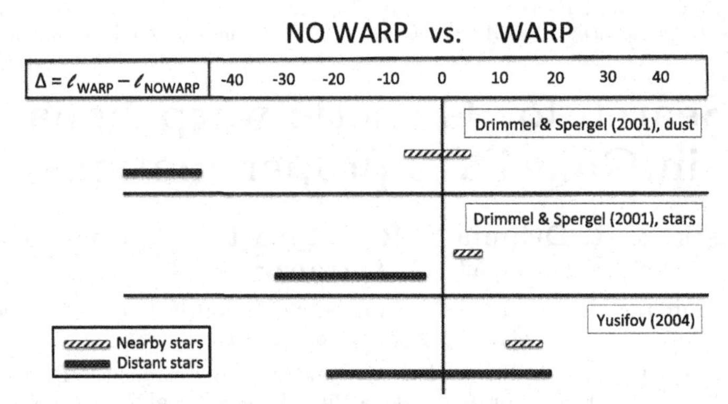

Figure 1. The difference of the loglikelihoods $\Delta = \ell_{WARP} - \ell_{NOWARP}$ (i.e. the likelihood ratio) for three different warp models according to the nearby ($1 < \varpi < 2$ mas) and distant ($\varpi < 1$ mas) stars of the sample. Positive (negative) values of Δ favour the warp (nowarp) model. The length of each bar is proportional to the uncertainty in Δ, calculated using bootstrap resamples.

The model. The modelled spatial distribution consists of four major spiral arms (Georgelin & Georgelin 1976, Taylor & Cordes 1993) and one local arm, with a gaussian density profile in the Galactic plane and an exponential vertical profile. Some spatial parameters are taken from the literature, while others are tuned to reproduce the longitude, latitude and apparent magnitude distribution observed in the data (see Paper I for the details). The kinematics is described by a simple model which includes solar motion from Schönrich *et al.* (2010), Galactic rotation from Bland-Hawthorn & Gerhard (2016) and velocity dispersions from Dehnen & Binney (1998). Astrometric errors are included in the model, together with the selection function of both Hipparcos catalogue and Hipparcos subset in *Gaia* DR1. Finally, the warp can be incorporated as a vertical displacement $z_W(R, \phi)$ in the z spatial coordinate, while its kinematic signal has a systematic offset $v_{z,W}(R, \phi)$ in the vertical velocities v_z.

Results. Depending on the warp spatial parameters, different kinematic signals are expected. Here we consider the three different sets of warp parameters, from Drimmel & Spergel (2001) (both dust and stars) and Yusifov (2004). The kinematic signal predicted by each of them is compared to the alternative model – the *nowarp* model, i.e. the absence of warp signal –, by constructing the expected probability distribution function in the proper motions μ_b as a function of galactic longitude l. Figure 1 summarizes our results, showing the likelihood of the various warp models with respect to the no warp model for our TGAS-Hipparcos dataset, divided into nearby ($1 < \varpi < 2$ mas) and distant ($\varpi < 1$ mas) stars. Our model for a long-lived warp predicts that the kinematic signal becomes stronger for larger distances, while the data do not show any evidence of warp signal for the most distant stars, consistent with the previous works of Smart *et al.* (1998) and Drimmel *et al.* (2000).

Interpretations. The absence of the warp kinematic signal in our dataset can be explained by several different scenarios. The first possible explanation is that our sample of OB stars in the Hipparcos subset of *Gaia* DR1 (approximately 0.5-3 kpc from the Sun) is not sufficiently sampling the Galactic warp. Indeed, there is no consensus about where the warp starts, although most studies find that the warp starts inside or close to the Solar circle (Momany *et al.* 2006; Reylé *et al.* 2009). Another possible interpretation is that the warp signal is overwhelmed by other perturbations, such as vertical waves in the disk (Gómez *et al.* 2013). Finally, it might be that our model of a long-lived stable warp is not appropriate, and additional effects should be taken into account, like precession

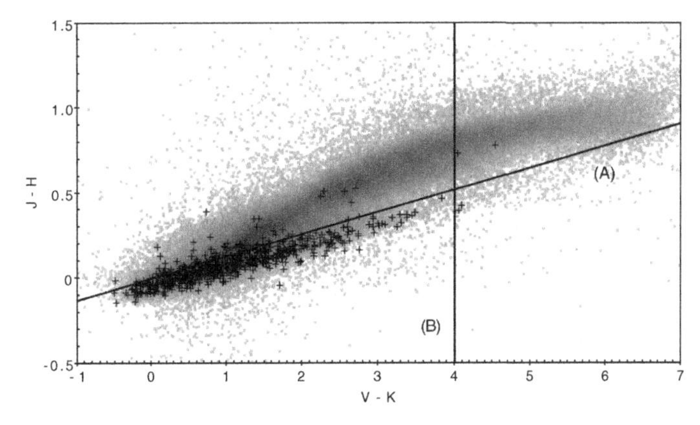

Figure 2. The TGAS stars with 2MASS and APASS photometry are represented by the grey dots, with the grey scale showing their density. Known O-B3 stars from the Tycho-2 Spectral Type Catalogue (black crosses) are located in the bottom-left part of the plot. Lines (A) and (B) are used for the selection process (see text).

or an amplitude varying with time. To shed further light on the nature of the Galactic warp it will be necessary to consider a larger dataset that samples a larger volumn of the Galactic disk.

3. TGAS in Gaia DR1

The selection of OB stars from the TGAS catalog is not trivial. Indeed, spectral classifications or parameter estimates such as $\log(g)$ and T_{eff} are available in literature only for a small fraction of TGAS stars, and not for the all sky. We therefore developed a selection criterium which combines astrometric with photometric measurements from the 2MASS (Skrutskie *et al.* 2006) and APASS (Henden *et al.* 2016) surveys, available for 1 824 237 TGAS stars. The first step of the method takes advantage of the fact that different stellar populations lie in different regions of the color-color plot shown in Figure 2. Known O-B3 stars from the Tycho-2 Spectral Type Catalogue (Wright *et al.* 2003, black crosses) are overplotted with the TGAS stars (grey density map). Extinction moves the OB stars to the right, but separated from the majority of giants and stars on the main-sequence. We begin by selecting all the TGAS stars below line (A) and bluer than line (B).

In order to reduce the fraction of contaminants that remain in our selected region of color-color space (mostly early A stars), we perform a second selection based on estimating the absolute magnitude of the stars using ϖ, σ_{ϖ} and G magnitude, assuming an exponentially decreasing prior for the heliocentric distance (similar to Bailer-Jones (2015)) and for the height from the Galactic plane. Taking extinction into account via the $(V-K)$ colors, we select as candidates those objects that have at least 75% probability of being brighter than B3 stars on the main sequence, resulting in $\approx$ 37000 candidate OB stars. Figure 3 shows the distribution on the sky of our selected sample. Applying our selection criterium to the Tycho-2 Spectral Type Catalogue, we estimate the amount of contamination from non-OB stars being about 40% for stars with $\varpi < 2$ mas and 30% for $\varpi < 1$ mas. However, the presence of late OB stars (i.e. with spectral type later than B3) is relevant ($\approx$ 40% of the selected stars). The relatively high fraction of contaminants is expected to be reduced with better parallax measurements in future Gaia releases.

Future works. We are developing a tool aimed at determining the likelihood of a model, given an observed sample selected as above. The objective is to perform a parameter

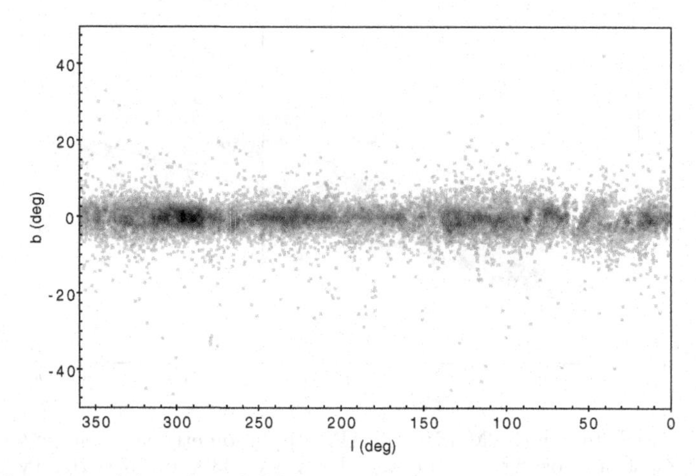

Figure 3. Location of the selected OB star candidates in the sky. The grey scale indicates stellar density.

adjustment for relevant kinematic warp parameters, such as the warp precession or possible amplitude variations.

This work has made use of data from the European Space Agency (ESA) mission *Gaia* (https://www.cosmos.esa.int/gaia), processed by the *Gaia* Data Processing and Analysis Consortium (DPAC, https://www.cosmos.esa.int/web/gaia/dpac/consortium). Funding for the DPAC has been provided by national institutions, in particular the institutions participating in the *Gaia* Multilateral Agreement.

References

Bailer-Jones, C. A. L. 2015, *PASP*, 127,994
Bland-Hawthorn, J. & Gerhard, O. 2016, *ARAA*, 54, 529
Dehnen, W. & Binney, J. J. 1998, *MNRAS*, 298, 387
Drimmel, R., Smart, R. L., & Lattanzi, M. G. 2000, *A&A*, 354, 67
Drimmel, R. & Spergel, D. N. 2001, *ApJ*, 556, 181
Gaia Collaboration (Brown, A. G. A., *et al.*) 2016, *A&A*, 595, A2
Georgelin, Y. M. & Georgelin, Y. P. 1976, *A&A*, 49, 57
Gómez, F. A., Minchev, I., O'Shea, B. W., Beers, T. C., Bullock, J. S., & Purcell, C. W. 2013, *MNRAS*, 429, 159
Henden, A. A., Templeton, M., Terrell, D., Smith, T. C., Levine, S., & Welch, D. 2016, *VizieR Online Data Catalog*, 2336
Momany, Y., Zaggia, S., Gilmore, G., Piotto, G., Carraro, G., Bedin, L. R., & de Angeli, F. 2006, *A&A*, 451, 515
Poggio, E., Drimmel, R., Smart, R. L., Spagna, A., & Lattanzi, M. G. 2017, *A&A*, 601, A115
Reylé, C., Marshall, D. J., Robin, A. C., & Schultheis, M. 2009, *A&A*, 495, 819
Schönrich, R., Binney, J., & Dehnen, W. 2010, *MNRAS*, 403, 1829
Skrutskie, M. F. *et al.* 2006, *AJ*,131,1163
Smart, R. L. & Lattanzi, M. G. 1996, *A&A*, 314, 104
Smart, R. L., Drimmel, R., Lattanzi, M. G., & Binney, J. J. 1998, *Nature*, 392, 471
Taylor, J. H. & Cordes, J. M. 1993, *ApJ*, 411, 674
van Leeuwen, F. 2008, *VizieR Online Data Catalogue*, 1311, 0
Wright, C. O., Egan, M. P., Kraemer, K. E., & Price, S. D. 2003, *AJ*, 125, 359
Yusifov, I. 2004, in: B. Uyaniker, W. Reich, & R. Wielebinski (eds.), *The Magnetized Interstellar Medium*, 165

Astronomy and Astrophysics in the Gaia sky
Proceedings IAU Symposium No. 330, 2017
A. Recio-Blanco, P. de Laverny, A.G.A. Brown
& T. Prusti, eds.

doi:10.1017/S1743921317005695

Can we detect Galactic spiral arms? 3D dust distribution in the Milky Way

Sara Rezaei Kh., Coryn A. L. Bailer-Jones, Morgan Fouesneau and Richard Hanson

Max Planck Institute for Astronomy (MPIA),
Königstuhl 17, 69117 Heidelberg, Germany
email: sara@mpia.de

Abstract. We present a model to map the 3D distribution of dust in the Milky Way. Although dust is just a tiny fraction of what comprises the Galaxy, it plays an important role in various processes. In recent years various maps of dust extinction have been produced, but we still lack a good knowledge of the dust distribution. Our presented approach leverages line-of-sight extinctions towards stars in the Galaxy at measured distances. Since extinction is proportional to the integral of the dust density towards a given star, it is possible to reconstruct the 3D distribution of dust by combining many lines-of-sight in a model accounting for the spatial correlation of the dust. Such a technique can be used to infer the most probable 3D distribution of dust in the Galaxy even in regions which have not been observed. This contribution provides one of the first maps which does not show the "fingers of God" effect. Furthermore, we show that expected high precision measurements of distances and extinctions offer the possibility of mapping the spiral arms in the Galaxy.

Keywords. Stars: Parallaxes – Stars: Distances – Galaxy: Dust Map – Galaxy: Milky Way – ISM: Dust – ISM: Extinction

---

1. Introduction

Dust is a tiny fraction of the mass of a galaxy; yet it plays an important role in different processes in the Galaxy; from star and planet formation and evolution to obscuring stellar lights via extinction. Therefore it is key to studying the different properties of this Interstellar Medium (ISM) component, especially its 3D distribution in the Galaxy.

Many studies have mapped (partially) the distribution of the extinction in the Galaxy using various approaches; from early 2D maps (e.g. Schlegel *et al.* 1998) to more recent 3D maps (e.g. Hanson & Bailer-Jones 2014, Hanson *et al.* 2016, Marshall *et al.* 2006, Sale *et al.* 2014, Green *et al.* 2015, Lallement *et al.* 2014). Looking in more detail, most of these maps show a common caveat: there are some lines-of-sight (l.o.s) traces, so-called "fingers of god", arising from the fact that each l.o.s is treated independently.

As published in Rezaei Kh. *et al.* 2017, we present a new method, which explicitly takes into account the correlations between close by l.o.s; therefore removing any l.o.s effect. Furthermore, we map the dust density distribution in the Galaxy rather than the extinction; while extinction is the integral over the dust densities along the l.o.s towards a star, dust density is the local property of the ISM. This way we present one of the first 3D dust maps in the Galaxy with no "fingers of god" effect.

2. Testing the method with mock data

Our method uses 3D positions and extinctions of stars within the Galaxy to infer the dust density at arbitrary positions in the 3D space. To do so, we divide each l.o.s into

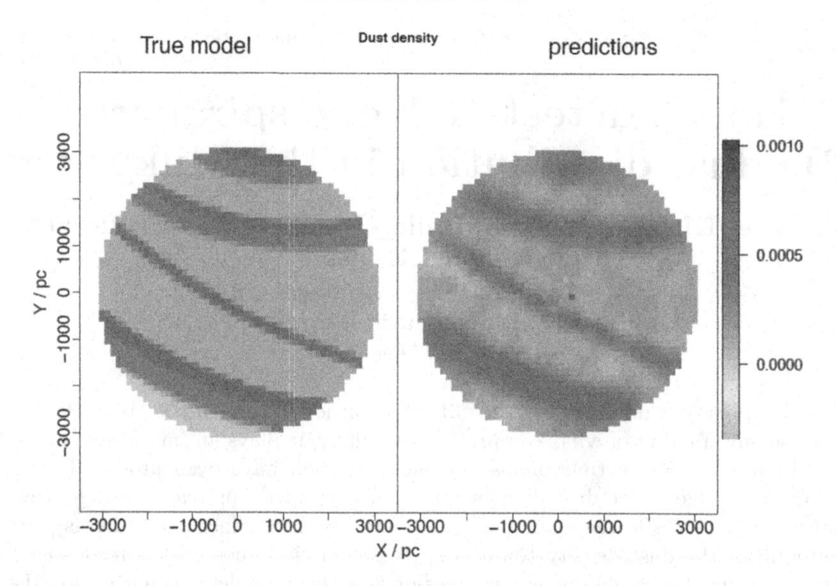

Figure 1. Comparing mock distribution (left) with model predictions (right) of dust density in the Galactic disk ($b < 2°$). Sun is at (0,0) and from the Sun upwards is toward the Galactic centre. Color indicates the dust density values in units of degree of attenuation per parsec.

small 1D cells and combine them with a Gaussian process, which exploits the correlation between these cells: the closer two points in the Galaxy in 3D, the more correlated they are. Through this method plus some linear algebra (see Rezaei Kh. *et al.* 2017), we find the posterior PDF of the dust density at any position in the Galaxy.

We test our model on a fiducial dust density similar to the disk of our Galaxy, i.e. over-densities at the locations of the spiral arms. We simulate 8000 stars at random positions in the disk of the Galaxy (2 degrees in latitude) and out to 3 kpc, generate their extinctions based on our mock model and add random Gaussian extinction noise with $\sigma = 0.05$ mag. We use these mock dataset as our input data and ask our model to find the dust density in the disk which describes best these noisy extinction measurements.

Figure 1 compares the predicted dust density (right) with the "true" distribution (left) used to generate these data. Qualitatively, the predictions recover all intrinsic structures, and the model precisely predicts the location and over-densities of the spiral arms.

3. Application on the real data

In this section, we infer the dust density in the Galaxy based on two different sets of observations: first, the APOGEE red clump (RC) star sample from Bovy *et al.* 2014, and second, the Gaia TGAS catalogue (Gaia Collaboration *et al.* 2016).

3.1. *Dust map in the disk with APOGEE RC stars*

Bovy *et al.* 2014 provides a catalogue of Red Clump stars based on APOGEE data. From their sample we select $\sim$ 8000 stars in the disk of the Galaxy (disk height = $\pm$ 500 kpc) and out to 5 kpc from the Sun with distances and K-band extinction (A_k) values. Their typical uncertainty in extinction is $\sigma_{A_k} = 0.05$ mag which we use in our model. However, we neglect the 5% uncertainty in the distance.

We compare dust features that may be associated with the spiral arms in our Galaxy described by the model of Reid *et al.* 2014, which assign high-mass star forming regions (masers) to the spiral arms based on their longitude-velocity diagram. We over-plot their

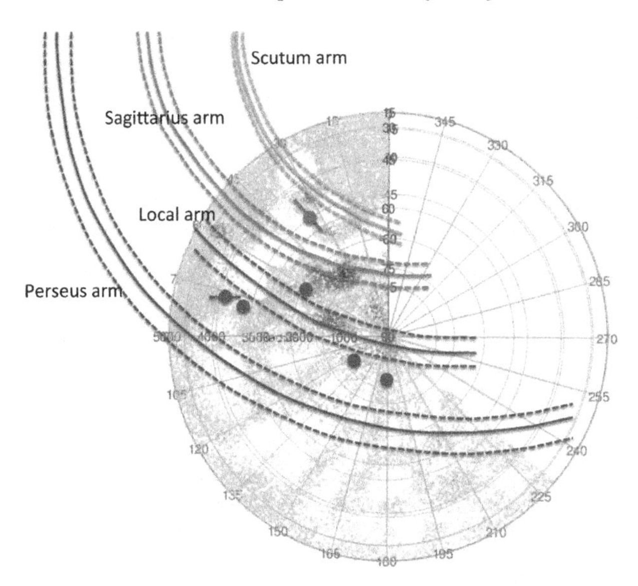

Figure 2. Dust density predictions in the Galactic disk, over-plotted with the spiral arm model and maser emissions from Reid *et al.* 2014. Color coding is as in fig. 2. Blue/pink circles are masers.

model of spiral arms plus some of their associated masers on our dust map as seen in Fig. 2. Some of our dust clouds (red) lie within 1σ of the arms; especially near the Sagittarius and the Local arms. However, some other dust clouds and masers from Reid *et al.* 2014 don't seem associated with the current model, for instance we do not find high dust density near the Perseus arm. We need more data with higher precision for further discussion on the Galactic spiral arms. Looking at fig. 2, there are some negative dust densities which are unphysical but due to the Gaussian nature of our model. This can be improved by truncating the Gaussian distribution or using other positive distributions.

3.2. *local 3D dust map with Gaia TGAS*

The higher the data quality, the more precisely can we map the dust. Therefore we explored the Gaia TGAS data. TGAS distances are provided by Astraatmadja and Bailer-Jones 2016 using Bayesian inference to infer distances from Gaia parallaxes, and we derive extinctions based on Rayleigh Jeans Color Excess (RJCE; Majewski *et al.* 2011) similarly to the extinctions in APOGEE (Zasowski *et al.* 2013), though restricting ourselves to giant stars where it was calibrated. As we currently neglect distance uncertainties in our model, we select stars with distance/standard error > 10. This gives us ~ 4000 stars and only out to 500 pc as input to map the dust density for a volume of 500 pc radius.

Figure 3 shows a top-down view of the 3D plot of dust density predictions, where the Sun is at the centre. We see some over-density between longitudes of 100° to 115° in the Galactic disk and at around $300 - 450$ pc. There are also some other over-densities like the one at negative latitude of $\sim -15°$ and longitude of $\sim 200°$ at ~ 400 pc which could be associated with the Orion cloud (Schlafly *et al.* 2014).

As discussed already, we need to use more data to better constrain the 3D structure of the dust in our Galaxy, which will eventually come from the next Gaia data releases and would enable us to build a precise 3D map of dust in the Milky way with no l.o.s. effect and will try to find some traces of spiral arms in our Galaxy.

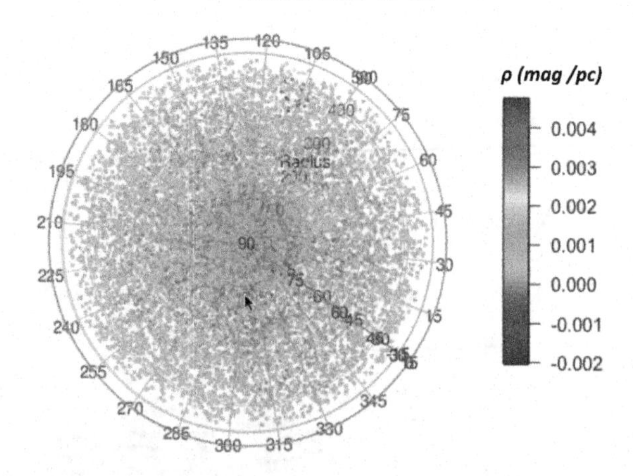

Figure 3. Top-down view of the 3D dust density predictions in a local 500 pc volume using TGAS data. Sun is in the centre and red numbers define the longitudes.

Acknowledgments

This work has made use of data from the European Space Agency (ESA) mission Gaia (http://www.cosmos.esa.int/gaia), processed by the Gaia Data Processing and Analysis Consortium (DPAC, http://www.cosmos.esa.int/web/gaia/dpac/consortium). Funding for the DPAC has been provided by national institutions, in particular the institutions participating in the Gaia Multilateral Agreement.

References

Astraatmadja, T. L. & Bailer-Jones, C. A. L. 2016, *ApJ*, 833, 119
Bovy, J., Nidever, D. L., Rix, H.-W. *et al.* 2014, *ApJ*, 790, 127
Gaia Collaboration *et al.* 2016, *A&A*, 595, A2
Green, G. M., Schlafly, E. F., Finkbeiner, D. P. *et al.* 2015, *ApJ*, 810, 25
Hanson, R. J. & Bailer-Jones, C. A. L. 2014, *MNRAS*, 438, 2938
Hanson, R. J., Bailer-Jones, C. A. L., Burgett, W. S. *et al.* 2016, *MNRAS*, 463, 3604
{Lallement, R., Vergely, J.-L., Valette, B., Puspitarini, L., Eyer, L., & Casagrande, L. 2014, *A&A*, 561, A91
Majewski, S. R., Zasowski, G., & Nidever, D. L. 2011, *ApJ*, 739, 25
Marshall, D. J., Robin, A. C., Reyl, C., Schultheis, M., & Picaud, S. 2006, *A&A*, 453, 635
Reid, M. J., Menten, K. M., Brunthaler, A. *et al.* 2014, *ApJ*, 783, 130
Rezaei Kh., S., Bailer-Jones, C. A. L., Hanson, R. J., & Fouesneau, M. 2017, *A&A*, 598, A125
Sale, S. E., Drew, J. E., Barentsen, G. *et al.* 2014, *MNRAS*, 443, 2907
Schlafly, E. F., Green, G., Finkbeiner, D. P. *et al.* 2014, *ApJ*, 786, 29
Schlegel, D. J., Finkbeiner, D. P., & Davis, M. 1998, *ApJ*, 500, 525
Zasowski, G., Johnson, J. A., & Frinchaboy, P. M. 2013, *AJ*, 146, 81

Astronomy and Astrophysics in the Gaia sky
Proceedings IAU Symposium No. 330, 2017
A. Recio-Blanco, P. de Laverny, A.G.A. Brown & T. Prusti, eds.

doi:10.1017/S1743921317006093

Galactic Disk Structure and Metallicity from Mono-age Stellar Populations of LAMOST

Maosheng Xiang[1,4], Xiaowei Liu[2], Jianrong Shi[1], Haibo Yuan[3], Yang Huang[2,4], Bingqiu Chen[2] and Chun Wang[2]

[1]National Astronomical Observatories, Chinese Academy of Science msxiang@nao.cas.cn
[2]Department of Astronomy, Peking University, China
[3]Department of Astronomy, Beijing Normal University, China
[4]LAMOST Fellow

Abstract. The LAMOST Galactic surveys provide robust stellar atmospheric parameters, abundances, masses and ages of millions of stars, allowing a unprecedented mapping of matter distribution, spatial structure, star formation rate, chemistry and kinematics of the Galaxy. In this proceeding we present structure and metallicity of the Galactic disk revealed by mono-age stellar populations within a few kilo-parsec of the solar neighborhood.

Keywords. Galaxy: general; Galaxy: disk; Galaxy: structure, Galaxy: abundances

1. Introduction

Characterizing the assemblage and evolution history of the Galactic disk is a fundamental task that becomes practicable in principle only until recently owing to the implementation of a few large-scale sky surveys, including photometric, spectroscopic and astrometric ones, that work together to deliver multi-dimensional parameters (3D positions, 3D motions, age, mass, metallicity and elemental abundances) of a huge, statistically meaningful number of stars. Among those parameters, stellar age is particularly a key to seek answers for many specific issues of the disk assemblage and evolution history, such as characteristic epochs of the disk formation, star formation and chemical enrichment histories, temporal evolution of disk morphology, etc. Nevertheless, delivering reliable stellar ages from the existing huge but usually inaccurate datasets are still challenging. As a consequence, robust and accurate age estimates for a large sample of stars were essentially absent in the past.

Due to fruitful efforts in deriving stellar atmospheric parameters, absolute magnitudes and abundances from low-resolution spectra of the LAMOST Galactic surveys, remarkable improvements have been achieved in both precision and accuracy of the stellar parameter estimates for millions of stars (Xiang *et al.* 2017a), which further leads to reliable age estimates of a million main sequence turn-off (MSTO) and subgiant stars of the Galactic disk. With this huge stellar sample, disk structure and stellar mass distribution, as well as metallicity and kinematics of mono-age stellar populations are explored. The results provide plenty of details and insights on the matter distribution, structure evolution, chemical and dynamical history of the Galactic disk. A combination of the LAMOST stellar parameters with parallax and proper motions from the coming data release (DR2+) of *Gaia* mission (Prusti *et al.* 2016) will provide better estimates of age, distance and tangential velocities for most LAMOST stars, thus is expected to promote significantly our knowledge of the assemblage and evolution history of our Galaxy.

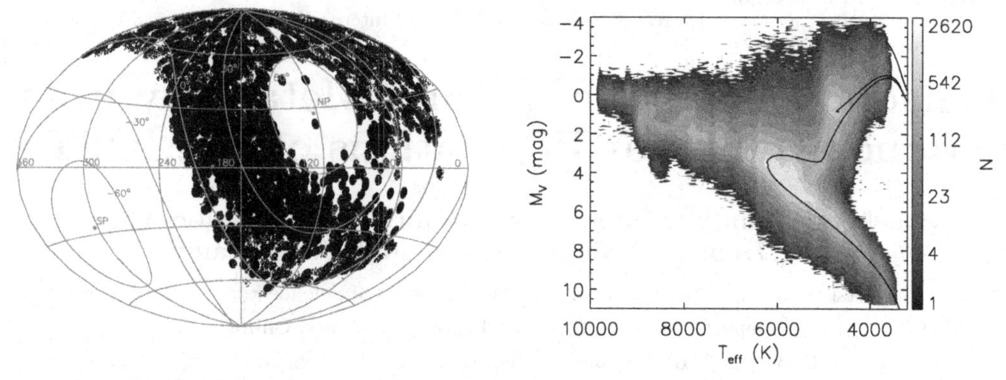

Figure 1. LAMOST sky coverage (left) in Galactic coordinate system and HR diagram (right) yielded by LSP3. Stellar isochrone of 4.5 Gyr and solar metallicity from Rosenfield *et al.* (2016) is shown in the HR diagram.

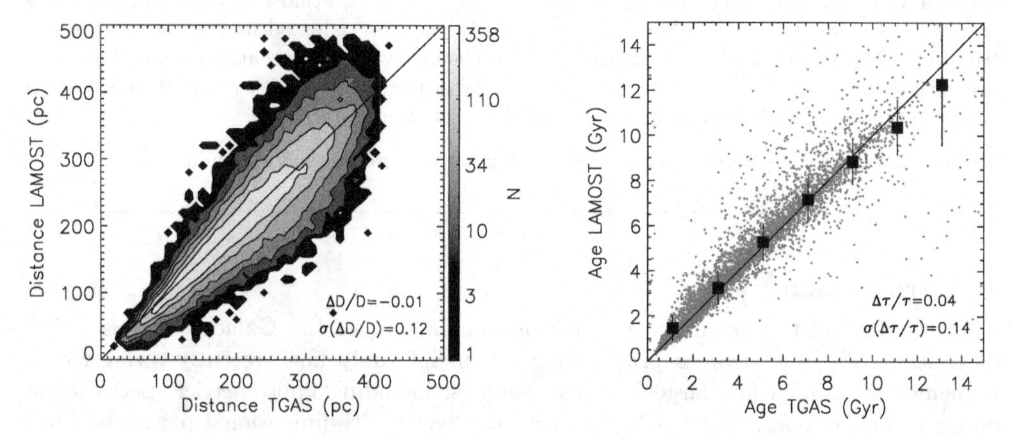

Figure 2. Comparisons of LAMOST stellar distance (left) and age (right) estimates with values derived based on TGAS parallax. Mean and dispersion of the relative differences are marked.

2. The LAMOST data

By June 2016, about 6.5 million low resolution ($R \sim$1800) stellar spectra for 4.4 million unique stars of 9–18.5 mag in r-band are collected by LAMOST Galactic surveys (Zhao *et al.* 2012, Deng *et al.* 2012) with a spectral signal-to-noise ratio (SNR) higher than 10. The left panel of Fig. 1 shows the sky coverage. Stellar parameters, including radial velocity $V_{\rm r}$, effective temperature $T_{\rm eff}$, surface gravity ($\log g$), absolute magnitudes M_V, M_{K_s}, metallicity [Fe/H], α-element to iron abundance ratio [α/Fe], carbon and nitrogen abundances [C/H], [N/H], have been derived from the spectra with the LAMOST Stellar Parameter Pipeline developed at Peking University (LSP3; Xiang *et al.* 2015a, Li *et al.* 2016, Xiang *et al.* 2017a). Extensive examinations with high-resolution spectroscopy, asteroseismology, open clusters as well as duplicate observations, indicate that given a SNR higher than 50, LSP3 yields stellar parameters with uncertainties of only 5 km/s, 100 K, 0.3 mag, 0.15 dex, 0.1 dex and 0.05 dex for $V_{\rm r}$, $T_{\rm eff}$, M_V, $\log g$, [Fe/H] and [α/Fe],respectively, and that LSP3 provides realistic parameter error estimates (Xiang *et al.* 2017b). The LSP3 stellar parameters, as well as interstellar extinction, distance and orbital parameters derived based on the LSP3 stellar parameters for stars targeted by the LAMOST Spectroscopic Survey of Galactic Anticentre (LSS-GAC; Liu *et al.* 2014, Yuan *et al.* 2015), a component of the LAMOST Galactic surveys, have been re-

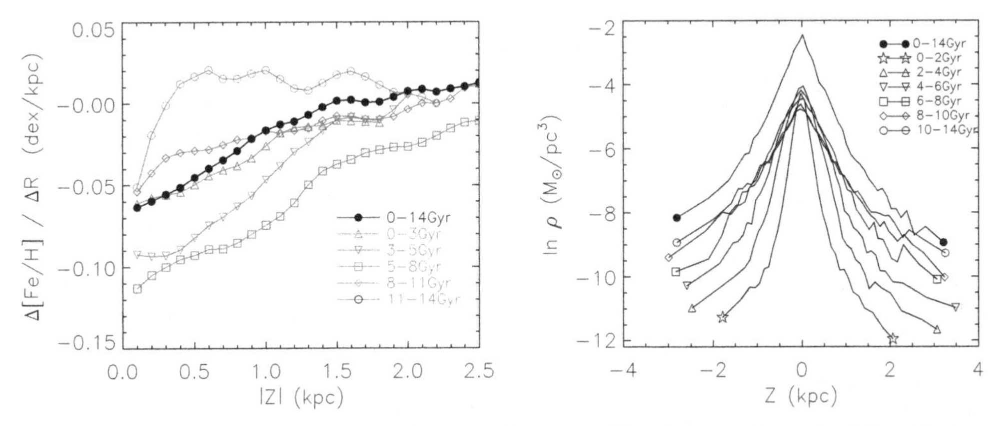

Figure 3. *Left*: radial metallicity gradient as a function of height away from the disk mid-plane for stellar populations of different ages. *Right*: vertical stellar mass density profile of different age populations in a Galactocentric radial annulus of $8.5 < R < 9.0$ kpc.

leased as LSS-GAC value-added catalogues (Yuan *et al.* 2015, Xiang *et al.* 2017b), and can be downloaded via *http://lamost973.pku.edu.cn/site/data*. The right panel of Fig. 1 shows the distribution of stars in the $T_{\rm eff}$–M_V diagram yielded by LSP3. The figure shows clear main sequence and red-giant branch, as well as a compact red clump. The diagram is in good agreement with stellar isochrones. The left panel of Fig. 2 shows a comparison of distance derived using the LSP3 $M_{\rm V}$ with those derived using *Gaia* TGAS parallax (Lindegren *et al.* 2016) for 50 000 stars that have a TGAS-based absolute magnitude error smaller than 0.2 mag. The figure illustrates that systematic error of the LAMOST distance estimates using LSP3 $M_{\rm V}$ is negligible, and the random error is smaller than 12 per cent.

From the LSP3 stellar parameters, we have selected a sample of a million main sequence turn-off and subgiant (MSTO-SUB) stars of the disk ([Fe/H] > -1), which have a minimal SNR of 20, and a median SNR of about 60. Stellar age and mass of the MSTO-SUB stars are estimated by matching stellar isochrones with a Bayesian algorithm using $T_{\rm eff}$, M_V, [Fe/H] and [α/Fe]. Extensive examinations indicate that half of the whole sample stars have an age uncertainty of only 20–30 per cent, and typical uncertainty of the mass estimates is a few per cent. The sample is going to be public available soon (Xiang *et al.* submitted). The right panel of Fig. 2 shows a comparison of stellar ages estimated using the LSP3 parameters with those using the TGAS parallax. The systematic difference is small, and the dispersion is 14 per cent only.

3. Disk structure and metallicity from mono-age stellar populations

The left panel of Fig. 3 shows the radial stellar metallicity gradients as a function of height above the disk mid-plane for populations of different ages. The gradients exhibit significant temporal evolution. The oldest stars have slightly positive (consistent with zero given an error of 0.01–0.02 dex/kpc) gradients at almost all heights except the very thin disk region. Younger stars have negative gradients, which flatten with height. Stars of 5–8 Gyr exhibit the steepest gradients, while both younger and older stars have smaller gradients. This confirms the findings of Xiang *et al.* (2015b), who utilized a smaller sample with less accurate age estimates. The different behaviors of metallicity gradients between the oldest and the younger populations strongly suggest the existence of different phases in the assemblage history of the Galactic disk. The results provide important constrains on the disk formation and evolution scenarios.

A three-dimensional stellar mass density distribution of the disk within a few kpc of the solar neighborhood for each mono-age population has been mapped using the MSTO-SUB stars. In doing so, selection functions of the LAMOST Galactic surveys are corrected, and the initial mass function of Kroupa (2001) is used to transfer the mass of MSTO-SUB stars to the total stellar mass of the whole underlying stellar population. The right panel of Fig. 3 shows the vertical stellar mass density profiles of different age populations in the disk of $8.5 < R < 9.0$ kpc. The figure shows a clear increasing trend of disk thickness with increasing stellar age. For almost all populations, the profiles can not be simply described by a single exponential function, which presents as a linear function in logarithm space, indicating that there are young thick disk and old thin disk, whose origins are particular interesting to further investigate.

Acknowledgements

This work is supported by National Key Basic Research Program of China 2014CB845700 and Joint Funds of the National Natural Science Foundation of China (Grant No. U1531244 and U1331120). Guoshoujing Telescope (the Large Sky Area Multi-Object Fiber Spectroscopic Telescope LAMOST) is a National Major Scientific Project built by the Chinese Academy of Sciences. Funding for the project has been provided by the National Development and Reform Commission. LAMOST is operated and managed by the National Astronomical Observatories, Chinese Academy of Sciences. The LAMOST FELLOWSHIP is supported by Special Funding for Advanced Users, budgeted and administrated by Center for Astronomical Mega-Science, Chinese Academy of Sciences (CAMS). This work has made use of data from the European Space Agency (ESA) mission *Gaia* (https://www.cosmos.esa.int/gaia), processed by the *Gaia* Data Processing and Analysis Consortium (DPAC, https://www.cosmos.esa.int/web/gaia/dpac/consortium). Funding for the DPAC has been provided by national institutions, in particular the institutions participating in the *Gaia* Multilateral Agreement.

References

Deng, L.-C., Newberg, H. J., Liu, C., *et al.*, 2012, *Research in Astronomy & Astrophysics*, 12, 735

Gaia Collaboration, Prusti, T., de Bruijne, J. H. J., Brown, A. G. A., *et al.*, 2016, *A&A*, 595, 1

Haywood, M., Di Matteo, P., Lehnert, M. D., Katz, D., & Gómez, A., 2013, *A&A*, 560, 109

Li, J., Han, C., Xiang, M.-S., *et al.*, 2016, *Research in Astronomy & Astrophysics*, 16, 110

Lindegren, L., Lammers, U., Bastian, U., *et al.*, 2016, *A&A*, 595, 4

Liu, X.-W., Yuan, H.-B., Huo, Z.-Y., *et al.*, 2014, in: S. Feltzing, G. Zhao, N. A. Walton & P. A. Whitelock (eds.), *Setting the scene for Gaia and LAMOST*, Proc. IAU Symposium No. 298, p. 310

Kroupa, P., 2001, *MNRAS*, 322, 231

Rosenfield, P., Marigo, P., Girardi, L., *et al.*, 2016, *ApJ*, 822, 73

Xiang, M.-S., Liu, X.-W., Yuan, H.-B., *et al.*, 2015, *MNRAS*, 448, 822

Xiang, M.-S., Liu, X.-W., Yuan, H.-B., *et al.*, 2015, *Research in Astronomy & Astrophysics*, 15, 1209

Xiang, M.-S., Liu, X.-W., Shi, J.-R., *et al.*, 2017, *MNRAS*, 464, 3657

Xiang, M.-S., Liu, X.-W., Yuan, H.-B., *et al.*, 2017, *MNRAS*, 467, 1890

Yuan, H.-B., Liu, X.-W., Huo, Z.-Y., *et al.*, 2015, *MNRAS*, 448, 855

Zhao, G., Zhao, Y.-H., Chu, Y.-Q., Jing, Y.-P., & Deng, L.-C., 2012, *Research in Astronomy & Astrophysics*, 12, 723

Astronomy and Astrophysics in the Gaia sky
Proceedings IAU Symposium No. 330, 2017
A. Recio-Blanco, P. de Laverny, A.G.A. Brown
& T. Prusti, eds.

doi:10.1017/S1743921317005622

Mapping young stellar populations towards Orion with *Gaia* DR1

Eleonora Zari and Anthony G. A. Brown

Leiden Observatory, Niels Bohrweg 2, 2333 CA Leiden, the Netherlands
email: zariem@strw.leidenuniv.nl

Abstract. OB associations are prime sites for the study of star formation processes and of the interaction between young massive stars with the interstellar medium. Furthermore, the kinematics and structure of the nearest OB associations provide detailed insight into the properties and origin of the Gould Belt. In this context, the Orion complex has been extensively studied. However, the spatial distribution of the stellar population is still uncertain: in particular, the distances and ages of the various sub-groups composing the Orion OB association, and their connection to the surrounding interstellar medium, are not well determined. We used the first *Gaia* data release to characterize the stellar population in Orion, with the goal to obtain new distance and age estimates of the numerous stellar groups composing the Orion OB association. We found evidence of the existence of a young and rich population spread over the entire region, loosely clustered around some known groups. This newly discovered population of young stars provides a fresh view of the star formation history of the Orion region.

Keywords. Stars: distances - stars: formation - stars: pre-main sequence - stars: early-type

1. Introduction

OB stars are not distributed randomly in the sky, but cluster in loose, unbound groups, which are usually referred to as OB associations (Blaauw 1964). In the solar vicinity, OB associations are located near star-forming regions (Bally 2008), hence they are prime sites for large scale studies of star formation processes and of the effects of early-type stars on the interstellar medium. The Orion star forming region is the nearest ($d \sim 400\,\mathrm{pc}$) giant molecular cloud complex. All stages of star formation can be found here, from protoclusters, to OB associations (Brown at al. 1994; Bally 2008; Briceño 2008; Muench *et al.* 2008; Da Rio *et al.* 2014). The different modes of star formation occurring here (isolated, distributed, and clustered) allow us to study the effect of the environment on star formation processes in great detail. Moreover, the Orion region is an excellent nearby example of the effects that young, massive stars have on the surrounding interstellar medium (Ochsendorf *et al.* 2015, Schlafly *et al.* 2015). The Orion OB association (Ori OB1) consists of several groups, with different ages, partially superimposed along our line of sight (Bally 2008) and extending over an area of $\sim 30^\circ \times 25^\circ$ (see Fig. 1, left). We use the first *Gaia* data release (Gaia Collaboration 2016a,b), hereafter *Gaia* DR1, to explore the three dimensional arrangement and the age ordering of the many stellar groups towards Orion, with the overall goal to construct a new classification and characterization of the stellar population in the region. Our approach is based on the parallaxes provided in the *Tycho-Gaia Astrometric Solution* (TGAS, Michalik *et al.* 2015, Lindegren *et al.* 2015), a sub-set of the *Gaia* DR1 catalogue, and on the combination of *Gaia* DR1 and 2MASS photometry. We find evidence for the presence of an extended young (age $<$ 20 Myr) population, loosely clustered around some known groups: 25 Ori, ϵ Ori and σ Ori, and NGC 1980 and the ONC. We derive distances to these sub-groups and (relative) ages. Our

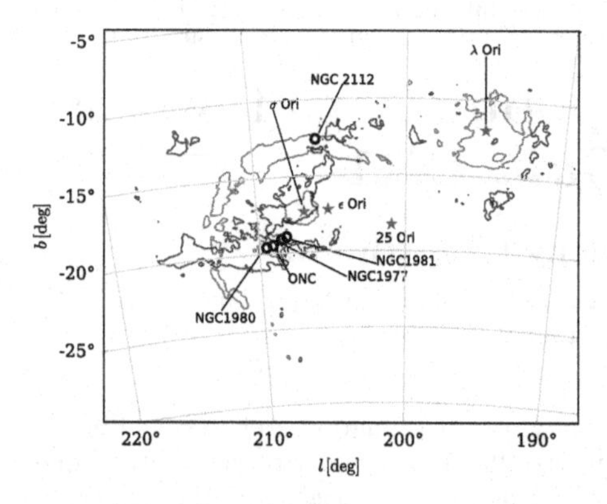

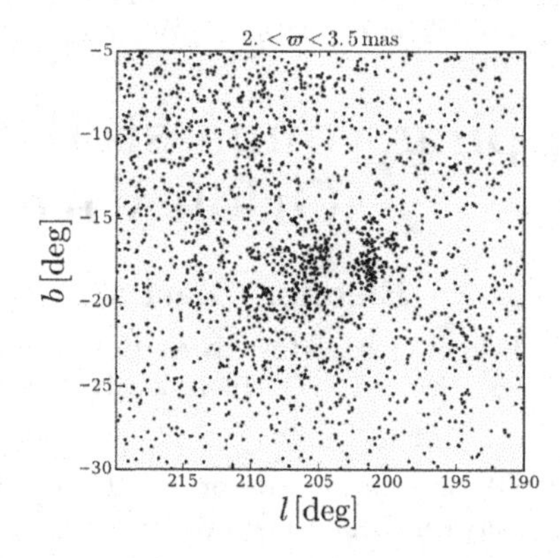

Figure 1. Left: Schematic representation of the field. The black contours correspond to the regions where $A_V > 2.5$ mag (Planck Collaboration, 2014), while the blue contours show the H_α structures (Finkbeiner, 2003): Barnard's loop and the λ Ori bubble. The positions of some known groups and stars are indicated with black circles and red stars, respectively. Right: Positions in the sky of the TGAS sources selected with Eq. 2.1 with parallax $2 < \varpi < 3.5$ mas.

results are the first step to unveil the complex star formation history of Orion towards obtaining a general overview of the episodes and the duration of the star formation processes in the entire region.

2. Orion in *Gaia* DR1

We first consider all the TGAS sources in the field. Since the motion of Orion OB1 is mostly directed radially away from the Sun, the observed proper motions are small. For this reason, a rough selection of the TGAS sources can be made requiring:

$$(\mu_{\alpha*} - 0.5)^2 + (\mu_\delta + 1)^2 < 25\,\mathrm{mas}^2\,\mathrm{yr}^{-2}, \tag{2.1}$$

where $\mu_{\alpha*}$ and μ_δ are the proper motions in right ascension and declination. Fig. 1 (right) shows the distribution in the sky of the sources with parallax $2 < \varpi < 3.5$ mas, which corresponds to a distance $285 < d < 500$ pc. Some source over-densities towards the center of the field, $(l, b) \sim (205°, -18°)$, are clearly visible, and they are not due to projection effects but are indicative of real clustering in three dimensional space. The stars within the density enhancements of Fig. 1 (right) also show a small gradient in the parallax - galactic longitude plane. In particular, the stars associated with 25 Ori have slightly larger parallaxes than those in the direction towards the ONC. We combine *Gaia* and 2MASS photometry to construct color-magnitude diagrams of the sources within the density enhancements. These sources define a sequence at the bright end of the color-magnitude diagram (black big dots in Fig. 2, left). This prompts us to look further at the entire field, and to use the entire *Gaia* DR1 catalogue to find evidence of the faint counterpart of the concentration reported above. Fig. 2 shows a G vs. $G - J$ color magnitude diagram of the central region of the field. A dense, red sequence is visible between $G = 14$ mag and $G = 18$ mag. This sequence might indicate the presence of a population of young stars, since it is situated above the main sequence at the distance of Orion. We decided to eliminate the bulk of the field stars by requiring the following conditions to hold (orange line in Fig. 2):

$$G < 2.5\,(G - J) + 10.5 \ \text{ for } G > 15\,\mathrm{mag}, \ G < 2.9\,(G - J) + 9.9 \ \text{ for } G < 15\,\mathrm{mag}. \tag{2.2}$$

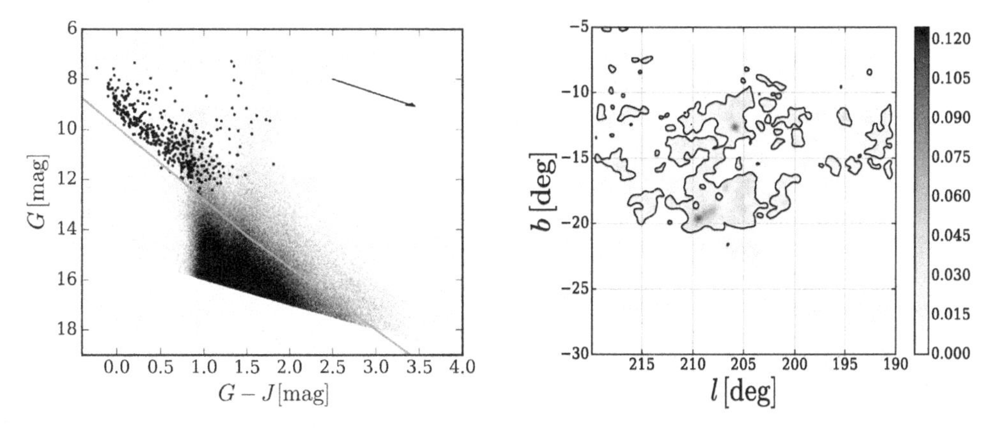

Figure 2. Left: Colour magnitude diagram of the *Gaia* DR1 sources cross matched with 2MASS. The sources we focus on are those responsible for the dense, red sequence in the lower part of the diagram. The orange line is defined in Eq. (2.2), and was used to separate the bulk of the field stars from the population we intended to study. The big black points represent the sources within the TGAS density enhancements. The arrow shows the reddening vector corresponding to $A_V = 1$ mag. Right: Background subtracted normalized probability density function of the stars selected with with Eq. (2.2). The density enhancements visible in the centre of the field correspond to the TGAS density enhancements (cf. Fig. 1, right). The peak at $(l, b) \sim (206, -12.5)$ deg corresponds to the open cluster NGC 2112.

We analyse the distribution in the sky of the sources selected with Eq. (2.2) using a multivariate normal kernel, with isotropic bandwidth $= 0.03°$. Fig. 2 (right) shows the background subtracted normalized density function of the source distribution. The groups clearly separate from the field stars. We selected all the sources within the contour levels shown in Fig. 2. To determine the age(s) of the population(s) we identified, we perform a Bayesian isochrone fit using a method similar to the one described in Jorgensen & Lindegren (2005) and Valls-Gabaud (2014). We compare the observed G magnitude and $G-J$ color to those predicted by the PARSEC (PAdova and TRieste Stellar Evolution Code, Bressan *et al.* 2012, Chen *et al.* 2014, Tang *et al.* 2014) library of stellar evolutionary tracks. We applied an extinction correction of $A_V = 0.5$ mag and we fixed the metallicity to $Z = 0.02$, following Brown *et al.* (1994). We applied the fitting procedure to all the stars within the contour levels of Fig. 2 (right), fixing the parallax to $\varpi = 2.65$ mas. Fig. 3 shows the density (Gaussian kernel, with bandwidth $= 0.05°$) of the source sky distribution as a function of their age, t. The coordinates of the density enhancements change with time. This means that the groups we identified have different relative ages. The last panel shows the stars with estimated ages > 20 Myr. These are field stars: their distribution is almost uniform, and increases towards the Galactic plane.

3. Conclusions

We studied the stellar population towards Orion and we found evidence for the presence of a young stellar population, at parallax $\varpi \sim 2.65$ mas, loosely distributed around some known clusters: 25 Ori, ϵ Ori and σ Ori, and NGC 1980 and the ONC. We also found hints of the presence of a parallax gradient going from 25 Ori to the ONC. We estimated the ages of the populations, and we found an age gradient corresponding to the parallax gradient. In particular, the closest stars to the Sun are also the oldest ones.

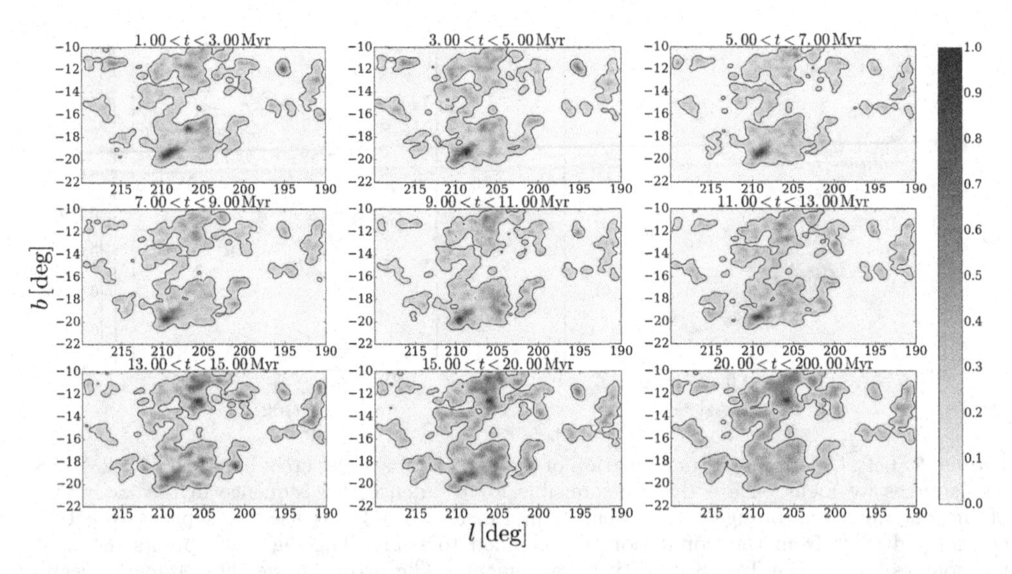

Figure 3. Distribution in the sky of the sources in different age intervals. The contours represent the 0.05 density level and are shown only for visualization purposes. The first eight panels show stars with estimated ages < 20 Myr, while the last one shows older sources. The young stars are not coeval, in particular the age distribution shows a gradient, going from 25 Ori and ϵ Ori towards the ONC and NGC 1980.

Acknowledgements. This project was developed in part at the 2016 NYC Gaia Sprint, hosted by the Center for Computational Astrophysics at the Simons Foundation in New York City. This work has made use of data from the European Space Agency (ESA) mission *Gaia* (`https://www.cosmos.esa.int/gaia`), processed by the *Gaia* Data Processing and Analysis Consortium (DPAC, `https://www.cosmos.esa.int/web/gaia/dpac/consortium`). Funding for the DPAC has been provided by national institutions, in particular the institutions participating in the *Gaia* Multilateral Agreement.

References

Blaauw, A. 1964, *ARA&A*, 2, 213
Bally, J. 2008, *Handbook of Star Forming Regions, Volume I*, ed. B. Reipurth, 459
Bressan, A., Marigo, P., Girardi, L. *et al.* 2012, *MNRAS*, 427, 127
Briceño, C. 2008, *Handbook of Star Forming Regions, Volume I*, ed. B. Reipurth, 838
Brown, A. G. A., de Geus E. J. & de Zeeuw P. T. 1994, *A & A*, 289, 101
Chen, Y., Girardi L. & Bressan, A. 2014, *MNRAS*, 444, 2525
Da Rio N., Tan J. C. & Jaehnig, K. 2014, *ApJ*, 795, 55
Finkbeiner, D. P. 2003, *ApJS*, 146, 407
Gaia Collaboration, Brown, A. G. A., Vallenari A. *et al.* 2016a, *A & A*, 595, A2
Gaia Collaboration, Prusti, T. de Bruijne, J. H. J., *et al.* 2016b, *A & A*, 595, A1
Jørgensen, B. R. & Lindegren, L. 2005, *A&A*, 436, 127
Lindegren L., Lammers, U., Bastian U. *et al.* 2016, *A&A*, 595, A4
Michalik, D., Lindegren L. & Hobbs, D. 2015, *A&A*, 574, A115
Muench, A., Getman K., Hillenbrand L. & Preibisch, T. 2008, *Handbook of Star Forming Regions, Volume I*, ed. B. Reipurth, 483
Ochsendorf, B. B., Brown A. G. A., Bally J. & Tielens, A. G. G. M. 2015, *ApJ*, 808, 111
Planck Collaboration, Abergel, A., Ade, P. A. R., *et al.* 2014, *A & A*, 571, A11
Schlafly, E. F., Green G., & Finkbeiner D. P. 2015, *ApJ*, 799, 116
Tang, J., Bressan A., Rosenfield P. *et al.* 2014, *MNRAS*, 445, 4287
Valls-Gabaud, D. 2014, *EAS Publication Series, Vol. 65*, 225, 265

Astronomy and Astrophysics in the Gaia sky
Proceedings IAU Symposium No. 330, 2017
A. Recio-Blanco, P. de Laverny, A.G.A. Brown & T. Prusti, eds.

doi:10.1017/S1743921317006597

The kinematics of the white dwarf population from the SDSS DR12

B. Anguiano, A. Rebassa-Mansergas, E. García-Berro, S. Torres, K. Freeman and T. Zwitter

Department of Astronomy, University of Virginia,
Charlottesville, VA 22904-4325, USA
email: ba7t@virginia.edu

Abstract. We use the Sloan Digital Sky Survey Data Release 12, which is the largest available white dwarf catalog to date, to study the evolution of the kinematical properties of the population of white dwarfs in the Galactic disc. We derive masses, ages, photometric distances and radial velocities for all white dwarfs with hydrogen-rich atmospheres. For those stars for which proper motions from the USNO-B1 catalog are available the true three-dimensional components of the stellar space velocity are obtained. This subset of the original sample comprises 20,247 objects, making it the largest sample of white dwarfs with measured three-dimensional velocities. Furthermore, the volume probed by our sample is large, allowing us to obtain relevant kinematical information. In particular, our sample extends from a Galactocentric radial distance $R_{\rm G} = 7.8$ kpc to 9.3 kpc, and vertical distances from the Galactic plane ranging from $Z = -0.5$ kpc to 0.5 kpc. We examine the mean components of the stellar three-dimensional velocities, as well as their dispersions with respect to the Galactocentric and vertical distances. We confirm the existence of a mean Galactocentric radial velocity gradient, $\partial\langle V_{\rm R}\rangle/\partial R_{\rm G} = -3 \pm 5$ km s^{-1} kpc^{-1}. We also confirm North-South differences in $\langle V_{\rm z}\rangle$. Specifically, we find that white dwarfs with $Z > 0$ (in the North Galactic hemisphere) have $\langle V_{\rm z}\rangle < 0$, while the reverse is true for white dwarfs with $Z < 0$. The age-velocity dispersion relation derived from the present sample indicates that the Galactic population of white dwarfs may have experienced an additional source of heating, which adds to the secular evolution of the Galactic disc.

Keywords. (stars:) white dwarfs; Galaxy: general; Galaxy: evolution; Galaxy: kinematics and dynamics; (Galaxy:) solar neighborhood; Galaxy: stellar content

1. Introduction

The ensemble properties of the white dwarf population are recorded in the white dwarf luminosity function, which therefore carries crucial information about the star formation history, the initial mass function, or the nature and history of the different components of our Galaxy — see the recent review of García-Berro & Oswalt (2016) for an extensive list of possible applications, as well as for updated references on this topic. Among these applications perhaps the most well known of them is that white dwarfs are frequently used as reliable cosmochronometers. The need of a complete sample with accurate measurements of true space velocities, distances, masses and ages, is crucial for studying the evolution of our Galaxy. In this sense, it is worth emphasizing that little progress has been done to use the Galactic white dwarf population to unravel the evolution of the Galactic disc studying the age-velocity relationship (AVR). Since white dwarfs are excellent natural clocks we use them to compute accurate ages and in this way we determine the AVR in the solar vicinity. All this allows us to investigate the kinematic evolution of the Galactic disc (see Anguiano *et al.* (2017) for details).

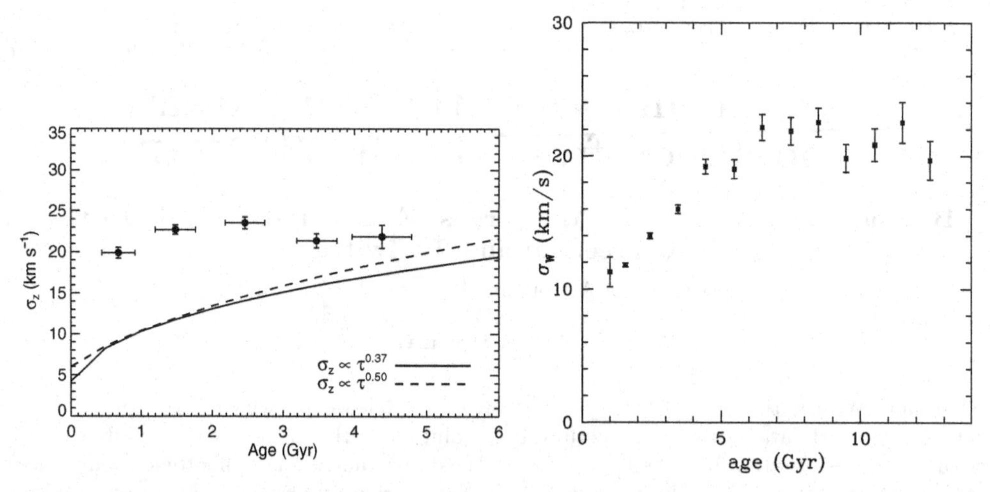

Figure 1. Left: Age-velocity dispersion relation for the vertical component of the velocity using the ages derived using the IFMR of Catalán *et al.* (2008). Dynamical heating functions following a power law, $\sigma_z \propto \tau^{\alpha}$, with $\alpha = 0.37$ (solid line) and 0.50 (dashed line) are also represented (Anguiano *et al.* (2017)). Right: Vertical velocity dispersion σ_W *vs* stellar age for the solar neighborhood, for stars with [Fe/H] > -0.3; data from Casagrande *et al.* (2011).

2. Age-velocity dispersion relation

The age-velocity dispersion relation derived from the present WD sample indicates that the Galactic population of white dwarfs may have experienced an additional source of heating, which adds to the secular evolution of the Galactic disc (Fig. 1). The GCS stars in the age range of our WDs (< 4 Gyr) have velocity dispersions that are clearly less than the 23 km s^{-1} found for the colder component of the WDs. The origin of this heating mechanism remains unclear. One possibility is that some thick disc-halo stars may have contaminated the younger bins of the AVR, thus making the age-velocity dispersion hotter. Another possibility is that there is an intrinsic dispersion for these stars. Given that white dwarfs are the final products of the evolution of stars with low and intermediate masses they might have experienced a velocity kick of ~ 10 km s^{-1} during the final phases of their evolution Davis *et al.* (2008). It is also important to keep in mind that, although we have excluded in our analysis all white dwarfs with masses smaller than 0.45 $M_{\odot}$, because they are expected to be members of close binaries, a possible contamination by close pairs — of the order of $\sim$10 per cent Badenes & Maoz (2012), Maoz & Hallakoun (2017) — may influence the age-velocity dispersion relation. Finally, our sample of DA white dwarfs is not homogeneously distributed on the sky, but instead is drawn from the SDSS. This means that, possibly, important selection effects could affect the results. All these alternatives need to be carefully explored employing detailed population synthesis models. We postpone this study for a future publication.

References

Anguiano, B., Rebassa-Mansergas, A., Garcia-Berro, E., *et al.* 2017, arXiv:1703.09152
Badenes, C. & Maoz, D. 2012, *ApJ*, 749, L11
Casagrande, L., Schönrich, R., Asplund, M., *et al.* 2011, *A&A*, 530, A138
Catalán, S., Isern, J., García-Berro, E., & Ribas, I. 2008, *MNRAS*, 387, 1693
Davis, D. S., Richer, H. B., King, I. R., *et al.* 2008, *MNRAS*, 383, L20
García-Berro, E. & Oswalt, T. D. 2016, *NewAR*, 72, 1
Maoz, D. & Hallakoun, N. 2017, *MNRAS*, 467, 1414

Astronomy and Astrophysics in the Gaia sky
Proceedings IAU Symposium No. 330, 2017
A. Recio-Blanco, P. de Laverny, A.G.A. Brown & T. Prusti, eds.

doi:10.1017/S1743921317006184

Accurate atomic data for Galactic Surveys

Maria Teresa Belmonte, Juliet C. Pickering, Christian Clear, Florence Liggings and Anne P. Thorne

Dept. Physics, Imperial College London, South Kensington Campus, London SW7 2BW, UK
email: m.belmonte-sainz-ezquerra@imperial.ac.uk

Abstract. Fourier Transform spectroscopy is able to provide high accuracy atomic parameters needed by many ongoing galactic surveys. Our laboratory has carried out a study of the neutral iron spectrum over the last years to measure oscillator strengths much needed for the calculation of chemical abundances. The main aim of this contribution is to encourage further dialogue with astronomers regarding their current necessities of spectroscopic data, as this would help spectroscopists to prioritise present-day needs within the field.

Keywords. atomic data, techniques: spectroscopic, stars: abundances

1. Introduction

The analysis of stellar spectra is vital in the determination of chemical abundances, the understanding of galaxy formation and evolution or the synthesis of the different elements. However, despite the large investment of time and money done to record spectra of astrophysical objects at unprecedented resolution, this work is being hindered by the lack of accurate atomic data, the Achilles' heel of stellar parameter determination (Bigot & Thévenin 2006). As stellar models are strongly dependent on parameters such as transition probabilities (Heiter *et al.* 2015) and their accuracy, this shortage of data or its poor quality leads to mistaken values of the chemical abundances and stellar ages.

The Fourier Transform Spectroscopy (FTS) Laboratory at Imperial College London has been conducting a very fruitful collaboration with the National Institute of Standards and Technology (NIST), the University of Wisconsin and the University of Lund over the last years to obtain accurate oscillator strengths (transition probabilities) for many spectral lines needed in surveys such as Gaia-ESO or APOGEE. The results can be found in Ruffoni *et al.* (2013), Ruffoni *et al.* (2014) and Den Hartog *et al.* (2014). The $\log(gf)$-values are obtained by combining branching fractions obtained in a Fourier Transform Spectrometer with upper energy level radiative lifetimes measured in a time-resolved laser-induced fluorescence experiment.

2. Atomic data needs

Due to the strong influence of atomic data on the algorithms used to model stellar atmospheres, a big effort is being made to compile critically reviewed line lists with reliable atomic data for different surveys (Heiter *et al.* 2015, Shetrone *et al.* 2015) that can be used as a standard input for the different models. This assures some homogeneity in the final results (Smiljanic *et al.* 2014) and allows the comparison of the different techniques (Hinkel *et al.* 2016). However, the existing experimental atomic data is very scarce. This obliges astronomers to chose values from a very limited selection which, on many occasions, contains atomic data with very high uncertainties.

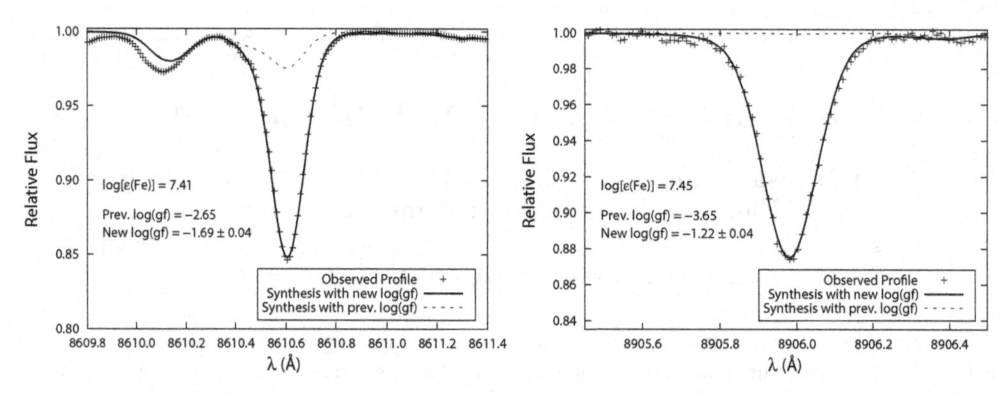

Figure 1. Sample line profiles used to measured Fe abundance.

The uncertainty of the oscillator strength used for the calculation of chemical abundances has a direct impact on the accuracy of the abundance obtained. Data needs have been previously discussed in some detail in Pickering *et al.* (2011), but the fast development of new surveys using high resolution spectra suggests that a new revision of the current atomic data necessities should be undertaken. Hence the appeal we would like to launch to all the Gaia community to start a fluid exchange of ideas to determine what data might be needed for future data releases.

3. Progress in new atomic data

The acute need for new accurately measured atomic parameters within the field of astronomy can be tackled by using high-resolution Fourier Transform spectroscopy. Transition probabilities, for example, can be obtained experimentally by combining branching fractions, BF, measured from high resolution emission spectra with upper level lifetimes obtained from time-resolved laser-induced fluorescence:

$$A_{ul} = \frac{BF_{ul}}{\tau_u} \tag{3.1}$$

with u and l representing the upper and lower energy level, respectively, and τ the radiative lifetime. Oscillator strengths can be obtained from transition probabilities by using the expression:

$$\log(g_l f) = \log\left[A_{ul} g_u \lambda^2 \times 1.499 \times 10^{-14}\right] \tag{3.2}$$

where g_l and g_u are the statistical weights of the lower and upper energy level and λ is the wavelength of the spectral line expressed in nm. Fourier transform spectroscopy has evolved dramatically over the past thirty years and nowadays, it is able to provide values of transition probabilities with uncertainties as low as 5% (0.02 dex in $\log(gf)$) for strong transitions.

The iron spectrum is of vital importance to obtain stellar metallicity. We have provided urgently needed Fe I oscillator strengths for two different surveys: APOGEE, where several tens of new Fe I $\log(gf)$s were measured in the H-band (1.5 - 1.7 μm) (Ruffoni *et al.* 2013) and the Gaia-ESO (GES) Survey (Ruffoni *et al.* 2014, Den Hartog *et al.* 2014). Studies of the Fe I spectrum within the GES spectral range revealed over 500 lines that were strong and unblended in stellar spectra. Around 50 of them had no previously published data and were urgently needed by GES. Fig 1, taken from Ruffoni *et al.* (2014), illustrates two sample line profiles used to measure the Solar Fe abundance

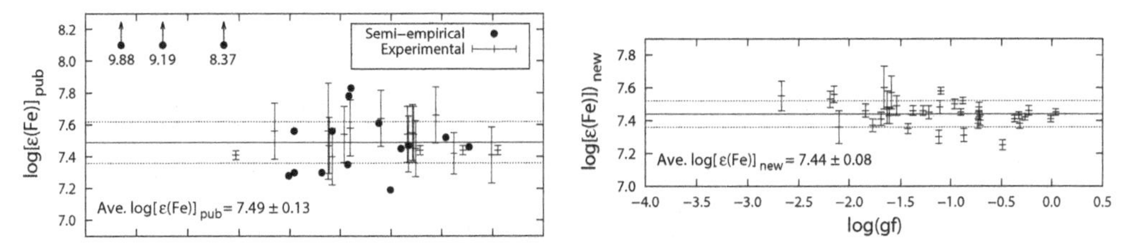

Figure 2. Determination of line-by-line solar Fe abundances using unblended solar lines.

($\log[\varepsilon(Fe)]$). The value of the Fe abundance provided in the plot was obtained by fitting the spectral lines with our new oscillator strengths. The dotted line shows the profiles that would have been obtained for these abundances with the best previously published $\log(gf)$-values.

To assess the impact of our new results on stellar spectral syntheses, we determined line-by-line solar Fe abundances for those that are unblended in the Sun and have good broadening parameters and continuum placement. It can see from Fig 2 (Ruffoni *et al.* 2014) how our new laboratory measured $\log(gf)$s have smaller uncertainties and the mean abundance calculated with them agrees well with recent values taken from the literature.

Collaboration with astronomers has been very fruitful, especially regarding the preparation of target line lists needed by different surveys. Our group has also published transition probabilities of neutral vanadium (Holmes *et al.* 2016) and is currently working on other elements such as manganese, nickel or scandium. Other data available from laboratory spectra are line wavelengths, used in cosmology to study the variation of fundamental constants, atomic energy levels needed to improve the theoretical modelling of lines and hyperfine splitting parameters to improve the synthesis of stellar spectra.

4. Conclusions

We would like to launch an appeal to collaborate with all those astronomers and research groups working on Gaia or any of the ongoing galactic surveys and who need accurate atomic data. Working closely with astronomers to know their data necessities, we will be able to provide very accurate atomic parameters that are much needed within the field.

This work has been funded by STFC of UK, Royal Society and Leverhulme Trust.

References

Bigot, L. & Thévenin, F. 2006, *Mon. Not. R. Astron. Soc.*, 372, 609

Den Hartog, E. A., Ruffoni, M. P., Lawler, J. E., Pickering, J. C., Lind, K., & Brewer, N. R. 2014, *ApJS*, 215, 23

Heiter, U., Lind, K., Asplund, M., *et al.* 2015, *Phys. Scr.*, 90, 054010.

Hinkel, N. R., Young, P. A., Pagano, M. D., *et al.* 2016, *ApJS*, 226:4.

Holmes, C. E., Pickering, J. C., Ruffoni, M. P., Blackwell-Whitehead, R., Nilsson, H., Engström, L., Hartman, H., Lundberg, H., & Belmonte, M. T. 2016, *ApJS*, 224:35.

Pickering, J. C., Blackwell-Whitehead R., Thorne, A. P., Ruffoni M. P., & Holmes C. E. 2011, *Can. J. Phys.*, 89, 387

Ruffoni M. P., Allende-Prieto C., Nave G., & Pickering J. C. 2013, *ApJ*, 779, 17

Ruffoni, M. P., Den Hartog, E. A., Lawler, J. E., Brewer, N. R., Lind, K., Nave, G., & Pickering, J. C. 2014, *Mon. Not. R. Astron. Soc.*, 441, 3127

Shetrone, M., Bizyaev, D., Lawler, J. E., *et al.* 2015, *ApJS*, 221:24.

Smiljanic, R., Korn, A. J., Bergemann, M., *et al.* 2014, *A&A*, 570, A122.

Astronomy and Astrophysics in the Gaia sky
Proceedings IAU Symposium No. 330, 2017
A. Recio-Blanco, P. de Laverny, A.G.A. Brown
& T. Prusti, eds.

doi:10.1017/S1743921317005865

Age dependence of metallicity gradients in the Galactic disc from astrometry and asteroseismology

Luca Casagrande

Research School of Astronomy & Astrophysics, Mount Stromlo Observatory
The Australian National University, ACT 2611, Australia
email: luca.casagrande@anu.edu.au

Abstract. Asteroseismology allows us to determine stellar parameters (distances, masses and ages) independently from Gaia astrometry, and it provides us with a new and complementary tool for studying stellar populations in the Galaxy. The prospects and synergies that asteroseismic and astrometric space-borne missions reserve to the field of Galactic archaeology are marvellous, and results have already started to emerge. For example, the study of metallicity gradients as function of age will provide powerful constraints to understand the evolution of the Milky Way disc at high-redshift.

Keywords. Galaxy: stellar content, Galaxy: disc, Galaxy: evolution, stars: fundamental parameters, stars: oscillations, surveys, techniques: photometric

---

1. Asteroseismology: the new kid on the block

Red giant stars are ideal targets to decipher the formation history of the Milky Way: on the HR diagram they span a vastly different range of gravities and luminosities, thus probing a large range of distances. Their ages essentially cover the entire history of the Universe, thus making them fossil remnants from different epochs of the formation of the Galaxy. The cold surface temperatures encountered in red giants are the realm of interesting atomic and molecular physics shaping their emergent spectra. This temperature regime is also dominated by convection, which is the main driver of the oscillation modes that we are now able to detect in several thousands of stars thanks to space borne asteroseismic missions such as *CoRoT*, *Kepler/K2*, and soon to be *TESS*. By measuring oscillation frequencies in stars, asteroseismology allows us to measure fundamental physical quantities, masses and radii in particular, which otherwise would be inaccessible in single field stars. Masses and radii can then be used to obtain information on stellar distances and ages (e.g., Hekker & Christensen-Dalsgaard 2016, for a recent review).

Asteroseismic ages are independent from astrometric distances (although the two methods can, and should, be combined to place tighter observational constraints), and thus provide a new and powerful tool for all photometric and spectroscopic stellar surveys. Even when accurate astrometric distances are available to allow comparison of stars with isochrones, the derived ages are still uncertain, and statistical techniques are required to avoid biases. Furthermore, isochrone dating is meaningful only for stars in the turnoff and subgiant phase, where stars of different ages are clearly separated in the HR diagram. This is in contrast, for example, to stars on the red giant branch, where isochrones with vastly different ages can fit equally well observational constraints such as effective temperatures, metallicities and surface gravities within their errors (e.g., Soderblom 2010, for a review). Thus, asteroseismology is the only way forward to determine ages of red giant stars.

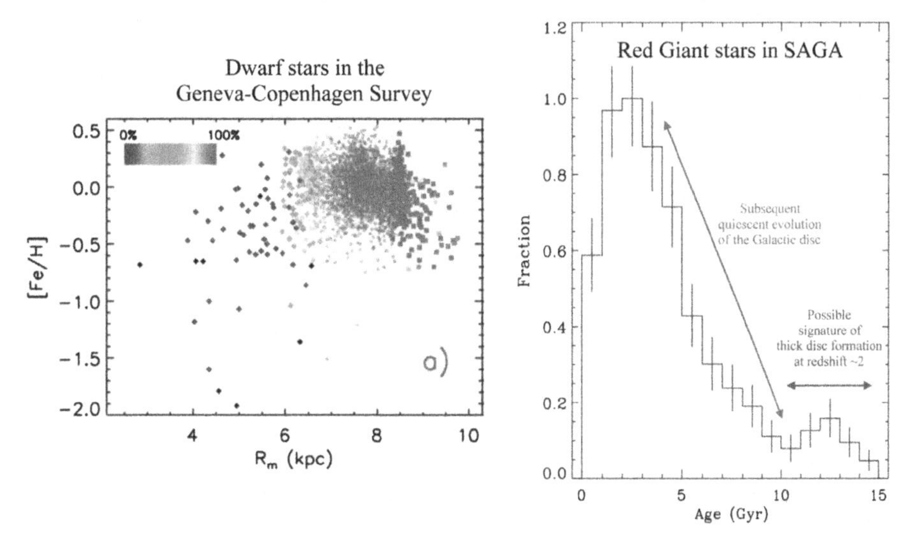

Figure 1. *Left panel:* metallicities of stars in the Geneva-Copenhagen Survey as function of their orbital radii. The probability of a star to belong to the thin disc is represented by colour (adapted from Casagrande *et al.* 2011). *Right panel:* age distribution of red giant stars in the Kepler field and SAGA (adapted from Casagrande *et al.* 2016).

2. Vertical & radial gradients from asteroseismology and astrometry

In Casagrande *et al.* (2011) we have exploited the power of combining Strömgren photometry with *Hipparcos* astrometry for the sake of studying stellar populations in the Solar Neighbourhood. We found that the radial metallicity gradient changes as function of time. Thick disc stars are preferentially older, and with a positive radial metallicity gradient (diamonds in Figure 1, left panel. See also discussion of Figure 18 in Casagrande *et al.* 2011). This positive metallicity gradient for thick disc stars might hold clues on the formation of the Milky Way thick disc at high-redshift (Kawata *et al.* 2017).

As part of the Strömgren survey for Asteroseismology and Galactic Archaeology (SAGA, Casagrande *et al.* 2014), we have then combined Strömgren photometry with *Kepler* asteroseismology. A direct measurement of the vertical age gradient of the Milky Way disc returns approximately 4 Gyr/kpc, though characterised by a large dispersion of ages at all heights. Once target selection effects are taken into account (e.g., Casagrande 2016), the smooth age distribution of red giant stars over the last 10 Gyr is consistent with that of a stellar population having a constant star formation rate; this is indicative of a rather quiescent evolution for the Milky Way disc since a redshift $z \simeq 2$ (modulo shifts on the absolute scale of stellar ages). More interestingly, the bump in the distribution at older ages could be indicative of thick disc formation (Figure 1, right panel. See Casagrande *et al.* 2016, for a more detailed discussion).

References

Casagrande, L., Schönrich, R., Asplund, M., *et al.* 2011, *A&A*, 530, A138
Casagrande, L., Silva Aguirre, V., Stello, D., *et al.* 2014, *ApJ*, 787, 110
Casagrande, L., Silva Aguirre, V., Schlesinger, K. J., *et al.* 2016, *MNRAS*, 455, 987
Casagrande, L. 2016, *Astronomische Nachrichten*, 337, 889
Hekker, S. & Christensen-Dalsgaard, J. 2016, arXiv:1609.07487
Kawata, D., Allende Prieto, C., Brook, C. B., *et al.* 2017, *MNRAS*, submitted.
Soderblom, D. R. 2010, *ARA&A*, 48, 581

Astronomy and Astrophysics in the Gaia sky
Proceedings IAU Symposium No. 330, 2017
A. Recio-Blanco, P. de Laverny, A.G.A. Brown
& T. Prusti, eds.

doi:10.1017/S1743921317006214

Galactic disk structure as revealed by LAMOST A stars

B.-Q. Chen,[1,4] X.-W. Liu,[1] H.-B. Yuan,[2] Y. Huang,[1,4] M.-S. Xiang,[3,4] C. Wang,[1] Z.-J. Tian[1,4] and H.-W. Zhang[1]

[1]Department of Astronomy, Peking University, Beijing 100871, P. R. China
email: bchen@pku.edu.cn
[2]Department of Astronomy, Beijing Normal University, Beijing 100875, P. R. China
[3]National Astronomy Observatories, Chinese Academy of Sciences, Beijing 100012, P. R. China
[4]LAMOST fellow

Abstract. Using the spectroscopic distances of over 0.12 million A-type stars selected from the LAMOST Spectroscopic Survey of the Galactic Anti-center (LSS-GAC), we map their three-dimensional number density distributions in the Galaxy. These stellar number density maps allow an investigation of the Galactic young age thin disk structure with no a priori assumptions about the functional form of its components. The data show strong evidence for a significant flaring young disk. A more detail analysis show that the stellar flaring have different behaviours between the Northern and the Southern Galactic disks. The maps also reveal spatially coherent, kpc-scale stellar substructure in the thin disk. Finally, we detect the Perseus arm stellar overdensity at R $\sim$ 10 kpc.

Keywords. Galaxy: disk, Galaxy: structure

---

1. Overview

The Milky Way is the only galaxy that we can study its structure using the stellar number densities. However the study is hampered by the heavy interstellar dust extinction and the difficulty of obtaining accuratedistances to individual stars. To study the Galactic thin disk, which has the strongest extinction and owns stars of a large range of ages, is particularly difficult.

Samples of stars with limited age range and accurate distance estimates can be used to avoid the problems discussed above. In this work we select a sample of A stars from the LSS-GAC (Liu *et al.* 2014) value-added catalogue, which contains data observed by LAMOST since June 2016 (Xiang *et al.* 2017b). In total, over 0.12 million A stars are selected. Absolute magnitude and stellar atmospheric parameters of these stars are estimated with the LAMOST Stellar Parameter Pipeline at Peking University (LSP3, Xiang *et al.* 2017a). Values of the interstellar extinction are obtained for individual stars with the standard pairing technique (Yuan *et al.* 2015). The selection effects of LSS-GAC are corrected similarly to the work of (Chen *et al.* 2017). To test the spectroscopic distances of stars in our sample, we compare them with with the Gaia-TGAS parallaxes (Gaia Collaboration *et al.* 2016). Our spectroscopic distances are in good agreement with the Gaia TGAS parallax, with a TGAS-based distance error smaller than 20%.

2. The Galactic thin disk (sub)structure

We have obtained the averaged stellar number density maps of our sample. The stellar density distributions for different ranges of ϕ show similar features. They do not show simplicity in agreement with a single exponential disk model. Significant stellar

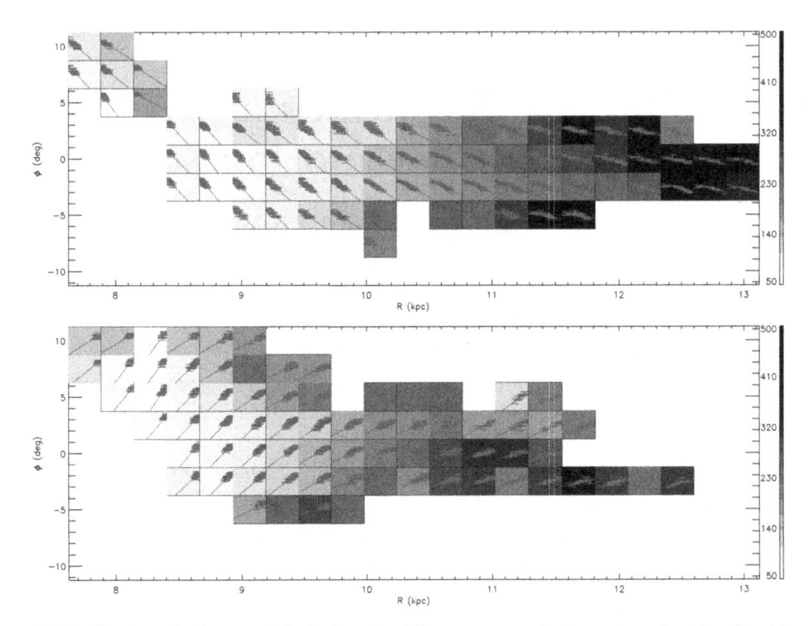

Figure 1. Disk flaring (color scale) derived with our sample for stars in the Northern (upper panel) and Southern (bottom panel) Galactic disks, respectively. Each sub-panel plots stellar density as a function of Z (points with error bars) along with a linear fit to the data (straight lines). The vertical axis of each sub-panel ranges ϕ from -10.5 to 10.5 deg and the horizontal axis R from 7 to 13 kpc.

flaring is clearly visible in the outer Galactic disk. There are also moderate North-South asymmetry in the outer part of the disk, starting at $R \sim 9$ kpc.

Detail analysis show that the stars flare differently between the Northern disk and the Southern disk (Fig. 1). Comparing to the Southern disk, the Northern disk has smaller solar neighbourhood scale height but larger flaring scale. We also find that for $R <$ 9.5 kpc, the Southern disk have more stars than the Northern disk, while for $R > 9.5$ kpc the opposite. This may be a result of the bending-mode perturbations recently identified across the disk plane (Sun *et al.* 2015). Finally, we detect an overdensity which may be refer to the Perseus Arm at $R \sim 10$ kpc.

Acknowledgements

This work is supported by National Key Basic Research Program of China 2014CB845700 and China Postdoctoral Science Foundation 2016M590014.

References

Chen, B.-Q., *et al.* 2017, MNRAS, 464, 2545

Gaia Collaboration, Prusti, T., de Bruijne, J. H. J., *et al.* 2016, *A&A*, 595, A1

Liu, X.-W., *et al.* 2014, in IAU Symposium, Vol. 298, IAU Symposium, ed. S. Feltzing, G. Zhao, N. A. Walton, & P. Whitelock, 310–321

Sun, N.-C., *et al.* 2015, *Research in Astronomy and Astrophysics*, 15, 1342

Xiang, M.-S., *et al.* 2017a, *MNRAS*, 464, 3657

Xiang, M.-S., *et al.* 2017b, *MNRAS*, 467, 1890

Yuan, H.-B., *et al.* 2015, *MNRAS*, 448, 855

Astronomy and Astrophysics in the Gaia sky
Proceedings IAU Symposium No. 330, 2017
A. Recio-Blanco, P. de Laverny, A.G.A. Brown
& T. Prusti, eds.

doi:10.1017/S1743921317005750

Using ground based data as a precursor for *Gaia* in getting proper motions of satellites

Tobias K. Fritz [1], Sean T. Linden[1], Paul Zivick[1], Nitya Kallivayalil[1] and Jo Bovy[2]

[1]Department of Astronomy, University of Virginia,
Charlottesville, 530 McCormick Road, VA 22904-4325, USA
email: tkf4w@virginia.edu

[2]Department of Astronomy & Astrophysics at University of Toronto,
50 St. George Street M5S 3H4 Toronto, Ontario, Canada

Abstract. We present our effort to measure the proper motions of satellites in the halo of the Milky Way with mainly ground based telescopes as a precursor on what is possible with Gaia. For our first study, we used wide field optical data from the LBT combined with a first epoch of SDSS observations, on the globular cluster Palomar 5 (Pal 5). Since Pal 5 is associated with a tidal stream it is very useful to constrain the shape of the potential of the Milky Way. The motion and other properties of the Pal 5 system constrain the inner halo of the Milky Way to be rather spherical. Further, we combined adaptive optics and HST to get an absolute proper motion of the globular cluster Pyxis. Using the proper motion and the line-of-sight velocity we find that the orbit of Pyxis is rather eccentric with its apocenter at more than 100 kpc and its pericenter at about 30 kpc. The dynamics excludes an association with the ATLAS stream, the Magellanic clouds, and all satellites of the Milky Way at least down to the mass of Leo II. However, the properties of Pyxis, like metallicity and age, point to an origin from a dwarf of at least the mass of Leo II. We therefore propose that Pyxis originated from an unknown relatively massive dwarf galaxy, which is likely today fully disrupted. Assuming that Pyxis is bound to the Milky Way we derive a 68% lower limit on the mass of the Milky Way of 9.5×10^{11} $M_{\odot}$.

Keywords. astrometry, globular clusters: individual (Palomar 5, Pyxis), Galaxy: halo, kinematics and dynamic

---

1. Introduction

The properties of the dark halo of our galaxy are some of the most poorly-constrained properties of the Milky Way, see e.g., Bland-Hawthorn & Gerhard (2016) for a review. Even its mass is uncertain, as different measurements obtain inconsistent values (e.g. Gibbons *et al.* (2014) and Boylan-Kolchin *et al.* 2013). This is problematic because many key properties of galaxies at about the mass of the Milky Way (like the number of faint satellites and characterization of the stellar halo) can be best observed in the Milky Way. Additionally, there are several potential problems for ΛCDM in near field cosmology, like for example 'Too big to fail' (Zavala *et al.* (2009), Boylan-Kolchin *et al.* 2011) and the missing satellites problem (Klypin *et al.* 1999). Whether these are really a problem depends on the exact mass of the Milky Way (Wang *et al.* 2012). To avoid error by extrapolation the mass is best measured at a large distance from the center.

The shape of the halo is another property where there might be a conflict between ΛCDM and observations. Law & Majewski (2010) obtained from detailed modeling of Sagittarius stream that the halo is triaxial. That would not be so surprising on its own, but the most minor axis of this nearly oblate halo is misaligned with the minor axis of the Galactic disc by nearly 90°. This configuration is surprising because it is unstable to torques (Debattista *et al.* 2013). Confirmation of this shape is necessary, and best done with measurements from new targeted observations of Milky Way halo objects.

Here we present proper motions and interpretation for two globular clusters. Firstly, the inner halo globular cluster Palomar 5 (Pal 5) (D$\approx$20 kpc) which is especially interesting because it has a thin tidal stream (Odenkirchen *et al.* 2001). Secondly, we concentrate on the proper motion of a more distant cluster, Pyxis (d$\approx$ 40 kpc, Da Costa 2000).

2. Measurements

Our proper motion measurement procedures for Pal 5 (using SDSS and LBC at LBT) and Pyxis (using HST and GeMS/GSAOI at Gemini South) are explained in detail in Fritz & Kallivayalil (2015) and Fritz *et al.* (2017). Here we summarize the method. Most procedures are similar in both cases although there are differences in the details. We work on single images not coadds of images, because stacks are difficult to correct for distortion. The distortion correction relies mainly on the fact that the distortion of one of the two data sets (i.e. either SDSS or HST) has a well known distortion solution. The object positions are corrected for differential chromatic refraction when it is relevant compared to the errors. The target stars are mainly selected photometrically, and in the case of Pal 5, the distance to the cluster center is also used. For Pyxis we also used the relative proper motions of stars to identify cluster members. The proper motions are measured relative to faint background galaxies. They are chosen as references because they can be found in all images. Their low SNR is the main error source of these measurements.

For Pal 5 we measure motions of $\mu_\alpha = -2.296 \pm 0.186$ mas/yr and $\mu_\delta = -2.257 \pm 0.181$ mas/yr, and for Pyxis, motions of $\mu_\alpha = 1.09 \pm 0.31$ mas/yr, $\mu_\delta = 0.68 \pm 0.29$ mas/yr.

3. Halo shape

Since Pal 5 is at most 20 kpc from the Galactic Center, it is not useful to obtain the full mass of the Milky Way. In contrast, Pal 5's associated stream is very useful to constrain the shape of the halo at the current distance of the cluster. Pearson *et al.* (2015) used the spatial properties and radial velocities of the Pal 5 cluster and stream to predict the expected proper motion for two different halos. For a spherical halo -2.35 mas/yr in both components is expected, and for the triaxial halo of Law & Majewski (2010), μ_α/μ_δ =-5.0/-3.7 mas/yr is expected. Our proper motion measurement better supports the spherical model. Further, Pearson *et al.* (2015) find that the L&M halo model would produce a fanned out stream, different from the observed stream of Pal 5. Combined, this suggests that if the halo is non-spherical, the symmetry plane lies in the Galactic plane, at least at the radial range probed by Pal 5 ($\approx$20 kpc). However, Pearson *et al.* (2015) only test one non-spherical model, which leaves the elongation of the third axis unconstrained with prolate, oblate, or indeed spherical shapes all possible.

We investigate this uncertainty further in Bovy *et al.* (2016). To model proper disruption of the globular cluster we use *galpy* (Bovy 2015). We also allow the other potential parameters to vary within the observational uncertainties. The fit obtains a halo axis ratio of $c/a = 0.9 \pm 0.2$. There is no relevant degeneracy with any other potential parameters. We then fit, in addition, the GD-1 stream measurements from Koposov *et al.* (2010). Our combined fit obtains $c/a = 1.05 \pm 0.14$, strictly seen prolate but consistent with spherical. A halo so close to spherical is in slight tension with the prediction of numerical cosmological simulations, where c/a is 0.7 to 0.8 (e.g., Kazantzidis *et al.* 2010), and likely 0.8 for the Milky Way as its disk is close to maximal (e.g., Bovy & Rix 2013).

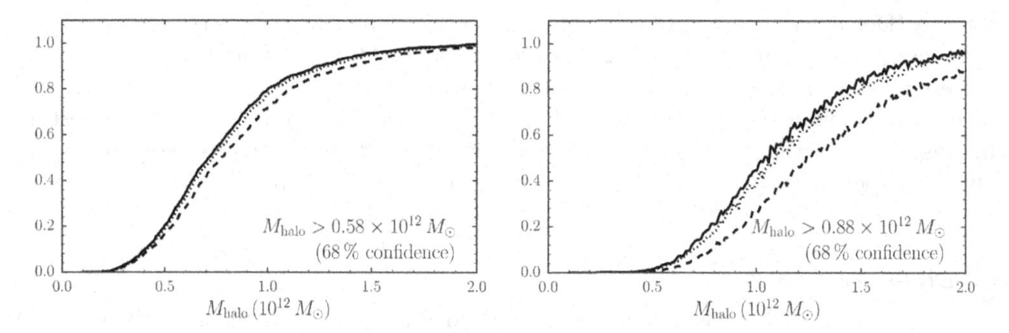

Figure 1. Lower limit constrain on the mass of the halo of the Milky Way. The left plots assumes that Pyxis is bound, the right assumes in addition also that Pyxis had a peripassage in the past. The three curves stand for different concentrations, the solid one uses $c = 15.3$, the short dashed one $c = 12$, the long one $c = 6$. The values are for the solid curves.

4. Origin of Pyxis

The proper motion of Pyxis and its radial velocity (Palma *et al.* 2000) results in a rather eccentric orbit. The pericenter is at about 30 kpc and the apocenter at more than 100 kpc. The apocenter is not well constrained because it is very sensitive to the total mass of the halo. Koposov *et al.* (2014) suggested that Pyxis could be the progenitor of the ATLAS stream. Our proper motion shows clearly that the two cannot be associated. Because Pyxis is somewhat younger (11.5 ± 1 Gyrs) than other globular clusters of its metallicity ([Fe/H]$= -1.45 \pm 0.1$) it is considered a young halo cluster (Irwin *et al.* 1995). These clusters (Zinn 1993) did not form in situ, and instead were once satellites of dwarf galaxies. Newer data as well as our orbit strengthen this classification. The mass of the former host dwarf galaxy can be estimated from metallicity and age. Assuming globulars are at most as metal rich as the host leads to a host of at least the mass of Leo II. Matching the age and metallicity of Pyxis with the age-metallicity relations of star clusters in different dwarf galaxies leads to an LMC like galaxy.

The Large Magellanic Cloud (LMC) was already proposed as a possible host by Irwin *et al.* (1995), also because Pyxis lies on the continuation of the Magellanic stream. We test this hypothesis using our proper motion and an LMC analog from a cosmological simulation (Sales *et al.* 2017), which is matched to the observed proper motion of the LMC (Kallivayalil *et al.* 2013). We find that while the tangential velocity of Pyxis matches, the radial velocity of formally bound particles at the position of Pyxis on the sky differ by more than 300 km/s from the measured velocity. Further, all other dwarf galaxies with known proper motions do not match the orbit of Pyxis. Since this includes all galaxies from Leo I mass upwards, the former host of Pyxis is not known. Because the host is rather massive, we assume that it would be detectable today, when it would be star forming, as usual for galaxies on first approach. Thus, we can conclude that Pyxis is not on first approach. Its former host is maybe hiding in second approach behind the Galactic Plane, or more likely was already disrupted, probably to a shell since the orbit of Pyxis is very eccentric.

5. Halo Mass

Since nearly all subhalos in simulations (Boylan-Kolchin *et al.* 2013) are bound to their halos, satellites of subhalos are bound to their subhalos. Thus, we can use the fact that Pyxis is very likely bound to the Milky Way to constrain the mass of the Galaxy.

When we require that Pyxis is 'just' bound we obtain that the halo mass of the Milky Way is to 68% probability larger than 0.58×10^{12} $M_{\odot}$ (Fig. 1). When we additionally require that Pyxis had a pericenter approach in the past, we obtain a halo mass which is to 68% probability larger than 0.88×10^{12} $M_{\odot}$. This mass depends only weakly on the concentration of the halo, in contrast to most mass estimates. Adding in the mass of disk and bulge, we get a total Milky Way mass which is with 68% probability larger than 0.95×10^{12} $M_{\odot}$.

6. Conclusion

As we have shown, proper motions of satellites can be great tools for constraining the properties of the halo of the Milky Way. With *Gaia* similar measurements will be easily possible for many targets with very high precision, although the precision is lower in DR2 since proper motions profit with $t^{1.5}$ from longer time baselines. However, there is a horizon of *Gaia*, beyond which its precision drops. There, measurements with other instruments are important and also require longer baselines, like with HST, see for example HST GO14734 (Kallivayalil *et al.* 2015) for an ongoing effort. At the distance of M31 old stellar populations are not detectable with *Gaia*, and thus proper motion measurements with other instruments are essential.

References

Bland-Hawthorn, J. & Gerhard, O. 2016, *ARAA*, 54, 529

Bovy, J. & Rix, H.-W. 2013, *ApJ*, 779, 115

Bovy, J. 2015, *ApJS*, 216, 29

Bovy, J., Bahmanyar, A., Fritz, T. K., & Kallivayalil, N. 2016, *ApJ*, 833, 31

Boylan-Kolchin, M., Bullock, J. S., & Kaplinghat, M. 2011, *MNRAS*, 415, L40

Boylan-Kolchin, M., Bullock, J. S., Sohn, S. T., Besla, G., & van der Marel, R. P. 2013, *ApJ*, 768, 140

Da Costa, G. S. 1995, *PASP*, 107, 937

Debattista, V. P., Roškar, R., Valluri, M., *et al.* 2013, *MNRAS*, 434, 2971

Fritz, T. K. & Kallivayalil, N. 2015, *ApJ*, 811, 123

Fritz, T. K., Linden, S. T., Zivick, P., *et al.* 2017, *ApJ*, 840, 30

Gibbons, S. L. J., Belokurov, V., & Evans, N. W. 2014, *MNRAS*, 445, 3788

Irwin, M. J., Demers, S., & Kunkel, W. E. 1995, *ApJ* (Letters), 7453, L21

Kallivayalil, N., van der Marel, R. P., Besla, G., Anderson, J., & Alcock, C. 2013, *ApJ*, 764, 161

Kallivayalil, N., Wetzel, A. R., Simon, J. D., *et al.* 2015, *arXiv*, 1503.01785

Kazantzidis, S., Abadi, M. G., & Navarro, J. F. 2010, *ApJ* (Letters), 720, L62

Klypin, A., Kravtsov, A. V., Valenzuela, O., & Prada, F. 1999, *ApJ*, 522, 82

Koposov, S. E., Rix, H.-W., & Hogg, D. W. 2010, *ApJ*, 712, 260

Koposov, S. E., Irwin, M., Belokurov, V., *et al.* 2014, *MNRAS* (Letters), 442, 85

Law, D. R. & Majewski, S. R. 2010, *ApJ*, 714, 229

Odenkirchen, M., Grebel, E. K., Rockosi, C. M., *et al.* 2001, *ApJ* (Letters), 548, L165

Palma, C., Kunkel, W. E., & Majewski, S. R. 2000, *PASP*, 112, 1305

Pearson, S., Küpper, A. H. W., Johnston, K. V., & Price-Whelan, A. M. 2015, *ApJ*, 799, 28

Sales, L. V., Navarro, J. F., Kallivayalil, N., & Frenk, C. S. 2017, *MNRAS*, 465, 1879

Wang, J., Frenk, C. S., Navarro, J. F., Gao, L., & Sawala, T. 2012, *MNRAS*, 424, 2715

Zavala, J., Jing, Y. P., Faltenbacher, A., *et al.* 2009, *ApJ*, 700, 1779

Zinn, R. 1993, *APC*, 48, 38

Astronomy and Astrophysics in the Gaia sky
Proceedings IAU Symposium No. 330, 2017
A. Recio-Blanco, P. de Laverny, A.G.A. Brown
& T. Prusti, eds.

doi:10.1017/S1743921317005762

What we learn from TGAS about the moving groups of the Solar neighbourhood

B. Goldman[1,2], E. Schilbach[3,1], S. Röser[3], P. Schöfer[4], A. Derekas[5], A. Moor[5], W. Brandner[1] and T. Henning[1]

[1]Max Planck Institute for Astronomy, Königstuhl 17, D–69117 Heidelberg, Germany
email: goldman,brandner,henning@mpia.de
[2]Observatoire astronomique de Strasbourg, Université de Strasbourg, CNRS, UMR 7550, 11 rue de l'Université, F-67000 Strasbourg, France
[3]Astronomisches Rechen-Institut, Zentrum für Astronomie der Universität Heidelberg, Möchhofstrasse 12–14, D–69120 Heidelberg, Germany
email: roeser,schilbach@ari.uni-heidelberg.de
[4]Institut für Astrophysik, Friedrich-Hund-Platz 1, 37077, Göttingen, Germany
schoefer@astro.physik.uni-goettingen.de
[5]Konkoly Observatory, Research Centre for Astronomy and Earth Sciences, Hungarian Academy of Sciences, P.O. Box 67, H-1525 Budapest, Hungary
moor,derekas@konkoly.hu

Abstract. We use the TGAS proper motions and parallaxes as well as published and new radial velocities to study the dynamics of nearby moving groups. In particular we try to determine their age using backtracing of the individual members to a common origin. We find that the current data, probably the radial velocities, do not allow to reach a successful conclusion.

Keywords. Galaxy: kinematics and dynamics, open clusters and associations: general

1. Introduction

Moving groups were originally detected as groups of stars sharing a common motion, generally showing sign of youth making them easier to distinguish against the stellar background (Zuckermann & Song 2004). In recent years, moving groups have become popular because membership allows to assign an age to nearby stars (useful to study stellar evolution, search for young, still warm extrasolar planets) (Gagné *et al.* 2015); and because they represent the end stage of open cluster evolution and dispersion of young stars out of their stellar nursery into the field. Comoving stars may also be the result of resonances within the Milky Way and trace its heterogeneous potential (Famaey *et al.* 2008). The ages of moving groups are based on evolutionary models, so it would be useful to obtain an independent determination, e.g. based on the dynamics of the groups.

2. Photometric selection

TGAS parallaxes allow to update the colour-magnitude diagrams of our moving groups, with a fraction of candidates shifting significantly. We reject the red points from the further kinematic analysis (Fig.1 *left*). The green dots nicely fall between the 50 and 100-Myr CIFIST isochrones (Allard, 2016).

3. Kinematic aging

Propagating back in time the motions of the group members until a point of minimum extension would allow to kinematically measure the group age, giving a

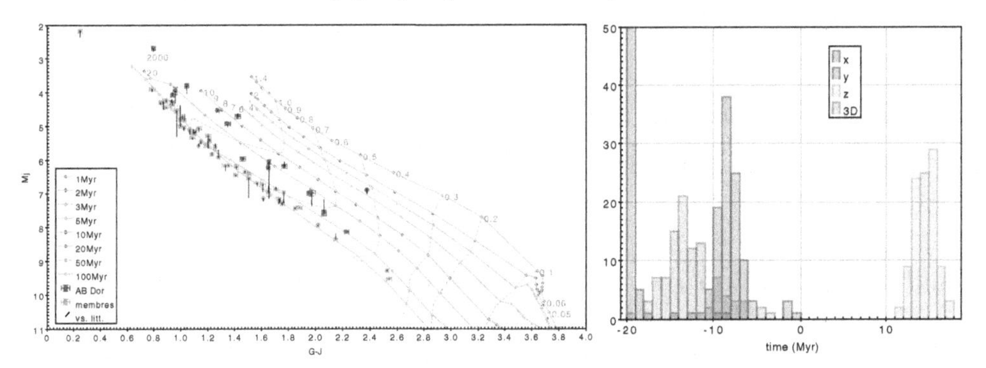

Figure 1. *left)* M_G vs. $G_{\mathrm{Gaia}}J_{\mathrm{2MASS}}$ for AB Dor members. We reject the red points from the kinematic analysis. *right)* Histograms of the times of minimal extension along the 3 directions and the 3-D RMS extension. Along the y direction most realisations have a minimal extension more than 20 Myr ago. The histograms clearly disagree.

model-independent age and a strong constraint on the stellar evolutionary models. For instance Ducourant *et al.* (2014) used 30 members of the TWA moving group to recover its birth place and age. This methodology requires good radial velocities as the 1km/s uncertainty translates in 10 Myr into a 10 pc positional uncertainty, a distance comparable to the expected size of the original cluster.

We selected members and member candidates from the literature of various moving groups with presumed age ranging from 8 Myr (TW Hydrae) to 100 Myr (AB Dor), and adopted TGAS proper motions and parallaxes as well as radial velocities from the literature or obtained by us using the CAFE spectrograph in Calar Alto. We propagate back in time using a epicyclic orbit the group member positions, for a number of realisations according to the current measurement uncertainties. We derive histograms of time of minimal extension, which can vary strongly in all three directions (Fig.1 *right*).

This analysis, even after removing the discrepant members, returns large minimal extension and incompatible times at maximal compression. We obtain similar inconsistent results for other moving groups, including for TWA (Ducourant *et al.* (2014) only present a 3-D analysis, which does not reveal these inconsistencies).

This work has made use of data from the European Space Agency (ESA) mission *Gaia* (`https://www.cosmos.esa.int/gaia`), processed by the *Gaia* Data Processing and Analysis Consortium (DPAC, `https://www.cosmos.esa.int/web/gaia/dpac/consortium`). Funding for the DPAC has been provided by national institutions, in particular the institutions participating in the *Gaia* Multilateral Agreement.

It was supported by the Sonderforschungsbereich SFB 881 "The Milky Way System" (subprojects B5 and B6) of the German Research Foundation (DFG).

References

Allard, F. 2016, in SF2A-2016: Proceedings of the Annual meeting of the French Society of Astronomy and Astrophysics, ed. C. Reylé, J. Richard, L. Cambrésy, M. Deleuil, E. Pécontal, L. Tresse, & I. Vauglin, 223–227

Ducourant, C., Teixeira, R., Galli, P. A. B., *et al.* 2014, *A&A*, 563, A121

Famaey, B., Siebert, A., & Jorissen, A. 2008, *A&A*, 483, 453

Gagné, J., Faherty, J. K., Cruz, K. L., *et al.* 2015, *ApJS*, 219, 33

Zuckerman, B., Song, I., & Bessell, M. S. 2004, *ApJL*, 613, L65

Astronomy and Astrophysics in the Gaia sky
Proceedings IAU Symposium No. 330, 2017
A. Recio-Blanco, P. de Laverny, A.G.A. Brown & T. Prusti, eds.

doi:10.1017/S1743921317006263

The AMBRE Project: *r*-process element abundances in the Milky Way thin and thick discs

Guillaume Guiglion[1], Patrick de Laverny[2], Alejandra Recio-Blanco[2] and C. Clare Worley[3]

[1]Leibniz-Institut für Astrophysik Potsdam (AIP) An der Sternwarte 16, 14482 Potsdam
email: gguiglion@aip.de

[2]Université Côte d'Azur, Observatoire de la Côte d'Azur, CNRS, Laboratoire Lagrange, France

[3]Institute of Astronomy, University of Cambridge, Madingley Rise, Cambridge CB3 0HA, UK

Abstract. Chemical evolution of *r*-process elements in the Milky Way disc is still a matter of debate. We took advantage of high resolution HARPS spectra from the ESO archive in order to derive precise chemical abundances of 3 *r*-process elements Eu, Dy & Gd for a sample of 4 355 FGK Milky Way stars. The chemical analysis has been performed thanks to the automatic optimization pipeline GAUGUIN. Based on the [α/Fe] ratio, we chemically characterized the thin and the thick discs, and present here results of these 3 *r*-process element abundances in both discs. We found an unexpected Gadolinium and Dysprosium enrichment in the thick disc stars compared to Europium, while these three elements track well each other in the thin disc.

Keywords. stars: abundances. techniques: spectroscopic. Galaxy: abundances. Galaxy: stellar content

---

1. Motivations, observations and chemical abundances

The synthesis of heavier elements than Iron is suppose to take place in stellar interiors, via capture of neutrons. Two main families of such elements are known, from *slow*-neutron capture (*s*-) or *rapid*- neutron capture (*r*-) compared to the timescale for β^- decay. In this study we will focus on 3 *r*-process elements: Eu, Gd and Dy. These elements are supposed to be formed via neutrino-induced winds from SN II (Woosley *et al.* 1994), merging of neutron stars (Freiburghaus 1999) or polar jets from SN II via pure MHD explosion (Nishimura *et al.* 2006), but current yields suffer from large uncertainties. On the observational side, there is a current lack of massive homogeneous catalog of *r*-process abundances in order to put constraints on their chemical evolution in the Milky Way. In this study, we took advantage of the ESO spectroscopic archive, focusing of the high resolution ($R \sim 110\,000$) HARPS data. Our working sample is composed of 4 355 individual stars with high quality atmospheric parameters ($T_{\rm eff}$, log(g), [Fe/H], [α/Fe]) derived by the AMBRE Project (De Pascale *et al.* 2014). We built a line-list based on VALD3 (Kupka *et al.* 1999) atomic blends, molecular blends and most recent laboratory and theoretical $\log gf$ measurements for the Eu (Lawler *et al.* 2001), Gd (Den Hartog *et al.* 2006) and Dy (Wickliffe *et al.* 2000). We used the optimization pipeline GAUGUIN (Guiglion *et al.* 2016) is order to derive the individual chemical abundances. We selected the best abundances for stars with $T_{\rm eff} > 4\,500$ K and $S/N > 50$, resulting in a sub-sample of 1 672 stars. Typical errors are 0.12 dex for Eu and Gd and 0.11 dex for Dy.

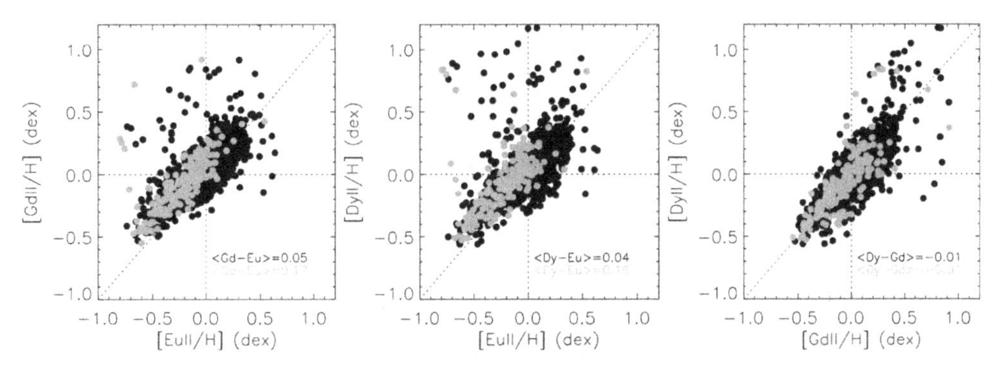

Figure 1. Gd *v.s.* Eu, Dy *v.s.* Eu and Dy *v.s.* Gd in the thin disc (black, filled) and the thick disc (grey, filled). Average differences are noted.

2. Looking for r-process enhancement in the thin & thick discs

We first chemically characterized the thin and the thick discs, taking advantage of the [α/Fe] ratio, following the same procedure as in (Guiglion *et al.* 2016). We identified 1 469 stars in the thin disc, 182 in the thick disc. We showed that the [Eu,Gd,Dy/Fe] ratios decrease with [M/H], following typical α-element trend. The thick discs shows higher [Eu,Gd,Dy/Fe] ratios with respect to the thin disc. It is the first time that the Gd and Dy contents are characterized in the thin and the thick disc stars. The Eu pattern is in agreement with recent observations (ex Battistini *et al.* 2016) and chemical evolution model (Travaglio *et al.* 2009). We note that Gd and Dy show very similar patterns, in both discs, suggesting a common chemical evolution. In addition, we observe that [Eu,Gd,Dy/Fe] ratios are strongly correlated to [α/Fe]. Indeed, Eu, Gd & Dy increase as a function of [α/Fe] in both discs, suggesting a higher chemical enrichment in r-process at early time.

A presented in Figure 1, we first observe that in the thin disc, Eu, Gd and Dy track well each other, without showing significant offset. On the other hand, we clearly see in the thick disc that Gd and Dy are enhanced compared to Eu, respectively of 0.17 and 0.16 dex (*left* and *middle panel*), while Gd and Dy track each other well (*right panel*). We think that the offset is not related to any line-list calibration issues because no significant offsets is observed in the thin disc stars. It is unexpected that Gd and Dy show an enrichment compared to Eu in the thick disc, because these three elements should originate from the same sources and sites of production. Up to now, no galactic chemical evolution model is able to explain such enrichment for these two r-process elements. Further investigations, especially thanks to the stellar ages, will be needed to understand such observations.

References

Battistini, C. & Bensby, T., 2016, *A&A*, 586, A49
Den Hartog, E. A., Lawler, J. E., Sneden, C., & Cowan, J. J., 2006, *ApJS*, 167, 292.
De Pascale, M., Worley, C. C., de Laverny, P., *et al.* 2014, *A&A*, 570, A68
Freiburghaus, C., Rosswog, S., & Thielemann, F.-K. 1999, *ApJ*, 525, L121
Guiglion, G, de Laverny, P., Recio-Blanco, A., Worley, C. C., *et al.* *A&A*, 2016, 595, A18
Kupka, F., Piskunov, N., Ryabchikova, T. A., Stempels, H. C., *et al.* *A&A*, 1999, 138, 119
Lawler, J. E., Wickliffe, M. E., den Hartog, E. A., & Sneden, C. 2001, *ApJ*, 563, 1075.
Nishimura, S., Kotake, K., Hashimoto, M., *et al.* 2006, *ApJ*, 642, 410
Travaglio, C., Galli, D., Gallino, R., *et al.* 1999, *ApJ*, 521, 69
Wickliffe, M. E., Lawler, J. E., & Nave, G., 2000, *J. Quant. Spec. Radiat. Transf.*, 2000, 66, 363
Woosley, S. E., Wilson, J. R., Mathews, G. J., Hoffman, R. D., *et al.* 1994, *ApJ*, 433, 229

Astronomy and Astrophysics in the Gaia sky
Proceedings IAU Symposium No. 330, 2017
A. Recio-Blanco, P. de Laverny, A.G.A. Brown & T. Prusti, eds.

doi:10.1017/S1743921317005774

The thick disc according to Gaia-ESO

Louise M. Howes and Thomas Bensby

Dept. of Astronomy & Theoretical Physics, Lund University,
Box 43, SE-221 00 Lund, Sweden
email: [louise; tbensby]@astro.lu.se

Abstract. In the era of large spectroscopic surveys, it is vital that selection effects are taken into account when making conclusions about the stellar populations of the Galaxy. Here we use the Galactic disc sample of stars from the Gaia-ESO Survey internal data release 4 (GES iDR4), applying the published selection function to characterise the vertical extent of the chemically defined thick and thin discs.

Keywords. Galaxy: disk, Galaxy: structure, Galaxy: stellar content, Galaxy: abundances

---

Gaia-ESO (Gilmore *et al.* 2012) has observed and analysed more than 12 000 field stars so far in the Milky Way disc using FLAMES/GIRAFFE on the VLT. The majority of these stars are out of the Galactic plane, making the sample an excellent probe of the vertical gradients in the thin and thick discs. Here we have used the measurements for Mg and Fe to examine the differences between the chemically defined disc, where alpha-rich are considered thick disc, and alpha-poor are considered thin disc. As can be seen in Figure 1, the sample covers both regions of abundance space, without a clear separation between the two. The separating line is taken from similar past studies (e.g., Bensby *et al.* 2014; Recio-Blanco *et al.* 2014; Kordopatis *et al.* 2015).

In creating Figure 1, crucially each star has been given a weight, taken from Stonkutė *et al.* (2016), to counter the effect that the survey's selection may have had on the trends. Distances are calculated using the BeSPP (Serenelli *et al.* 2013). Each star is represented by a probability distribution, calculated based on the likelihood of the star falling in a certain region of space (e.g., $6.0 < R < 7.5$ and $|Z| < 0.3$ kpc), the weight given to the star by the selection function, and the size of the error bars in [Fe/H] and [α/Fe] (represented as Gaussians). Each plot has more than 150 stars.

Close to the plane, nearly all stars are metal-rich, alpha-poor. Only at a height of ~ 0.5 kpc does an alpha-rich population start to grow. This transition occurs more quickly in the inner disc (blue), which also has a much larger variation in [Fe/H] as one moves above the plane. The sample does not seem to include many thin disc stars at metallicities lower than Solar, perhaps due to the targeting of fields out of the plane. Thick disc stars make up the majority of the population in the inner disc region above 0.5 kpc, but for $R > 8.5$ kpc, the thin disc forms a significant population at all heights, including $|Z| > 2.0$ kpc. These results agree with previous work by Bensby *et al.* (2011), and the APOGEE survey (Hayden *et al.* 2015).

The lack of separation between the thin and thick disc components could be real, however the large uncertainties in the distances to these stars mean that it is impossible to tell. Parallaxes from *Gaia* DR2 (*Gaia* Collaboration *et al.* 2016) will allow us to reduce these significantly, and so resolve the crucial detail in these abundance plots.

References

Bensby, T, Alves-Brito, A., Oey, M. S., Yong, D., & Meléndez, J 2011, *ApJ*, 735, L46
Bensby, T., Feltzing, S., & Oey M. S. 2014, *A&A*, 562, A71

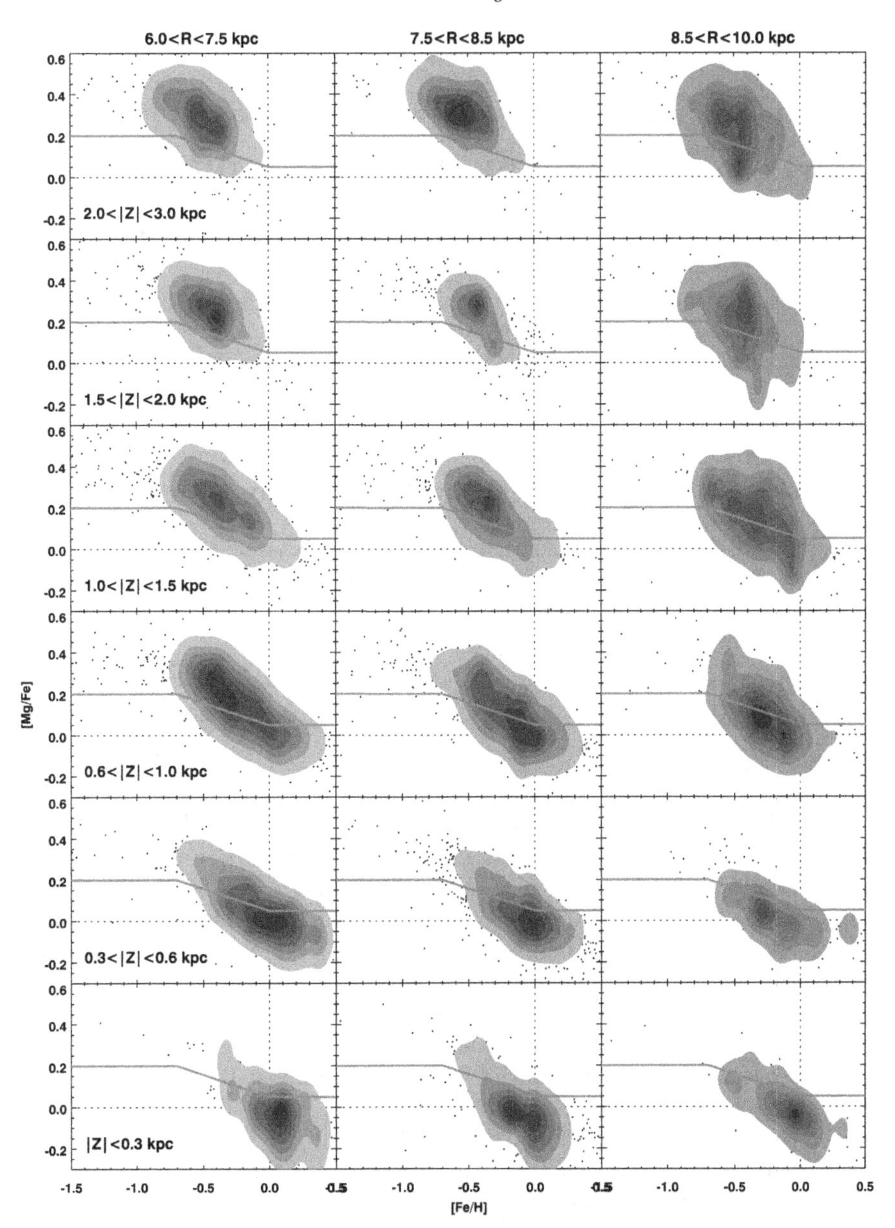

Figure 1. Abundance plot split into bins according to position in the Galaxy. Each star is represented by a probability distribution function of its distance from the Sun; each bin contains the fraction of the PDF that falls in that bin, summed over all stars.

Gaia Collaboration, Prusti, T., de Bruijne, J. H. J., *et al.* 2016, *A&A*, 595, A1
Gilmore, G., Randich, S., Asplund, M., *et al.* 2012, *ESO Messenger*, 147, 25
Hayden, M. R., Bovy, J., Holtzman, J. A., *et al.* 2015, *ApJ*, 808, 132
Kordopatis, G., Wyse, R. F. G., Gilmore, G., *et al.* 2015, *A&A*, 582, A122
Recio-Blanco, A., de Laverny, P., Kordopatis, G., *et al.* 2014, *A&A*, 567, A5
Serenelli, A. M., Bergemann, M., Ruchti, G., & Casagrande, L. 2013, *MNRAS*, 429, 3645
Stonkutė, E., Koposov, S. E., Howes, L. M., *et al.* 2016, *MNRAS*, 460, 1131

Astronomy and Astrophysics in the Gaia sky
Proceedings IAU Symposium No. 330, 2017
A. Recio-Blanco, P. de Laverny, A.G.A. Brown & T. Prusti, eds.

doi:10.1017/S1743921317006378

The Galactic mass distribution from the LAMOST Galactic spectroscopic surveys

Y. Huang[1,4], X.-W. Liu[1], H.-B. Yuan[2], H.-W. Zhang[1], M.-S. Xiang[3,4], B.-Q. Chen[1,4] and C. Wang[1]

[1]Department of Astronomy, Peking University, Beijing 100871, China
email: yanghuang@pku.edu.cn

[2]Department of Astronomy, Beijing Normal University, Beijing 100875, China

[3]National Astronomical Observatories, Chinese Academy of Sciences, Beijing 100012, China

[4]LAMOST Fellow (Supported by Special fund budgeted and administrated by CAMS)

Abstract. Using the data from the LAMOST Galactic spectroscopic surveys and some other surveys, we have started a series of work to measure the mass distribution of our Galaxy. As a result of the first-stage, we have constructed the Galactic rotation curve out to 100 kpc and the Galactic escape velocity curve between 5 and 14 kpc. From the two curves, we have built parametrized mass models for our Galaxy, respectively. Both models yield a similar result for the Milky Way's virial mass: $\sim 0.9 \times 10^{12}$ M$_{\odot}$.

Keywords. Galaxy: fundamental parameters – Galaxy: kinematics and dynamics – dark matter

---

1. The project: the Milky Way mass by LAMOST Survey

Modelling the mass distribution of our Galaxy is one of the fundamental tasks of Galactic astronomy. An accurate measurement of the mass of the Milky Way (MW) is of vital importance for understanding the Galaxy formation and evolution. For example, the number of massive subhalos are highly decided by the virial mass of their host galaxies (e.g. Wang *et al.* 2012). Therefore, the key to understanding the so-called *too big to fail* problem (Boylan-Kolchin *et al.* 2011) for the MW is to accurately estimate the mass of our Galaxy. However, the current measurements of the MW mass are quite uncertain and the results could differ by a factor of 5, from 0.5 to 2.5×10^{12} M$_{\odot}$.

With the huge stellar spectra database collected by LAMOST (Cui *et al.* 2012), combined with other survey data (e.g. Gaia, SEGUE, APOGEE), we attempt to obtain an accurate measurement for the mass distribution of our Galaxy by a series of studies, including the Galactic rotation curve, the Galactic escape velocity curve, the cold stellar streams and the mass surface density. The Galactic rotation and escape velocity curves could be used to measure the overall mass of the MW, while the mass surface density could provide vital constraint on deriving the mass distribution of different Galactic components (i.e. the bulge, the disc and the dark matter). The study of cold stream could tell the shape information (e.g. spherical, oblate or prolate) of the dark matter. At present, we have finished the work of constructions of the Galactic rotation curve and the escape velocity curve and both studies give a similar result for the MW's virial mass: 0.9×10^{12} M$_{\odot}$ (Huang *et al.* 2016; Huang *et al.* to be submitted). In what follows, we briefly introduce the constructions of the Galactic rotation curve and escape velocity curve, respectively.

First, using ~ 6000 primary red clump giants (PRCGs) in the outer disc selected from the LSS-GAC (Liu *et al.* 2014) and the SDSS-III/APOGEE survey, combined with ~ 5700

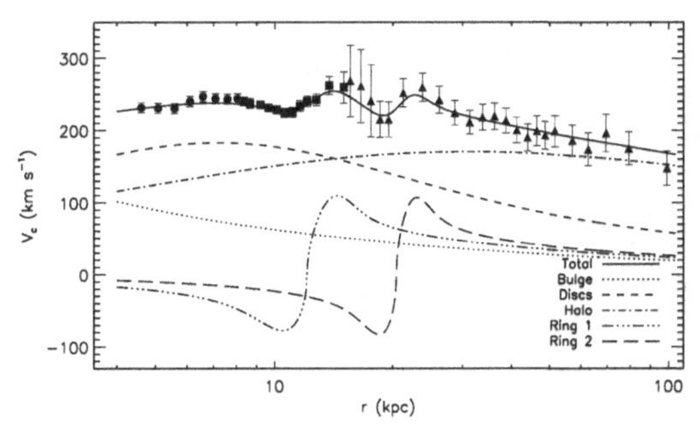

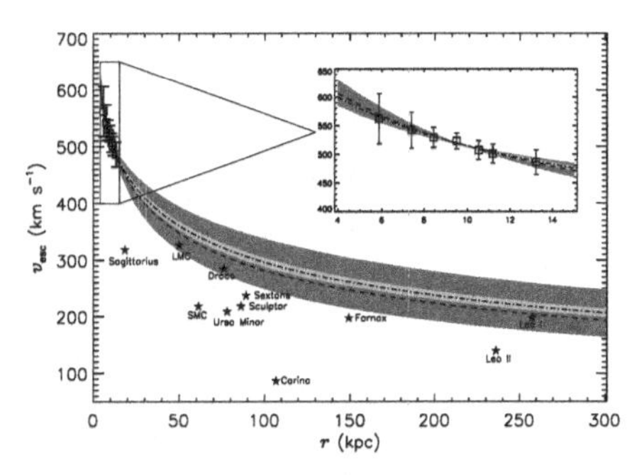

Figure 1. *Left panel:* Final combined rotation curve of the MW to 100 kpc derived from H I data (dots), PRCGs (squares) and HKGs (triangles). Lines of different styles as labelled in the bottom-right corner of the diagram represent the best-fitting rotation curves contributions to the components of the Milky Way, with the solid line representing the sum of contributions from all the mass components. For details, please see Huang *et al.* (2016). *Right panel:* Galactic escape velocity $v_{\rm esc}$ as a function of Galactocentric radius r. Black boxes represent Galactic escape velocities newly derived from our sample of 527 high velocity halo stars for r between 5 and 14 kpc. The dashed line and deep gray shade indicate our best-fit model and 1σ uncertainty area without a prior on the local circular speed, while the dash-doted line and the shallow gray shade indicate those with a prior on the local circular speed. The inset in the top-right is to zoom in the region of the newly observed Galactic escape velocity curve for clarity. Black stars represent the total velocities of the named MW classical satellite galaxies. For details, please see Huang *et al.* (to be submitted).

halo K giants (HKGs) selected from the SDSS/SEGUE survey, we have constructed the rotation curve of the MW out to ~ 100 kpc. The newly constructed rotation curve has a generally flat value of 240 km s^{-1} within r of 25 kpc and then decreases steadily to 150 km s^{-1} at $r \sim 100$ kpc (See the *left panel* of Fig. 1).

Secondly, using a sample of 527 high velocity halo stars ($|v_r| \geqslant 300$ km s^{-1} and [Fe/H] < -1) selected from the LAMOST Galactic spectroscopic surveys, we have measured the Galactic escape velocities at r between 5 and 14 kpc. The Galactic escape velocity at the solar position $v_{\rm esc}\,(R_0)$ is found to 529 ± 29 km s^{-1} (90 per cent confidence). As the *right panel* of Fig. 1 shows, the newly constructed Galactic escape velocity curve decreases steadily from 562 km s^{-1} at $r \sim 5.9$ kpc to 486 km s^{-1} at $r \sim 13.2$ kpc.

Finally, from the constructed rotation and escape velocity curves, we have built parametrized mass models for our Galaxy, respectively. Both models yield a similar results of the virial mass of the MW's dark matter halo of $0.9 \pm 0.1 \times 10^{12}$ M$_\odot$. The local dark matter density is also found to be $\sim 0.008 \pm 0.001$ M$_\odot$ pc^{-3} from both models.

Acknowledgements

This work is supported by the National Key Basic Research Program of China 2014CB845700.

References

Boylan-Kolchin, M., Bullock, J. S., & Kaplinghat, M. 2011, *MNRAS*, 415, L40

Cui, X.-Q., Zhao, Y.-H., Chu, Y.-Q., *et al.* 2012, *RAA*, 12, 1197

Huang, Y., Liu, X.-W., Yuan, H.-B., *et al.* 2016, *MNRAS*, 463, 2623

Liu X. -W., *et al.*, 2014, in Feltzing S., Zhao G., Walton N., Whitelock P., eds, Proc. IAU Symp. 298, Setting the scene for Gaia and LAMOST, Cambridge University Press, pp. 310-321, preprint (arXiv: 1306.5376)

Wang, J., Frenk, C. S., Navarro, J. F., Gao, L., & Sawala, T. 2012, *MNRAS*, 424, 2715

Astronomy and Astrophysics in the Gaia sky
Proceedings IAU Symposium No. 330, 2017
A. Recio-Blanco, P. de Laverny, A.G.A. Brown & T. Prusti, eds.

doi:10.1017/S1743921317006251

Modelling the Milky Way with *Gaia*-TGAS

Jason A. S. Hunt†,

Dunlap Institute for Astronomy & Astrophysics, University of Toronto, 50 St. George Street, Toronto, ON M5S 3H4, Canada

Abstract. I summarize two recent projects involving the *Gaia*-TGAS data. Firstly, I discuss a detection of a lack of disc stars in the Solar neighbourhood with velocities close to zero angular momentum. We use predictions of this effect to make a measurement of the Solar rotation velocity around the Galactic centre, and also of R_0. Secondly, I discuss a detection of a group of stars with systematically high Galactic rotation velocity. We propose that it may be caused by the Perseus arm and compare the data with simulations.

Keywords. Galaxy: kinematics and dynamics, Galaxy: fundamental parameters

---

1. Introduction

The European Space Agency's *Gaia* mission (Gaia Collaboration *et al.* 2016a) gives us a new window on the dynamics of the Solar neighbourhood. *Gaia* DR1 (Gaia Collaboration *et al.* 2016b) contains the Tycho-Gaia Astrometric Solution (Michalik *et al.* 2015), a catalogue of $\sim 2 \times 10^6$ stars in common between *Gaia* and Tycho-2, the star mapper from *Hipparcos*, with positions on the sky, parallaxes and proper motions. We are entering an exciting era for Milky Way astronomy and astrophysics, and even from the first data release we are able to explore new facets of our Galaxy. In this work, I summarize two recent projects involving the *Gaia*-TGAS data. Please note that both works are described more fully in their respective publications.

2. Measuring the Solar motion with *Gaia*-TGAS and RAVE

In this Section, I summarize our recent work reporting on a detection of a dearth of stars with zero angular momentum in the Solar neighbourhood (Hunt *et al.* 2016), and describe how we used this feature to make a measurement of the Solar rotation velocity around the Galactic centre, $V_{\odot}$, and the distance to the Galactic centre, R_0.

The TGAS data provides only 5D phase space information, (α, δ, π, μ_α, μ_δ). However, we can cross match TGAS with ground based surveys, such as the Radial Velocity Experiment (RAVE; e.g. Steinmetz *et al.* 2006) or the Large sky Area Multi-Object fibre Spectroscopic Telescope (LAMOST; e.g. Zhao *et al.* 2012) to add line-of-sight velocities.

We cross matched TGAS with RAVE DR5, resulting in over 200,000 stars with the full 6D phase space information, allowing us to calculate tangential velocities, v_Y (km s^{-1}. The left panel of Fig. 1 shows the distribution of v_Y for stars within 700 pc, with a 'dip' visible around $v_Y = -240$ km s^{-1}. A feature like this was predicted by Carlberg & Innanen (1987), who suggest that a likely explanation is that the missing stars, which have zero angular momentum, plunge into the Galactic centre and are scattered onto highly chaotic orbits by interaction with the Galactic nuclear potential.

We modelled this effect by integrating test particles in a Milky Way-like potential and observing their orbits. We use the MWPotential2014 from galpy (Bovy 2015)

† jason.hunt@dunlap.utoronto.ca

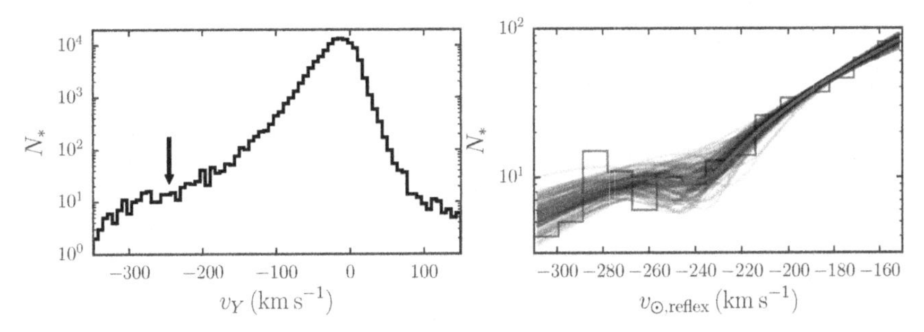

Figure 1. Left: Distribution of tangential velocities, v_Y (km s^{-1}), with a 'dip' visible around $v_Y = -240$ km s^{-1}, marked with an arrow. **Right:** Tail of the v_Y distribution overlaid with the best-fit model (red), and 100 MCMC samples (black). As shown in Hunt *et al.* (2016).

combined with a Plummer potential representing the Galactic nuclear potential. We created a function from the resulting distribution of stars which remained upon non-chaotic orbits (shown in Figure 3 in Hunt *et al.* 2016) by smoothly interpolating the model and combining it with an exponential.

We then fit this function to the distribution of disc stars in the range -310 km s^{-1} $\leqslant v_Y < -150$ km s^{-1}. We also fit the exponential component alone and compared the likelihoods of the respective models. We determined that the data prefers the dip model at a significance of 2.7σ. We explored the uncertainties on the model by performing a Markov Chain Monte Carlo (MCMC) analysis with EMCEE (Foreman-Mackey *et al.* 2013). We find $v_\odot = 239 \pm 9$ km s^{-1}, which also gives us $R_0 = 7.9 \pm 0.3$ kpc when combined with the proper motion of Sgr A*. The right panel of Fig. 1 shows the tail of the v_Y distribution overlaid with the best fit model (red) and the MCMC samples (black). We also performed the same analysis using TGAS+RAVE+LAMOST data. Here we use the bayesian distance estimates from Astraatmadja & Bailer-Jones (2016), whereas previously we used the RAVE spectrophotometric distance estimates. For this data set we find $v_\odot = 236 \pm 9$ km s^{-1} at 3σ. It is encouraging that the measurements are consistent.

Gaia DR2 will have the power to measure this feature, if real, with substantially increased precision. Assuming the errors decrease on the order of $\sqrt{N}$, we expect *Gaia* to enable us to constrain $v_\odot$ to within approximately 1 km s^{-1}. In turn, the error on R_0 would become dominated by the systematics. As such, any future exploration of this feature should take into account factors such as the Galactic bar, spiral arms, and giant molecular clouds, although the effects are expected to be small.

3. Searching for the Perseus arm with *Gaia*-TGAS

In this Section, I summarize our recent work reporting on a detection of a small group of stars in the TGAS data with systematically high Galactocentric rotation velocity as described fully in Hunt *et al.* (2017).

In Kawata *et al.* (2014) we showed that the distribution of v_ϕ is notably different for models which treat spiral structures as density waves, compared to N-body simulations which give rise to transient winding spirals arms, and in Hunt *et al.* (2015) we show that the final *Gaia* data release will easily enable us to observe the kinematic signatures of a co-rotating spiral arm, if present. Thus, we analyze the Galactocentric rotation velocity, v_ϕ, of stars in TGAS. Here we wish to explore the distribution of velocities for stars in the direction of the Perseus spiral arm. In the direction of the Galactic anti-centre we can cross-match TGAS with LAMOST, but this results in very few stars. Thus we

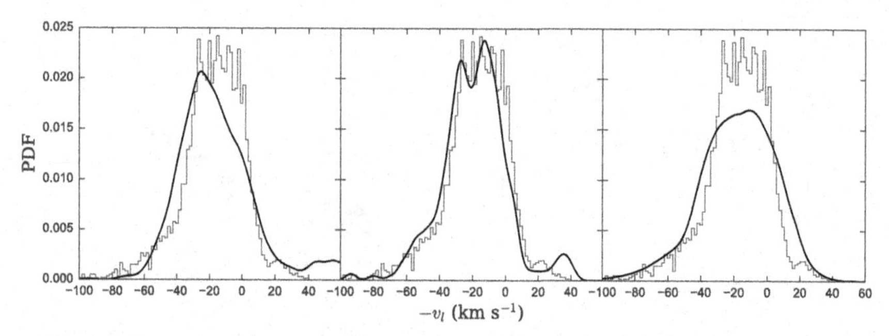

Figure 2. Distribution of v_l (km s^{-1}) at $(l, b) = (180, 0) \pm 5$ deg (red) overlaid with three model predictions (black), originally shown in Hunt *et al.* (2017).

convert the proper motions (μ_α, μ_δ) to (μ_l, μ_b), because $v_l = 4.74\mu_l/\pi$ is analogous to Galactocentric rotation velocity when observing the line-of-sight $(l, b) = (180, 0)$ deg.

Fig. 2 shows the distribution of v_l (km s^{-1}) at $(l, b) = (180, 0) \pm 5$ deg (red) overlaid with predictions from a N-body model with a strong spiral arm (left), an N-body model with a weak spiral arm (centre) and a test particle model with density wave like arms (right). The left and right panels of Fig. 2 show the strong arm model and the density wave model are not a good fit to the data. The centre panel is closer, but the high velocity 'bump' caused by the co-rotation of the N-body arm is still too strong compared with the data. We propose that a still weaker arm may reproduce the data nicely, as the strength of the arm effects the acceleration of the stars in the high velocity tail. Alternatively, the bump may be a resonance feature of a density wave like arm. Future *Gaia* data should enable us to better distinguish between the models.

Acknowledgements

This work has made use of data from the European Space Agency (ESA) mission *Gaia* (https://www.cosmos.esa.int/gaia), processed by the *Gaia* Data Processing and Analysis Consortium (DPAC, https://www.cosmos.esa.int/web/gaia/dpac/consortium). Funding for the DPAC has been provided by national institutions, in particular the institutions participating in the *Gaia* Multilateral Agreement.

References

Allende-Prieto, C,. Kawata, D. & Cropper, M. 2016 *A&A*, 596, 98
Astraatmadja, T. L., & Bailer-Jones, C. A. L. 2016 *ApJ*, 833, 119
Bovy, J. 2015 *ApJS*, 216, 29
Carlberg, R. & Innanen, K. A. 1987 *AJ*, 94, 666
Foreman-Mackey, D., Hogg, D. W., Lang, D., & Goodman, J. 2013 *PASP*, 125, 306
Gaia Collaboration *et al.* 2016 *A&A*, 595, 1
Gaia Collaboration *et al.* 2016 *A&A*, 595, 2
Hunt, J. A. S. *et al.* 2015 *MNRAS*, 450, 2132
Hunt, J. A. S., Bovy, J., & Carlberg, R. 2016 *ApJL*, 832, L25
Hunt, J. A. S., *et al.* 2017 *MNRAS*, 467, L21
Kawata, D., *et al.* 2014 *EAS*, 67, 247
Michalik, D., Lindegren, L. & Hobbs, D. 2015 *A&A*, 574, 115
Monari, G., Kawata, D., Hunt, J. A. S., & Fameay, B. 2017 *MNRAS*, 466, L113
Steinmetz, M., Zwitter, T. & Siebert, A. 2006 *AJ*, 132, 1645
Zhao, G., Zhao, Y-H., Chu, Y-Q., Jing, Y-P., & Deng, L-C. 2012 *RAA*, 12, 723

Astronomy and Astrophysics in the Gaia sky
Proceedings IAU Symposium No. 330, 2017
A. Recio-Blanco, P. de Laverny, A.G.A. Brown & T. Prusti, eds.

doi:10.1017/S1743921317006330

Identification of binary and multiple systems in TGAS using the Virtual Observatory

F. Jiménez-Esteban[1,2,3] and E. Solano[1,2]

[1]Centro de Astrobiología (INTA-CSIC), Madrid, Spain
email: fran.jimenez-esteban@cab.inta-csic.es

[2]Spanish Virtual Observatory
[3]Suffolk University

Abstract. Binary and multiple stars have long provided an effective method of testing stellar formation and evolution theories. In particular, wide binary systems with separations > 20,000 au are particularly challenging as their physical separations are beyond the typical size of a collapsing cloud core (5,000 - 10,000 au). We present here a preliminary work in which we make use of the TGAS catalogue and Virtual Observatory tools and services (Aladin, TOPCAT, STILTS, VOSA, VizieR) to identify binary and multiple star candidate systems. The catalogue will be available from the Spanish VO portal (http://svo.cab.inta-csic.es) in the coming months.

Keywords. Astronomical data bases, catalogues, virtual observatory tools, parallaxes, proper motions, binaries: visual.

---

1. Search of co-moving candidates

We selected TGAS (Gaia Collaboration *et al.* 2016) sources with positive parallaxes and errors < 10% in both proper motion components and parallaxes (250,466 sources). Co-moving systems were defined as any group of sources with the same parallaxes and proper motions within 3 times the sum of the individual errors (∼150,000 candidate pairs). We set a conservative limit of 500,000 au (∼2.5 pc) in the projected sky separation to discard systems whose components are too far to be physically bound. 15,645 sources grouped in almost 7,000 systems remained.

2. Radial velocity confirmation

We searched radial velocity (RV) data at RAVE-DR5 (Kunder *et al.* 2017) and LAMOST-DR2 (Luo *et al.* 2015) catalogues, both available at Vizier. RVs were found for 2,243 sources. The components of 339 systems (∼5%) showed the same RV within 3 times the sum of the individual errors. 459 sources were rejected because of their discrepant RV values. This left 15,186 sources in 6,703 systems (339 confirmed and 6364 candidate).

3. Binding Energies

We used the Gaia-2MASS (G-K_s) color and the absolute magnitude M_G obtained using the TGAS distance to place our candidate systems in a H-R diagram (Fig. 1). According to their position 804 objects (∼5%) were classified as subgiants or giants and 13,400 as dwarfs.

Spectroscopic determinations of the effective temperatures (T_{eff}) of 1,972 stars were obtained from RAVE-DR5, LAMOST-DR2, and PASTEL (Soubiran *et al.* 2016). For the rest of objects, T_{eff} were estimated using VOSA (Bayo *et al.* 2008). (Fig. 2).

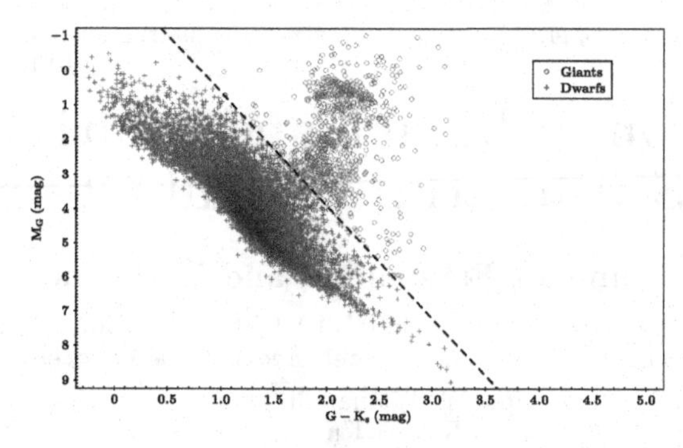

Figure 1. H-R diagram showing the location of our candidate systems. Dashed line separates dwarfs (below) from giants (above).

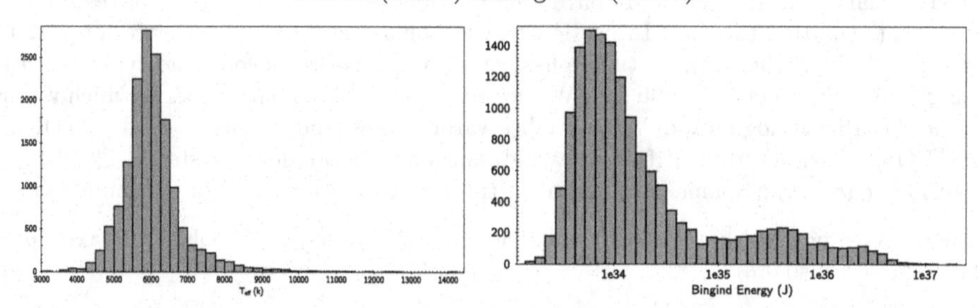

Figure 2. Left: T_{eff} distribution. The mayority of our objects show effective temperaturs typical of F-G spectral types; Right: Binding energy distribution. The bi-modal shape reported in Dhital *et al.* (2010) is also found here.

Stellar masses were derived by interpolating T_{eff} in the tables B1 and B2 (Gray 2008) using the dwarf/giant information. Masses and projected physical separation were used to estimate the binding energy $U = -GM_1M_2/r$ (Fig. 2). Following Dhital *et al.* (2010), we set a lower limit for the binding energy of 10^{33} J.

An analysis of the distribution of binding energies and the identification of interesting systems (weakly bound, FGK-M pairs,...) is currently on-going.

Acknowledgements

This work has made use of data from the European Space Agency (ESA) mission *Gaia* (https://www.cosmos.esa.int/gaia), processed by the *Gaia* Data Processing and Analysis Consortium (DPAC, https://www.cosmos.esa.int/web/gaia/dpac/consortium). Funding for the DPAC has been provided by national institutions, in particular the institutions participating in the *Gaia* Multilateral Agreement.

References

Bayo, A., Rodrigo, C., Barrado Y Navascués, D., *et al.* 2008, *A&A*, 492, 277

Dhital, S., West, A. A., Stassun, K. G., & Bochanski, J. J. 2010, *AJ*, 139, 2566

Gaia Collaboration *et al.* 2016, *A&A*, 595, A2

Gray, D. F. 2008, *The Observation and Analysis of Stellar Photospheres, by David F. Gray, Cambridge, UK: Cambridge University Press*

Kunder, A., Kordopatis, G., Steinmetz, M., *et al.* 2017, *AJ*, 153, 75

Luo, A.-L., Zhao, Y.-H., Zhao, G., *et al.* 2015, *Research in Astronomy and Astrophysics*, 15, 1095

Soubiran, C., Le Campion, J.-F., Brouillet, N., & Chemin, L. 2016, *A&A*, 591, A118

Astronomy and Astrophysics in the Gaia sky
Proceedings IAU Symposium No. 330, 2017
A. Recio-Blanco, P. de Laverny, A.G.A. Brown & T. Prusti, eds.

doi:10.1017/S1743921317005920

Open star clusters and Galactic structure

Yogesh C. Joshi

Aryabhatta Research Institute of Observational Sciences, Nainital, India - 263002
email: yogesh@aries.res.in

Abstract. In order to understand the Galactic structure, we perform a statistical analysis of the distribution of various cluster parameters based on an almost complete sample of Galactic open clusters yet available. The geometrical and physical characteristics of a large number of open clusters given in the MWSC catalogue are used to study the spatial distribution of clusters in the Galaxy and determine the scale height, solar offset, local mass density and distribution of reddening material in the solar neighbourhood. We also explored the mass-radius and mass-age relations in the Galactic open star clusters. We find that the estimated parameters of the Galactic disk are largely influenced by the choice of cluster sample.

Keywords. open clusters: general; Galaxy: evolution – Galaxy: structure, – method: statistical – astronomical data bases

---

1. Introduction

Open clusters are distributed throughout the Galactic disk and span a wide range in ages hence they are useful tool to study the effects of dynamical evolution of the Galactic disk (Carraro *et al.* 1998). Even though much progress has been made in understanding the general properties of open star clusters and Galactic structure but continuous increase of the cluster sample and refinement of cluster parameters pave way for a better understanding of the Galactic structure (Joshi *et al.* 2016). The observed distribution of cluster parameters and correlations between various parameters offer important empirical constraints not only on the cluster formation, but also on the history of the Galaxy. However, any meaningful study of star clusters is very much dependent on the accurate determination of the cluster parameters as well as the homogeneity and degree of completeness of the cluster sample.

2. Data

To understand the Galactic structure, statistical analysis of a large number of star clusters in the Galaxy is very important because individual cluster parameters may be affected by the considerable uncertainty in their determination. We used MWSC catalogue that gives the physical parameters of 3208 star clusters (Kharchenko *et al.* 2013, Schmeja *et al.* 2014, Scholz *et al.* 2015) which were determined in homogeneous manner since parameters of all the clusters are estimated through the same technique. Our analysis found that MWSC catalogue is complete only up to a distance of 1.8 kpc which is similar to Dias catalogue† on open clusters (Joshi 2005, 2007). After rejecting the dubious clusters in the MWSC catalogue within the completeness limit of 1.8 kpc, we used 1218 open clusters to study the Galactic structure.

† http://www.astro.iag.usp.br/ocdb

3. Results and Conclusion

Although a detailed study to understand the Galactic structure in the solar neighbourhood is reported in Joshi *et al.* (2016) based on MWSC survey catalogue, a brief summary of the results is given below.

(*a*) Assuming a uniform density model for the distribution of clusters, the catalogue of open clusters is found to be complete only up to 1.8 kpc distance.

(*b*) We found a huge deficiency of old open clusters (> 700 Myr) within 500 pc from the Sun, having only 6 out of a total 91 clusters (6.6%) are located within this distance. It is understood that the most of the massive clusters in the solar neighbourhood would dissolve even before reaching the age of 1 Gyr causing the deficiency of old clusters in the solar neighbourhood.

(*c*) The distribution of clusters in various longitude bins revealed that the maximum clusters are located in the region of around 125° (Cassiopeia), 210 deg (Monoceros), 240 deg (Canis Major), and 285 deg (Carina) while deep minima is primarily found in two longitudes, around 50 deg (Sagitta) and 150 deg (Perseus).

(*d*) The maximum Galactic absorption is seen towards $l \sim 40^o$ and minimum towards $l \sim 220^o$.

(*e*) The solar offset is found to be in the range of 6 to 15 kpc, however, it strongly depends on the choice of stellar population, data sampling and estimation method.

(*f*) The disk scale height is found to be $z_h = 64 \pm 2$ pc, however, it decreases to $z_h = 60 \pm 2$ pc when we only consider clusters younger than 700 Myr. Although it is close to generally accepted value of z_h deduced from the cluster sample but it is noticed that the difference in measurement of z_h is strongly influenced by the selection criteria of the cluster sample.

(*g*) It is found that scale height increases with mean age of the clusters and, in general, z_h increases from ~40 pc at 1 Myr to ~75 pc at 1 Gyr. Furthermore, z_h is found to be larger in the direction of Galactic anti-center than the Galactic center and, on an average, z_h is more than twice as large as in the outer region than in the inner region of the solar circle, except for the youngest population of clusters.

(*h*) We estimated a local mass density of $\rho_0 = 0.090 \pm 0.005\ M_\odot/\mathrm{pc}^3$ but did not find any significant contribution of dark matter in the solar neighborhood.

(*i*) The reddening in the direction of clusters suggests a strong correlation with their vertical distance from the Galactic plane with a respective slope of $dE(B-V)/dz = 0.40\pm0.04$ and 0.42 ± 0.05 mag/kpc below and above the Galactic Plane.

(*j*) We observed a linear mass-radius and mass-age relations and derived slopes of $\frac{dR}{d(logM)} = 2.08 \pm 0.10$ and $\frac{d(logM)}{d(logT)} = 0.36 \pm 0.05$ for the Galactic open clusters.

Acknowledgments

I wish to thank IAU for providing financial support to attend the conference.

References

Joshi, Y. C. 2005, *MNRAS*, 362, 1259
Joshi, Y. C. 2007, *MNRAS*, 378, 768
Joshi, Y. C., Dambis, A. K., Pandey, A. K., & Joshi, S. 2016, *A&A*, 593, 116
Kharchenko, N. V., Piskunov, A. E., Schilbach, E., *et al.* 2013, *A&A*, 2013, 558, 53
Carraro, G., Ng, Y. K., & Portinari, L. 1998, *MNRAS*, 296, 1045
Schmeja, S., Kharchenko, N. V., Piskunov, A. E., *et al.* 2014, *A&A*, 568, 51
Scholz, R.-D., Kharchenko, N. V., Piskunov, A. E., *et al.* 2015, *A&A*, 581, 39

Astronomy and Astrophysics in the Gaia sky
Proceedings IAU Symposium No. 330, 2017
A. Recio-Blanco, P. de Laverny, A.G.A. Brown & T. Prusti, eds.

doi:10.1017/S1743921317006135

The time evolution of gaps in tidal streams in axisymmetric potentials

Helmer H. Koppelman and Amina Helmi

Kapteyn Instituut, University of Groningen,
P.O. Box 800, 9700 AV, Groningen, the Netherlands
email: h.h.koppelman@rug.nl

Abstract. Our goal is to understand the evolution and properties of gaps produced by dark matter subhalos in stellar tidal streams. Here we explore how gaps grow in spherical potentials in comparison to axisymmetric potentials. We develop a model that uses the divergence of two orbits, one on each side of the gap, to describe the size of the gap and how this varies with time and depends on the characteristics of the encounter with the dark subhalo. To this end we use a formalism based on action-angle variables.

Keywords. Galaxy: halo, Galaxy: structure, Galaxy: kinematics and dynamics

---

1. Introduction & Motivation

ΛCDM simulations typically predict thousands of satellites orbiting the Milky Way to be completely dark matter dominated (Springel *et al.* 2008). Possibly interesting objects to probe this population of (otherwise invisible) dark satellites are tidal streams. We thus aim to infer the properties of the dark matter satellites from the properties of the gaps formed as a dark subhalo comes sufficiently close to a tidal stream. Modeling the evolution of a gap as the divergence of two orbits has shown good agreement with N-body experiments carried out in a spherical host potential (Helmi & Koppelman, 2016). We present here the results obtained for a more realistic Stäckel Galactic potential, now including a disk and a halo component.

2. Model

Building on Helmi & Koppelman (2016) we model the size of the gap as the separation of two orbits A & B which correspond to particles in the stream that have been perturbed the most because of the interaction with the subhalo. To obtain the initial separation of these two orbits we use the impulse approximation (Erkal & Belokurov, 2015). By assuming a Plummer profile for the dark matter we can obtain analytic expressions for their separations in phase-space, ΔX_0 and ΔV_0. This initial separation can be transformed to a separation in action-angle space with a matrix: $M_0 = \partial(\Theta, J)/\partial(X, V)$, which is the Jacobian of the coordinate transformation. In action-angle space, time evolution is simple: $\Delta\Theta(t) = \Delta\Theta_0 + \Omega(J)t$, where $\Omega(J)$ are the frequencies of motion, and $J(t) = J_0$. We encode this evolution in a matrix $\Omega' = \begin{bmatrix} I_3 & \partial\mathbf{\Omega}/\partial\mathbf{J}t \\ [0.3em]0 & I_3 \end{bmatrix}$. At each timestep the separation vector in action-angle space is transformed back locally to Cartesian coordinates with matrix M_t^{-1}. For a schematic overview of the model see Fig.1a.

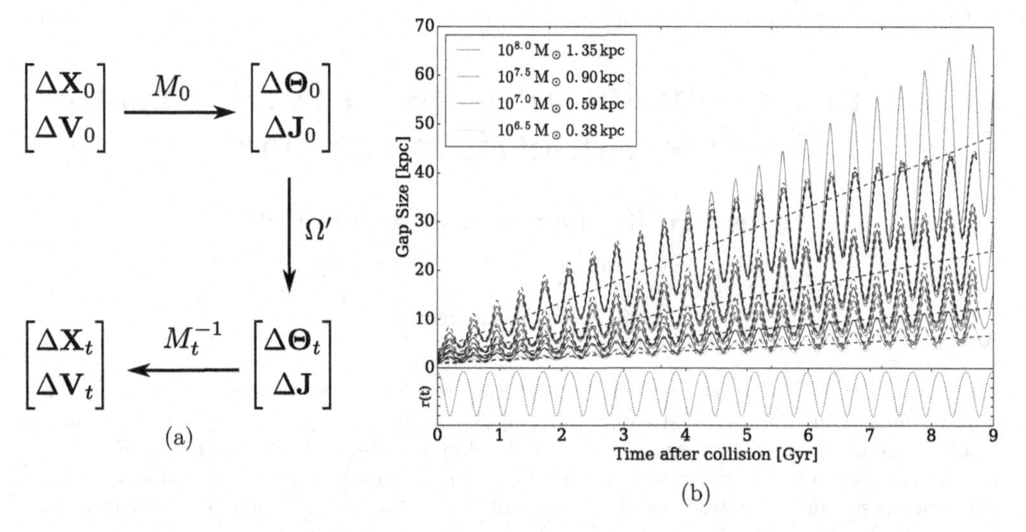

Figure 1: Panel (a): A schematic overview of the model. First we calculate the initial size of the gap from the impulse approximation, then we evolve this separation in time using actions and angles. Panel (b): Gaps growing in tidal streams. The colored lines show the predicted sizes of the gaps from our model, the solid black lines show the size measured in N-body experiments. The fitted dashed lines show the linear rate of growth of the size of the gap with time for these experiments run in a realistic (axisymmetric) Galactic potential.

3. Results & Conclusions

Fig. 1b shows the evolution in time of the four different gaps produced by encounters with different mass subhalos. The solid colored lines show the prediction for the size of the gap from our model, the solid black lines are the measured size of the gaps in the corresponding N-body experiments. The fitted dashed lines show that a linear fit to the growth of the gap size in time works well on average. The oscillations seen depend on the orbital phase of the gap in the stream.

Our model thus shows very nice agreement with gap sizes measured in the N-body experiments. The rate of growth is dependent on the properties of the dark matter subhalo, e.g. its mass and scale radius, on the geometry of the impact, and on the age of the stream. The latter two have not been varied for the results shown in Fig. 1b. The evolution in time of the size of a gap is thus linear, also in the case of an axisymmetric potential.

References

Helmi, A. & Koppelman, H. H. 2016, *ApJ* (Letters), 828, L10
Helmi, A. & White, S. D. M. 1999, *MNRAS*, 307, 3
Erkal, D. & Belokurov, V. 2015, *MNRAS*, 450, 1
Springel, V., Wang, J., Vogelsberger, M., *et al.* 2008, *MNRAS*, 391, 1685

Astronomy and Astrophysics in the Gaia sky
Proceedings IAU Symposium No. 330, 2017
A. Recio-Blanco, P. de Laverny, A.G.A. Brown & T. Prusti, eds.

doi:10.1017/S1743921317006007

On-sky verification of the 6-h periodic basic angle variations of the Gaia satellite

Shilong Liao[1,2,*], Mario G. Lattanzi[2], Alberto Vecchiato[2], Zhaoxiang Qi[1], Mariateresa Crosta[2] and Zhenghong Tang[1]

[1]Shanghai Astronomical Observatory, Chinese Academy of Sciences, 80 Nandan Road, 200030 Shanghai, China
*email: shilongliao@shao.ac.cn

[2]INAF-Osservatorio Astrofisico di Torino, Strada Osservatorio 20, I-10025 Pino Torinese, TO, Italy

Abstract. A Basic Angle (BA) of 106.5° separates the view directions of Gaia's two fields of view (FoV). A precise determination of the BA variations (BAV) is essential to guarantee a correct reconstruction of the global astrometric sphere, as residual systematic errors would result in, e.g., a bias in the parallaxes of the final Gaia catalog. The Basic Angle Monitoring (BAM) device, which provides a reliable and accurate estimation of BAV, shows that there exists a ~1 mas amplitude, 6-h period BA oscillation. It's essential to verify to what extent this signal is caused by real BAV, or is at least in part an effect of the BAM device itself. Here, we propose an astrometric on-sky approach to re-determine the 6-h periodic BAV. The results of this experiment, which treated a full day (17 Oct 2016) of Gaia astrometric data, recover a value for the 6-h oscillation of 1.856±0.857 mas. This is consistent, within the errors, with the BAM finding for that day.

Keywords. periodic basic angle variation, astrometric method, validation

1. Introduction

An early discovery in the Gaia data was that there exists a ~1 mas amplitude, 6-h period oscillation in the BAM data (Mora *et al.* 2014). An immediate question was whether this signal is caused by a real variation of the BA, or an effect in the BAM device itself. Here, we discuss a method based solely on astrometric measurements of transiting stars to validate whether that 6-h periodic BAV is seen on sky.

2. Astrometric Method

Gaia measures the crossing time of the target image center transiting the CCD fiducial lines (Prusti *et al.* 2016). The proposed method is to analyze the difference of the time required for two targets, coming from the two FoVs and quasi-simultaneously entering the common astrometric focal plane, to cross two consecutive fiducial lines. The quasi-simultaneous entering ensures that the two targets share the same attitude, while the BAV effect will be left in the time difference since they come from different FoVs.

As seen from the left panel of Fig. 1, the transiting time interval of star S_1 from PFoV and star S_2 from FFoV to cross two consecutive fiducial lines can be written as:

$$\Delta t = t_2 - t_1 = \Delta t_\eta + \delta t_{att}^{S_1} + \delta t_{noise}^{S_1} + \Delta t_{f_0} \quad \text{and} \quad \Delta T = T_2 - T_1 = \Delta T_\eta + \delta T_{att}^{S_2} + \delta T_{noise}^{S_2} + \Delta T_{f_1} \tag{2.1}$$

The differential of their transiting time intervals can be written as:

$$\Delta t - \Delta T = (\Delta t_\eta - \Delta T_\eta) + (\delta t_{att}^{S_1} - \delta T_{att}^{S_2}) + (\Delta t_{f_0} - \Delta T_{f_1}) + (\delta t_{noise}^{S_1} - \delta T_{noise}^{S_2}) \tag{2.2}$$

where $(\Delta t_\eta - \Delta T_\eta)$ is the effect of the proper distance along scan difference between two fiducial lines (that we consider constant within 24h); $(\delta t_{att}^{S_1} - \delta T_{att}^{S_2})$ is the effect of differential attitude

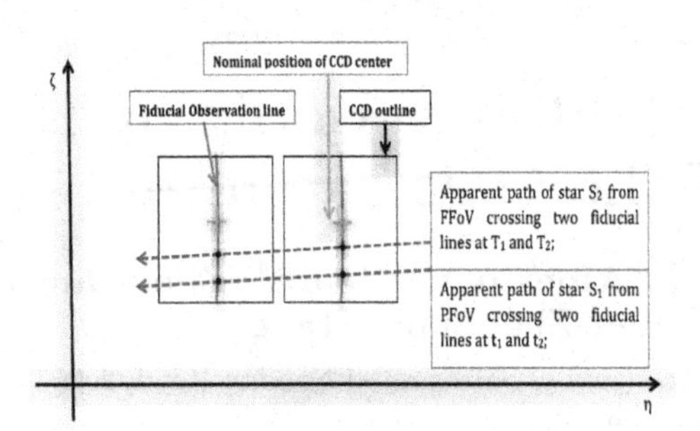

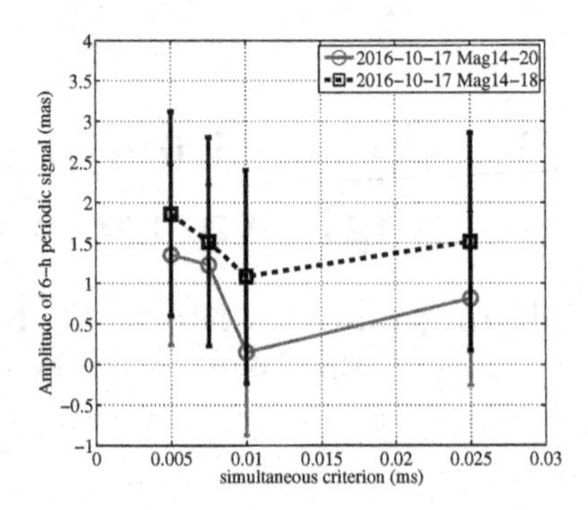

Figure 1. Left: Schematic illustration of quasi-simultaneous observations. The dashed lines are the apparent path of two targets from the two FoVs crossing two fiducial lines at (t_1,t_2) and (T_1,T_2), respectively. Right: The results of the amplitude of the 6-h periodic signal under different simultaneity criterions for day 17 Oct 2016; two magnitude ranges, ($14 \leqslant G \leqslant 20$) and ($14 \leqslant G \leqslant 18$), are compared. The results show that with a larger quasi-simultaneity criterion, the signals become less significant, and, with more bright stars, the SNR is better.

noise. Quasi-simultaneous observations ensure the two targets S_1 and S_2 share the same attitude, so that $(\delta t_{att}^{S_1} - \delta T_{att}^{S_2})$ can be eliminated, at least the low frequency terms; $(\Delta t_{f_0} - \Delta T_{f_1})$ is the effect of the two lines of sight change, which is treated as the basic angle variation effect between t_1 and t_2, that is $\frac{\Delta\Gamma(t_2)-\Delta\Gamma(t_1)}{\Omega} = (\Delta t_{f_0}^{S_1} - \Delta T_{f_1}^{S_2})$. $\delta t_{noise}^{S_1}$ and $\delta T_{noise}^{S_2}$ are the measurement noise effect of the image location estimation of S_1 and S_2, respectively. Of course, the brighter the stars, the better the noise. To least-square the data, we use the following functional form:

$$\Delta\Gamma(t) = \sum_{k=1}^{8} [A_k \cos(k\Omega t) + B_k \sin(k\Omega t)] \tag{2.3}$$

Where A_k and B_k are the coefficients of the periodic variation terms, Ω is the satellite spin angular velocity. $\Delta\Gamma(t)$ represents only the time variations caused by the periodic BAV components. A secular term is present in the data, and was taking into account in the reduction, but it's not shown here.

3. Results and Conclusions

The right panel of Fig. 1 shows the fitting results of the selected data set under different quasi-simultaneity criterions. The results show that the optimal simultaneity criterion is 0.005ms, and the best magnitude range is to G magnitude 18. A 6-h periodic component with an amplitude of 1.856±0.857 mas is found, which is compatible with the BAM results within the errors for that day. The SNR of the result is low and we'll try to improve it by using a data set with a higher density and larger number of bright stars (G magnitude $\leqslant$ 16), such as when the satellite is scanning the disk of the Milky Way.

However, the results might be telling a more interesting story, as the lines of sight change measured by this approach represents an average over the whole focal plane; by contrast, the BAM results only represent a specific part of the focal plane. Also, our results are affected by the focal length changes and optical distortion in the two different telescopes.

Acknowledgements

This work was supported by grants from the National Science Foundation of China (NSFC) 11573054, ASI under contracts Gaia Mission- The Italian Participation to DPAC, I/058/10/0-1 and 2014-025-R.1.2015. This work has made use of data from the European Space Agency (ESA) mission Gaia, processed by the Gaia Data Processing and Analysis Consortium (DPAC).

References

Gaia collaboration, Prusti, T., *et al.* 2016, *A&A*, A1, 595
Mora, A., Biermann, M., *et al.* 2014, *SPIE*, 91430X

Astronomy and Astrophysics in the Gaia sky
Proceedings IAU Symposium No. 330, 2017
A. Recio-Blanco, P. de Laverny, A.G.A. Brown
& T. Prusti, eds.

doi:10.1017/S1743921317006123

Open Cluster Dynamics via Fundamental Plane

Chien-Cheng Lin[1,2] and Xiao-Ying Pang[3] †

[1]Shanghai Astronomical Observatory, 80 Nandan Rd. 200030 Shanghai, China
[2]Max Planck Institute for Astronomy, Königstuhl 17, 69117 Heidelberg, Germany
email: `cclin@mpia.de`
[3]Shanghai Institute of Technology, 100 Haiquan Rd., 201418 Shanghai, China
email: `xypang@bao.ac.cn`

Abstract. Open clusters (OCs) are important objects for stellar dynamics studies. The short survival timescale of OCs makes them closely related to the formation of Galactic field stars. We motivate to investigate the dynamical evolution of OCs on the aspect of internal effect and the external influence. Firstly, we make use of the known OC catalog to obtain OCs masses, effective radii. Additionally, we estimate OCs kinematics properties by OC members cross-matched with radial velocity and metallicity from SDSSIV/APOGEE2. We then establish the fundamental plane of OCs based on the radial velocity dispersion, the effective radius, and average surface brightness. The deviation of the fundamental plane from the Virial Plane, so called the tilt, and the r.m.s. dispersion of OCs around the average plane are used to indicate the dynamical status of OCs. Parameters of the fitted plane will vary with cluster age and distance.

Keywords. (Galaxy:) open clusters and associations: general

---

1. Introduction

The fundamental plane (FP) is represented with the effective radius (r_e), average surface brightness (μ_e), and central velocity dispersion (σ_0) of normal elliptical galaxies (Djorgovski & Davis 1987). And the FP of elliptical galaxies can be described as $\log(r_e) = a{\cdot}\log(\sigma_0) + b{\cdot}\log(\mu_e) + c$, where a, b, and c are fitted parameters. Similar idea has also been applied to 56 globular clusters (GCs) and the derived FP imply that they are virialized-systems with constant mass-to-light ratios, i.e., $a = 2$ and $b = -1$ (Djorgovski 1995). Moreover, Bonatto & Bica (2005)'s 11 nearby open clusters (OCs) sample indicates an FP of OCs. Thanks to the modern sky surveys, we can homogeneously enlarge the sample of clusters to study the dynamical evolution of OCs via their tilt of FP.

2. Data and Data Analysis

The initial OC sample was mainly based on the collection of Kharchenko *et al.* (2013), due to their homogeneously study of about 3000 OCs. We then crossed matched the member stars of the OC sample with the 13th data release from the Sloan Digital Sky Survey (SDSS Collaboration *et al.* 2016) for APOGEE2 stars in order to obtain the dispersion of radial velocity for each OC. We also combined 10 OCs from Gaia-ESO survey to enlarge the sample. The OCs members were considered: 1. stars are along with the isochrone of cluster age, 2. stars are located within apparent radius r_2 of each OC, 3. stars are clustering on radial velocity and metallicity diagram of each OC (Fig. 2 left panel is selected demonstration of step 3 results). The final sample includes 36 OCs with sufficient radial velocities members, i.e., > 5 stars.

† Further questions please address to the corresponding author: Xiaoying Pang.

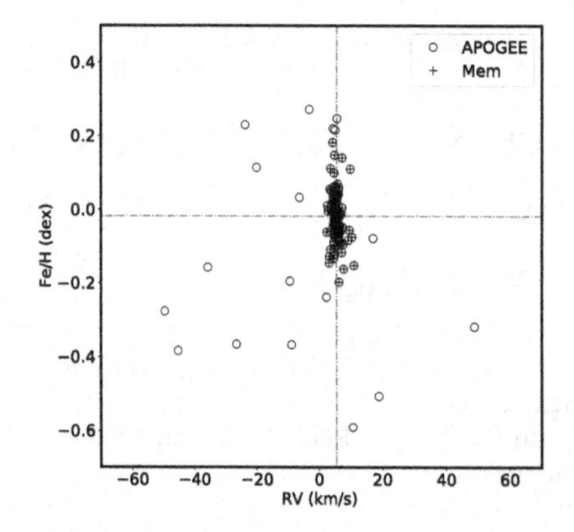

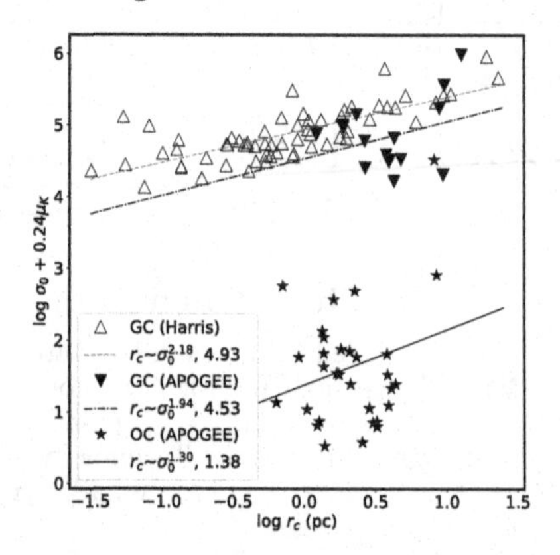

Figure 1. Left: The radial velocity and metallicity diagram. Black empty circles are stars matched with APOGEE data, while black crosses are satisfied with the three criteria in session 2. Right: The correlation between the core radius (r_c), surface brightness (μ_K) in K band, and central velocity dispersion (σ_0). The empty triangles are GCs from Harris (1996), filled triangles are those matched with APOGEE2 data, and the filled asterisks are OC sample. The dashed line, dashed-dotted line, and solid lines are the least-square fitting results by the equation in session 1 of GCs from Harris (1996), GCs with APOGEE2 data, and OCs with APOGEE2 and Gaia-ESO data, respectively.

3. Results and Discussions

Fig. 2 right panel shows comparison of the three parameters correlation of GCs between Harris (1996) and APOGEE2 data. Both groups illustrate the FP slopes are 2.18 and 1.94, which are in a good agreement with the virialized-systems. A total of 31 OCs (filled asterisks) with ages from 0.1 to 4 Gyr (intermediate and old ages) show the FP is tilted with about 10 degrees. The tilt of these relative old-OCs FP may be mainly due to environmental effect, such as their locations on the disk make them suffer from larger tidal effect due to spiral arm or giant molecular clouds. The intercept of GC FP is different from OC FP, which might owning to different mass to light between these two groups. N-body simulations will be used to investigate the physical processes that response for the parameter variations. Gaia DR2 may help us enlarge the OC sample including younger ones to make comparisons of FP between young and old ages.

4. Acknowledgement

This work is supported by the National Science Council of Taiwan under grants MOST 106-2917-I-564-042 (Chien-Cheng Lin) and by National Natural Science Foundation of China, grant No: 11503015 (Xiaoying Pang).

References

Bonatto, C. & Bica, E. 2005, *A&A*, 437, 483
Djorgovski, S. & Davis, M. 1987, *ApJ*, 313, 59
Djorgovski, S. 1995, *ApJ*, 438, L29
Harris, W. E. 1996, *AJ*, 112, 1487
Kharchenko, N. V., Piskunov, A. E., Schilbach, *et al.*, 2013, *A& A*, 558, A53
SDSS Collaboration, Albareti, F. D., Allende Prieto, C., *et al.* 2016, *arXiv:1608.02013*

Astronomy and Astrophysics in the Gaia sky
Proceedings IAU Symposium No. 330, 2017
A. Recio-Blanco, P. de Laverny, A.G.A. Brown & T. Prusti, eds.

doi:10.1017/S1743921317006536

A 3D-Study of the residual vector field

Francisco J. Marco[1], **María J. Martínez**[2] **and Jose A. López**[1]

[1]Universidad Jaume I. Dept. Matemáticas, IMAC.
Avda. Vicent Sos Baynat s/n, 12071, Castellón, Spain
email: marco@mat.uji.es, lopez@mat.uji.es

[2]Universidad Politécnica de Valencia. Dept. Matemática Aplicada. IUMPA.
Camino de Vera s/n. 46022 Valencia. Spain
email: mjmartin@mat.upv.es

Abstract. One of the important challenges that Gaia imposes on the Astrometric Catalogs, is a careful study in everything affected by parallax. A particularly important case is the necessary linkage Gaia - HCRF - ICRF2, which require methods of analysis that are accurate enough so that the provided results are at the same precision level as the work data.

Keywords. astronomical data bases: miscellaneous, catalogs, reference systems

---

1. Introduction

We will consider the usual discrete points for which we know the Right Ascension, the Declination, and the trigonometric parallax p. We denote $r = 1/p$. We will use Hipparcos2 and UCAC4 catalogs. Regarding the limitations in some measurements in both catalogues, this paper represents an initial approximation to a further work that will be done using the new Gaia catalogue. We consider the residuals $\Delta\alpha^*$, $\Delta\delta$ together with their corresponding r values. These residuals can be seen as unidimensional random variables (r.v.) A regression on r shows that there exist dependencies $\Delta\alpha^*(r)$, $\Delta\delta(r)$. For a r.v., we calculate the regression as:

$$E\left[X|r\right] = \frac{\int xf\left(x,r\right)dx}{\int f\left(x\right)dx}$$

where each f represents different density functions, depending on the context. Our final aim is to consider the vector field $[\Delta\alpha^*(r), \Delta\delta(r)]$ for the N = 82446 selected stars- Due to low density of the 3D-discrete, we have decided to work in "slices" of 200 pcs. Here, we focus on a particular slice centered at 400 pcs.

2. Methods

Adjusting a three-dimensional spherical function

$$\widehat{m}\left(\alpha,\delta,r\right) = \sum_{i=1}^{n}\varpi_i z_i$$

where $z_i = \Delta\alpha_i^*\left(r_i\right)$ or $\Delta\delta_i\left(r_i\right)$ and ϖ_i are the weights for the Nadaraya-Watson method in the three-dimensional case. The low three-dimensional density recommends the use of a relatively high bandwidth in r *Vector spherical harmonics (VSH) development*

Denoting by Y_{nm}, $n \geqslant 0$, $|m| \leqslant n$, the usual spherical surface harmonics, the VSH are:

$$\vec{R}_{nm} = \vec{r}Y_{nm}, \vec{S}_{nm} = \vec{\nabla Y}_{nm}, \vec{T}_{nm} = -\vec{r}\times\vec{\nabla Y}_{nm}$$

They are an orthogonal and complete base in the Hilbert space of the vector fields $L^2(S^2)$.

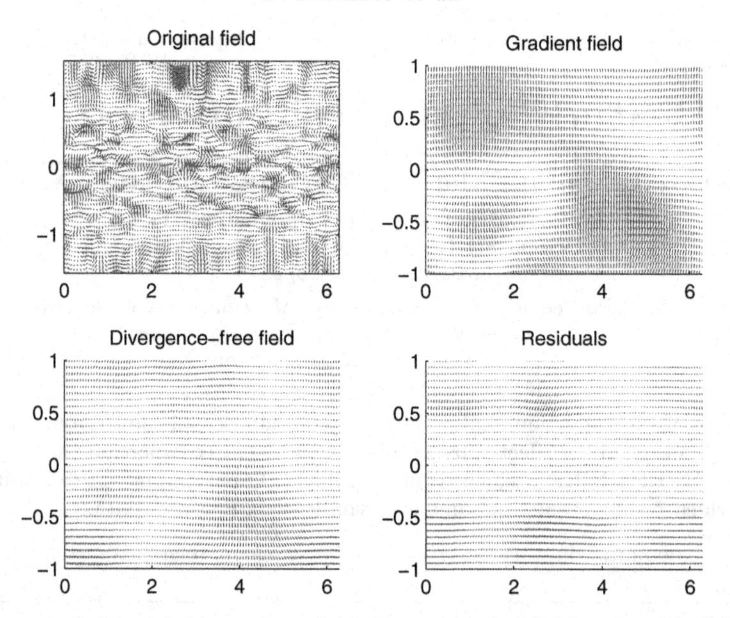

Figure 1. Original field, gradient field, divergence-free field and residual field.

Table 1. Div-free and curl-free components for the residual vector fields around $r = 400$ pcs

r (pcs)		$s_{1,0}$	$s_{1,1}$	$s_{1,-1}$	$t_{1,0}$	$t_{1,1}$	$t_{1,-1}$
400	Curl-free	8.72	7.90	1.50	−−	−−	−−
	Div-free	−4.04	−4.20	−0.09	3.38	−0.08	3.61
	Sum	4.68	3.70	1.41	3.38	−0.08	3.61

<u>*Helmholz Decomposition*</u> We know that we can obtain a divergence-free vector from a noise vector. This is a potential source of errors and that is why we propose, as a method of analysis of coefficients and errors, to perform previously a Helmholtz decompositions for each especially difficult vector fields.
Let us consider a field on the sphere $\overrightarrow{X}$, Helmholtz theorem provides a (unique) decomposition: $\overrightarrow{X} = \overrightarrow{\nabla\phi} + \overrightarrow{\nabla} \times \overrightarrow{u} + \overrightarrow{\varepsilon}$ where the gradient of the potential, $\overrightarrow{\nabla\phi}$, is curl-free, the curl $\overrightarrow{\nabla} \times \overrightarrow{u}$ is divergence-free and the residual field $\overrightarrow{\varepsilon}$ verifies both properties. From this point, we recompute the VSH coefficients again.

3. Results

Certain 3D regions should be studied in more detail, due to the variation in the values of certain coefficients. We have taken the neighbourhood of 400 pcs. Here, the distances between residuals in RA and Dec are relatively large. For RA, the residuals increase near a local maximum and decrease near a local minimum for Dec. This must reflect in the vector field. In Table 1 we show the results for $r = 400pcs$. The complete, curl-free, divergence free and remnant ($\overrightarrow{\varepsilon}$) vector fields may be seen in Figure 1.
We should point out that UCAC4 is not independent of the Hipparcos catalogue, as Tycho-2 stars are used as reference for UCAC4. This makes more difficult the interpretations of the residuals, which will be discussed in a further paper.

Acknowledgements

This work was supported by a grant UJI-B2016-18.

Astronomy and Astrophysics in the Gaia sky
Proceedings IAU Symposium No. 330, 2017
A. Recio-Blanco, P. de Laverny, A.G.A. Brown
& T. Prusti, eds.

doi:10.1017/S1743921317006457

Impact on the Hipparcos2-UCAC4 geometric relation from some physical properties of the stars

Francisco J. Marco[1], **María J. Martínez**[2] **and Jose A. López**[1]

[1]Universidad Jaume I. Dept. Matemáticas, IMAC.
Avda. Vicent Sos Baynat s/n, 12071, Castellón, Spain
email: marco@mat.uji.es, lopez@mat.uji.es

[2]Universidad Politécnica de Valencia. Dept. Matemática Aplicada. IUMPA.
Camino de Vera s/n. 46022 Valencia. Spain
email: mjmartin@mat.upv.es

Abstract. The aim of this paper is the study of the impact that the consideration of different physical properties as magnitude and spectral type of stars has on the geometric relations between Hipparcos2 and UCAC4. In this sense, the pairs of residuals $\Delta\alpha^*$ and $\Delta\delta$ can be considered as functions of (α, δ, r) and for each fixed r, we can fit a vector field on the sphere from which to obtain its components in the VSH basis. The same can be done by grouping the stars considering their magnitudes, spectral types (or mixing them) and then studying the variations in the mentioned geometry. We must not forget that $\Delta\alpha^*$ and $\Delta\delta$ are numerical random variables whose regression on the magnitude m, for example, can be estimated. The results will be computed taking into account r as well as the physical mentioned properties. So, we avoid the assumption that the harmonic coefficients depend only on m.

Keywords. astronomical data bases: miscellaneous, catalogs, reference systems

1. Magnitudes and distances

We have selected two sets of common stars from Hipparcos and UCAC4 regarding their magnitudes. We call these sets m5 (23282 stars with magnitudes ranging [7.153, 8.556]) and m6 (31753 stars ranging [8.556, 9.959]), and distances up to 800 pcs. We study the different behaviors in $\Delta\alpha^*$ and $\Delta\delta$. Both magnitudes cover the 80% of the total population that has been considered (79591 non-double stars and with strictly positive parallaxes) If we proceed as in Marco *et al.* (2017), we see that around $r = 400$ pcs, for both m5 and m6 populations, there are possible problems for the vector field of the residuals (possible saddle point for m5 and almost coincident, on a long range of r, for m6, which could imply a possible degeneration).

Table 1 shows that the behavior of the coefficients is different for both populations, as expected. Globally s_{10} and t_{1-1} decrease with r, being similar for the two consecutive distances 300 and 400pcs. We also find this behavior in s_{11}. On the other hand, t_{10} decreases slightly. Finally, s_{1-1} and t_{11} have unparalleled behaviors. We also notice that formal errors are generally high, although a more detailed study shows that the stars collected in m6 have a more stable behavior.

2. Spectral type and distances. Mixing properties

The spectral type, by itself, does not provide relevant information, but this changes if we consider the data together with the J-magnitude. We used stars up to 800 pcs

Table 1. Results for m5, m6 and weighted m5+m6 sets

m_5 (pcs)	$s_{1,0}$	$s_{1,1}$	$s_{1,-1}$	$t_{1,0}$	$t_{1,1}$	$t_{1,-1}$
S^2	9.68 ± 0.02	2.42 ± 0.02	0.24 ± 0.02	4.16 ± 0.02	0.08 ± 0.02	3.50 ± 0.02
[200, 400]	5.08 ± 0.80	4.44 ± 0.50	0.16 ± 0.01	5.15 ± 0.40	-1.09 ± 0.40	5.56 ± 0.12
[300, 500]	3.28 ± 0.14	4.04 ± 0.25	-1.43 ± 0.35	5.63 ± 0.30	-1.32 ± 0.11	3.53 ± 0.03
m_6 (pcs)	$s_{1,0}$	$s_{1,1}$	$s_{1,-1}$	$t_{1,0}$	$t_{1,1}$	$t_{1,-1}$
S^2	8.83 ± 0.01	2.88 ± 0.01	0.79 ± 0.01	2.65 ± 0.01	1.77 ± 0.01	3.44 ± 0.01
[200, 400]	6.36 ± 0.31	3.30 ± 0.10	1.48 ± 0.43	2.11 ± 0.01	0.44 ± 0.38	2.99 ± 0.18
[300, 500]	4.62 ± 0.63	2.90 ± 0.01	1.91 ± 0.25	1.17 ± 0.02	0.96 ± 0.05	1.77 ± 0.35
$m_5 + m_6$ (pcs)	$s_{1,0}$	$s_{1,1}$	$s_{1,-1}$	$t_{1,0}$	$t_{1,1}$	$t_{1,-1}$
S^2	9.19 ± 0.03	2.69 ± 0.03	0.56 ± 0.03	3.29 ± 0.03	1.06 ± 0.03	3.47 ± 0.03
[200, 400]	5.81 ± 0.40	2.10 ± 0.02	0.92 ± 0.30	5.86 ± 0.30	3.40 ± 0.20	4.08 ± 0.10
[300, 500]	4.05 ± 0.60	3.38 ± 0.30	0.50 ± 0.07	3.63 ± 0.04	0.00 ± 0.30	2.51 ± 0.12

Table 2. Results for m5 and m6 sets mixed with KM and No-KM stars

$KM - m_6$ (pcs)	$s_{1,0}$	$s_{1,1}$	$s_{1,-1}$	$t_{1,0}$	$t_{1,1}$	$t_{1,-1}$
S^2	5.85 ± 0.05	2.54 ± 0.01	-0.23 ± 0.01	6.58 ± 0.01	2.99 ± 0.02	2.06 ± 0.01
[200, 400]	4.16 ± 0.80	2.99 ± 0.20	0.34 ± 0.30	5.36 ± 0.12	3.34 ± 0.30	1.51 ± 0.25
[300, 500]	3.21 ± 0.50	3.74 ± 0.30	0.13 ± 0.01	5.33 ± 0.15	3.06 ± 0.04	1.45 ± 0.03
$No - KM - m_6$ (pcs)	$s_{1,0}$	$s_{1,1}$	$s_{1,-1}$	$t_{1,0}$	$t_{1,1}$	$t_{1,-1}$
S^2	11.43 ± 0.02	3.70 ± 0.10	1.56 ± 0.02	0.97 ± 0.02	1.23 ± 0.02	3.92 ± 0.04
[200, 400]	9.87 ± 0.90	3.75 ± 0.10	1.75 ± 0.60	-0.48 ± 0.21	-1.00 ± 0.73	4.08 ± 0.05
[300, 500]	10.43 ± 0.04	1.85 ± 0.10	2.96 ± 0.16	-1.57 ± 0.06	-2.72 ± 0.60	1.63 ± 0.35
$KM - m_5$ (pcs)	$s_{1,0}$	$s_{1,1}$	$s_{1,-1}$	$t_{1,0}$	$t_{1,1}$	$t_{1,-1}$
S^2	5.52 ± 0.08	1.52 ± 0.01	-1.04 ± 0.01	7.89 ± 0.01	2.88 ± 0.02	0.97 ± 0.01
[200, 400]	4.18 ± 0.94	1.82 ± 0.20	-1.10 ± 0.04	7.05 ± 0.50	2.48 ± 0.30	1.62 ± 0.40
[300, 500]	1.90 ± 2.23	2.24 ± 0.50	-1.18 ± 0.15	7.05 ± 0.45	2.80 ± 0.04	0.07 ± 0.75
$No - KM - m_5$ (pcs)	$s_{1,0}$	$s_{1,1}$	$s_{1,-1}$	$t_{1,0}$	$t_{1,1}$	$t_{1,-1}$
S^2	10.24 ± 0.01	2.62 ± 0.10	1.14 ± 0.02	2.05 ± 0.01	-0.08 ± 0.02	4.56 ± 0.02
[200, 400]	8.07 ± 0.14	4.12 ± 0.38	2.04 ± 0.60	3.40 ± 0.90	-1.43 ± 0.70	5.84 ± 0.20
[300, 500]	7.92 ± 0.07	3.48 ± 0.10	2.43 ± 0.43	4.09 ± 0.70	-2.72 ± 0.80	5.16 ± 0.16

with magnitudes in the m5 and m6 groups (already candidates per se, because of the behavior observed in the previous study considering magnitudes) and both KM and no-KM. The obtained results are given in Table 2. The results of Marco *et al.* (2017) for a slice around 400 pcs are similar to the expected results for the m_6 or $m_5 + m_6$ set from table 1. The difference in the $t_{1,-1}$ coefficient is due to the bias in δ in that slice. Further considerations and interpretations of the coefficients will be provided in a later paper, now in preparation.

Acknowledgements

This work was supported by a grant UJI-B2016-18.

References

Marco, F. J., Martínez, M. J. & López J. A. 2017, *Astrometry and Astrophysics in the Gaia Sky*, In Press.

Astronomy and Astrophysics in the Gaia sky
Proceedings IAU Symposium No. 330, 2017
A. Recio-Blanco, P. de Laverny, A.G.A. Brown & T. Prusti, eds.

doi:10.1017/S1743921317006354

How far away and how old are these stars?

Paul J. McMillan and the RAVE Collaboration

Lund Observatory, Lund University, Department of Astronomy and Theoretical Physics,
Box 43, SE-22100, Lund, Sweden
email: paul@astro.lu.se

Abstract. RAVE is the spectroscopic survey with the largest overlap with TGAS (around 200 000 stars). Since RAVE's fourth data release, it has contained distance estimates based on a Bayesian estimation scheme. Here we compare these estimates to TGAS's parallaxes, to determine the strengths and weaknesses of each. We also combine the two datasets together to find more precise distance estimates for all these stars.

Keywords. surveys, Galaxy: structure, Galaxy: kinematics and dynamics, methods: statistical

1. Introduction

The RAVE survey (Kunder *et al.* 2017) is a spectroscopic survey that has taken spectra for ~500 000 stars. This provides the radial velocity for all of these stars, along with structural parameters of the stars such as their effective temperature, $T_{\rm eff}$, surface gravities, $\log g$, and metallicities $[M/H]$. These can be used, in combination with photometry (in our case from 2MASS: Skrutskie *et al.* 2006) to derive the distances to stars, and since RAVE's fourth data release these have been provided by the Bayesian method introduced by Burnett & Binney (2010). This method inevitably gives insight on other stellar properties (such as age) while finding distance estimates.

The arrival of parallaxes from the Tycho-Gaia astrometric solution (Lindegren *et al.* 2016) means that we now have parallaxes estimates for ~ 200 000 of these stars. In this note, we compare the parallaxes found by TGAS and those found from the RAVE spectrophotometric pipeline. We then incorporate the TGAS parallaxes into the Bayesian distance estimation. This provides us with distance (or parallax) estimates that are significantly smaller than those using either alone.

2. Results

Figure 1 shows the median difference between the TGAS parallaxes ϖ_T and those from RAVE ϖ_{Sp} for the same stars, binned by position on sky (left panel). There is a clear problem region (running to the ecliptic pole) corresponding to $\varpi_T > \varpi_{Sp}$. This seems to be a problem with the TGAS parallaxes (related to the scanning law), and was also seen by Arenou *et al.* (2017). The right-hand panel of figure 1 shows the median value of the parallax difference divided by uncertainty as a function of $\log g$ for giant stars, suggesting that RAVE parallaxes are systematically underestimated for stars with low $\log g$ values in RAVE DR5. In addition, we find the combined uncertainties are overestimated. We suspect that this is due to an overestimate of the TGAS uncertainties by ~0.2 mas.

In Figure 2 we show the improvement in precision using TGAS parallaxes in the RAVE distance estimates. In the left-hand panel we show the improvement in distance accuracy, divided into that for dwarfs and giants. The median fractional distance uncertainty improves by 200% for dwarfs and 61 % for giants (corresponding to 107% as the median improvement for all stars). For giants (which tend to be more distant in these magnitude-limited samples), the improvement in parallax uncertainty over TGAS alone is 55%.

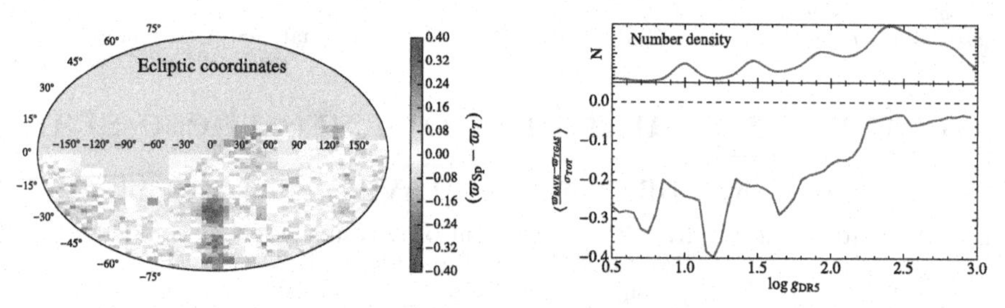

Figure 1. Two plots illustrating problems with TGAS parallaxes (left) and RAVE parallax estimates (right). The left hand plot shows the median difference between the two parallax estimates, binned by position on the sky. The right hand plot shows the median difference between the two estimates divided by the combined uncertainty, as a function of RAVE $\log g$ for giants. The upper panel shows the number density in $\log g$, for reference.

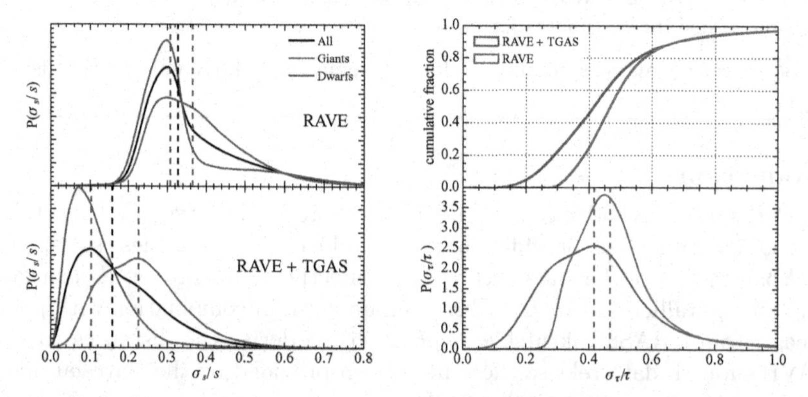

Figure 2. Plots illustrating the improvement in distance uncertainties (left) and age uncertainties (right) when TGAS parallaxes are used. The distance uncertainties are divided in to those for dwarfs ($\log g \geqslant 3.5$) and giants ($\log g < 3.5$). Vertical dashed lines indicate median values.

The right-hand panel of Figure 2 shows the improvement in age uncertainty when using TGAS parallaxes as input. Previously almost no stars had age uncertainties smaller than 30%, but with TGAS around 25% (mostly near the main sequence turn-off) do.

Publication of results.

Full results will be published in McMillan *et al.* (2017, submitted), and released publicly.

Acknowledgement

This work has made use of data from the European Space Agency (ESA) mission *Gaia* (www.cosmos.esa.int/gaia), processed by the *Gaia* Data Processing and Analysis Consortium (DPAC, www.cosmos.esa.int/web/gaia/dpac/consortium). Funding for the DPAC has been provided by national institutions, in particular the institutions participating in the *Gaia* Multilateral Agreement. Funding for RAVE (www.rave-survey.org) has been provided by institutions of the RAVE participants and by their national funding agencies.

References

Arenou, F., *et al.*, 2017, *A&A*, 599, A50
Burnett, B., Binney J., 2010, *MNRAS* 407, 339
Kunder, A., *et al.*, 2017, *AJ*, 153, 75
Lindegren, L., *et al.*, 2016, *A&A*, 595, A4
Skrutskie, M. F., *et al.*, 2006, *AJ*, 131, 1163

Astronomy and Astrophysics in the Gaia sky
Proceedings IAU Symposium No. 330, 2017
A. Recio-Blanco, P. de Laverny, A.G.A. Brown & T. Prusti, eds.

doi:10.1017/S1743921317005749

CNO distributions in the Solar neighborhood with Gaia data

Šarūnas Mikolaitis, Gražina Tautvaišienė, Renata Ženovienė, Arnas Drazdauskas, Erika Pakštienė, Rimvydas Janulis, Vilius Bagdonas and Lukas Klebonas

Institute of Theoretical Physics and Astronomy, Vilnius University,
Saulėtekio al. 3, LT-10257, Vilnius, Lithuania
email: Sarunas.Mikolaitis@tfai.vu.lt

Abstract. A spread of lifebuilding elements, such as carbon, nitrogen, and oxygen in the Galactic discs is yet not well investigated. In this study, we use spectra from the UVES spectrograph (Gaia-ESO survey) and the VUES spectograph (SPFOT-PLATO survey) and determine the carbon, nitrogen, and oxygen abundances in FGK stars using the same technique. For some of our target stars the Gaia space observatory has already presented accurate distances, thus we overview the first results of radial and vertical CNO abundance distributions in the Galactic thin and thick disc populations.

Keywords. Stars:abundances, galaxy:abundances

1. Introduction

Distributions of C, N, and O are particularly interesting in a context of the Galactic chemo-dynamical evolution studies. These elements participate in many nucleosynthetic reactions during the stellar evolution. Stars can experience dredge-ups and extra mixing processes that alter abundances of these elements in stellar atmospheres. Thus, the analysis of C, N, and O is quite complex, especially in the context of Galactic abundance distributions. Accurate stellar distances are also very important. The large astrometric *Gaia* mission is a very important source of stellar distances. In this work, we use the Tycho-Gaia astrometric solution (Michalik *et al.* 2015, Arenou *et al.* 2017) as a source of distances and the C, N, and O abundances from the Gaia-ESO survey (Gilmore *et al.* 2012) fouth internal data release and the SPFOT-PLATO survey which we are running at the Molėtai Astronomical observatory of the Vilnius University (Ženovienė *et al.* 2017).

2. Overview

Carbon and nitrogen abundances are affected by stellar evolution; however, the summed abundance of carbon and nitrogen (C+N) conserves the initial conditions of stellar formation. Thus, in order to avoid the evolutionary effects, we analyse gradients of the summed [C+N/Fe] ratios. We taged the thin and thick disc populations using alpha-element abundance-to-iron ratios similarly as in Mikolaitis *et al.* (2014) and Masseron & Gilmore (2015). We find that the radial [C+N/Fe] gradients are slightly negative for both discs (Fig. 1). Therefore, the thin disc [C+N/Fe] vertical abundance gradient is slightly negative, but the thick disc vertical [C+N/Fe] abundance gradient is slightly positive.

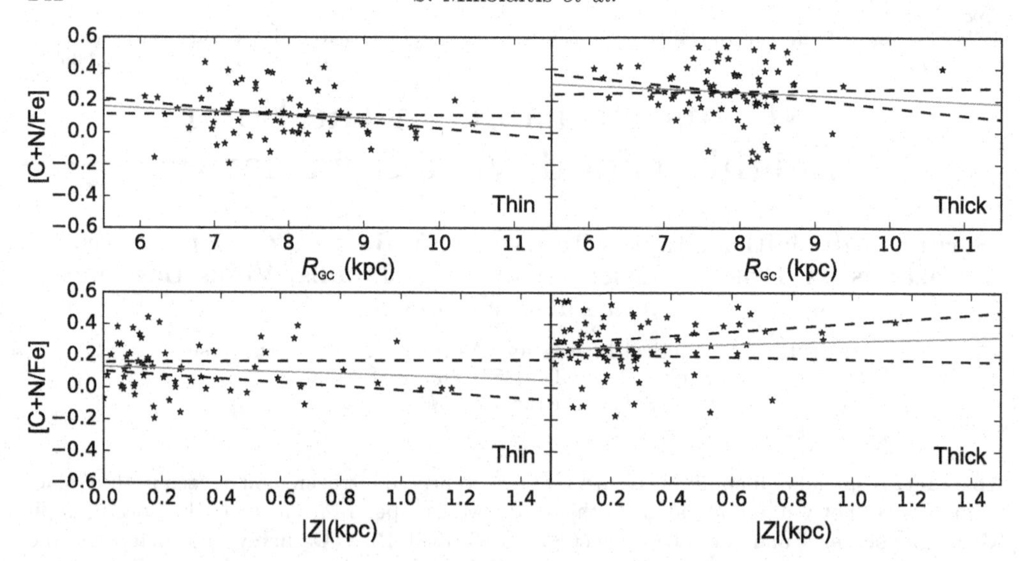

Figure 1. [C+N/Fe] radial (upper panel) and vertical (lower panel) gradients of thin and thick disc stars.

3. Implications

A good model of the Galactic chemical evolution should explain radial and vertical chemical element abundance gradients. For example, our oxygen results agree with the models of Cescutti *et al.* (2007), which assumed an inside-out build-up. However, the thick disc formation is still unclear. The data set is still insufficient to make clear conclusions, however in the future when new releases of high-resoluton spectroscopic surveys will be opened and the *Gaia* distances from DR2 will be provided, C, N and O abundance distributions will be studied with much higher precision.

Acknowledgements

This work has made use of data from the European Space Agency (ESA) mission *Gaia* (https://www.cosmos.esa.int/gaia), processed by the *Gaia* Data Processing and Analysis Consortium (DPAC, https://www.cosmos.esa.int/web/gaia/dpac/consortium). Funding for the DPAC has been provided by national institutions, in particular the institutions participating in the *Gaia* Multilateral Agreement. Based on data products from observations made with ESO Telescopes at the La Silla Paranal Observatory (programme ID 188.B-3002) and from observations at the Molėtai Astronomical Observatory of Vilnius University (funded by the Research Council of Lithuania, LAT-16019).

References

Arenou, F., Luri, X., Babusiaux, C., *et al.* 2017, *A&A*, 599, A50
Cescutti, G., Matteucci, F., François, P., & Chiappini, C. 2007, *A&A*, 462, 943
Gaia Collaboration, Prusti, T., de Bruijne, J. H. J., *et al.* 2016, *A&A*, 595, A1
Gilmore, G., Randich S., Asplund, M. *et al.* 2012, *The Messenger*, 147, 25
Masseron, T. & Gilmore, G. 2015, *MNRAS*, 453, 1855
Michalik, D., Lindegren, L., & Hobbs, D. 2015, *A&A*, 574, A115
Mikolaitis, Š., Hill, V., Recio-Blanco, A., *et al.* 2014, *A&A*, 572, A33
Recio-Blanco, A., de Laverny, P., Kordopatis, G., *et al.* 2014, *A&A*, 567, A5
Ženovienė, R., Mikolaitis, Š., Gražina Tautvaišienė, *et al.* 2017, *IAUS 330, in press*

Astronomy and Astrophysics in the Gaia sky
Proceedings IAU Symposium No. 330, 2017
A. Recio-Blanco, P. de Laverny, A.G.A. Brown & T. Prusti, eds.

doi:10.1017/S1743921317005981

Developing Automated Spectral Analysis Tools for Interstellar Features Extraction to Support Construction of the 3D ISM Map

L. Puspitarini[1a], R. Lallement[2b], A. Monreal-Ibero[2], H.-C. Chen[3], H. L. Malasan[1], Aprilia[1], M. I. Arifyanto[1] and M. Irfan[1]

[1]Department of Astronomy and Bosscha Observatory, FMIPA Institut Teknologi Bandung, Jalan Ganesha 10 Bandung 40132 Indonesia ([a]email: lucky.puspitarini@as.itb.ac.id)

[2]GEPI, Observatoire de Paris, CNRS UMR8111, Université Paris Diderot, Place Jules Janssen, 92190 Meudon, France

[3]Institute of Astronomy, National Central University, Chungli, Taiwan

Abstract. One of the ways to obtain a detailed 3D ISM map is by gathering interstellar (IS) absorption data toward widely distributed background target stars at known distances (line-of-sight/LOS data). The radial and angular evolution of the LOS measurements allow the inference of the ISM spatial distribution. For a better spatial resolution, one needs a large number of the LOS data. It requires building fast tools to measure IS absorption. One of the tools is a global analysis that fit two different diffuse interstellar bands (DIBs) simultaneously. We derived the equivalent width (EW) ratio of the two DIBs recorded in each spectrum of target stars. The ratio variability can be used to study IS environmental conditions or to detect DIB family.

Keywords. ISM: structure, abundances, extinction, bands

1. Introduction

To construct a detailed 3D ISM map through the LOS data requires building automated tools to measure the IS absorption features (Vergely *et al.* 2010, Lallement, *et al.* 2014). The tools must disentangle the stellar continuum and all features presented in the observed spectrum. In principle, the spectrum is fitted by a combination of polynomial function or synthetic stellar model, IS profile model, and telluric transmission model (see Puspitarini *et al.* 2013, Puspitarini *et al.* 2015, Chen *et al.* 2013, Monreal-Ibero & Lallement 2017).

2. Method, Data, and Analysis

One of the spectral analysis tools is global analysis to fit simultaneously different IS tracers with multi-fitting functions while linking some of the parameters (Puspitarini *et al.* 2015). We applied the tool to extract diffuse interstellar bands (DIBs). Despite their unknown carriers, they are a promising tool to trace IS cloud, in particular at large distance as they are not easily saturated. We show here the two-DIBs global analysis in late-type star spectra. During the fit, radial velocities of the two DIBs are linked, but remain a free parameter.

We applied the tool to 12 target stars from Gaia-ESO-Survey (GES)/UVES spectra (Gilmore *et al.* 2012) in $(l, b \simeq 213°, -2°)$ and $(l, b \simeq 37°, -7°)$ fields. When the tool were applied to measure narrow and broad DIBs, 6614 and 6283 Å resp. (Fig. 1), the two fields differ in terms of EW ratios. It can be explained by the effect of radiation field that may destroy or favor the DIB carrier(s) (Vos *et al.* 2011). When we applied to a strong and weak DIBs, 6614 and 6196 Å resp., they do not show significant systematic difference between the fields. It confirms that the two DIBs might originate from the same carrier

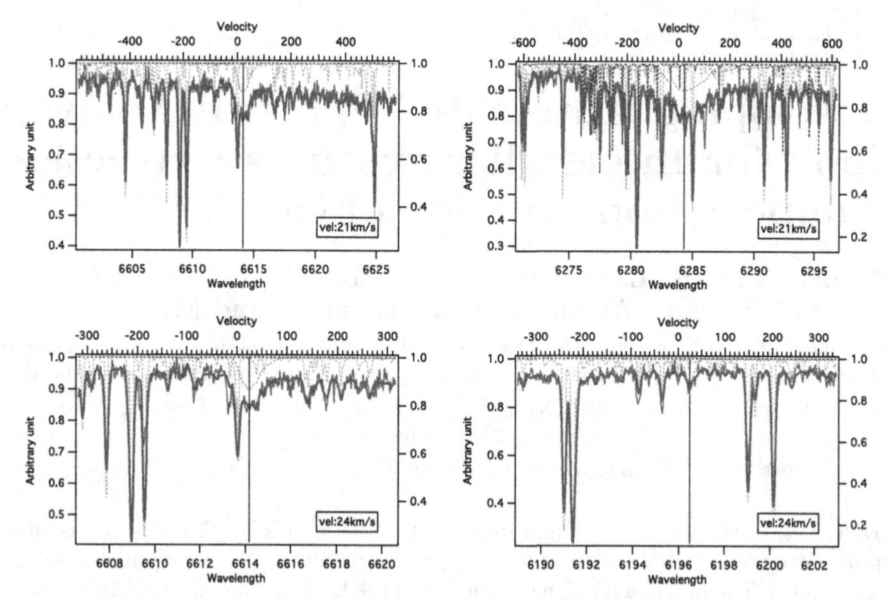

Figure 1. **Upper:** An example of the global analysis: the 6614 Å DIB (left) and 6283 Å DIB (right). Each spectrum is fitted by a combination of synthetic stellar model, DIB model, and telluric transmission model. The thick line shows the fitting. Models are in dotted lines. Vertical lines show the IS radial velocity (v_r) which is linked. The global analysis helps constraining v_r for a very wide DIB from a narrower DIB. **Lower:** Same as upper figure, but for the 6614Å DIB (left) and 6196 Å DIB (right). The method helps detecting a weaker DIB (6196 Å) from the constraint on its radial velocity that is linked to a stronger DIB (6614Å).

(family of DIBs). The global analysis allows to understand relationship between the two DIBs and to derive their kinematic at the same time. The EW ratio can be used to study IS environmental conditions or to detect DIB family. The global analysis method can also be a strong advantage for DIB detection.

3. Discussions and Future Works

We have briefly discussed the principle of the tools and shown the two-DIBs global analysis. The EW measurements will increase the LOS data for the 3D ISM map. The EW can be converted into A_0 by using the EW-A_0 relationship with caution. Together with precise stellar distances from Gaia, these can be used to infer ISM distribution. We will improve and apply the tools to more datasets, e.g., GES, Bosscha Compact Spectrograph spectra (Malasan *et al.* 2001), etc.

Acknowledgement

We thank financial support from the LKBF and the IAU grant.

References

Chen, H.-C., Lallement, R., Babusiaux, C., *et al.* 2013, *A&A* 550, A62

Gilmore, G., Randich, S., Asplund, M., *et al.* 2012, *The Messenger*, 147, 25

Lallement, R., Vergely, J.-L., Valette, B., *et al.* 2014, *A&A*, 561, A91

Malasan, H. L., Yamamuro, T., Takeyama, N., *et al.* 2001, *Proc. of the Indonesia-German Conf. on Instrumentation, Measurements and Communication for the Future*, p. 159

Monreal-Ibero, A. & Lallement, R. 2017, *A&A*, 599, A74

Puspitarini, L., Lallement, R., & Chen, H.-C. 2013, *A&A*, 555, A25

Puspitarini, L., Lallement, R., Babusiaux, C., *et al.* 2015, *A&A*, 573, AA35

Vergely, J.-L., Valette, B., Lallement, R., & Raimond, S. 2010, *A&A*, 518, A31

Vos, D. A. I., Cox, N. L. J., Kaper, L., Spaans, M., & Ehrenfreund, P. 2011, *A&A*, 533, A129

Astronomy and Astrophysics in the Gaia sky
Proceedings IAU Symposium No. 330, 2017
A. Recio-Blanco, P. de Laverny, A.G.A. Brown
& T. Prusti, eds.

doi:10.1017/S1743921317005890

Finding evolved stars in the inner Galactic disk with *Gaia*

L. H. Quiroga-Nuñez[1,2], H. J. van Langevelde[2,1], Y. M. Pihlström[3], L. O. Sjouwerman[4] and A. G. A. Brown[1]

[1]Leiden Observatory, Leiden University,
P.O. Box 9513, 2300 RA Leiden, The Netherlands.
emails: `quiroganunez@strw.leidenuniv.nl`; `brown@strw.leidenuniv.nl`

[2]Joint Institute for VLBI ERIC (JIVE),
Postbus 2, 7990 AA Dwingeloo, The Netherlands.
email: `langevelde@jive.eu`

[3]Department of Physics and Astronomy, University of New Mexico,
MSC07 4220, Albuquerque, NM 87131, USA.
email: `ylva@unm.edu`

[4]National Radio Astronomy Observatory,
P.O. Box 0, Lopezville Road 1001, Socorro, NM 87801, USA.
email: `lsjouwer@nrao.edu`

Abstract. The Bulge Asymmetries and Dynamical Evolution (BAaDE) survey will provide positions and line-of-sight velocities of $\sim 20,000$ evolved, maser bearing stars in the Galactic plane. Although this Galactic region is affected by optical extinction, BAaDE targets may have *Gaia* cross-matches, eventually providing additional stellar information. In an initial attempt to cross-match BAaDE targets with *Gaia*, we have found more than 5,000 candidates. Of these, we may expect half to show SiO emission, which will allow us to obtain velocity information. The cross-match is being refined to avoid false positives using different criteria based on distance analysis, flux variability, and color assessment in the mid- and near-IR. Once the cross-matches can be confirmed, we will have a unique sample to characterize the stellar population of evolved stars in the Galactic bulge, which can be considered fossils of the Milky Way formation.

Keywords. Galaxy: bulge, stars: AGB and post-AGB, masers, astrometry.

---

1. Motivation

The characterization of the stellar population of the Galactic bulge represents a key piece to understand the morphology and dynamical evolution of the inner Galaxy. This stellar population is dynamically affected by a massive bar (e.g. Dwek *et al.* (1995)) and recent studies have shown an X-shaped structure (e.g. Wegg & Gerhard (2013)), similar to what it is seen in extragalactic edge-on boxy bulges. Optical surveys —notably *Gaia*— are limited due to optical extinction, and are not able to make unhindered stellar astrometric measurements in the Galactic bulge, which complicates the characterization of this stellar population.

Radio campaigns are not affected by extinction and can therefore provide complementary information to optical surveys, especially at low latitudes. The Bulge Asymmetries and Dynamical Evolution (BAaDE) project surveys red giant stars for SiO maser emission at 43 and 86 GHz with the VLA and ALMA, eventually providing positions and radial velocities of approximately 20,000 targets along the Galactic plane (Sjouwerman *et al.* (2016)). The BAaDE survey aims to significantly improve the dynamical models

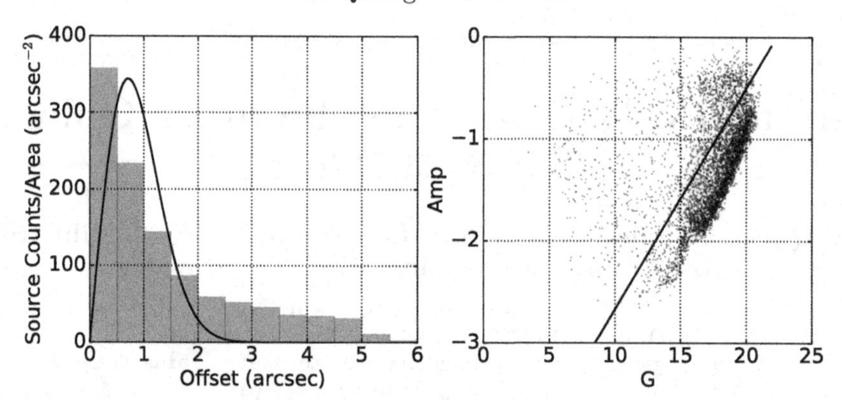

Figure 1. Left: Offset distance between BAaDE and *Gaia* sources. The solid line shows the offset distribution for 2 arcsec source position uncertainty, implying that sources with larger offsets may be false positives. **Right**: Amplitude-magnitude diagram for the cross-matches obtained between BAaDE and *Gaia*. Higher amplitudes can be associate with pulsating AGB stars.

using radio sources in regions not reachable with optical surveys. The BAaDE survey is expanding the currently known stellar tracers in the inner Galaxy by a large number.

Tests for dynamical models of the Galaxy require large samples of stars with accurate positions and velocities. More details can be derived if distances are tied to stellar velocities. Therefore, we present an initial attempt to cross-match BAaDE targets with *Gaia* DR1, resulting in more than 5,000 matches. However, since BAaDE targets were selected based on mid-IR colors measured with the MSX mission (Sjouwerman *et al.* (2009), (2016)) —where the positional uncertainty is up to 2 arcsec— the cross-matched sample could be contaminated by false positives. After confirming the matches, we will have a sample with optical, IR and radio information that can be used to characterize the stellar populations in the inner Galaxy, as well as to test dynamical models. In particular, we can obtain the positions, proper motions, parallaxes, colors and periods from *Gaia* DR2 (April 2018). Until then, we can use the *Gaia* DR1 positions for cross-matching.

2. Cross-matching description

The BAaDE target selection was based on MSX colors, which in turn were based on IRAS color-color diagrams (see Sjouwerman *et al.* (2009)). Van der Veen & Habing (1988) developed an IRAS color-color diagram to study dust/gas envelopes (DGE) of Asymptotic Giant Branch (AGB) stars. They found that DGE stars appear in a sequence in the IRAS color-color diagram, perhaps associated with an evolutionary track with an increasing mass-loss rate. In this color-color diagram, SiO maser stars are found within a specific color regime, allowing a stellar selection based on the IRAS colors. Later on, Sjouwerman *et al.* (2009) were able to transform parts of this IRAS color-color diagram onto colors in the mid-IR, using MSX data. With the improved angular resolution provided by MSX, red giant stars (with envelopes likely to harbor SiO maser emission) can be efficiently selected in the Galactic plane.

To positionally match the BAaDE targets with other surveys, we consider a circular area with 5 arcsec radius around the BAaDE targets, based on the MSX positional accuracy (2 arcsec). Although the cross-match can be done directly with *Gaia*, we initially cross-match BAaDE targets and 2MASS, because of three different reasons. Firstly, we do not expect that a target displaying both mid-IR emission (MSX) and optical emission (*Gaia*) would not have emission in the near-IR (2MASS). Hence, by initially

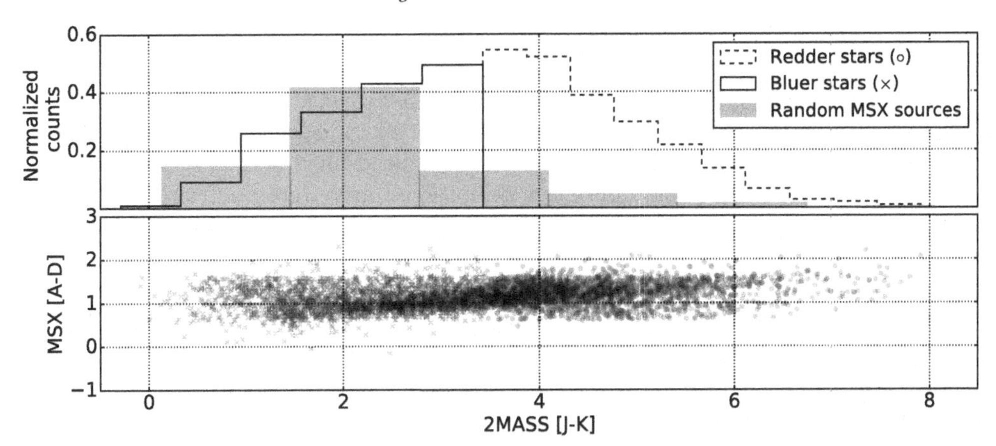

Figure 2. Lower: Color-color diagram for the cross-matches obtained between BAaDE targets and *Gaia*. The sample was split in bluer (crosses) and redder sources (circles), see Sect. 3.2. **Upper**: Histogram distributions for bluer and redder sources. The gray histogram represents a random sample of MSX sources showing that most of them are part of the bluer sample.

cross-matching with 2MASS, we are already avoiding some false positives. Secondly, the cross-match between 2MASS and *Gaia* was already made by Marrese *et al.* (in preparation) using the best neighbor algorithm, finding more than 90% coincidences. Finally, 2MASS contributes with useful near-IR information to characterize the stellar population.

The cross-match between BAaDE targets and 2MASS produced more than 90% coincidences within 5 arcsec. However, looking at these sources in the *Gaia* catalog, out of 5,674 coincidences seem to have a counter part at optical wavelengths. From those, 4,814 sources have only one *Gaia* match and 860 sources have 2 or more *Gaia* matches within the search radius. We will focus on the 4,814 sources that have a unique *Gaia* match.

3. False positive filters

To refine the cross-matching by avoid false positives, several filtering methods have been considered. Below, we outline the most successful methods that we have applied.

3.1. *Distance analysis*

The distribution of matches between the BAaDE targets and *Gaia* shows Gaussian distributions for both components ($\Delta\alpha \times cos(\delta), \Delta\delta$) with absolute mean values < 0.2 arcsec. This is 2D Gaussian distribution can be converted to a function of the distance offset, which is a first-order Bessel function assuming the same standard deviations in both components. The left panel of Fig. 1 shows the expected distribution for a radius of 2 arcsec (representative as the typical MSX positional uncertainty) as a solid line. Excess sources at offsets above ~ 2 arcsec may be considered false positives.

3.2. *Color filters in the mid- and near-IR*

Since the cross-match was made through 2MASS, the near-IR filters (J, H, K) and the mid-IR (MSX bands) can be used for color-color diagrams. The lower panel of Fig. 2 shows the color-color diagram between [A-D] MSX bands and [J-K] 2MASS filters for the matches between BAaDE targets and *Gaia*. We calculated the mean value for the 2MASS colors and we split the sample in two different subsamples, i.e., $[J-K] < 3.6$ (bluer stars) and $[J-K] > 3.6$ (redder stars). AGB stars are expected to have redder colors (represented by a steeper slope in their SED), and therefore we expect that the redder stars are more likely to be correct cross-matches.

Moreover, the upper panel of Fig. 2 shows the histograms for the bluer and redder stars respectively, plotted on top of the distribution for random subset of MSX sources. The plot shows that most of the MSX sources are indeed bluer stars, in agreement with that redder stars (representing half of our sample) are more rare and could more easily be associated with pulsating AGB stars.

3.3. *Variability of evolved stars*

The observed variability of the optical g-band can be quantified with an amplitude measure, defined as $\mathrm{Amp} = \log_{10}(\sqrt{\mathrm{N_{obs}}}\frac{\sigma_g}{\mathrm{g}})$, where N_{obs} is the number of observations. Belokurov *et al.* (2017) calculated the amplitude for different stellar populations in the LMC and SMC, and localized Mira variables in the upper region of the amplitude-magnitude plot. This implies for a given range in G these variable stars have a higher value of Amp than non-variable source of the same brightness. The right panel of Fig. 1 shows an amplitude-magnitude plot for the matches between BAaDE targets and *Gaia*. The solid line represent the typical behavior for most of the *Gaia* sources. Stars with amplitudes higher than -1 are highly related with pulsating stars and hence could be confirmed as properly matched. In contrast, Stars with amplitudes lower than -1 must be carefully reviewed by an alternative criterion for false positives.

3.4. *Statistical arguments*

Assuming an uniform distribution of sources in the bulge for the *Gaia* detections and for the BAaDE targets, one could calculate the number of sources that randomly will match given the resolution of each survey. We estimate that the number of random matches should be less 1,200, which is low compared with our finding of 5,674 matches. Moreover, in the statistical calculation we have assumed that there is no optical extinction that could limit the number of *Gaia* sources. Therefore, the actual number of random matches should be much lower than 1,200, confirming that our cross-match is not a consequence of random matches of unrelated sources.

Acknowledgements

This work has made use of data from the European Space Agency mission *Gaia*, processed by the *Gaia* Data Processing and Analysis Consortium (DPAC). Funding for the DPAC has been provided by national institutions, in particular the institutions participating in the *Gaia* Multilateral Agreement. This material is based upon work supported by the National Science Foundation under Grant Number 1517970.

References

Belokurov, V., Erkal, D., Deason, A. J., Koposov, S. E. *et al.* 2017, *MNRAS*, 466, 4711

Dwek, E., Arendt, R. G., Hauser, M. G., Kelsall, T., Lisse, C. M. *et al.* 1995, *ApJ*, 445, 716

Sjouwerman, L. O., Capen, S. M., & Claussen, M. J. 2009, *ApJ*, 705, 1554

Sjouwerman, L. O., Pihlström, Y. M., Rich, R. M., Morris, M. R. *et al.* 2016, in: R.M. Crocker, S.N. Longmore & G.V. Bicknell (eds.), *The Multi-Messenger Astrophysics of the Galactic Centre*, Proc. IAU Symposium No. 332, p. 103

van der Veen, W. E. C. J. & Habing, H. J. 1988, *A&A*, 194, 125

Wegg, C. & Gerhard, O. 2013, *MNRAS*, 435, 1874

Astronomy and Astrophysics in the Gaia sky
Proceedings IAU Symposium No. 330, 2017
A. Recio-Blanco, P. de Laverny, A.G.A. Brown & T. Prusti, eds.

doi:10.1017/S1743921317005580

Magellanic Clouds Proper Motion and Rotation with Gaia DR1

J. Sahlmann and R. van der Marel

Space Telescope Science Institute, 3700 San Martin Drive, Baltimore, MD 21218, USA
email: jsahlmann@stsci.edu

Abstract. We used the Gaia data release 1 to study the proper motion fields of the Large and Small Magellanic Clouds (LMC, SMC) on the basis of the Tycho-Gaia Astrometric Solution (van der Marel & Sahlmann 2016). The Gaia LMC and SMC proper motions have similar accuracy and agree to within the uncertainties with existing HST proper motion measurements. Since Gaia probes the young stellar population and uses different methods with different systematics, this provides an external validation of both data sets and their underlying approaches.

Keywords. Magellanic Clouds, Local Group, galaxies: kinematics and dynamics, astrometry

1. Introduction

Precise proper motions can inform us on the dynamics of galaxies, in particular those of the relatively nearby Local Group. The Hipparcos survey was not sufficiently precise for detailed studies, thus most proper motions studies of Local Group dynamics relied on measurements with the Hubble Space Telescope (HST, e.g. van der Marel & Kallivayalil 2014) and VLBI. This picture is changing with the Gaia survey and we studied the dynamics of the LMC and SMC using the first Gaia Data Release (DR1, Gaia Collaboration et al. 2016a,b) proper motions of individual bright stars (van der Marel & Sahlmann 2016). The highest-precision Gaia proper motions available in DR1 were obtained for Hipparcos stars as part of the Tycho-Gaia Astrometric Solution (TGAS) catalog. We retrieved TGAS proper motions for 29 Hipparcos stars in the LMC and for 8 Hipparcos stars in the SMC, mostly B and A supergiants, a sample that was studied by Kroupa & Bastian (1997).

2. LMC proper motion and rotation

The LMC center-of-mass proper motions measured with Gaia and HST have similar accuracy and agree to within the uncertainties. In Figure 1, clockwise stellar motion is clearly evident. This qualitatively validates the accuracy of the TGAS data, and confirms that the stars belong to the LMC. We modelled the LMC proper motion field to derive its kinematic and geometric parameters. Although the Gaia and HST studies probe different stellar populations, we find excellent agreement in terms of the derived LMC parameters.

3. SMC results

The SMC center-of-mass proper motions measured with Gaia and HST have similar accuracy and agree to within the uncertainties. No rotation in the plane of the sky is evident, because the SMC is smaller than the LMC, and the TGAS stars are closer to the galaxy center then they are for the LMC.

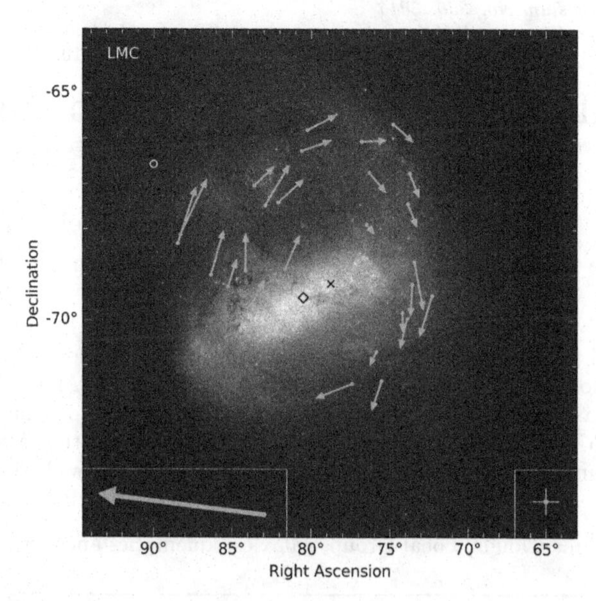

Figure 1. Spatially variable component of the observed Gaia TGAS proper motion field for the LMC, overlaid on a representation of the Gaia DR1 source density. Solid dots show the positions of the 29 sample stars and arrows correspond to the individual Gaia proper motions minus the best-fit center-of-mass motion (bottom left inset). Clockwise rotation is clearly evident. The bottom right inset shows the median proper motion uncertainty. The panel is centered on the dynamical center (cross). The locations of the southern ecliptic pole (open circle) and the JWST astrometric calibration field (diamond, see Sahlmann *et al.* in this volume) are indicated.

4. Conclusions

We have used the Gaia DR1 to obtain new insights into the motions and internal kinematics of the Magellanic Clouds. Within the uncertainties, the results inferred from Gaia are consistent with existing HST studies. Since Gaia probes the young stellar population and uses different methods with different systematics, this provides an external validation of both data sets and their underlying approaches. Both Gaia and HST (van der Marel & Kallivayalil 2014) confidently detect and quantify the rotation of the LMC disk. In a follow-up search for LMC runaway stars, Lennon et al. (2016) discovered a candidate hypervelocity star based on its TGAS proper motion.

Acknowledgements

This work has made use of data from the European Space Agency (ESA) mission *Gaia* (`https://www.cosmos.esa.int/gaia`), processed by the *Gaia* Data Processing and Analysis Consortium (DPAC, `https://www.cosmos.esa.int/web/gaia/dpac/consortium`). Funding for the DPAC has been provided by national institutions, in particular the institutions participating in the *Gaia* Multilateral Agreement.

References

Gaia Collaboration, Brown, A. G. A., Vallenari, A., *et al.* 2016a, *A&A*, 595, A2
Gaia Collaboration, Prusti, T., de Bruijne, J. H. J., *et al.* 2016b, *A&A*, 595, A1
Kroupa, P. & Bastian, U. 1997, New Astronomy, 2, 77
Lennon, D. J., van der Marel, R. P., Ramos Lerate, M., *et al.* 2016, ArXiv e-prints
van der Marel, R. P. & Kallivayalil, N. 2014, *ApJ*, 781, 121
van der Marel, R. P. & Sahlmann, J. 2016, *ApJL*, 832, L23

Astronomy and Astrophysics in the Gaia sky
Proceedings IAU Symposium No. 330, 2017
A. Recio-Blanco, P. de Laverny, A.G.A. Brown
& T. Prusti, eds.

doi:10.1017/S1743921317006111

Proper motions of stars in the globular clusters using WFI@2.2 m telescope

Devesh P. Sariya[1], Ing-Guey Jiang[1] and R. K. S. Yadav[2]

[1]Department of Physics and Institute of Astronomy, National Tsing-Hua University, Hsin-Chu, Taiwan
email: deveshpath@gmail.com, jiang@phys.nthu.edu.tw

[2]Aryabhatta Research Institute of Observational Sciences, Manora Peak, Nainital 263 002, India
email: rkant@aries.res.in

Abstract. We present results of our studies for a sample of Galactic globular star clusters with the aim of deriving relative proper motions. We used CCD archival data observed with Wide Field Imager (WFI) mounted on ESO 2.2 m telescope at La Silla, Chile. Astrometric software designed by Anderson *et al.* is used to derive relative proper motions. The vector point diagrams show clear separation of field stars from the cluster stars. We used proper motions to determine membership probabilities and to produce color-magnitude diagrams with most probable cluster member stars. Our membership catalogue can be used to study the membership status of the peculiar stars including various variables reported in the literature.

Keywords. (Galaxy:) globular clusters: general, stars: kinematics

---

1. Introduction

Proper motions in Galactic globular clusters are of great interest due to their application to interpret the kinematics and dynamics of the Milky Way. Proper motions can be used to decide whether a particular star belongs to the cluster or it is more likely a field star. Ground based data holds importance in proper motion work by virtue of covering wider region of the clusters and complements the HST which can resolve the central regions of globular clusters. CCDs have revolutionized the proper motion work as the required epoch gap has reduced and one can go to much fainter magnitudes. Wide field imaging CCDs made with mosaicing techniques cover a wide region of globular clusters.

2. Data and reduction procedures

The images used for proper motion studies of globular clusters were obtained with the Wide Field Imager (WFI) camera mounted on 2.2 m ESO/MPI telescope located at La Silla, Chile. The epoch gaps for the studied globular clusters are in the range of $\sim$3-14 yrs. WFI is a mosaic of 8 CCDs, 2k$\times$4k pixels each. Mosaic CCDs of the WFI enable a wide field of view observation (34$\times$33 arcmin^2) of the globular clusters. Typical seeing during the observations was $\sim$1-1.5 arcsec.

To reduce the WFI images, the procedure discussed by Anderson *et al.* (2006, A06) was used. The first main aspect of this procedure is to make multiple PSFs per CCD chip to capture the variation of the PSF across the mosaic system. Usually 3$\times$5 PSFs are made for each CCD, thus 120 total PSFs for an image. The second main thing is to solve for the geometric distortion effect using the solution provided by A06. Finally, the local transformation approach is used to remove any uncorrected geometric distortion effect.

Reference stars to calculate proper motions were selected using the position of stars in the color-magnitude diagrams (CMD) and then, the program was iterated to include good proper motion. The detailed procedures to obtain the proper motion has been discussed by A06; Bellini *et al.* (2009); Yadav *et al.* (2008, 2013); Sariya *et al.* (2012, 2015, 2017). A centroid around the cluster proper motion center in the vector point diagrams gives the cluster members. Then, the method given by Balaguer-Núñez *et al.* (1998) was used to determine membership probabilities of the stars.

3. Results and Discussion

Our research group derived proper motions and membership probabilities for the stars in the area of globular clusters NGC 6809, NGC 6366 and NGC 3201 (Sariya *et al.* 2012, 2015, 2017). Number of the studied stars is ~12600, 2530 and 8322 for NGC 6809, NGC 6366 and NGC 3201 consecutively. The electronic catalogues contain equatorial coordinates, proper motions, photometric magnitudes and membership probabilities. Using most probable cluster members, we produced CMDs showing clean sequences having only a little field stars contamination. The membership catalogues were also used to discuss membership status of known variable stars, blue stragglers and X-ray sources reported in literature. Our studies exhibit the superiority of CCDs over photographic plates with proper motion determinations possible using data having only a few years epoch gap.

Acknowledgement

Devesh P. Sariya and Ing-Guey Jiang acknowledge the grant from the Ministry of Science and Technology (MOST), Taiwan. The grant numbers are MOST 103-2112-M-007-020-MY3, MOST 104-2811-M-007-024, and MOST 105-2811-M-007-038. This work is based on data obtained from the ESO Science Archive Facility.

References

Anderson, J., Bedin, L. R., Piotto, G., Yadav, R. K. S., & Bellini, A., 2006, *A&A*, 454, 1029, [A06]
Balaguer-Núñez, L., Tian, K. P., & Zhao, J. L., 1998, *A&AS*, 133, 387
Bellini, A., Piotto, G., Bedin, L. R., *et al.*, 2009, *A&A*, 493, 959
Sariya, D. P., Yadav, R. K. S., & Bellini, A., 2012, *A&A*, 543, A87
Sariya, D. P. & Yadav, R. K. S., 2015, *A&A*, 584, A59
Sariya, D. P. & Jiang, Ing-Guey, Yadav, R. K. S., 2017, *AJ*, 153, 134
Yadav, R. K. S., Bedin, L. R., Piotto, G. *et al.*, 2008, *A&A*, 484, 609
Yadav, R. K. S., Sariya, D. P., & Sagar, R., 2013, *MNRAS*, 430, 3350

Astronomy and Astrophysics in the Gaia sky
Proceedings IAU Symposium No. 330, 2017
A. Recio-Blanco, P. de Laverny, A.G.A. Brown & T. Prusti, eds.

doi:10.1017/S1743921317006020

Finding the stars that reionized the Universe

Mahavir Sharma, Tom Theuns and Carlos Frenk

Institute for Computational Cosmology, Department of Physics, Durham University, South Road, Durham, UK, DH1 3LE, email: `mahavir.sharma@durham.ac.uk`

Abstract. We study the abundance of the remnants of stars that reionized the Universe in galaxies in the present day Universe using the EAGLE cosmological hydrodynamical simulation. High mass galaxies contain most of these 'reionizers'. The fractional number of galaxies that do not host reionizers increases with decreasing stellar mass, $M_\star$. For the galaxies that host reionizers, the fraction of mass of the galaxy in reionizers increases with decreasing $M_\star$, such that the fraction is low ($\sim 10^{-4}$) for high mass galaxies and can be as high as 0.1 in low mass galaxies, $M_\star \leqslant 10^7$ $M_\odot$. In Milky-Way like galaxies, the distribution of reionizers is spatially more extended than that of normal stars.

Keywords. cosmology: theory, galaxies: evolution, Galaxy: formation, stars: statistics

---

1. Introduction

Metal poor stars allow us to track the assembly history of our galaxy, and allow us to constrain the nature and abundances of stars that reionized the Universe. It has been a challenge to identify the descendants of reionizers and first stars in metal poor stellar population in the Milky Way, though considerable progress has been made in this direction at the observational front in recent years (Frebel & Norris 2015).

We use the EAGLE cosmological hydrodynamical simulation to track descendants of reionizers. EAGLE is a suite of Smoothed Particle Hydrodynamic (SPH) simulations that follow the formation and evolution of galaxies (Schaye *et al.* 2015; Crain *et al.* 2015). In this work we use the high resolution run 'L0025N0752', that simulates a cubical region of the Universe with comoving size of 25 Mpc.

In a recent study (Sharma *et al.* 2017a) we developed a model of reionization in which we used the idea that winds from starbursts open up channels through which the ionizing photons escape and ionize the Universe. In particular, we assumed that when the local surface density of star formation, $\dot{\Sigma}_\star > \dot{\Sigma}_{\star,\mathrm{crit}}$, a wind is launched that opens up a channel. Following on that work, we define reionizers as the stars that formed before redshift 6 in regions of high $\dot{\Sigma}_\star$. The later criterion is automatically satisfied for most of the star particles at $z > 6$ since $\dot{\Sigma}_\star$ is typically very high during those early times. Galactic winds played a major role in reionization and might also have led to peculiar distribution and abundance patterns in the metal poor stars that formed during such episodes and may still be visible today (Sharma *et al.* 2016, 2017b).

In this work, we study the present day distribution of the descendants of reionizers in galaxies and in particular in the Milky Way type galaxies, which can be compared with existing catalogues of metal poor stars and with upcoming data from GAIA (Gaia Collaboration *et al.* 2016).

2. Results: distribution of reionizers

In the hierarchical structure formation galaxies grow via mergers and accretion. For example the Milky Way is formed of a number of progenitors that merged with the main

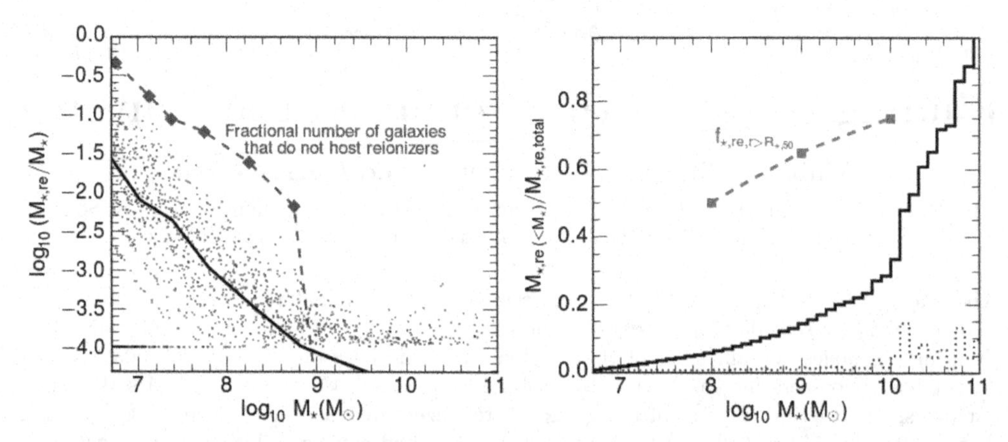

Figure 1. Left panel: fraction of reionizers to normal stars for an individual galaxy is shown as a function of the stellar mass, $M_\star$, for galaxies from the EAGLE simulation. Each dot represents a galaxy, and, the solid curve is the median of galaxies that host reionizers. The galaxies that do not host reionizers have been assigned a value of 10^{-4} and their numerical fraction increase with decreasing stellar mass (dashed blue curve). Right panel: fraction of the total stellar mass of reionizers in the Universe that resides in galaxies below a given $M_\star$ is plotted as a solid line. The corresponding differential histogram is shown as a dotted line. The dashed red line shows the fraction of reionizers in a galaxy that reside outside the stellar half mass radius ($f_{\star,re,r>R_{\star,50}}$).

progenitor during its lifetime. Therefore, the descendants of first stars and reionizers may end up in a galaxy that is not necessarily the galaxy in which they were born. In fact we find that 70 percent of the reionizers live in high mass galaxies with stellar mass, $M_\star > 10^{10}$ M$_\odot$, in the present day Universe (Fig. 1, right panel), however, they formed in low mass galaxies in the early Universe.

For Milky Way type galaxies the fraction of stars that are reionizers is typically 10^{-4}–10^{-3}, while for a low mass galaxy ($M_\star \sim 10^7$M$_\odot$) the fraction can be in the range 0–0.1. This fraction has a considerable scatter from galaxy to galaxy, particularly at the low mass end. The fractional number of galaxies that do not host reionizers increases with decreasing stellar mass (dashed blue curve in the left panel of Fig. 1)

We also find that the reionizers have a shallower spatial distribution when compared to normal stars and the shallowness increases with $M_\star$ (dashed red line in the right panel, see also Sharma *et al.* 2017b), which implies that the relative abundance of reionizers to normal stars is higher in the Galaxy outskirts (right panel). Therefore it may be easier to detect reionizers in the solar neighbourhood rather than in the bulge of the Galaxy.

We thank the EAGLE collaboration and STFC for the capital grant ST/K00042X/1, ST/H008519/1 and DiRAC Operations grant ST/K003267/1.

References

Crain, R. A., Schaye, J., Bower, R. G., *et al.* 2015, *MNRAS*, 450, 1937
Frebel, A. & Norris, J. E. 2015, *ARA&A*, 53, 631
Gaia Collaboration, Prusti, T., de Bruijne, J. H. J., *et al.* 2016, *A&A*, 595, A1
Schaye, J., Crain, R. A., Bower, R. G., *et al.* 2015, *MNRAS*, 446, 521
Sharma, M., Theuns, T., Frenk, C., *et al.* 2017a, *MNRAS*, 468, 2176
Sharma, M., Theuns, T., Frenk, C., & Cooke, R. 2016, *arXiv*: 1611.03868
Sharma, M., Theuns, T., & Frenk, C. 2017b, *MNRAS in prep.*

Astronomy and Astrophysics in the Gaia sky
Proceedings IAU Symposium No. 330, 2017
A. Recio-Blanco, P. de Laverny, A.G.A. Brown & T. Prusti, eds.

doi:10.1017/S1743921317006044

Determining the Local Dark Matter Density with SDSS G-dwarf data

Hamish Silverwood[1], Sofia Sivertsson[2], Justin Read[3], Gianfranco Bertone[4] and Pascal Steger[5]

[1]Institut de Ciències del Cosmos (ICCUB), Universitat de Barcelona (IEEC-UB), Martì Franquès 1, E08028 Barcelona, Spain
email `hamish.silverwood@icc.ub.edu`

[2]The Oskar Klein Centre for Cosmoparticle Physics, Department of Physics, Stockholm University, AlbaNova, SE-106 91 Stockholm, Sweden

[3]Department of Physics, University of Surrey, Guildford, GU2 7XH, Surrey, UK

[4]GRAPPA, University of Amsterdam, Science Park 904, 1098 XH Amsterdam, The Netherlands

[5]Institute for Astronomy, Department of Physics, ETH Zürich, Wolfgang-Pauli-Strasse 27, CH-8093 Zürich, Switzerland

Abstract. We present a determination of the local dark matter density derived using the integrated Jeans equation method presented in Silverwood *et al.* (2016) applied to SDSS-SEGUE G-dwarf data processed by Büdenbender *et al.* (2015). For our analysis we construct models for the tracer density, dark matter and baryon distribution, and tilt term (linking radial and vertical motions), and then calculate the vertical velocity dispersion using the integrated Jeans equation. These models are then fit to the data using MultiNest, and a posterior distribution for the local dark matter density is derived. We find the most reliable determination to come from the α-young population presented in Büdenbender *et al.* (2015), yielding a result of $\rho_{\rm DM} = 0.46^{+0.07}_{-0.09}\,{\rm GeV\,cm^{-3}} = 0.012^{+0.001}_{-0.002}{\rm M_\odot\,pc^{-3}}$. Our results also illuminate the path ahead for future analyses using Gaia DR2 data, highlighting which quantities will need to be determined and which assumptions could be relaxed.

Keywords. dark matter, Galaxy: kinematics and dynamics, Galaxy: disk, etc.

---

1. Introduction

The proper interpretation of results from in-laboratory searches for cosmological Dark Matter (DM) is reliant on a high quality determination of the local DM density $\rho_{\rm DM}$ from astrometry and astrophysics. The detection rates of these experiments feature a degeneracy between the local DM density, and the coupling of DM to Standard Model particles. This latter quantity is crucial in investigating the particle physics nature of DM and its role in Beyond-the-Standard-Model theories, and so systematic problems in the determination of $\rho_{\rm DM}$ can flow from astrometry through experimental physics to theoretical physics. See Read (2014) for a review of the local DM density.

In this work we give a synopsis of results presented at the IAU Symposium 330 in Nice, France (24-28 April 2017), and published in full in Sivertsson *et al.* (2017).

2. Method

The method used for this analysis is an evolution of that presented in Silverwood *et al.* 2016. Broadly speaking we use a Jeans equation based method to analyse the vertical motions of stars in the Milky Way disc plane and determine the total mass distribution in the vicinity of the Sun. The inclusion of a model for the baryon distribution then allows us to make a determination of the local DM density.

2.1. *Mathematical Details*

The starting point for the method is the z-direction Jeans equation with the assumption of axisymmetry and dynamical equilibrium (Binney & Tremaine (2008), Silverwood *et al.* (2016)):

$$\frac{1}{\nu}\frac{\partial}{\partial z}(\nu\sigma_z^2) + \underbrace{\frac{1}{R\nu}\frac{\partial}{\partial R}(R\nu\sigma_{Rz})}_{\text{tilt term: } \mathcal{T}} = -\frac{\partial \Phi}{\partial z}, \tag{2.1}$$

where ν is the tracer density, σ_z^2 is the z-direction velocity dispersion, σ_{Rz} is the (R, z) cross term of the velocity dispersion tensor, and Φ is the total gravitational potential. From measurements of the positions, z- and R-velocities, we could derive all the terms on the LHS of equation 2.1 and derive the z-gradient of the total potential Φ from which we could derive the surface density profile $\Sigma_z(z)$. However this would require differentiating numerical data, which would amplify the noise present in the data. Thus we instead integrate equation 2.1, to yield (Sivertsson *et al.* (2017)):

$$\nu(z)\sigma_z^2(z) = \nu(z_0)\sigma_z^2(z_0) - \int_{z_0}^{z} \nu(z')[2\pi G\Sigma(z') + \mathcal{T}(z')]\mathrm{d}z'. \tag{2.2}$$

The surface density $\Sigma(z)$ is a sum of both baryons and DM:

$$\Sigma(z) = 2\int_{z=0}^{z} dz\,[\rho_{\text{baryon}}(z') + \rho_{\text{DM}}(z')]\,. \tag{2.3}$$

We then build models for the tracer density profile $\nu(z)$, the baryon density $\rho_{\text{baryon}}(z)$, the DM density $\rho_{\text{DM}}(z)$, and the tilt term $\mathcal{T}(z)$. Using equation 2.2 we can then calculate the vertical velocity dispersion $\sigma_z^2(z)$. Finally we fit these models to data using Bayesian nested sampling as implemented by MultiNest (Feroz & Hobson (2007), Feroz *et al.*(2008), Feroz *et al.*(2013)). The comparison between model and data is made with a χ^2 test on $\nu(z)$, $\sigma_z^2(z)$, and $\sigma_{Rz}(z)$. From the output of the MultiNest we can derive a posterior distribution on the local DM density.

The link between potential Φ and the density ρ includes a rotation curve term $\mathcal{R}$:

$$\frac{\partial^2\Phi}{\partial z^2} + \underbrace{\frac{1}{R}\frac{\partial}{\partial R}\left(R\frac{\partial\Phi}{\partial R}\right)}_{\mathcal{R}} = 4\pi G\rho, \tag{2.4}$$

For this study we assume a flat rotation curve, and thus $\mathcal{R} = 0$, but any deviation from this would manifest itself as a shift in the local DM density (Silverwood *et al.* (2016), Sivertsson *et al.* (2017)).

2.2. *Input Data*

The data we use to derive the local DM density here is taken from Büdenbender *et al.* (2015). In this paper they process observations of G-type dwarf stars from SDSS/SEGUE into tracer density $\nu(z)$ and velocity dispersions $\sigma_z^2(z)$ and $\sigma_{Rz}(z)$. The extract two populations from the sample using metallicity cuts: an α-old population, with $0.3 <$ [α/Fe] and $-1.2 <$ [Fe/H] < -0.3, and an α-young population with [α/Fe] < 0.2 and $-0.5 <$ [Fe/H]. We take their derived data for $\nu(z)$, $\sigma_z^2(z)$ and $\sigma_{Rz}(z)$ and feed it into our method.

2.3. *Parameterised Models*

For the tracer density we assume an exponential distribution, as this was also assumed by Büdenbender *et al.* (2015) and fits the data well. The DM density is assumed to be constant in height above the disc plane, and so is described by a single parameter.

The baryon model is derived from data drawn from a number of sources, primarily McKee *et al.*(2015). We use a total surface density of $46.85\,\mathrm{M_\odot\,pc^{-2}} \pm 13\%$. We only model the profiles of components with significant contributions to the mass density above the minimum z-value for which we have data. All other components can be modelled as a constant surface density term, i.e. modelling them as concentrated in the disc plane. The components we model are a thick disc component of all stars, the thin disc component of the dwarfs (consisting of M dwarfs, brown dwarfs, white dwarfs, plus a smaller contribution from neutron stars and black holes), and the HII gas component.

The tilt term links the vertical and radial velocity dispersions. Any potential that is symmetric in z is separable up to second order at the midplane ($z = 0$). Thus the radial and vertical motions will decouple at $z = 0$, and the tilt term will vanish. Though it can increase rapidly with z the tilt term is generally small at low z and has an increasing impact further from the midplane. From equation 2.2 the tilt term is given by (Sivertsson *et al.* (2017))

$$\mathcal{T}(R_\odot, z) \equiv \frac{1}{R\nu}\frac{\partial}{\partial R}(R\nu\sigma_{Rz}) \tag{2.5}$$

The radial variation of the tracer density and (R, z) cross term of the velocity dispersion tenor are modelled as

$$\nu(R,z) = \nu(R_\odot,z)\exp(-k_0(R-R_\odot)) \quad \text{and} \quad \sigma_{Rz}(R,z) = \sigma_{Rz}(R_\odot,z)\exp(-k_1(R-R_\odot)), \tag{2.6}$$

which yields a tilt term of

$$\mathcal{T}(R_\odot, z) = \left(\frac{1}{R_\odot} - k\right)\sigma_{Rz}(z) = \left(\frac{1}{R_\odot} - k\right)Az^n, \tag{2.7}$$

where $k \equiv k_0 + k_1$, and $R_\odot \simeq 8$ kpc is our distance to the galactic center. While the A and n parameters of the $\sigma_{Rz}(z)$ vertical profile model can be fit to the data from Büdenbender *et al.* (2015), the values of the k_1 and k_2 parameters controlling the radial variation of ν and σ_{Rz} must be imposed via a prior. Based on results from Bovy *et al.* (2016) we set a flat prior on k with range $-0.5 \leqslant k \leqslant 1.5\,\mathrm{kpc^{-1}}$ for the α-old population and $-1.3 \leqslant k \leqslant 1\,\mathrm{kpc^{-1}}$ for the α-young population. We take the galactocentric distance to be $R_\odot = 8$ kpc, though any uncertainty on this would be effectively marginalised over along with k.

3. Results

We perform five different analysis runs: α-young data only, with and without a tilt term included in the model; α-old, again both with and without tilt; and a combined analysis with tilt, using a common DM and baryon model but different tracer density and tilt term models for each population. The results are summarised in Table 1.

As expected the inclusion or exclusion of the tilt term has only a minor effect on the outcome of the α-young results. This is because the α-young population populates a canonical 'thin'-disc and is concentrated close towards the midplane, thus not experiencing a strong impact from the tilt term. The thick disc results, on the other hand, undergo a large shift in ρ_{DM} values, with the median shifting from $0.019\,\mathrm{M_\odot\,pc^{-3}} = 0.73\,\mathrm{GeV\,cm^{-3}}$ when including tilt to $0.012\,\mathrm{M_\odot\,pc^{-3}} = 0.46\,\mathrm{GeV\,cm^{-3}}$ when neglecting it.

Table 1. Limits on the credible regions (CRs) of the marginalised posterior of $\rho_{\rm DM}$. The five different analysis runs are presented here: α-young and α-old analyses, done with tilt and without tilt, and a combined α-young and α-old analysis performed with tilt. The α-young with tilt analysis is considered the most reliable, and is shown in bold face.

		α-young		α-old		Combined analysis
		Tilt	No Tilt	Tilt	No Tilt	Tilt
68% CR upper	$M_\odot\,pc^{-3}$	**0.013**	0.014	0.021	0.013	0.012
	$GeV\,cm^{-3}$	**0.53**	0.53	0.79	0.48	0.43
Median	$M_\odot\,pc^{-3}$	**0.012**	0.013	0.019	0.012	0.011
	$GeV\,cm^{-3}$	**0.46**	0.48	0.73	0.46	0.40
68% CR lower	$M_\odot\,pc^{-3}$	**0.0098**	0.011	0.017	0.012	0.0097
	$GeV\,cm^{-3}$	**0.37**	0.42	0.68	0.44	0.37

The thick disc results with tilt are also anomalous compared to the other four results. Several factors lead us to distrust this result and instead favour the thin disc with tilt result. As commented on earlier the thick disc is more dependent on the tilt term, and so is more dependent on the model and assumptions we use to describe the tilt term. The posterior for the k parameter describing the radial variation of ν and σ_{Rz} appears to suffer from a prior dependency in the thick disc case, with tensions in the data driving the k to the edges of the prior (Sivertsson *et al.* (2017)). In the thin disc case however the posterior is generally flat within the prior range. Additionally the thick disc stars are further away with potentially greater errors, and are more susceptible to halo contamination. Furthermore the assumption of a flat rotation curve is informed by knowledge of galaxy close to the midplane - this assumption may be erroneous in the regions high above the midplane probed by the thick disc population.

These issues inform us as to how to proceed with Gaia data. For the stellar samples analysed with this method we must not only find the vertical distribution of the tracer density and velocity dispersions, but also their radial dependence, rather than relying on a prior for the latter. The assumption of the flat rotation curve must be tested at a range of z values. Beyond these immediately apparent avenues, Gaia data may also allow us to dispense with the assumption of axisymmetry, and the the associated assumption of time independence.

References

Sivertsson, S., Silverwood, H., Read, J. I., Bertone, G., & Steger, P., 2017, *in preparation*

Silverwood, H., Sivertsson, S., Steger, P., Read, J. I., & Bertone, G., 2016, *Mon.Not.Roy.Astron.Soc.* 459 (2016) no.4, 4191-4208

Binney, J. & Tremaine, S., 2008, *Galactic Dynamics*, Princeton University Press, 2008

Feroz, F. & Hobson, M. P., 2007, *Mon.Not.Roy.Astron.Soc.*, 384 (2007), 449?463

Feroz, F., Hobson, M. P., & and Bridges, M., 2008, *Mon.Not.Roy.Astron.Soc.*, 398 (2009), 1601-1614

Feroz, F., Hobson, M. P., Cameron, E., & Pettitt, A. N. 2013, arXiv:1306.2144

Büdenbender, A., van de Ven, G., Watkins, L. L. 2015 *Mon.Not.Roy.Astron.Soc.*, 452 (2015), 956-968

McKee, C. F., Parravano, A., Hollenbach, D.J. 2015 *Astrophys. J.*, 814 (2015), 13-36

Bovy, J., Rix, H-W., Schlafly, E. F., Nidever, D.L, Holtzman, J. A., Shetrone, M., Beers, T. C., 2016 *Astrophys. J.*, 823 (2016), 30-50

Read, J. I., 2014 *J. Phys. G*, 41 (2014), 063101

Astronomy and Astrophysics in the Gaia sky
Proceedings IAU Symposium No. 330, 2017
A. Recio-Blanco, P. de Laverny, A.G.A. Brown & T. Prusti, eds.

doi:10.1017/S1743921317005944

The relation between velocity dispersions and chemical abundances in RAVE giants

Rodolfo Smiljanic[1] and Rafael Silva de Souza[2]

[1]Nicolaus Copernicus Astronomical Center, Polish Academy of Sciences, Bartycka 18, 00-716, Warsaw, Poland
email: rsmiljanic@camk.edu.pl

[2]MTA Eötvös University, EIRSA "Lendulet" Astrophysics Research Group, Budapest 1117, Hungary
email: rafael.2706@gmail.com

Abstract. We developed a Bayesian framework to determine in a robust way the relation between velocity dispersions and chemical abundances in a sample of stars. Our modelling takes into account the uncertainties in the chemical and kinematic properties. We make use of RAVE DR5 radial velocities and abundances together with Gaia DR1 proper motions and parallaxes (when possible, otherwise UCAC4 data is used). We found that, in general, the velocity dispersions increase with decreasing [Fe/H] and increasing [Mg/Fe]. A possible decrease in velocity dispersion for stars with high [Mg/Fe] is a property of a negligible fraction of stars and hardly a robust result. At low [Fe/H] and high [Mg/Fe] the sample is incomplete, affected by biases, and likely not representative of the underlying stellar population.

Keywords. Stars: abundances – Stars: kinematics – Galaxy: stellar content

1. Introduction

In a sample of giants from RAVE DR4, Minchev *et al.* (2014) discovered a decrease in the velocity dispersion of old stars (low [Fe/H] and [Mg/Fe] > +0.4). Comparing with a chemo-dynamical model (from Minchev *et al.* 2013), this decrease was interpreted as evidence of the migration of old inner disk stars with cool kinematics. The migration was likely triggered by an early merger. In this work, we re-address the problem of analyzing the relation between chemical abundances and velocity dispersions. We introduce a hierarchical Bayesian approach that models the velocity dispersions without the need of arbitrary data binning (Smiljanic & de Souza 2017, in prep).

2. Stellar sample

We use RAVE DR5 (Kunder *et al.* 2017). RAVE is a stellar spectroscopic survey that provides radial velocities, atmospheric parameters and abundances for half million stars. We selected RAVE giants ($T_{\rm eff}$ > 4000 K, 1.5 < log g < 3.5) with good-quality distances (either based on Gaia parallaxes or from RAVE itself; Binney *et al.* 2014, Astraatmadja & Bailer-Jones 2016, Lindegren *et al.* 2016), good-quality proper motions, and radial velocities (22242 giants). Here, we present results of the preliminary analysis of a subsample with high-signal-to-noise spectra (> 65) and volume restricted ($|Z| < 0.5$ kpc, 7 kpc < X < 9 kpc, and $|Y| < 1$ kpc), i.e., about 4500 stars.

Comparing the atmospheric parameters with isochrones, we noticed that the RAVE giants with [Fe/H] < −0.50 tend to be cooler and brighter than expected. This suggests accuracy problems in the analysis of metal-poor stars in RAVE.

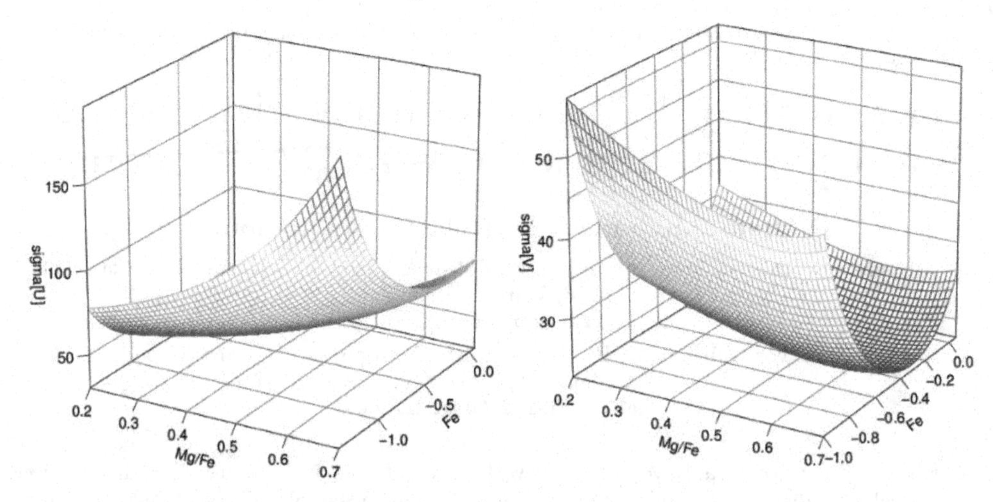

Figure 1. Dispersion of the U (left) and V (right) velocities for giants with [Mg/Fe] > +0.20.

3. Discussion

We fit the variance of each Galactic velocity component assuming that the logarithm of the variance is a linear function of the velocity. This ensures that the variance is always positive. The MCMC integration is done within R, using JAGS. The uncertainties of the velocities and abundances are fully taken into account in the modelling.

Our preliminary analysis suggests that, for low-[Fe/H] high-[Mg/Fe] stars (Fig. 1), the velocity dispersions are uncertain. The model does not converge well and the credibility intervals are wide. The small number of stars in this regime is likely not representative of the underlying population and does not constrain well the velocity dispersions. Moreover, the low accuracy of the atmospheric parameters also raises questions about the quality of the abundances (Fe and Mg) for these stars. Therefore, we caution that conclusions based on these stars are uncertain and should be seen with care.

Acknowledgement

We acknowledge E. E. O. Ishida for useful discussions. R. Smiljanic acknowledges support from NCN (grant 2014/15/B/ST9/03981) and from the Polish Ministry of Science and Higher Education. This work has made use of data from the European Space Agency (ESA) mission *Gaia* (`https://www.cosmos.esa.int/gaia`), processed by the *Gaia* Data Processing and Analysis Consortium (DPAC, `https://www.cosmos.esa.int/web/gaia/dpac/consortium`). Funding for the DPAC has been provided by national institutions, in particular the institutions participating in the *Gaia* Multilateral Agreement.

References

Astraatmadja, T. L. & Bailer-Jones, C. A. L. 2016, *ApJ*, 883, 119
Binney, J., Burnett, B., & Kordopatis, G. 2014, *MNRAS*, 437, 351
Kunder, A., Kordopatis, G., & Steinmetz, M. 2017, *AJ*, 153, 75
Lindegren, L., Lammers, U., Bastian, U. *et al.* 2016, *A&A*, 595, A4
Minchev, I., Chiappini, C., & Martig, M. 2013, *A&A*, 558, A9
Minchev, I., Chiappini, C., Martig, M., *et al.* 2014, *ApJ*, 781, L20

Astronomy and Astrophysics in the Gaia sky
Proceedings IAU Symposium No. 330, 2017
A. Recio-Blanco, P. de Laverny, A.G.A. Brown & T. Prusti, eds.

doi:10.1017/S1743921317006056

HST Proper Motions of Distant Globular Clusters: Constraining the Formation & Mass of the Milky Way

S. Tony Sohn[1], Roeland P. van der Marel[1], Alis Deason[2], Andrea Bellini[1], Gurtina Besla[3] and Laura Watkins[1]

[1]Space Telescope Science Institute,
3700 San Martin Drive, Baltimore 21218, USA
email:tsohn@stsci.edu

[2]Institute for Computational Cosmology,
Department of Physics, Durham University,
South Road, Durham DH1 3LE, UK

[3]Steward Observatory, University of Arizona,
933 North Cherry Avenue, Tucson, AZ 85721, USA

Abstract. Proper motions (PMs) are required to calculate accurate orbits of globular clusters (GCs) in the Milky Way (MW) halo. We present our HST program to create a PM database for 20 GCs at distances of R_{GC} = 10–100 kpc. Targets are discussed along with PM measurement methods. We also describe how our PM results can be used for Gaia as an external check, and discuss the synergy between HST and Gaia as astrometric instruments in the coming years.

Keywords. astrometry, Galaxy: evolution, Galaxy: formation, Galaxy: globular clusters: general, Galaxy: halo

1. Introduction

The globular cluster (GC) system of the Milky Way (MW) provides important information on the MW's present structure and past evolution. Clusters in the halo are particularly useful tracers; because of their long dynamical timescales, their orbits retain imprints of the origin and accretion history. Full 3d motions are required to calculate orbits of GCs. While most GCs in the MW halo have known line of sight velocities, accurate proper motion (PM) measurements have only been available for a few halo GCs. Our HST program GO-14235 is designed to remedy this situation by creating a high-quality PM database for several distant GCs in the MW halo. Orbit calculations based on our PMs will provide important clues to the origins of individual halo GCs. Our PMs will also yield the best handle yet on the velocity anisotropy profile of any tracer population in the halo, and subsequently provide an improved estimate of the MW mass.

2. Target Clusters and Proper Motions

Table 1 lists our 20 target GCs along with their distances, metallicities, and GC "class" based on Mackey & van den Bergh (2005). We included clusters with a wide range of properties to select a representative sample of the halo population. To measure PMs of the GCs, we used multi-epoch HST ACS/WFC and WFC3/UVIS data. First-epoch data for most of the targets were obtained through the two survey programs by Sarajedini *et al.* (2007) and Dotter *et al.* (2011), except for Pal 13 and NGC 2419. We obtained second-epoch data for all target clusters through this program using the same detectors, telescope

Table 1. Target globular clusters and their properties

Cluster	R_{GC} (kpc)	$R_\odot$ (kpc)	[Fe/H]	Class	Cluster	R_{GC} (kpc)	$R_\odot$ (kpc)	[Fe/H]	Class
NGC 6101	11.1	15.3	−1.98	Old	NGC 5024	18.3	17.8	−2.10	Old
NGC 6934	12.8	15.7	−1.47	Young	Rup 106	18.5	21.2	−1.68	Young
NGC 6426	14.6	20.7	−2.15	Young	Terzan 8	19.1	26.0	−2.16	Sgr
IC 4499	15.7	18.9	−1.53	Young	NGC 4147	21.3	19.3	−1.80	Sgr
NGC 2298	15.7	10.7	−1.92	Old	Arp 2	21.4	28.6	−1.75	Sgr
Pal 12	15.9	19.1	−0.85	Sgr	Pal 13	26.7	25.8	−1.88	Young
Terzan 7	16.0	23.2	−0.32	Sgr	Pal 15	37.9	44.6	−2.07	Old
NGC 5466	16.2	15.9	−1.98	Young	NGC 7006	38.8	41.5	−1.52	Young
NGC 5053	16.9	16.4	−2.27	Young	Pyxis	41.7	39.7	−1.20	Young
NGC 1261	18.2	16.4	−1.27	Young	NGC 2419	91.5	84.2	−2.15	Old

pointings and orientations as in the first-epoch observations. The PM measurements were carried out following the same methodology we used for measuring PMs of M31 (Sohn *et al.* 2012) and Leo I (Sohn *et al.* 2013). In short, we measured the bulk motions of stars in our target GCs with respect to the background galaxies found in the same HST field of each cluster. We have carefully tested and verified our method to minimize possible systematic errors by correcting for local effects around each background galaxy. The final 1-D PM errors are in the range 3–20 $\mathrm{km\,s^{-1}}$ at the distances of targets, with a median uncertainty of 6 $\mathrm{km\,s^{-1}}$.

3. Prospects of Proper Motions with HST and Gaia

The average velocity dispersion of the MW halo is measured to be $\sim 100\,\mathrm{km\,s^{-1}}$ (e.g., Battaglia *et al.* 2005), and so a 1-D PM error lower than $\sim 50\,\mathrm{km\,s^{-1}}$ is required to get a meaningful dynamical measure of halo objects. Beyond 50 kpc, objects with required PM qualities are considerably lacking. Together with the Gaia PM results, our current and future HST programs will significantly increase the number of samples that will help better constrain the MW halo mass profile out to and beyond 100 kpc.

With the end-of-mission data, Gaia will measure systemic PMs and parallaxes of GCs to 1% or better out to $\sim$ 15 kpc (Pancino *et al.* 2017). For classical dwarf spheroidal galaxies, Gaia will be able measure PMs using plenty of stars brighter than its detection limits out to $\sim$ 200 kpc. Therefore, for halo objects in the range $R_{GC} \lesssim 200$ kpc, the current and upcoming PM results from HST will be used as important external checks for the Gaia results. Beyond $R_{GC} = 200$ kpc, stars in satellite objects are too faint to be detected by Gaia, so HST will continue to be a unique platform for distant objects. Meanwhile, it may be possible to use Gaia astrometry against HST or JWST for bright stars to measure PMs.

References

Battaglia, G., Helmi, A., Morrison, H. *et al.* 2005 *MNRAS*, 364, 433
Dotter, A., Sarajedini, A. & Anderson, J. 2011 *ApJ*, 738, 74
Mackey, A. D. & van den Bergh, S. 2005, *MNRAS*, 360, 631
Pancino, E., Bellazzini, M., Giuffrida, G. & Marino, S. 2017 *MNRAS*, 467, 412
Sarajedini, A., Bedin, L. R., Chaboyer, B. *et al.* 2007 *AJ*, 133, 1658
Sohn, S. T., Anderson, J., & van der Marel, R. P. 2012, *ApJ*, 753, 7
Sohn, S. T., Besla, G., van der Marel, R. P. *et al.* 2013, *ApJ*, 768, 139

Astronomy and Astrophysics in the Gaia sky
Proceedings IAU Symposium No. 330, 2017
A. Recio-Blanco, P. de Laverny, A.G.A. Brown & T. Prusti, eds.

doi:10.1017/S1743921317006019

Chemo-dynamical signatures in simulated Milky Way-like galaxies

Alessandro Spagna[1], Anna Curir[1], Marco Giammaria[2], Mario G. Lattanzi[1], Giuseppe Murante[3] and Paola Re Fiorentin[1]

[1]INAF – Osservatorio Astrofisico di Torino
via Osservatorio 20, 10025 Pino Torinese, Italy
email: spagna@oato.inaf.it

[2]Universitá degli Studi di Torino, Italy

[3]INAF – Osservatorio Astronomico di Trieste, Italy

Abstract. We have investigated the chemo-dynamical evolution of a Milky Way-like disk galaxy, AqC4, produced by a cosmological simulation integrating a sub-resolution ISM model. We evidence a global inside-out *and* upside-down disk evolution, that is consistent with a scenario where the "thin disk" stars are formed from the accreted gas close to the galactic plane, while the older "thick disk" stars are originated *in situ* at higher heights. Also, the bar appears the most effective heating mechanism in the inner disk. Finally, no significant metallicity-rotation correlation has been observed, in spite of the presence of a negative [Fe/H] radial gradient.

Keywords. Galaxy: abundance, evolution, structure, kinematics and dynamics

---

In order to study the chemo-dynamical signatures related to the galactic formation processes, we have analyzed a new ΛCDM cosmological simulation, AqC4, carried out with the GADGET-3 TreePM+SPH code, where star formation, chemical evolution and stellar feedback are described using a sub-grid Multi Phase Particle Integrator (MUPPI) model. The main parameters of this simulation are listed in Table 1, while further details on the model are described by Murante *et al.* (2015).

The morphology of AqC4 at redshift $z = 0$ is well represented by the three main stellar components of a Milky Way-like galaxy (i.e. bulge/spheroid, thin and thick disk). An extensive analysis of the spatial, kinematic, and chemical properties of AqC4 is presented by Giammaria (2017), who confirmed that the 6D $(\mathbf{x}, \mathbf{v})$ stellar distribution is similar to that observed in the Milky Way. The main difference is the presence of a more massive bulge (B/T=0.34), which also causes a faster rotation velocity ($V_\phi \sim 270$ km/s at $R \simeq 5-10$ kpc). Morever, the metallicity distribution of the stellar halo results a few dex higher than that of the Galactic halo.

However, the overall evolution of AqC4 appears quite consistent with similar studies based on independent simulations (e.g. Minchev *et al.* 2012, Bird *et al.* 2013, Martig *et al.* 2014, Ma *et al.* 2017).

Table 1. AqC4 parameters. $M_{\rm DM}$: mass of DM particles; $M_{\rm gas}$ initial mass of gas particle; ϵ = smoothing parameter; $M_{\rm vir}$ and $R_{\rm vir}$: DM mass and virial radius at redshift z=0; $N_{\rm DM}, N_{\rm gas}, N_{\rm star}$: number of DM particles, gas particles and stellar particles within $R_{\rm vir}$ at $z = 0$; Ω_i: density parameters; H_0: Hubble constant.

$M_{\rm DM}$ [M$_\odot$]	$M_{\rm gas}$ [M$_\odot$]	ϵ [kpc]	$M_{\rm vir}$ [M$_\odot$]	$R_{\rm vir}$ [kpc]	$N_{\rm DM}$	$N_{\rm gas}$	$N_{\rm star}$	$\Omega_{\rm m}$	Ω_Λ	$\Omega_{\rm b}$	H$_0$
$2.7 \cdot 10^5$	$5.1 \cdot 10^4$	0.163	$1.49 \cdot 10^{12}$	237	5 518 587	1 348 120	6 919 646	0.25	0.75	0.04	71

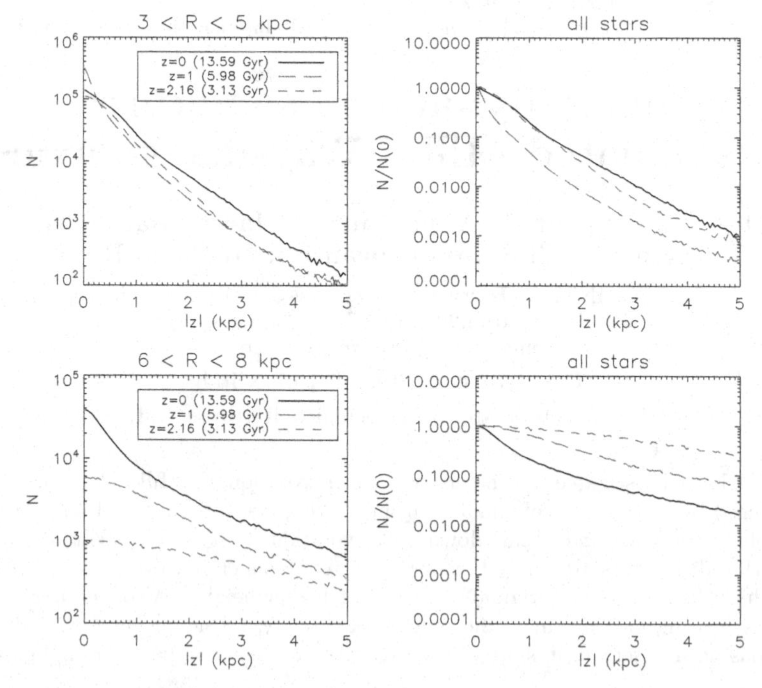

Figure 1. Disk $|z|$ distributions with R between 3-5 kpc and 6-8 kpc.

Figure 1 compares the evolution of the vertical distributions at redshift 0, 1, and 2.16, that clearly evidence an *upside-down* disk evolution in the "solar" annulus, 6 kpc $< R <$ 8 kpc. Such result supports a formation scenario that firstly generates the ancient "thick disk" stars *in situ*, from an initially turbulent ISM characterized by shorter scale-lenghts and higher scale-heights with respect to the younger "thin disk" stars (Haywood *et al.* 2013, Bird *et al.* 2013). Conversely, after about 6 Gyr ($z = 1$), a strong disk thickening is observed in the inner disk, 3 kpc $< R <$ 5 kpc, possibly due to the heating induced by the central bar (cfr. Grand *et al.* 2016).

Finally, although a typical negative [Fe/H] radial gradient is present in the disk of AqC4, no significant rotation-metallicity relation is present, as observed in our Milky Way (Spagna *et al.* 2010, Lee *et al.* 2011; see also Shönrich & McMillan 2017). This result deserves further investigation as it may be an effect of the specific formation history of this simulation.

References

Bird, J. C. *et al.* 2013, *ApJ*, 773, 43
Giammaria, M. 2017, *Tesi laurea magistrale in Fisica - Ind. Astrofisica*, Turin University (Italy)
Grand, R. J. J., Springel, V., Gómez, F. A. *et al.* 2016, *MNRAS*, 459, 199
Haywood, M., Di Matteo, P., Lehnert, M. D. *et al.* 2013, *A&A*, 560, A109
Lee, Y. S., Beers, T. C., An, D. *et al.* 2011, *ApJ*, 738, 187
Ma, X., Hopkins, P. F., Wetzel, A. R. *et al.* 2017, *MNRAS*, 467, 2430
Martig, M., Minchev, I., & Flynn, C. 2014, *MNRAS*, 442, 2474
Minchev, I., Famaey, B., Quillen, A. C. *et al.* 2012, *A&A*, 548, A127
Murante, G., Monaco, P., Borgani, S., *et al.* 2015, *MNRAS*, 447, 178
Schönrich, R. & McMillan, P. J. 2017, *MNRAS*, 467, 1154
Spagna, A., Lattanzi, M. G., Re Fiorentin, P., & Smart, R. L. 2010, *A&A*, 510, L4

Astronomy and Astrophysics in the Gaia sky
Proceedings IAU Symposium No. 330, 2017
A. Recio-Blanco, P. de Laverny, A.G.A. Brown
& T. Prusti, eds.

doi:10.1017/S1743921317005968

Revisiting TW Hydrae association in light of Gaia-DR1

R. Teixeira[1], E. R. Gonoretzky[1], C. Ducourant[2], P. A. B. Galli[1] and A. G. O. Krone-Martins[3]

[1]Universidade de São Paulo, IAG, São Paulo, Brazil
email: rama.teixeira@iag.usp.br

[2]Laboratoire d'Astrophysique de Bordeaux, Univ. Bordeaux, CNRS, France

[3]Faculdade de Ciencias, Universidade de Lisboa, Portugal

Abstract. TW Hydrae is a very young and nearby association with about 30 known members which is an excellent target for studies on stellar evolution since several of its members present a particular interest (planetary system, brown dwarfs, etc.). With the new data from TGAS and the Gaia DR1 eventually combined with others astrometric data we intend to improve our kinematic knowledge of this association.

Keywords. stellar association, TW Hydrae, proper motion, Gaia, TGAS

We started to investigate the impact of the new astrometric data reality as consequence of the first Gaia data release (Gaia Collaboration 2016) on the kinematics of the TW Hydrae association. Indeed until now, only dedicated small field astrometric data were available leading to imprecise kinematic membership determination. Recent papers (Weinberg *et al.* 2013, Ducourant *et al.* 2014) argued about the utilization of the trace back strategy to determine the age of stars belonging to TWA. This is a relevant point since the age so obtained is model independent but assumes all stars of the association were formed at the same time in the place.

Although proper motions from several astrometric catalog as PPMXL (Roser *et al.* 2010), SPM4 (Girard *et al.* 2011), UCAC4 (Zacharias *et al.* 2013) and more recently HSOY (Altmann *et al.* 2017) and UCAC5 (Zacharias *et al.* 2017) can be considered globally reliable, they can locally present some inconsistencies (Teixeira *et al.* 2014) that can perturb the membership determination or the trace back results. In Figure 1 we present the various published proper motions of the "well-behaved" star TWA07 from TW Hydrae. We notice huge discrepancies between the values that cannot be explained by the fact that UCAC5 and HSOY are not in the same reference system than the others.

This example is illustrative of what can be frequently be found when one gets interested in specific objects and it becomes evident that dedicated studies of individual targets (i.e., stars) are needed in punctual works. In view of this scenario, we have been determining new TWA proper motions using the various positions found in the literature, including Gaia-DR1. Figure 2 confirms the existence of large discrepancies that could, for example, change the membership status or the convergence in the trace back analysis.

We are grateful to the French and Brazilian organisms COFECUB, FAPESP, CAPES and CNPq for financial support. This work has made use of data from the European Space Agency (ESA) mission *Gaia* (https://www.cosmos.esa.int/gaia), processed by the *Gaia* Data Processing and Analysis Consortium (DPAC, https://www.cosmos.esa.int/web/gaia/dpac/consortium). Funding for the DPAC has been provided by national institutions, in particular the institutions participating in the *Gaia* Multilateral Agreement.

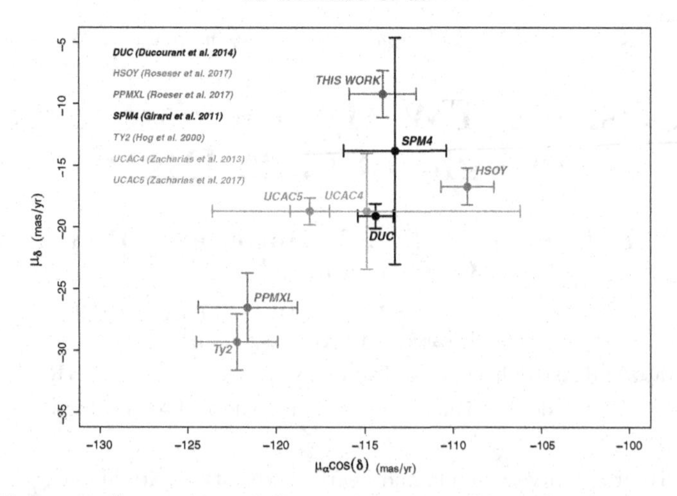

Figure 1. Comparison of the proper motions from several sources.

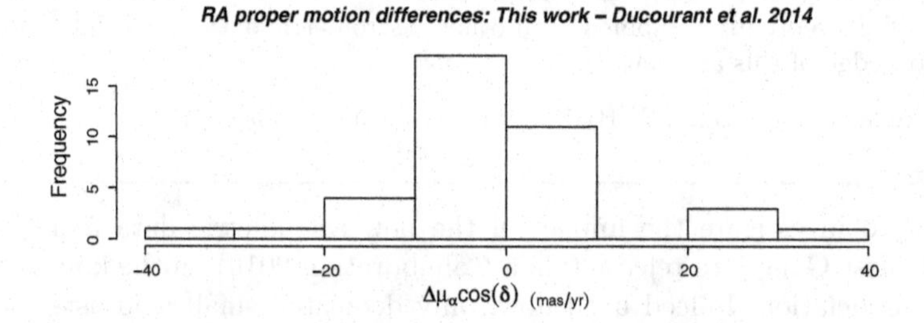

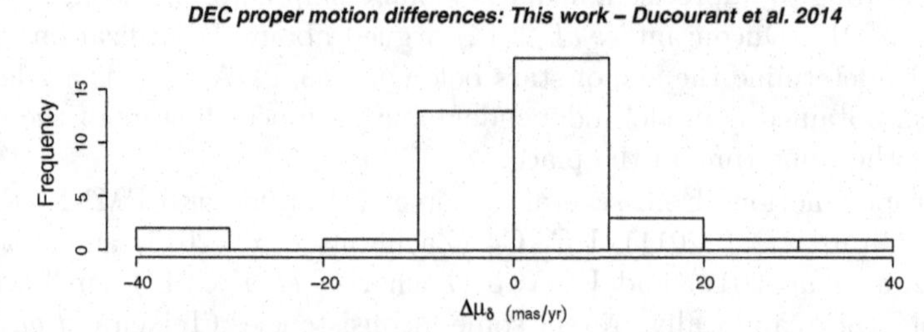

Figure 2. Differences between our proper motions and those from Ducourant *et al.* 2014.

References

Altmann M., Roeser S., Demleitner M., Bastian U. & Schilbach E. 2017, *A&A*, 600, L4
Ducourant C., Teixeira R., Galli P. A. B. *et al.* 2014, *A&A*, 563, A121
Gaia Collaboration, Brown, A. G. A., Vallenari, A., Prusti, T., de Bruijne, J. H. J., Mignard, F. *et al. A&A, 595A, 2G*
Girard, T. M., van Altena, W. F., Zacharias, N. *et al.* 2011, *AJ*, 142, 15
Høg, E., Fabricius, C., Makarov, V. V., *et al.* 2000 *A&A*, 355, L27
Roeser S., Demleitner M. & Schilbach E. 2010, *AJ*, 139, 2440
Teixeira, R., Galli, P. A. B., Le Campion, *et al.* 2014, RMxAC, 43, 55
Weinberger A. J., Anglada-Escudé G. & Boss A. P. 2013, *ApJ*, 762, 118
Zacharias N., Finch C. T., Girard T. M. *et al.* 2013, *AJ*, 145, 44
Zacharias N., Finch C. & Frouard J. 2017, *AJ*, 153, 166

Astronomy and Astrophysics in the Gaia sky
Proceedings IAU Symposium No. 330, 2017
A. Recio-Blanco, P. de Laverny, A.G.A. Brown & T. Prusti, eds.

doi:10.1017/S1743921317006780

[Y/Mg] stellar dating calibration

A. Titarenko[1], A. Recio-Blanco[1], P. de Laverny[1], M. Hayden[1], G. Guiglion[2] and C. Worley[3]

[1]Université Côte d'Azur,Observatoire de la Côte d'Azur, CNRS, Laboratoire Lagrange, France
[2]Leibniz-Institut fur Astrophysik Potsdam (AIP), 14482 Potsdam, Germany
[3]Institute of Astronomy, University of Cambridge (IOA), Cambridge, United Kingdom

Abstract. Gaia DR1 has opened a new era of stellar age dating, that is crucial for many astrophysical objectives. In addition, the Gaia based isochrone fitting ages can be compared to other chemical clocks like the [Y/Mg] one (Nissen *et al.* 2015). In our work we have used ESO archived data of the AMBRE project (de Laverny *et al.* 2013) for UVES spectra, in order to evaluate the age [Y/Mg] abundance correlation for turn off stars. 310 turn off stars of the UVES-archive (setups 564 and 580) are included in the TGAS database. Isochrone fitting ages were derived. We have applied the GAUGUIN procedure for those stars to derive the Mg and the Y abundances. As the result we present the [Y/Mg] vs stellar age dependence for ~40 TO-stars.

Keywords. stars: abundances, atmospheres

---

1. Overview

The AMBRE (Archeologie avec Matisse Basee sur les aRchives de l'ESO) project, established by ESO and the Observatoire de la Cote d'Azur in 2009 (de Laverny *et al.* 2013), is parameterizing stellar spectra archived at ESO. In the AMBRE project, stellar spectra collected by four high resolution spectrographs were analysed (FEROS, HARPS, UVES and GIRAFFE). According to the selected MgI lines in the range of wavelength 5167-6319 Å, we used data from 2 setups of UVES data: RED564 [4583-6686 Å, gap 5644-5654 Å] and RED580 [4726-6835 Å, gap 5804-5817 Å], that leads us to a final number of 4156 spectra. We used GAUGUIN code, developed in OCA (Guiglion *et al.* 2016). After polynomial normalization for the continuum (if needed), it applies a Gauss-Newton method to get the final solution for the abundances. The GAUGUIN method is used also for Gaia RVS data within the DPAC/Apsis pipeline. In order to make a calibration of the method we used [Mg/Fe] values for benchmark stars from Jofre *et al.* 2015. A scientific validation was performed thanks to the abundances [Mg/Fe] from Mikolaitis *et al.* 2017 for HARPS and FEROS instruments that are inside AMBRE project. Using stellar ages, calculated by Hayden *et al.* (in prep.), we selected a sample of 80 turn off stars for UVES data with reliable ages. We selected [Y/H] abundances for 345 stars from Guiglion *et al.* (in prep.) based on 6 lines of YII in 4980-5820 Å wave range. The total number of stars for which the values [Mg/Fe], [Y/Fe] and robust ages are available is ~40.

2. Conclusions

We determined the abundances of [Mg/Fe] for 590 stars (from the ESO archive data UVES setups 564 and 580) using the GAUGUIN code. On the other side, the cross match with TGAS gave us 310 stars with 80 of them with reliable values of ages. Using the strict selection of [Y/Fe] abundances calculated by 6 YII lines for these two UVES setups we got 345 stars. Collecting all the available data in order to get [Y/Mg] vs Age relation we find, that the final sample of ~40 turn off stars shows a similar correlation

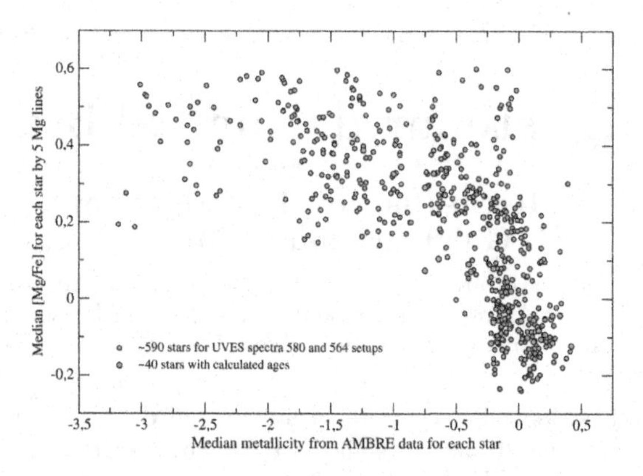

Figure 1. The relation [Mg/Fe] to metallicity for the final UVES sample. The stars for which we have reliable values of the ages are colored in orange.

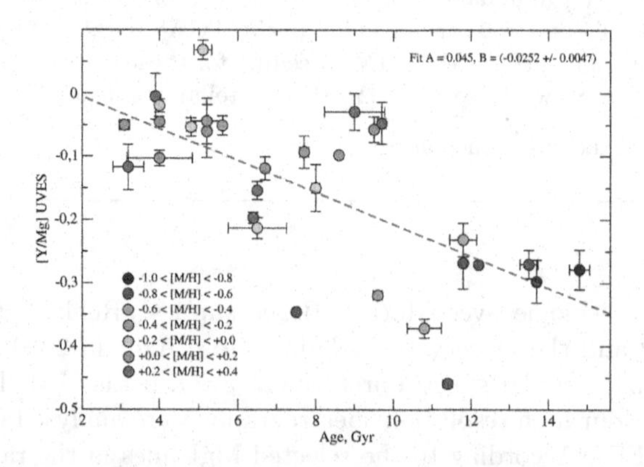

Figure 2. The [Y/Mg] vs stellar age relation for UVES TO-stars. Color code represents the stellar metallicity, the dashed line - linear fit with of the sample. The dached line represents the character of the one from Tucci *et al.* 2016.

that the one presented in Nissen 2015 and Tucci *et al.* 2016 for the solar twins. Our data allow to explore the character of the [Y/Mg] ratio as a stellar clock in a wide range of metallicities. Also we assume that the relation between [Y/Mg] and stellar age can give us a possibility to estimate the ages for the AMBRE stars that are not inside TGAS extending the studies of the disc outside the solar neighbourhood.

This work has made use of data from the European Space Agency (ESA) mission *Gaia* (`https://www.cosmos.esa.int/gaia`), here, processed by the *Gaia* Data Processing and Analysis Consortium (DPAC, `https://www.cosmos.esa.int/web/gaia/dpac/consortium`). Funding for the DPAC has been provided by national institutions, in particular the institutions participating in the *Gaia* Multilateral Agreement.

References

Nissen, P. E. 2015, *Astronomy & Astrophysics*, 579, A52
Tucci, *et al.* 2016, *Astronomy & Astrophysics*, 590, A32
Laverny, *et al.* 2013, *The Messenger*, 153
Guiglion, *et al.* 2016, *Astronomy & Astrophysics*, 595, A18
Jofre, *et al.* 2015, *Astronomy & Astrophysics*, 582, A81
Mikolaitis, *et al.* 2017, *Astronomy & Astrophysics*, 600, A22

Astronomy and Astrophysics in the Gaia sky
Proceedings IAU Symposium No. 330, 2017
A. Recio-Blanco, P. de Laverny, A.G.A. Brown
& T. Prusti, eds.

doi:10.1017/S1743921317006287

Dynamics of the Oort Cloud In the Gaia Era I: Close Encounters

S. Torres, S. Portegies Zwart and A. G. A. Brown

Leiden Observatory, Leiden University, PO Box 9513 Leiden, NL.2300 RA, the Netherlands.
email: storres@strw.leidenuniv.nl, spz@strw.leidenuniv.nl &
brown@strw.leidenuniv.nl

Abstract. Comets in the Oort cloud evolve under the influence of internal and external perturbations from giant planets to stellar passages, the Galactic tides, and the interstellar medium.Using the positions, parallaxes and proper motions from TGAS in Gaia DR1 and combining them with the radial velocities from the RAVE-DR5, Geneva-Copenhagen and Pulkovo catalogues, we calculated the closest encounters the Sun has had with other stars in the recent past and will have in the near future. We find that the stars with high proper motions near to the present time are missing in the Gaia-TGAS, and those to tend to be the closest ones. The quality of the data allows putting better constraints on the encounter parameters, compared to previous surveys.

Keywords. Oort cloud, Solar System

1. Introduction

The outer regions of the Solar System are populated by a large number of planetesimals. Further away, more than 1,000 times farther than the Sun, and almost extending to the nearest stars, is the Oort cloud. It was proposed in the late 1950s by the Dutch astronomer Jan Hendrik Oort at the Sterrewacht Leiden, by anticipating its existence when he realised that long term comets (a >40 AU) bound to the Sun must come from an area well beyond Neptune. Oort (1950) pointed out that a spike in the distribution of $1/a$ of the long period comets with $a<10^4$ AU (see for example Wiegert & Tremaine 1999), and isotropic distributions in $cos\ i$, ω, and Ω, would argue for the existence of a reservoir of objects in quasi-spherical symmetry surrounding the Solar System. The external perturbations such as galactic tides, stellar flybys and molecular clouds play an important role in the formation and evolution of the Oort cloud. In particular, the role of passing stars is to reshuffle the comet distribution in the Oort could and to refill the high inclination region where comets are pushed into the planetary region. The Gaia-DR1 (Gaia Collaboration *et al.* 2016) has opened a new window for the understanding of the Milky Way. In the particular case of the Solar System, Gaia detected the local star systems within 50pc from the Sun (compared to the 20% detected by Hipparcos (Perryman *et al.* 1997)). This allows for the stellar positions to be traced back far in time, which in turn allows reconstructing the history of stellar encounters and its implication for the comets of the Oort cloud.

2. Method

Using the positions, parallaxes and proper motions from TGAS (Gaia Collaboration *et al.* 2016) and combining them with the radial velocities from the RAVE-DR5 (Kunder *et al.* 2017), Geneva-Copenhagen (Nordström *et al.* 2004), and Pulkovo (Gontcharov 2006) catalogues, we computed the 6D elements for the stars in the immediate solar

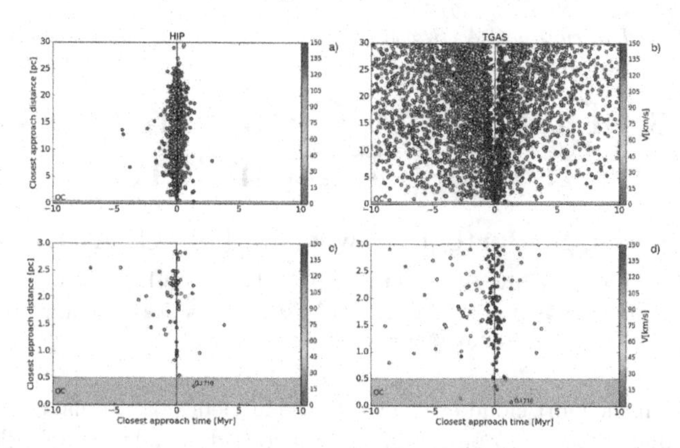

Figure 1. Closest approach of distance vs time for stars within 30pc (**a** -Hipparcos data-, **b** -TGAS data-) and 3 pc (**c** -Hipparcos data-, **d** -TGAS data-) with respect to the Sun; colorbar represents the velocity of the stellar approaches.

neighbourhood (<30 pc) backwards and forward in time ($\pm$10 Myr) to identify the past and future solar encounters. The orbital integration of the stars was performed by using an axisymmetric potential included in the module *Galaxia* in AMUSE (Pelupessy *et al.* 2013).

3. Results/Discussion

The closest encounter to the Sun has been studied for years by taken advantage of the Hipparcos catalogue (see for example García-Sánchez et al. 2001). It has been a general agreement of the potential perturbers to the Sun, particularly with the star HIP 89825 (Gliese 710) that according to García-Sánchez et al. 2001 will perturb the Sun in $1.36 Myr$ to within 0.33pc respect to the Sun (Figure 1 a,c); however the Hipparcos data is incomplete. Using TGAS, we recalculate the encounters history of the Sun and the stars in the solar neighbourhood (Figure 1 b,d) and we find that Gliese 710 will approach the Sun in $1.35 Myr$ to within 0.06pc (see also Berski and Dybczyński 2016). Due to the better completeness of the Gaia data, we increased the number of encounters within 30 pc, and we find a distribution of the closest stars with better accuracy. However, we are missing the recent encounters with high proper motions, which are missing from the TGAS catalogue.

This work has made use of data from the European Space Agency (ESA) mission *Gaia* (`https://www.cosmos.esa.int/gaia`).

References

Berski, F. & Dybczyński, P. A. 2016, *A&A*, 595, L10
García-Sánchez, J., Weissman, P. R., Preston, R. A., *et al.* 2001, *A&A*, 379, 634
Gaia Collaboration, Brown, A. G. A., Vallenari, A., *et al.* 2016, *A&A*, 595, A2
Gontcharov, G. A. 2006, *Astronomy Letters*, 32, 759
Kunder, A., Kordopatis, G., Steinmetz, M., *et al.* 2017, *AJ*, 153, 75
Nordström, B., Mayor, M., Andersen, J., *et al.* 2004, *A&A*, 418, 989
Oort, J. H. 1950, bain, 11, 91
Pelupessy, F. I., van Elteren, A., de Vries, N., *et al.* 2013, *A&A*, 557, A84
Perryman, M. A. C., Lindegren, L., Kovalevsky, J., *et al.* 1997, *A&A*, 323, L49
Wiegert, P. & Tremaine, S. 1999, *Icarus*, 137, 84

Astronomy and Astrophysics in the Gaia sky
Proceedings IAU Symposium No. 330, 2017
A. Recio-Blanco, P. de Laverny, A.G.A. Brown & T. Prusti, eds.

doi:10.1017/S1743921317006202

Stellar parameters with FASMA: a new spectral synthesis package

M. Tsantaki[1], D. T. Andreasen,[2] G. D. C. Teixeira,[2] S. G. Sousa,[2] N. C. Santos,[2] E. Delgado-Mena,[2] and G. Bruzual[1]

[1]Instituto de Radioastronomía y Astrofísica, UNAM,
Campus Morelia, A.P. 3-72, C.P. 58089, Michoacán, Mexico
email: mtsantaki@crya.unam.mx

[2]Instituto de Astrofísica e Ciências do Espaço, Universidade do Porto,
CAUP, Rua das Estrelas, Porto, 4150-762, Portugal

Abstract. Current Galactic surveys, including the Gaia mission, rely on the efficiency of the spectral analysis techniques to provide precise and accurate spectral information (i.e. effective temperature, surface gravity, metallicity, and chemical abundances) in the shortest computational time. In this work, we present a new package to preform complete spectral analyses based on the spectral synthesis technique (Tsantaki *et al.* 2017, submitted). We focus on deriving atmospheric parameters for FGK-type stars using both high and medium resolution (GIRAFFE) spectra. This method is implemented on the Gaia-ESO benchmark stars to confirm its validity, achieving similar accuracy for the two resolution setups.

Keywords. techniques: spectroscopic, stars: atmospheres, fundamental parameters

1. Introduction

In the last decades, due to the growing number of spectroscopic surveys dedicated to the study of the Galactic stellar populations, the number of high quality spectra has increased to several hundreds of thousands mainly owing to ground-based surveys, such as APOGEE, the Gaia-ESO Survey (GES), the GALAH survey, to name a few. The success of the above surveys depends on the efficiency of the spectral analysis techniques to provide precise and accurate spectral information in the shortest computation time.

Motivated by that, we developed a new package to derive the fundamental atmospheric parameters using the spectral synthesis technique. We named this package 'FASMA' which is built around the spectral synthesis code, MOOG (version 2014, Sneden 1973). FASMA includes other spectral functionalities, among them the analysis of spectra based on the equivalent width which is described in detail in Andreasen *et al.* (2017). In this work we describe a new additional driver of FASMA for the derivation of atmospheric parameters using the spectral synthesis technique. Among the advantages of this work is its applicability for both high and medium resolution regimes and the wide coverage of the stellar parameter space including high rotational velocities.

2. Methodology

Model atmospheres. The model atmospheres included in FASMA are the grid generated by the ATLAS program (Kurucz 1993), the grid of MARCS models (Gustafsson *et al.* 2008) and the grid for the APOGEE survey based on ATLAS9 (Mészáros *et al.* 2012).

Line list. The line list covers a wide range in the optical and therefore can be applied to spectra obtained by various spectrographs. It includes the regions of HR10 and HR15n set-ups of the GIRAFFE spectrograph (5399–5619 Å and 6470–6790 Å, respectively). We queried for all atomic and molecular lines inside intervals of ±2Å around mainly iron lines. From the 249 unique lines bigger than 10 mÅ (of the Sun) of our line list, 159 are iron. We used the NSO Atlas and the typical solar parameters to improve the transition probabilities ($\log gf$ values) in an inverted analysis.

Minimization procedure. FASMA performs local normalization for the adopted intervals and includes the parameter optimization procedure based on the Levenberg-Marquardt algorithm to solve the nonlinear least-squares problem, yielding the parameters that minimize the χ^2, namely T_{eff}, $\log g$, $[M/H]$, and $v \sin i$.

The code is written in python and the complete package is provided freely here: `https://github.com/MariaTsantaki/fasma-synthesis`. It is run either from the terminal or through a GUI interface for a more user-friendly approach.

3. Results

We explore how our parameters are affected by different characteristics: 1) the choice of initial conditions, 2) the different signal-to-noise values, 3) the different resolution set-ups, and 4) different rotational velocities. Moreover, to check the performance of our code, we select the GES benchmark stars as our comparison sample. The sample (excluding the M stars) contains 29 stars and is described in Heiter *et al.* (2015). We added the latest metal-poor benchmark stars suggested by Hawkins *et al.* (2016), reaching a total of 34 stars. The spectra in high resolution are taken from Blanco-Cuaresma *et al.* (2014) (95 spectra in total). For medium resolution, we query the GES archive for spectra taken with the GIRAFFE spectrograph.

4. Conclusion

With our spectral package, we provide stellar parameters for a wide range of spectral types and luminosity classes and can be used to analyse large samples in a reasonable amount of time. Our parameters show almost no dependence on the choice for initial parameters. Signal-to-noise deviations should be considered for very noisy spectra (below 50). The effects of rotational velocities become visible after 35 km s^{-1}. We compare our results with the GES benchmark stars using spectra both in high and medium resolution and we find very good agreement for metallicities for both resolutions. When we have better external estimations of surface gravities (e.g. using parallaxes), we can improve the temperature determinations.

References

Andreasen, D. T., Sousa, S. G., Tsantaki, M., *et al.* 2017, *A&A*, 600, A69
Blanco-Cuaresma, S., Soubiran, C., Jofré, P., & Heiter, U. 2014, *A&A*, 566, A98
Gustafsson, B., Edvardsson, B., Eriksson, K., *et al.* 2008, *A&A*, 486, 951
Hawkins, K., Jofré, P., Heiter, U., *et al.* 2016, *A&A*, 592, A70
Heiter, U., Jofré, P., Gustafsson, B., *et al.* 2015, *A&A*, 582, A49
Jofré, P., Heiter, U., Soubiran, C., *et al.* 2014, *A&A*, 564, A133
Kurucz, R. L. 1993, Kurucz CD-ROM, Cambridge, Smithsonian Astrophysical Observatory
Mészáros, S., Allende Prieto, C., Edvardsson, B., *et al.* 2012, *ApJ*, 144, 120
Sneden, C. 1973, *ApJ*, 184, 839
Tsantaki, M., Andreasen, D. T., Teixeira, G. D., Sousa, S. G., *et al.* 2017 (submitted), *MNRAS*

Astronomy and Astrophysics in the Gaia sky
Proceedings IAU Symposium No. 330, 2017
A. Recio-Blanco, P. de Laverny, A.G.A. Brown
& T. Prusti, eds.

doi:10.1017/S1743921317005993

Complex study of the open cluster NGC 2281

Jaroslav Velčovský and Jan Janík

Department of Theoretical Physics and Astrophysics,
Faculty of Science, Masaryk University,
Kotlářská 2, CZ-611 37, Brno, Czech Republic
email: `375641@mail.muni.cz`, `honza@physics.muni.cz`

Abstract. We present the complex study of the open cluster NGC 2281 where both traditional and newly developed methods for study of open clusters have been used. Morphological and dynamical parameters of the cluster were obtained from the accepted astrometric data. The new method "Superposition of Gaussian surfaces" along with proper motion of stars was used to determine membership probabilities which were helpful in selection of stars for further analysis. Metallicity and radial velocity of the cluster were obtained from spectroscopic measurements. Age, colour excess, and distance of the cluster were determined using absolute CCD photometry combined with previous results. The results were compared with those of previous studies.

Keywords. complex study, NGC 2281, membership probability

---

1. Membership probability

A new method "Superposition of Gaussian surfaces" (SGS) was developed for determination of membership probability p_μ of field stars of the cluster. The SGS method generates three-dimensional proper motion diagram of field stars of the cluster. For each point in diagram the surface density is given by equation:

$$f_s(\mu'_\alpha, \mu_\delta) = \sum_i^n R + \frac{K}{m_i^2} \exp\left[- \left(\frac{(\mu'_\alpha - \mu'_{\alpha\, i})^2}{2\sigma'^2_{\mu_{\alpha\, i}}} \frac{(\mu_\delta - \mu_{\delta i})^2}{2\sigma^2_{\mu_{\delta i}}} \right) \right], \tag{1.1}$$

where R and K are constants, m_i is brightness of i-th star in the V filter, $\mu_{\alpha i}$, $\mu_{\delta i}$, $\sigma_{\mu_\alpha i}$, and $\sigma_{\mu_\delta i}$ are proper motions of i-th star and their uncertainties. Both $\mu'_{\alpha i}$ and $\sigma'_{\alpha i}$ are equal to $\cos\delta_i \mu_{\alpha i}$ and $\cos\delta_i \sigma_{\alpha i}$ respectively to reduce a spherical projection.

Proper motions data from the PPMXL (Roeser *et al.*, 2010, reference positions), NOMAD (Zacharias *et al.*, 2004), UCAC4 (Zacharias *et al.*, 2012), IGSL (Smart *et al.*, 2013), URAT1 (Zacharias *et al.*, 2015), and TYCHO (Hog *et al.*, 2000) catalogues were averaged for each star up to 15th magnitudethe in the V filter inside a radius of half degree with the centre in cluster's catalogue coordinates. The averaged data were used for generate of the diagram by SGS method. A superposition of two Gaussian surfaces was used as a model for description of the resulting diagram, where the first surface is for background stars and the second surface for stars of the cluster. Membership probabilities of field stars were derived from a single Gaussian surface with determined parameters of the cluster.

2. Spectroscopy

Spectra of two stars were acquired in October 2014 with the Echelle spectrograph ($R = 18\,000$) on the 2.4m telescope (Ritchey-Chretien) at Doi Inthanon in Thailand. The SYNSPEC software (E1) with ATLAS9 grid models (Castelli *et al.*, 2003) and ROTIN3 software (E1) for rotational convolution were used for modelling of synthetic spectra.

The first star with $p_\mu = 59$ % has metallicity [Fe/H] $= 0.13 \pm 0.03$ dex and the second star with $p_\mu = 58$ % has [Fe/H] $= 1.30 \pm 0.05$ dex which is typical of chemically peculiar stars. Therefore, metallicity of the first star was accepted as metallicity of the cluster. Both stars have almost the same radial velocity (-17.7 and -17.1 km/s) and their average value is $v_r = -17 \pm 1$ km/s.

3. Photometry

Photometric CCD data were acquired in October 2013 and January 2014 using the 600mm reflector at the Suhora Observatory in Poland. Landolt's field (PG1633+099, Landolt *et al.*, 1992) as a season's primary standards and chosen stars from NGC 2281 and NGC 7142 as a night's secondary standards were used for transformation of the cluster stars into the standard photometric system.

Measured photometric data with $p_\mu > 10$ % were plotted into the colour–magnitude diagram where a well-defined evolving trace is created. Points were fitted by the PARSEC isochrone (Bressan *et al.*, 2012) with the known metallicity and colour excess. The distance modulus $m_V - M_V = 8.81$ mag and the logarithm of age $\log t = 8.70$ of the cluster were obtained from a vertical shift and a change of the shape of the isochrone towards points. A red isochrone for binary stars with the -0.75 mag shift in the y axis was added.

Final membership of field stars to the cluster was determined as intersection of stars with high membership probability and those which are close to isochrones. Photometric data from the 2MASS catalogue were used to cover the whole one degree diameter. Stars with $p_\mu > 50$ % and distance less then 0.07 mag from the isochrones were considered as real members of the cluster. Stars with 50 % $\geqslant p_\mu >$ 10 % and distance less then 0.05 mag were considered as candidates of membership to the cluster. There were found 69 real members and 42 candidates of membership of the cluster.

4. Results and discussions

Derived morphological and dynamical parameters of the cluster are almost the same as the current catalogue values. Measured radial velocity of the cluster is opposite to the current catalogue value +13.3 $\pm$ 4.1 km/s (Kharchenko *et al.*, 2013). The value -17 $\pm$ 1 km/s should be adopted as a radial velocity of the cluster because of good agreement of radial velocities of both chosen stars. The found distance d = 550 pc and age t = 500×10^6 yr of the cluster are slightly different from the catalogue values (Kharchenko *et al.*, 2013) because stars with high probability of membership were used in our method. The found metallicity (Z = +0.019) is the same as in the catalogue (Kharchenko *et al.*, 2013). The cluster should be classified as I3m instead I3p in Trumpler's system because of the amount of true star members of the cluster.

References

Bressan, A., *et al.* 2012, *MNRAS*, 427, 127
Castelli, F. & Kurucz, R. L. 2003, *IAUS* 210
Hog, E., *et al.* 2000, *A&A*, 355, L27
Kharchenko, N. V., *et al.* 2013, *A&A*, 558, A53
Landolt, A. 1992, *AJ*, 104, 340
Smart, R., Nicastro, L., VizieR, I/324, 11/2013
Roeser, S., *et al.* 2010, *AJ*, 139, 2440
Zacharias, N., *et al.* 2004, *BAAS*, 36, 1418
Zacharias, N., *et al.*, VizieR, I/322A, 07/2012
Zacharias, N., *et al.* 2015, *AJ*, 150, 13
E1: http://nova.astro.umd.edu/Synspec49/synspec.html

Astronomy and Astrophysics in the Gaia sky
Proceedings IAU Symposium No. 330, 2017
A. Recio-Blanco, P. de Laverny, A.G.A. Brown & T. Prusti, eds.

doi:10.1017/S1743921317007141

Unveiling the stellar halo with TGAS

Jovan Veljanoski, L. Posti, A. Helmi and M. A. Breddels

Kapteyn Astronomical Institute, University of Groningen
Landleven 12, 9747 AD Groningen, The Netherlands
email: jovan@astro.rug.nl

Abstract. The detailed study of the Galactic stellar halo may hold the key to unlocking the assembly history of the Milky Way. Here, we present a machine learning model for selecting metal poor stars from the TGAS catalogue using 5 dimensional phase-space information, coupled with optical and near-IR photometry. We characterise the degree of substructure in our halo sample in the Solar neighbourhood by measuring the velocity correlation function.

Keywords. Galaxy: halo, Galaxy: structure, Galaxy: kinematics and dynamics

1. Introduction

Selecting a reliable halo sample of stars is a challenging task. Helmi *et al.* (2017) used mostly a spectroscopic metallicity criterion to select halo stars from the cross-match between the TGAS and RAVE catalogues (Gaia Collaboration *et al.* 2016; Kunder *et al.* 2017). However, this limited that study to a sample covering only one hemisphere.

Here, we present the first attempt at creating a method based on a Support Vector Classifier (SVC) to select metal poor halo star candidates relying on the synergy between TGAS and ALLWISE, both all-sky surveys (Wright *et al.* 2010). This powerful technique can deal with large amounts of data, is not memory intensive, and performs classifications quickly. Most importantly, the SVC can handle the extreme imbalance driven by the fact that halo stars are expected to be $< 1\%$ of the total sample.

2. Training the SVC

To train the SVC, we label stars as halo by using the full phase-space and metallicity information available via TGAS and RAVE. We tag stars as halo if they have [m/H] $\leqslant -1.5$ dex, $\Delta\pi/\pi \leqslant 0.3$, distance > 100 pc, $\Delta v_{\rm los} \leqslant 10$ km/s and for which the astrophysical parameters have been reliably determined (S/N $\geqslant 20$, AlgoConv $\neq$ -1, and CorrCoeff $\geqslant 10$ in RAVE). The tagged stars also need to have a probability of $\geqslant 90\%$ of being halo according to a dynamical model fitted to the RAVE data (Piffl *et al.* 2014).

To train the SVC we use 10 features: 5D phase-space information (x, y, z, v_l, v_b) from TGAS, the values of the distribution functions for the thin disk and halo, marginalized over the $v_{\rm los}$ for each star from the dynamical model by Piffl *et al.* (2014), and the Gaia G together with the ALLWISE WISE1 and WISE2 magnitudes, whose colours are proxy for metallicity. Here we focus on the RGB stars (G-WISE1 $\geqslant 1.7$ and absolute G $\leqslant 4$).

After careful tuning, our model recovers at least 75% of the tagged halo stars, with a maximum contamination level of $\sim 65\%$. The two panels in Fig. 1 show the performance of the model on TGAS $\times$ RAVE data unused during the fitting process.

When appling our model to all RGB stars in the TGAS $\times$ ALLWISE data with $\Delta\pi/\pi \leqslant 0.3$, we classify 1036 stars as halo candidates. The left and middle panels in Fig. 2 show colour magnitude diagrams on which the halo star candidates are marked with open circles.

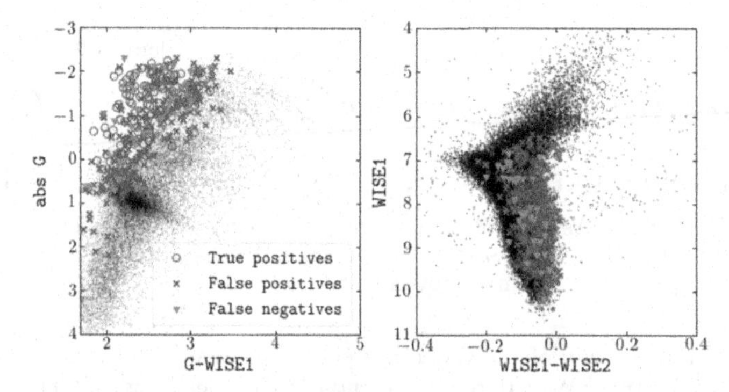

Figure 1. Evaluating the SVC on TGAS × RAVE data, unused during the model fitting and optimizing process. The open circles are the correctly predicted halo stars, the crosses are the false positives, and the solid triangles are the false negatives.

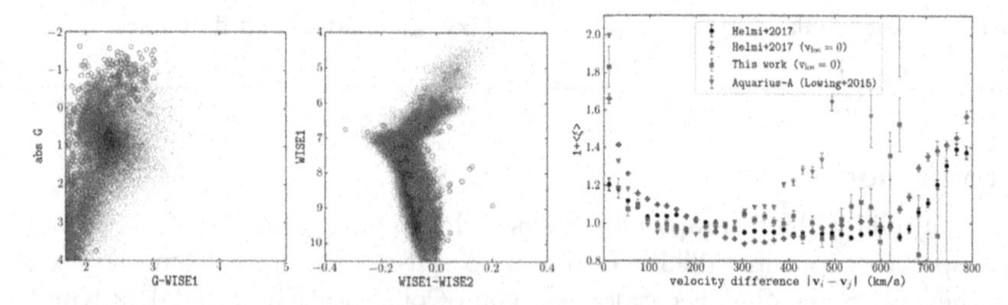

Figure 2. Left and middle panels: halo star candidates in TGAS (open circles). Right: velocity correlation function for the halo sample by Helmi *et al.* (2017), simulations, and this work which uses 5 of the 6 phase-space coordinates to compute the velocities. All samples are in qualitative agreement with each other, and exhibit significant clustering on small scales.

3. Velocity correlation function

To estimate the degree of substructure in our halo sample, we calculate the velocity correlation function using 5D phase-space information. On the third panel in Fig. 2 we compare it to the 6D phase-space sample from Helmi *et al.* (2017), together with the case obtained by assuming $v_{\rm los} = 0$. We also show the correlation function for an "equivalent" sample of stars from one of the Aquarius haloes (Lowing *et al.* 2015). We find significant clustering on small scales, in qualitative agreement with the metallicity selected halo sample from Helmi *et al.* (2017) and the simulations (Lowing *et al.* 2015).

Acknowledgement

This work has made use of data from the European Space Agency (ESA) mission *Gaia* (`https://www.cosmos.esa.int/gaia`), processed by the *Gaia* Data Processing and Analysis Consortium (DPAC, `https://www.cosmos.esa.int/web/gaia/dpac/consortium`). Funding for the DPAC has been provided by national institutions, in particular the institutions participating in the *Gaia* Multilateral Agreement.

References

Gaia Collaboration, Brown, A. G. A., Vallenari, A., *et al.* 2016, *A&A*, 595, A2
Helmi, A., Veljanoski, J., Breddels, M. A., Tian, H., & Sales, L. V. 2017, *A&A*, 598, A58
Kunder, A., Kordopatis, G., Steinmetz, M., *et al.* 2017, *AJ*, 153, 75
Lowing, B., Wang, W., Cooper, A., *et al.* 2015, *MNRAS*, 446, 2274
Piffl, T., Binney, J., McMillan, P. J., *et al.* 2014, *MNRAS*, 445, 3133
Wright, E. L., Eisenhardt, P. R. M., Mainzer, A. K., *et al.* 2010, *AJ*, 140, 1868

Astronomy and Astrophysics in the Gaia sky
Proceedings IAU Symposium No. 330, 2017
A. Recio-Blanco, P. de Laverny, A.G.A. Brown
& T. Prusti, eds.

doi:10.1017/S1743921317005956

Herbig Ae/Be stars with TGAS parallaxes in the HR diagram

Miguel Vioque[1,2*], René D. Oudmaijer[1] and Deborah Baines[2]

[1]School of Physics and Astronomy, University of Leeds,
Leeds, United Kingdom

[2]Isdefe, European Space Astronomy Centre (ESAC),
Madrid, Spain

*email: pymvdl@leeds.ac.uk

Abstract. The intermediate mass Herbig Ae/Be stars are young stars approaching the Main Sequence and are key to understanding the differences in formation mechanisms between magnetic low mass stars and non-magnetic high mass stars. A large fraction of known Herbig Ae/Be stars have TGAS parallaxes, which were used to derive luminosities and place 107 of these objects in the HR diagram, increasing the number of objects using directly determined parallaxes by a factor of 5. We also studied the characteristics of the infrared excesses of this set of Herbig Ae/Be stars and we linked our results to an evolutionary analysis.

Keywords. stars: formation, stars: fundamental parameters, Hertzsprung-Russell diagram, stars: pre–main-sequence, infrared: stars.

---

1. Methodology, Infrared Excesses and Evolutionary Analysis

We selected 254 Herbig Ae/Be star candidates (see Chen *et al.* 2016) and cross-matched them with TGAS, reducing the set to 107 sources. An atmosphere model from Castelli & Kurucz (2004) of the appropriate Teff, log(g) and metallicity was scaled to the dereddened Johnson V band point for each star. A total flux was obtained by integrating below the atmosphere model and by means of the parallax it was converted to luminosity (in a similar way to what was done by van den Ancker *et al.* 1998). Finally, as a control sample, a similar procedure was done for 73240 TGAS sources (selected from McDonald *et al.* 2012) whose parallaxes resulted in better than 3σ detections (see Fig. 1). We can study the infrared properties of this set of Herbig Ae/Be stars by grouping similar sources in colour-colour diagrams, colour-excess or even excess-excess diagrams and then observing how they are placed in the HR diagram. For example, in Fig. 1 we can appreciate how the majority of A type stars are mostly only present at high J-K_s and W1-W4 whilst many B type stars have little excess. This may indicate that they are more evolved and have already cleared most of their dust. Some may be misclassified Be stars. Another approach for studying the evolution of Herbig Ae/Be stars is through the SEDs. Fixing a mass value and picking several stars on the corresponding Pre-Main Sequence track (see Fig. 1) provides an evolutionary movie of a Herbig Ae/Be star of that mass.

2. Conclusions and Forthcoming Research

This work constitutes the largest to date homogeneous analysis of Herbig Ae/Be stars using directly determined distances. It is also an example of how useful the HR diagram can be for studying general properties of these stars. This work serves as an illustration for our Gaia based project to search, identify and analyse new Herbig Ae/Be stars.

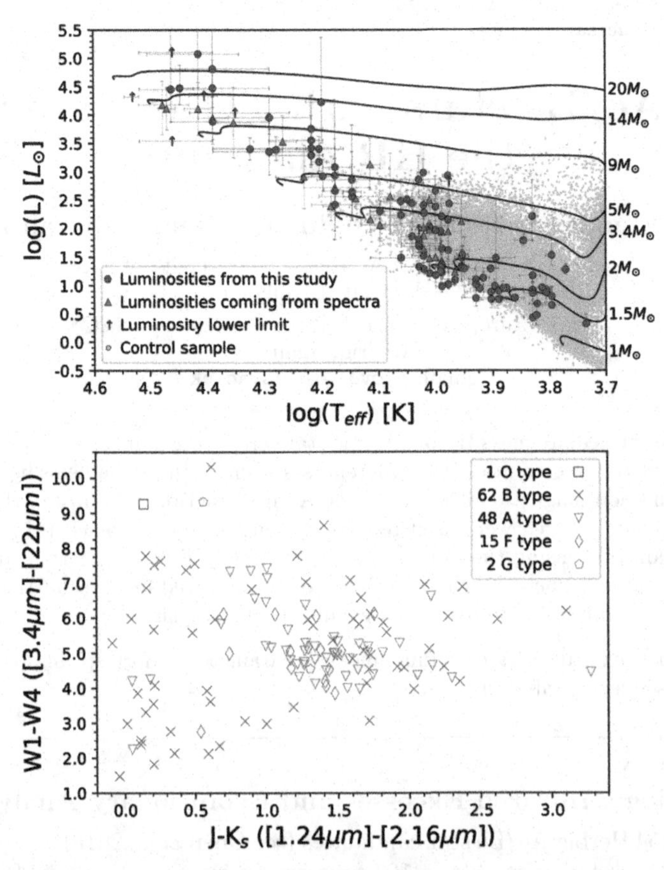

Figure 1. *Top:* 107 Herbig Ae/Be stars in the HR diagram. 33 additional Herbig Ae/Be stars whose luminosities are known from spectra are shown (from Fairlamb *et al.* 2015 and Montesinos *et al.* 2009). Vertical error bars are dominated by parallax uncertainties. The Pre-Main Sequence tracks are from Bressan *et al.* (2012). We can study SED evolution by picking sources on same Pre-Main Sequence tracks. *Bottom:* Colour-colour diagram of infrared excess. Note that most A stars have a very cool infrared excess.

The STARRY project has received funding from the European Union's Horizon 2020 research and innovation programme under MSCA ITN-EID grant agreement No 676036. This work used data from the European Space Agency (ESA) mission *Gaia* (`https://www.cosmos.esa.int/gaia`), processed by the *Gaia* Data Processing and Analysis Consortium (DPAC, `https://www.cosmos.esa.int/web/gaia/dpac/consortium`). Funding for the DPAC has been provided by national institutions, in particular the institutions participating in the *Gaia* Multilateral Agreement.

References

van den Ancker, M. E., de Winter, D., & Tjin A Djie, H. R. E. 1998, *A&A*, 330, 145
Bressan, A., Marigo, P., Girardi, L., Salasnich, B., Dal Cero, C., *et al.* 2012, *MNRAS*, 427, 127
Castelli, F. & Kurucz, R. L. 2004, *ArXiv Astrophysics e-prints*, arXiv:astro-ph/0405087
Chen, P. S., Shan, H. G., & Zhang, P. 2016, *New Astron.*, 44, 1
Fairlamb, J. R., Oudmaijer, R. D., Mendigutía, I., *et al.* 2015, *MNRAS*, 453, 976
McDonald, I., Zijlstra, A. A., & Boyer, M. L. 2012, *MNRAS*, 427, 343
Montesinos, B., Eiroa, C., Mora, A., & Merín, B. 2009, *A&A*, 495, 901

Astronomy and Astrophysics in the Gaia sky
Proceedings IAU Symposium No. 330, 2017
A. Recio-Blanco, P. de Laverny, A.G.A. Brown & T. Prusti, eds.

doi:10.1017/S1743921317006172

The age-metallicity relation with RAVE and TGAS

Jennifer Wojno[1], Georges Kordopatis[2], Matthias Steinmetz[1], Paul J. McMillan[3] and the RAVE collaboration

[1]Leibniz Institut für Astrophysik Potsdam, Potsdam, Germany
[2]Laboratoire Lagrange, Université Côte d'Azur, Observatoire de la Côte d'Azur, Nice, France
[3]Lund Observatory, Lund University, Lund, Sweden

Abstract. Using RAVE data release 5 (DR5), we explore the age and chemistry of a sample of ~25,000 FGK turnoff stars in the extended solar neighbourhood (7 < R < 9 kpc), by separating our sample into two chemical disc components, and investigating the nature of the age-metallicity relation for both. Overall, we find a flat trend in [Fe/H] as a function of age for our α-low disc, and a correlation between age and metallicity for the oldest α-high stars, confirming age-metallicity trends found in more local, high-resolution studies now for a larger volume. We also find a positive gradient in [Mg/Fe] as a function of age for our oldest stars. These results have implications for models which include dynamical evolutionary processes such as radial migration.

Keywords. stars: ages, stars: abundances, Galaxy: formation

1. Introduction

Determining the age of field stars has long remained a significant hurdle in relating stellar evolution with internal Galactic evolutionary processes to reconstruct the chemo-dynamical history of our Galaxy. In particular, understanding the local age-metallicity relation is a crucial ingredient in accurately modeling stellar populations in the solar neighbourhood. With improvements to isochrone fitting techniques, and parallaxes available via the Tycho-Gaia Astrometric Solution (TGAS, Lindegren *et al.* 2016), ages of field stars can be reasonably estimated for a large sample of solar neighbourhood stars. Ages are determined using an updated Bayesian method described in McMillan *et al.* (2017, this volume), taking TGAS parallaxes as a prior, together with T_{eff}, log g, [M/H] from RAVE (Steinmetz *et al.* 2006), and an underlying Galactic model.

2. Sample selection and chemical separation of disc components

For this study, we select a local (distance < 1 kpc), high-quality (SNR > 60) sample of turnoff stars from RAVE DR5 (Kunder *et al.* 2017). Our selection criteria in T_{eff}-log g space is shown in Fig. 1 by the dashed red lines. Our final sample consists of 25,017 stars. To our sample of turn-off stars, we apply the probabilistic chemical separation method described in Wojno *et al.* (2016). This method uses a model metallicity distribution function and [α/Fe]-distribution function for both the α-low (thin disc) and α-high (thick disc) to determine membership likelihood for each component.

3. Age-metallicity and age-α relations

Age-metallicity relation. We find different age-metallicity relations (AMRs) for our two chemical disc components. Our thin disc is consistent with a flat trend, i.e., no correlation between age and metallicity (Fig. 1). In contrast, we find a correlation between age and

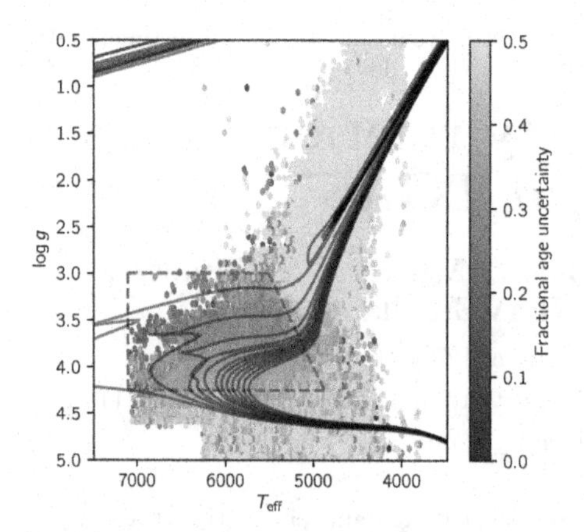

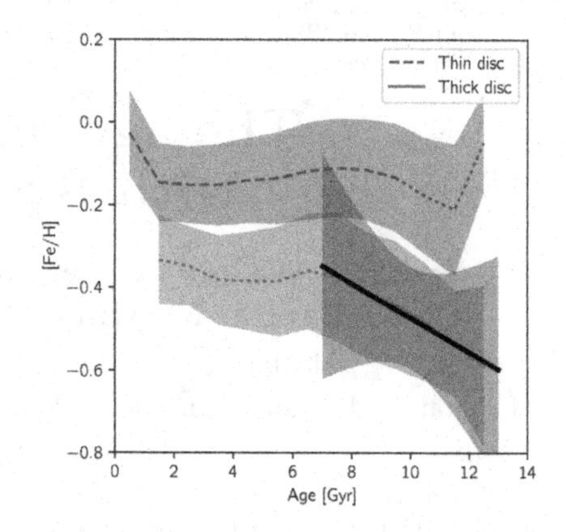

Figure 1: Left: $T_{\rm eff}$-$\log g$ diagram showing our parameter space selection (red dashed lines). Bins are colour-coded by their average fractional age uncertainties. Solar metallicity isochrones from 0 to 13 Gyr, in 1 Gyr steps, are plotted in black.
Right: The AMR for our selected thin (dashed blue) and thick (solid red) disc stars. Dotted lines indicate regions were we assume strong contamination between disc components. The black solid line shows the fit used to estimate the AMR for the oldest thick disc stars.

metallicity for our thick disc. When we consider the oldest stars ($\tau > 8$ Gyr) in the thick disc, we measure a gradient of -0.05 ± 0.08 dex $\rm Gyr^{-1}$. This falls between trends found by high-resolution studies (~ -0.2 dex $\rm Gyr^{-1}$, e.g. Haywood *et al.* 2013, Bensby *et al.* 2014), and the recent study by Fuhrmann *et al.* 2017 (~ -0.017 dex $\rm Gyr^{-1}$).

Age-[α/Fe] relation. For the age-[α/Fe] (here, [Mg/Fe]) relation, we find a flat trend for young stars ($\tau < 8$ Gyr), and a slight positive trend (0.02 ± 0.04 dex $\rm Gyr^{-1}$) for the oldest stars. This trend is more shallow than those found by Bensby *et al.* (2014) and Haywood *et al.* (2015) (~ 0.06 and ~ 0.05 dex $\rm Gyr^{-1}$, respectively), indicating a weaker correlation between age and [α/Fe] for our chemical thick disc, possibly due to contamination as a result of large age uncertainties. However, we do find a knee at ~ 8 Gyr as in the thick disc AMR, which clearly illustrates that the chemical thin and thick discs have experienced distinctly different enrichment histories (e.g. Haywood *et al.* 2013).

References

Bensby, T., Feltzing, S., & Oey, M. S. 2014, *A&A*, 562, A71
Binney, J., Burnett, B., Kordopatis, G., *et al.* 2014, *MNRAS*, 437, 351
Fuhrmann, K., Chini, R., Kaderhandt, L., & Chen, Z. 2017, *MNRAS*, 464, 2610
Haywood, M., Di Matteo, P., Lehnert, M. D., Katz, D., & Gómez, A. 2013, *A&A*, 560, A109
Haywood, M., Di Matteo, P., Snaith, O., & Lehnert, M. D. 2015, *A&A*, 579, A5
Kunder, A., Kordopatis, G., Steinmetz, M., *et al.* 2017, *AJ*, 153, 75
Lindegren, L., Lammers, U., Bastian, U., *et al.* 2016, *A&A*, 595, A4
Steinmetz, M., Zwitter, T., Siebert, A., *et al.* 2006, *AJ*, 132, 1645
Wojno, J., Kordopatis, G., Steinmetz, M., *et al.* 2016, *MNRAS*, 461, 4246

This work has made use of data from the European Space Agency (ESA) mission *Gaia* (`https://www.cosmos.esa.int/gaia`), processed by the *Gaia* Data Processing and Analysis Consortium (DPAC, `https://www.cosmos.esa.int/web/gaia/dpac/consortium`). Funding for the DPAC has been provided by national institutions, in particular the institutions participating in the *Gaia* Multilateral Agreement.

Astronomy and Astrophysics in the Gaia sky
Proceedings IAU Symposium No. 330, 2017
A. Recio-Blanco, P. de Laverny, A.G.A. Brown & T. Prusti, eds.

doi:10.1017/S1743921317006366

Reanalysis of 24 Nearby Open Clusters using Gaia data

Steffi X. Yen[1], Sabine Reffert[1], Siegfried Röser[1,2], Elena Schilbach[1,2], Nina V. Kharchenko[2,3] and Anatoly E. Piskunov[2,4]

[1]Zentrum für Astronomie der Universität Heidelberg, Landessternwarte, Königstuhl 12, 69117 Heidelberg, Germany
email: syen@lsw.uni-heidelberg.de

[2]Zentrum für Astronomie der Universität Heidelberg, Astronomisches Rechen-Institut, Mönchhofstraße 12-14, 69120 Heidelberg, Germany

[3]Main Astronomical Observatory, 27 Academica Zabolotnogo Str., 03680 Kiev, Ukraine

[4]Institute of Astronomy of the Russian Acad. Sci., 48 Pyatnitskaya Str., 109017 Moscow, Russia

Abstract. We have developed a fully automated cluster characterization pipeline, which simultaneously determines cluster membership and fits the fundamental cluster parameters: distance, reddening, and age. We present results for 24 established clusters and compare them to literature values. Given the large amount of stellar data for clusters available from Gaia DR2 in 2018, this pipeline will be beneficial to analyzing the parameters of open clusters in our Galaxy.

Keywords. open clusters and associations: general

Open clusters are gravitationally bound groups of young stars located in the disk of the Milky Way, making them excellent objects for studying stellar evolution and the Galactic disk. The fundamental cluster parameters can be determined from fitting isochrones to the cluster color-magnitude diagram (CMD). In anticipation of the large amount of stellar data from Gaia, we have developed an automated pipeline to simultaneously determine cluster membership and fit mean cluster parameters: distance d, reddening $E(B-V)$, and age $\log t$. We use a χ^2 minimization to fit isochrones to the photometric observations of candidate cluster members. Our data set is comprised of ASCC-2.5 BV-, 2MASS JHK_s-, and Gaia DR1 G-band photometry, as well as TGAS proper motions and parallaxes.

The pipeline first determines cluster membership through proper motion and parallax selections. The proper motion selection begins by computing the weighted mean cluster proper motion ($\bar{\mu}_{\rm cluster}$) using membership probabilities from the Milky Way Star Clusters (MWSC) catalog (Kharchenko *et al.* 2013). Guided by the astrometric accuracy of TGAS, all stars within 2 mas/yr of $\bar{\mu}_{\rm cluster}$ are selected as cluster members, as well as stars within 5 mas/yr, if $\bar{\mu}_{\rm cluster}$ lies within the 3σ proper motion error ellipse of the star. This selection is iterated with a new $\bar{\mu}_{\rm cluster}$ until membership is unchanged. Next, the weighted mean cluster parallax ($\bar{\pi}_{\rm cluster}$) is computed for the proper motion-selected members. Cluster members are defined as those stars for which $\bar{\pi}_{\rm cluster}$ lies within the 3σ parallax error of the star. Similarly, the parallax selection is iterated until membership is unchanged.

Next is a photometric selection, where the zero-age main sequence (ZAMS) is fitted to the photometric observations of the astrometrically-selected cluster stars to obtain initial cluster $E(B-V)$ and d. The evolved members of the cluster, if any, are identified and down weighted. With highly probable cluster members, all isochrones in our set (Bressan *et al.* (2012) with $Z = Z_{\odot}$) are then fitted to the cluster photometry , where stellar binaries are accounted for with an 0.1 mag offset in all passbands. The isochrone

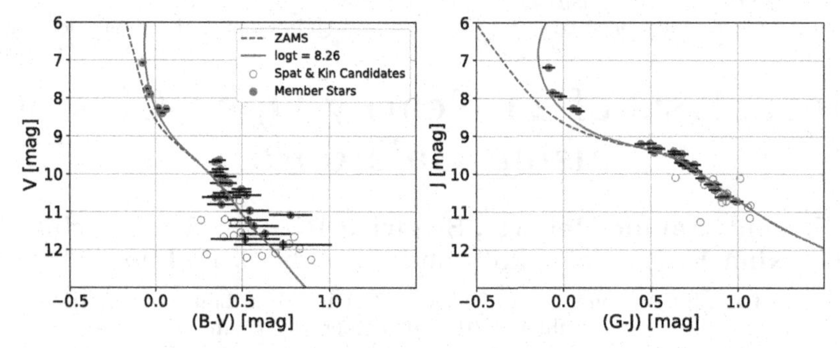

Figure 1. CMDs for Blanco 1 with fitted isochrone ($\log t = 8.26$, $E(B-V) = 0.044 \pm 0.012$ mag, $d = 258 \pm 3$ pc).

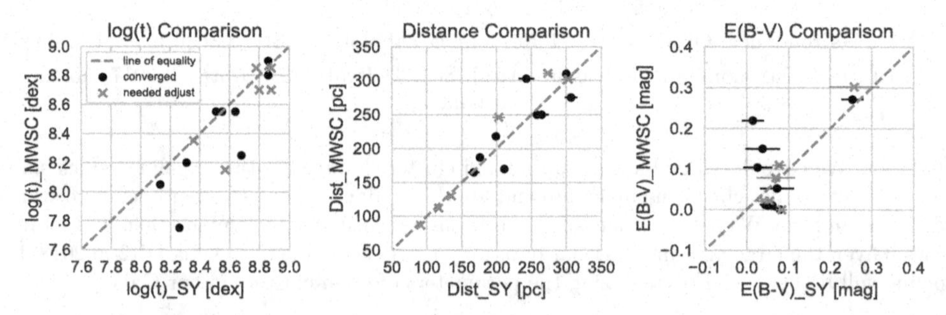

Figure 2. Comparison of cluster parameter results from this work (SY) to the MWSC.

yielding the minimum reduced χ^2 is selected and the star with the largest χ^2 contribution is removed. This process is repeated, starting with the photometric selection, until a reduced $\chi^2 < 5$ is achieved, yielding the cluster's final age, $E(B-V)$, and d. In Fig. 1, we illustrate the result of our fitting procedure with two CMDs for the cluster Blanco 1.

We have analyzed 24 nearby clusters with our automated pipeline. Parameter estimates were returned for 15 clusters, of which 9 converged unaided and 6 converged after individual adjustments. The remaining 9 clusters contained too few members to do a proper fit. In Fig. 2, we compare our fitted parameters for 15 clusters to those given in the MWSC. Six of our 15 clusters were also explored by the Gaia Collaboration, van Leeuwen *et al.* (2017). A comparison of the parallaxes of these six clusters gave a mean difference of 0.26 ± 0.15 mas. Our next steps include refining the selection criteria in the pipeline and adding HSOY (Altmann *et al.* 2017) proper motions and photometry.

Acknowledgements

This work has made use of data from the European Space Agency (ESA) mission *Gaia* (`https://www.cosmos.esa.int/gaia`), processed by the *Gaia* Data Processing and Analysis Consortium (DPAC, `https://www.cosmos.esa.int/web/gaia/dpac/consortium`). Funding for the DPAC has been provided by national institutions, in particular the institutions participating in the *Gaia* Multilateral Agreement.

References

Altmann, M., Röser, S., Demleitner, M., Bastian, U., & Schilbach, E. 2017, *A&A*, 600, 4A
Bressan, A., Marigo, P., Girardi, L., *et al.* 2012, *MNRAS*, 427, 127
Gaia Collaboration, van Leeuwen, F., Vallenari, A., Jordi, C., *et al.* 2017, *A&A*, 601, A19
Kharchenko, N. V., Piskunov, A. E., Schilbach, E., *et al.* 2013, *A&A*, 558, A53

Astronomy and Astrophysics in the Gaia sky
Proceedings IAU Symposium No. 330, 2017
A. Recio-Blanco, P. de Laverny, A.G.A. Brown & T. Prusti, eds.

doi:10.1017/S1743921317006317

Spectroscopic and Photometric Survey of Northern Sky for the ESA PLATO space mission

Renata Ženovienė, Vilius Bagdonas, Arnas Drazdauskas, Rimvydas Janulis, Lukas Klebonas, Šarūnas Mikolaitis, Erika Pakštienė, and Gražina Tautvaišienė

Institute of Theoretical Physics and Astronomy, Vilnius University,
Saulėtekio al. 3, LT-10257, Vilnius, Lithuania
email: Sarunas.Mikolaitis@tfai.vu.lt

Abstract. The ESA-PLATO 2.0 mission will perform an in-depth analysis of the large part of the sky-sphere searching for extraterrestrial telluric-like planets. At the Molėtai Astronomical Observatory of Vilnius University, we started a spectroscopic and photometric survey of the northern sky fields that potentially will be targeted by the PLATO mission. We aim to contribute in developing the PLATO input catalogue by delivering a long-duration stellar variability information and a full spectroscopic characterization of brightest targets. First results of this survey are overviewed.

Keywords. Stars:abundances, stars:oscillations

1. Objectives

The main objective of the scientific preparation of the PLATO 2.0 mission (Ricker *et al.* 2015) is to create the most promising input catalogue of targets. The main source for the input catalogue will be the ESA-Gaia mission results. However, a supplementary material will be collected from other catalogues and on ground instruments. Majority of the large spectroscopic surveys, e. g. Gaia-ESO (Gilmore *et al.* 2012), RAVE (Steinmetz *et al.* 2006), are performed on large telescopes situated in the southern hemisphere and do not contain brightest objects in their star-lists. We estimated that only up to 30 % of necessary information from high-resolution spectroscopy is available for the brightest stars. Another issue is a lack of information on stellar variability that can perturb signals of photometric observations and that may cause the false-positive detections of planets in the PLATO mission.

Thus, our objective is to contribute in preparing a dataset of brightest targets of the most-northern regions of the sky-sphere for the northern PLATO 2.0 fields employing the spectroscopic and photometric instruments of the Molėtai Astronomical Observatory of ITPA VU, taking advantage of their northern geographical location.

2. Observations

We conduct spectroscopic and photometric observations using instruments of the Molėtai Astronomical Observatory of ITPA VU: (1) the high-resolution VUES spectrograph that covers a range of 400–880 nm with the resolving power R=60 000 (Jurgenson *et al.* 2016, *http://mao.tfai.vu.lt*) that is installed on the 1.65 m Cassegrain-type telescope and (2) the CCD photometer installed on the 35/51 cm wide field Maksutov-type telescope.

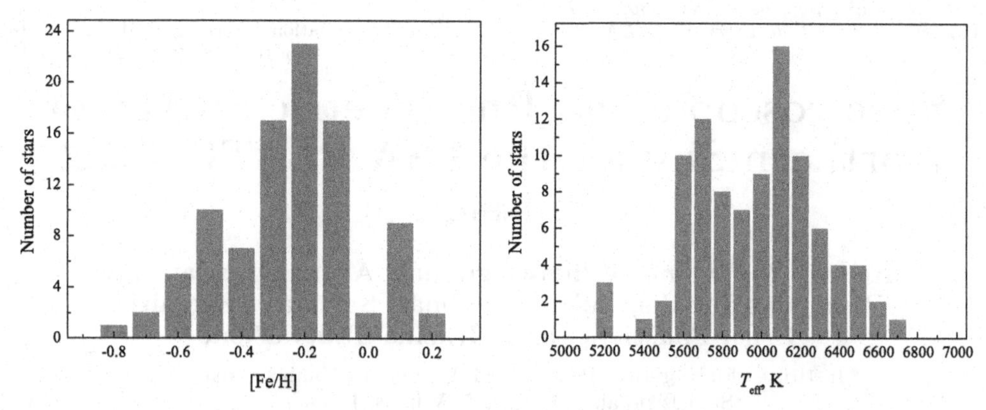

Figure 1. Metallicity (left) and effective temperature (right) distributions of sample stars.

Targets for the spectroscopic observations in the PLATO STEP02 field were selected from the Geneva-Copenhagen survey (Nordström *et al.* 2004) taking FGK-type dwarfs with V<8 mag. The photometric variability observations are going for stars with V<11 mag.

3. First results

We have already observed more than 200 stars spectroscopically and determined the main atmospheric parameters and chemical composition of iron group elements for about 130 stars with $v\sin i$<20 km s^{-1}. The study is done in accordance with the Gaia-ESO survey standards and techniques used by the Vilnius node (Smiljanic *et al.* 2014). In Fig. 1 we show the metallicity [Fe/H] and effective temperature $T_{\rm eff}$ distributions of sample stars.

The accomplished photometric observations already allowed us to discovered 20 previously unknown long-duration variables in several northern PLATO fields.

Using the information from spectroscopy and photometry, we will perform an in-depth analysis of the targets-of-interest. Results of this survey will be published in the upcoming papers by Mikolaitis *et al.* and Pakštienė *et al.* (in prep.).

Acknowledgements

This work was partly supported by the grant from the Research Council of Lithuania (LAT-16019).

References

Gilmore, G., Randich S., Asplund, M. *et al.* 2012, *The Messenger*, 147, 25

Jurgenson, C., Fischer, D., McCracken, T., Sawyer, D., Giguere, M., Szymkowiak, A., Santoro, F., & Muller, G. 2016, *Journal of Astronomical Instrumentation*, 5, id. 1650003-239

Nordström, B., Mayor, M., Andersen, J., Holmberg, J., Pont, F., Jorgensen, B. R., Olsen, E. H., Udry, S., & Mowlavi, N. 2004, *A&A*, 418, 989

Ricker, G. R., Winn, J. N., Vanderspek, R. *et al.* 2015, *Journal of Astronomical Telescopes, Instruments, and Systems*, 1, id. 014003

Smiljanic, R., Korn, A. J., Bergemann, M. *et al.* 2014, *A&A*, 570, A122

Steinmetz, M., Zwitter, T., Siebert, A. *et al.* 2006, *ApJ*, 132, 4

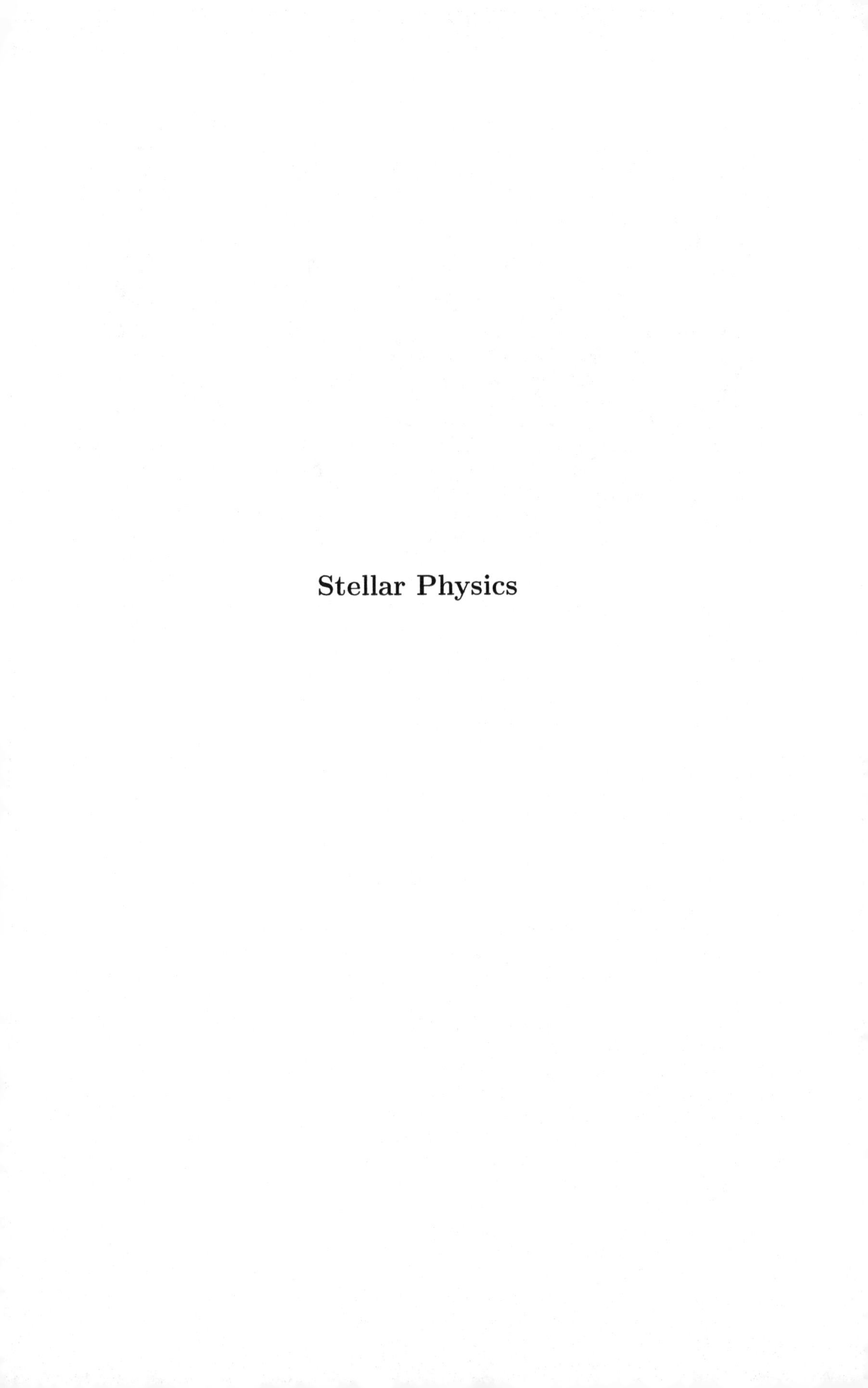

Stellar Physics

Some conference participants during a coffee break in the Palais de la Méditerrannée.

Astronomy and Astrophysics in the Gaia sky
Proceedings IAU Symposium No. 330, 2017
A. Recio-Blanco, P. de Laverny, A.G.A. Brown & T. Prusti, eds.

doi:10.1017/S1743921317005506

Variable stars in the Gaia era: Mira, RR Lyrae, δ and Type-II Cepheids

Martin A. T. Groenewegen

Koninklijke Sterrenwacht van België
Ringlaan 3, B-1180 Brussel, Belgium
email: martin.groenewegen@oma.be

Abstract. Classical variables like RR Lyrae, classical and Type-II Cepheids and Mira variables all follow period-luminosity relations that make them interesting as distance indicators. Especially the RR Lyrae and δ Cepheids are crucial in establishing the distance scale in the Universe, and all classes of variables can be used as tracers of galactic structure. I will present an overview of recent period-luminosity relations and review the work that has been done using the *Gaia* DR1 data so far, and discuss possibilities for the future.

Keywords. Cepheids, stars: variables: other, stars: distances, distance scale, stars: AGB and post-AGB, Magellanic Clouds

---

1. Introduction

Miras, RR Lyrae, classical and Type-II Cepheids belong to the oldest known variable stars, certainly in the literature of the west. The American Association of Variable Star Observers (AAVSO) website has interesting historical information about the prototypes *o* Ceti or Mira (discovered by Fabricius in 1596, see `www.aavso.org/vsots_mira`), RR Lyrae (discovered by Wilhelmina Fleming, and published in Pickering *et al.* 1901, see `www.aavso.org/vsots_rrlyr`), and δ Cephei (discovered by John Goodricke in 1784, see `www.aavso.org/vsots_delcep`).

Type-II Cepheids are subdivided in three classes, typically based on pulsation period: The BL Herculis variables (BLH; periods 1-4 days; discovery paper by Cuno Hoffmeister 1929), the W Virginis stars (WVir; periods 4-20 days; discovered by Eduard Schönfeld in 1866, see `www.aavso.org/vsots_wvir`), and the RV Tauri stars (RVT; periods 20-$\sim$70 days; discovered by Lidiya Tseraskaya (or Ceraski), published in Ceraski 1905, see `www.aavso.org/vsots_rvtau`).

In evolutionary terms, RR Lyrae (RRL) variables are evolved, metal poor, core He-burning stars at or slightly brighter than the zero-age horizontal branch (ZAHB). Marconi *et al.* (2015) provide recent nonlinear, time-dependent convective hydrodynamical models of RRL over a broad range in metal abundances (Z = 0.0001-0.02) and masses, ranging from 0.8 $M_\odot$ (for Z= 0.0001) to 0.54 $M_\odot$ (for Z= 0.02). They provide analytical relations for the edges of the instability strip (IS) as a function of Z. Period-radius-metallicity relations for fundamental and first-overtone pulsators are determined, as well as a large set of period-luminosity and period-Wesenheit relations.

Classical or δ Cepheids (CEPs) are evolved objects with initial masses in the range $\sim$ 2 to $\sim$ 15 $M_\odot$. Theoretical pulsation models have been calculated by Bono *et al.* (2000) and Anderson *et al.* (2014, 2016) who considered the effect of rotation on the evolution and pulsation. A Cepheid can cross the IS up to three times The, so-called, first crossing occurs when the star evolves from the main sequence to the red giant branch during a core contraction phase. This crossing is expected to be fast, and Cepheids in this phase should be rare. The majority of Cepheids are expected to be on the second and third crossings during the so-called "blue loops" experiencing core helium burning.

As mentioned above, the Type-II Cepheids (T2C) are subdivided in three classes, typically based on period, but they are thought to have different evolutionary origins. Evolutionary modelling of T2Cs has been pioneered by Gingold (1976, 1985) establishing the classical picture that T2Cs are low-mass stars, evolving from the blue HB through the IS to the asymptotic giant branch (AGB) for the short-period stars, blue loops off the AGB for the stars of intermediate period, and post-AGB (PAGB) evolution for the longest period, also see Wallerstein (2002) and Bono *et al.* (2016).

The anomalous Cepheids (ACs) are also pulsating stars which overlap in period range with RRL and the BLH stars. They form a separate PL relation clearly different from the RRL, classical Cepheids and T2Cs. They pulsate in the fundamental mode (FU) and first overtone (FO) mode (unlike T2C). Models have been calculated by Fiorentino & Monelli (2012). Their mean mass is around $1.2 \pm 0.2\ M_{\odot}$, and there is also discussion if ACs are the result of binary interaction.

Groenewegen & Jurkovic (2017a, b) recently studied the 335 T2C and ACs discovered by the OGLE-III survey in the Large and Small Magellanic Cloud. From fitting the spectral energy distribution (SED) they derived effective temperature and luminosity. In the 2017a paper the resulting Hertzsprung-Russell diagram was compared in a qualitative way to modern evolutionary tracks. In agreement with the findings cited above the BL Her can be explained by stars in the mass range $\sim 0.5 - 0.6\ M_{\odot}$ and the ACs by stars in the mass range $\sim 1.1 - 2.3\ M_{\odot}$. The origin of the (p)WVir is unclear however: tracks of $\sim 2.5 - 4\ M_{\odot}$ cross the IS at the correct luminosity, as well as (some) lower mass stars on the AGB that undergo a thermal pulse when the envelope mass is small, but the timescales make these unlikely scenarios to explain this class of objects as a whole. The peculiar W Vir have been suggested to be binaries, and in general, some of the phenomenon observed in T2C and ACs may well be linked to so called binary evolutionary pulsators (BEP; Karczmarek *et al.* 2016).

In the 2017b paper, pulsation models for RRL (Marconi *et al.* 2015) and Cepheids (Bono *et al.* 2000) were used to estimate the pulsation mass for all objects. Both estimates agreed best for the BLH ($M \sim 0.49\ M_{\odot}$) and the ACs ($M \sim 1.3\ M_{\odot}$). The masses of the W Vir appeared similar to the BL Her. The situation for the pWVir and RVT stars was less clear. For many RV Tau the masses are in conflict with the standard picture of (single-star) post-AGB evolution, the masses being either too large ($\gtrsim 1\ M_{\odot}$) or too small ($\lesssim 0.4\ M_{\odot}$).

Groenewegen & Jurkovic (2017a) found that $\sim$ 60% of the RVT showed an infrared excess in their SEDs, not unsurprising if RVT have indeed evolved of the AGB. Surprisingly however, $\sim$ 10% of the W Vir (including the pWVir) objects also showed an infrared excess, confirming the result of Kamath *et al.* (2016) that there exist stars with luminosities below that predicted from single-star evolution that show a clear infrared excess, and which they called dusty post-red giant branch stars, and suggested to have evolved off the RGB as a result of binary interaction.

AGB and super-AGB stars are intermediate mass stars (initial mass $\sim$0.8-12 $M_{\odot}$) in the last phase of active nuclear burning, that undergo double-shell burning, experience thermal pulses (or Helium shell flashes) that change the composition of the envelope, making it increasingly rich in carbon, so that S-stars (C/O ratio close to one, and that show increased abundances of s-process elements), and C-stars (Carbon stars with C/O>1) can form. They are cool giants, where dust can form close to the star that is driven outward in a slow stellar wind.

(S)AGB stars also pulsate, classically divided into irregular (Lb), semi-regular (SR) and Mira (M) variables. The SR and M are sometimes also called long-period variables (LPVs), as they are not so different. Miras are not necessarily less regular than the SR, and the definition that Miras should have an pulsation amplitude in the visual band larger than 2.5 magnitudes is arbitrary.

That Miras follow a *PL* relation is well known (Glass & Lloyd Evans 1981), and it should be noted that a 500 day Mira is ~0.5 mag brighter than a 50 day Cepheid in the near and mid-IR. The revolution came with advent of the microlensing surveys, MACHO and OGLE. Wood *et al.* (1999) and Wood (2000) showed that red giants in the LMC follow several sequences, 3 that define pulsating stars, a sequence that consists of binary systems, and one that is formed by the long secondary period (LSP) that occurs in many red giants. Subsequent works expanded on this in various ways (Ita *et al.* 2004, Soszyński *et al.* 2004, 2005, Fraser *et al.* 2008, Riebel *et al.* 2012, Soszyński & Wood 2013, Soszyński *et al.* 2013) and revealed many more (sub-)sequences, including those for RGB stars.

2. Cosmological connection

The small dispersion in the *PL* relation of classical Cepheids makes them the primary calibrator in the distance ladder, and ultimately in determining the Hubble constant (Freedman *et al.* 2001). Riess *et al.* (2016) find $H_0 = 73.24 \pm 1.74$ km/s/Mpc by using locally calibrated Cepheids (15 Cepheids with parallaxes in our MW, 8 detached eclipsing binaries (dEBs) in the LMC, 2 dEBs in M31, and the megamaser in NGC 4258), to determine the brightness of Type-I SNe in 19 galaxies that host Cepheids and SNIa, and then measure the distance to $\sim$ 300 SNIa in the Hubble flow with $z < 0.15$. This value for H_0 differs by more than 3σ from the $H_0 = 66.93 \pm 0.62$ km/s/Mpc determined by the Planck mission (Planck Collaboration *et al.* 2016). Whether this discrepancy is real is of obvious importance and requires that all steps in the stellar distance ladder are investigated and improved, and this includes the Cepheid *PL* relation.

In this line, the Carnegie-Chicago Hubble Program (Beaton *et al.* 2016; this volume) aims at a 3% measurement of H_0 using alternative methods to the traditional Cepheid distance scale. They aim to establish a completely independent route to the Hubble constant using RRL variables, and the tip of the red giant branch (TRGB) method. This requires a reassessment of the RRL *PL* relation.

3. Period-luminosity relations

Mostly recent empirical *PL* and *PLZ* relations for RRL, Type-II, ACs and classical Cepheids in selected filters and Wesenheit relations are compiled in Table 1. If a slope was adopted it is listed between parentheses. Large the table is, it is certainly not complete and the numbers hide important details in their derivation. Period-luminosity relations exist in other infrared filters than K, and in other Wesenheit combinations than V and I, or V and K. The original references should be consulted about solutions for other filter combinations or pulsation modes, the details of the filter(s) used, the details in the definition of the Wesenheit function, any cuts in pulsation period that were applied, or, mostly for the LMC, whether the effect of the orientation of the disc was taken out or not.

Table 1 includes solutions based on *Gaia* (Gaia Collaboration *et al.* 2016b) data release one (GDR1; Gaia Collaboration *et al.* 2016a). Clementini *et al.* (2016) derived *PL* relations in the *Gaia G* band based on data in the south ecliptic pole in the outskirts of the LMC. Gaia collaboration *et al.* (2017) contains several *PL* relations based on known RRL, T2C and CEP in our Galaxy based on the TGAS solution. They present solutions based on three approaches. The first is based on a least-square fit of absolute magnitude versus $\log P$ where the absolute magnitude is calculated from

$$M = m_0 + 5 \log \pi - 10, \tag{3.1}$$

with m_o the dereddened magnitude and the parallax is in milli-as. A simple application of this method requires a selection in parallax space ($\pi > 0$) and is therefore subject to Lutz-Kelker bias (Lutz & Kelker 1973, Oudmaijer *et al.* 1998, Koen 1992), which Gaia collaboration *et al.* (2017) did not correct for. They also present two methods that work in parallax space. In this case Eq. 3.1 is rewritten (for a *PL* relation $\alpha + \beta \log P$, or similarly for a M_V–[Fe/H] relation for RRL) as

$$10^{0.2\alpha} = \pi \cdot 10^{0.2(m_0 - \beta \log P - 10)}. \tag{3.2}$$

The first method is based on a Bayesian approach, and the other, on a weighted non-linear least squares solution of this equation, called the astrometric based luminosity (ABL), and these are the solutions listed in Table 1. They cite Arenou & Luri (1999), although the method was used in a classical paper by Feast & Catchpole (1997) to determine the zeropoint of the Cepheid *PL* relation based on *Hipparcos* data for 220 Cepheids. The method was shown to be free from bias by Koen & Laney (1998).

Table 1 also includes *PL* relations based on individual distances to Galactic and MC Cepheids based on the Baade-Wesselink method (Storm *et al.* 2011a,b, Groenewegen 2013). This method depends on the so-called projection factor, p, that translates the pulsational velocity to the radial velocity in the line-of-sight measured via spectroscopy. Both papers derive a p factor that depends quite strongly on period from the condition that the distance to the LMC should not depend on pulsation period (Storm *et al.* find $p = 1.550 - 0.186 \log P$; Groenewegen 2013 find $p = 1.50 - 0.24 \log P$). However, the most recent studies indicate that the data is consistent with a constant p factor of 1.29 $\pm$ 0.04 (Kervella, this volume; Kervella *et al.* 2017). The reason behind this discrepancy is currently unknown.

4. GCVS $\Longleftrightarrow$ GDR1

As a usefull exercise I cross correlated the latest edition of the General Catalog of Variable Stars (version 5.1; Samus *et al.* 2017, see `www.sai.msu.su/gcvs/gcvs/gcvs5/htm/`) with the GDR1 for several types of variables. The results are summarised in Table 2. The third column lists which identifiers were used in the search, the fourth column how many of those types are listed in the GCVS, the fifth column how many have a parallax listed in the GDR1 TGAS solution, and the last column how many of those have a relative parallax error less than 16%. The week after the conference, on May 1st, Gaia Collaboration *et al.* (2017) appeared that did a very similar search, to ultimately derive the *PL* relations listed in Table 1.

The RRL is the class with the largest number of (accurate) parallax data available. As they are intrinsically faint GDR1 is not hampered by the fact that many bright stars are not listed there. Table 3 lists the 18 RRL with $\sigma_\pi/\pi < 0.16$ plus SU Dra, sorted by relative parallax error. SU Dra is included as it is one of five RRLs which have the parallax determined using the *HST* by Benedict *et al.* (2011). For those five stars their is good agreement between the *HST* based parallaxes and GDR1. The last column gives the parallax listed by van Leeuwen (2007) based on the re-reduction of the *Hipparcos* data.

The next entries in Tables 2 and 3 are for the T2C, of the BLH & WVir, and RVT types. Only 2 have an accurate parallax determined in GDR1. κ Pav, one of two T2C with an HST based parallax in Benedict *et al.* (2011), is missing, probably because it is

Table 1. PLZ relations for RRL and different classes of Cepheids ($\text{mag} = \alpha + \beta \log P + \gamma\ [\text{Fe/H}]$).

Class	Band	α	β	γ	Sample	Reference
RRLab	V	19.385 ± 0.017	-	0.214 ± 0.047	LMC	Gratton *et al.* (2004)
RRLab	M_V	0.93 ± 0.12	-	0.23 ± 0.04	GCC	Chaboyer *et al.* (1999)
RRLab	M_V	0.82 ± 0.04	-	(0.214)	GAL	Gaia Collaboration *et al.* (2017)
RRLab	W(V,I)	17.172 ± 0.003	-2.933 ± 0.009	-	LMC	Jacyszyn-Dobrzeniecka *et al.* (2017)
RRLab	W(V,I)	17.492 ± 0.007	-3.001 ± 0.028	-	SMC	Jacyszyn-Dobrzeniecka *et al.* (2017)
RRLab	K	17.43 ± 0.01	-2.73 ± 0.25	0.03 ± 0.07	LMC	Murareva *et al.* (2015)
RRLab	K	13.28 ± 0.02	-2.33 ± 0.08	-	M5	Coppola *et al.* (2011)
RRLab	K	10.420 ± 0.024	-2.33 ± 0.07	-	M4	Braga *et al.* (2015)
RRLab	K	12.752 ± 0.054	-2.232 ± 0.044	0.141 ± 0.020	ω Cen	Navarrete *et al.* (2017)
RRLab	M_K	-1.16 ± 0.27	(-2.33)	-	GAL	Groenewegen & Salaris (1999)
RRLab	M_K	-1.05 ± 0.13	-2.38 ± 0.04	0.08 ± 0.11	GCC	Sollima *et al.* (2006)
RRLab	M_K	-0.95 ± 0.14	-2.53 ± 0.36	0.07 ± 0.04	GAL	Murareva *et al.* (2015)
RRLab	M_K	-1.17 ± 0.10	(-2.73)	0.07 ± 0.07	GAL	Gaia Collaboration *et al.* (2017)
RRLab	[3.6]	10.229 ± 0.010	-2.332 ± 0.106	-	M4	Neeley *et al.* (2015)
RRLab	[4.5]	10.192 ± 0.010	-2.336 ± 0.105	-	M4	Neeley *et al.* (2015)
RRLab	W1	-1.113 ± 0.013	-2.38 ± 0.20	-	GAL	Klein *et al.* (2014)
RRLab	W2	-1.111 ± 0.013	-2.39 ± 0.20	-	GAL	Klein *et al.* (2014)
T2C	G	18.640 ± 0.085	-1.650 ± 0.109	-	LMC	Clementini *et al.* (2016)
T2C	W(V,I)	17.365 ± 0.015	-2.521 ± 0.022	-	LMC	Matsunaga *et al.* (2009)
T2C	W(V,I)	17.554 ± 0.083	-2.304 ± 0.107	-	SMC	Matsunaga *et al.* (2011)
T2C	K	13.27 ± 0.10	-2.24 ± 0.14	-	GB	Groenewegen *et al.* (2008)
T2C	K	17.412 ± 0.029	-2.278 ± 0.047	-	LMC	Matsunaga *et al.* (2009)
T2C	K	17.600 ± 0.082	-2.113 ± 0.105	-	SMC	Matsunaga *et al.* (2011)
T2C	K	17.47 ± 0.02	-2.385 ± 0.030	-	LMC	Ripepi *et al.* (2015)
T2C	K	17.405 ± 0.038	-2.483 ± 0.089	-	LMC	Bhardwaj *et al.* (2017)
T2C	M_K	-1.58 ± 0.17	(-2.385)	-	GAL	Gaia Collaboration *et al.* (2017)
T2C	W(V,K)	17.33 ± 0.02	-2.49 ± 0.03	-	LMC	Ripepi *et al.* (2015)
T2C	W(V,K)	17.415 ± 0.012	-2.456 ± 0.025	-	LMC	Bhardwaj *et al.* (2017)
AC FU	G	18.00 ± 0.04	-2.95 ± 0.27	-	LMC	Clementini *et al.* (2016)
AC FU	K	16.74 ± 0.02	-3.54 ± 0.15	-	LMC	Ripepi *et al.* (2014)
AC FU	W(V,K)	16.58 ± 0.02	-3.58 ± 0.15	-	LMC	Ripepi *et al.* (2014)
CEP FU	G	17.361 ± 0.020	-2.818 ± 0.032	-	LMC	Clementini *et al.* (2016)
CEP FU	M_V	-1.43 ± 0.10	(-2.81)	-	GAL	Feast & Catchpole (1997)
CEP FU	M_V	-1.275 ± 0.023	-2.678 ± 0.076	-	GAL	Fouqué *et al.*(2007)
CEP FU	M_V	-1.54 ± 0.10	(-2.678)	-	GAL	Gaia Collaboration *et al.* (2017)
CEP FU	W(V,I)	16.375 ± 0.014	-3.314 ± 0.020	-	SMC	Ngeow *et al.* (2015a)
CEP FU	W(V,I)	15.897 ± 0.001	-3.327 ± 0.001	-	LMC	Inno *et al.* (2016)
CEP FU	W(V,I)	16.492 ± 0.002	-3.358 ± 0.005	-	SMC	Jacyszyn-Dobrzeniecka *et al.* (2017)
CEP FU	W(V,I)	15.888 ± 0.004	-3.313 ± 0.006	-	LMC	Jacyszyn-Dobrzeniecka *et al.* (2017)
CEP FU	$M_{W(V,I)}$	-2.60 ± 0.03	-3.32 ± 0.08	(0.0)	MC+G	Storm *et al.* (2011b)
CEP FU	$M_{W(V,I)}$	-2.414 ± 0.022	-3.477 ± 0.074	-	GAL	Fouqué *et al.* (2017)
CEP FU	$M_{W(V,I)}$	-2.82 ± 0.11	(-3.477)	-	GAL	Gaia Collaboration *et al.* (2017)
CEP FU	K	16.494 ± 0.026	-3.212 ± 0.033	-	SMC	Groenewegen (2000)
CEP FU	K	16.514 ± 0.025	-3.213 ± 0.032	-	SMC	Ngeow *et al.* (2015a)
CEP FO	K	15.941 ± 0.032	-3.132 ± 0.083	-	SMC	Bhardwaj *et al.* (2016b)
CEP FU	K	16.051 ± 0.050	-3.281 ± 0.040	-	LMC	Persson *et al.* (2004)
CEP FU	K	16.070 ± 0.017	-3.295 ± 0.018	-	LMC	Ripepi *et al.* (2012)
CEP FU	K	15.984 ± 0.017	-3.228 ± 0.004	-	LMC	Macri *et al.* (2015)
CEP FO	K	15.458 ± 0.014	-3.257 ± 0.023	-	LMC	Macri *et al.* (2015)
CEP FU	M_K	-2.282 ± 0.019	-3.365 ± 0.063	-	GAL	Fouqué *et al.* (2007)
CEP FU	M_K	-2.63 ± 0.10	(-3.365)	-	GAL	Gaia Collaboration *et al.* (2017)
CEP FU	M_K	-2.33 ± 0.03	-3.30 ± 0.06	(0.0)	MC+G	Storm *et al.* (2011b)
CEP FU	M_K	-2.49 ± 0.08	-3.07 ± 0.07	-0.05 ± 0.10	MC+G	Groenewegen (2013)
CEP FU	W(V,K)	15.870 ± 0.013	-3.325 ± 0.014	-	LMC	Ripepi *et al.* (2012)
CEP FU	W(V,K)	15.894 ± 0.002	-3.314 ± 0.002	-	LMC	Inno *et al.* (2016)
CEP FU	W(V,K)	15.837 ± 0.049	-3.287 ± 0.010	-	LMC	Bhardwaj *et al.* (2016a)
CEP FU	$M_{W(V,K)}$	-2.87 ± 0.10	(-3.32)	-	GAL	Gaia Collaboration *et al.* (2017)
CEP FU	$M_{W(V,K}$	-2.69 ± 0.08	-3.11 ± 0.07	$+0.04 \pm 0.10$	MC+G	Groenewegen (2013)
CEP FU	[3.6]	16.01 ± 0.02	-3.31 ± 0.05	-	LMC	Monson *et al.* (2012)
CEP FU	[4.5]	15.90 ± 0.02	-3.21 ± 0.06	-	LMC	Monson *et al.* (2012)
CEP FU	$M_{[24]}$	-2.46 ± 0.10	-3.18 ± 0.10	-	GAL	Ngeow *et al.* (2015b)

Table 2. Link between GCVS classes and GDR1.

Class	Type	GCVS	Number in GCVS	Number in GDR1	Number with $(\sigma_\pi/\pi) < 0.16$
RRL		RRab, RRc	6631	331	18
T2C	BLH/WVir	CW, CW:, CWA, CWA:, CWB, CWB:	271	44	2
T2C	RVT	RV, RV:, RVA, RVA:, RVB, RVB+EA, RVB:	159	52	0
AC		BLBOO	1	0	0
CEP		DCEP	632	289	1
M/SR		M, SRA, SRB	10491	732	1

Table 3. Data on RR Lyrae and Type-II Cepheids.

Name	Hipparcos	Type	Period (d)	G (mag)	$\pi \pm \sigma_\pi$ (mas, GDR1)	$\pi \pm \sigma_\pi$ (mas, HST)	$\pi \pm \sigma_\pi$ (mas, Hipp)
			RR Lyrae				
RR Lyr	95497	RRab	0.567	7.6	3.64 ± 0.23	3.77 ± 0.13	3.46 ± 0.64
FO CVn		RRc	0.284	10.8	3.15 ± 0.25		
RZ Cep	111839	RRc	0.309	9.2	2.65 ± 0.24	2.54 ± 0.19	0.59 ± 1.48
CS Eri	12199	RRc	0.311	8.9	2.16 ± 0.23		2.71 ± 1.10
X Ari	14601	RRab	0.651	9.5	2.02 ± 0.22		0.88 ± 1.32
UV Oct	80990	RRab	0.542	9.5	2.02 ± 0.23	1.71 ± 0.10	2.44 ± 0.81
AR Per	19993	RRab	0.425	10.3	1.99 ± 0.24		0.93 ± 1.45
DX Del	102593	RRab	0.472	9.9	1.66 ± 0.22		0.77 ± 1.38
EW Cam	36213	RRab	0.628	9.4	1.69 ± 0.23		2.13 ± 1.10
V1057 Cas		RRc	0.423	10.0	2.20 ± 0.31		
XZ Dra	94134	RRab	0.476	10.3	1.43 ± 0.21		2.26 ± 0.88
SW And	1878	RRab	0.442	9.6	1.77 ± 0.26		1.48 ± 1.21
XZ Cyg	96112	RRab	0.467	9.9	1.56 ± 0.23	1.67 ± 0.17	2.29 ± 0.84
AV Peg	107935	RRab	0.390	10.4	1.53 ± 0.23		2.28 ± 1.72
V4424 Sgr	97923	RRab	0.425	10.2	1.66 ± 0.25		0.92 ± 1.94
RX Eri	22442	RRab	0.587	9.7	1.83 ± 0.28		1.50 ± 1.12
BH Peg	112994	RRab	0.641	10.6	1.40 ± 0.22		0.31 ± 1.82
BN Vul	95702	RRab	0.594	10.7	1.45 ± 0.23		6.09 ± 2.24
SU Dra	56734	RRab	0.660	9.7	1.43 ± 0.28	1.42 ± 0.16	0.20 ± 1.13
			Type-II Cepheids				
VY Pyx	43736	BL Her	1.239	7.0	3.85 ± 0.28	6.44 ± 0.23	5.01 ± 0.44
KT Com	66179	W Vir	4.070	8.0	4.16 ± 0.66		5.50 ± 0.73
κ Pav	93015	W Vir	9.078	(5.0)		5.57 ± 0.28	6.52 ± 0.77

so bright. Interestingly, the parallax measurement for VY Pyx differs quite a bit from the HST and the *Hipparcos* based value.

The ACs are listed under the identifier "BLBOO" in the GCVS. There is only one, BL Boo, which is not listed in GDR1.

There is only one classical Cepheid with an accurate parallax in GDR1, CK Cam. Table 4 lists that star and the 12 stars which have an *HST* based parallax from Benedict *et al.* (2007), Riess *et al.* (2014) and Casertano *et al.* (2016). Most are too bright to be included in GDR1. The two fainter stars suggest that the parallaxes derived using the new WFC3 scanning technique will be competitive beyond GDR2.

There is a very large number of Mira and SR variables listed in the GCVS, but since these stars are intrinsically bright only one has an accurate parallax, the anonymous SRb variable V375 And. Whitelock & Feast (2000) and Whitelock *et al.* (2008) studied Miras and Mira-like variables and derived the K band PL relation. Table 5 lists 8 stars with relative parallax error < 0.16 in *Hipparcos* data. I also added R Dor, the star with the largest angular diameter on the sky (see column 6). This is a relevant factor for these very large giants and supergiants, that have large convective cells. Chiavassa *et al.* (2011) show that in a star like Betelgeuse the photocentre shifts by a noise characterised by a standard deviation of the order of 0.1 AU. They find that in the worst situation, the

Table 4. Data on classical Cepheids.

Name	V	$\pi \pm \sigma_\pi$ (mas, HST)	$\pi \pm \sigma_\pi$ (mas, Hipparcos)	$\pi \pm \sigma_\pi$ (mas, GDR1)
β Dor	3.5	3.14 ± 0.16	3.64 ± 0.28	
δ Cep	3.7	3.66 ± 0.15	3.81 ± 0.20	
FF Aql	4.7	2.81 ± 0.18	2.05 ± 0.34	1.64 ± 0.89
l Car	3.2	2.01 ± 0.20	2.06 ± 0.27	
RT Aur	5.3	2.40 ± 0.19	−0.23 ± 1.01	
T Vul	5.5	1.90 ± 0.23	2.31 ± 0.29	
Y Sgr	5.1	2.13 ± 0.29	3.73 ± 0.32	
X Sgr	4.0	3.00 ± 0.18	3.39 ± 0.21	
ζ Gem	3.8	2.78 ± 0.18	2.71 ± 0.17	
W Sgr	4.3	2.28 ± 0.20	2.59 ± 0.75	
SS CMa	9.9	0.348 ± 0.038		0.69 ± 0.23
SY Aur	9.1	0.428 ± 0.054		0.69 ± 0.25
CK Cam	7.6		−0.59 ± 1.13	1.56 ± 0.25

Table 5. Data on Mira and SR variables.

Name	Type	V (GCVS) (max - min)	$\pi \pm \sigma_\pi$ (mas, Hipparcos)	$\pi \pm \sigma_\pi$ (mas, GDR1)	θ (mas)	Reference for θ
V375 And	SRb	7.0 - 7.2	2.35 ± 0.54	2.91 ± 0.46		
o Cet	M	2.0 - 10.1	10.91 ± 1.22		33.6 ± 3.5	Whitelock & Feast (2000)
L_2 Pup	SRb	2.6 - 6.2	15.61 ± 0.99		17.9 ± 1.6	Kervella *et al.* (2014)
R Car	M	3.9 - 10.5	6.34 ± 0.81		~ 20	Ireland *et al.* (2004)
R Leo	M	4.4 - 11.3	9.01 ± 1.42		37.4 ± 2.3	Whitelock & Feast (2000)
R Hya	M	3.5 - 10.9	8.24 ± 0.92		28.7 ± 3.3	Whitelock & Feast (2000)
W Hya	SRa	7.7 - 11.6	9.59 ± 1.12		45 ± 4	Whitelock & Feast (2000)
W Cyg	SRb	6.8 - 8.9	5.72 ± 0.38		11.5 ± 0.4	Dyck *et al.* (1996)
R Cas	M	4.4 - 13.5	7.95 ± 1.03		24.9 ± 2.9	Whitelock & Feast (2000)
R Dor	SRb	4.8 - 6.6	16.02 ± 0.69		57 ± 5	Whitelock & Feast (2000)

degradation of the astrometric fit caused by this photocentric noise will be noticeable up to about 5 kpc for the brightest supergiants.

The effect could possibly be present in Cepheids as well but should be almost an order of magnitude smaller. The largest Cepheid is l Car with a mean angular diameter of ~ 3 mas (Kervella *et al.* 2004) comparable to its parallax. Others are smaller; see Table 12 in Groenewegen (2013) for predicted angular diameters and references to measured ones.

5. GDR1

Several papers have used GDR1 data in order to study the classical variables. Two important ones have already been mentioned, (1) Gaia collaboration *et al.* (2017) that analysed the parallax data in TGAS for known RRL, T2C, CEP and derived the zeropoint of various PL relations (see Table 1), and (2) Clementini *et al.* (2016) that analysed classical variables in the south ecliptic pole data.

Casertano *et al.* (2017) used the 212 Cepheids from van Leeuwen *et al.* (2007) with $VIJH$ data to construct the $m_{\rm H} = m_{160} - 0.3861(m_{555} - m_{814})$ magnitude and compare the TGAS parallax to the photometric parallax calculated from their adopted absolute calibration $M_{\rm H} = -2.77 - 3.26 \log P$. They find that "the parallaxes are in remarkably good global agreement with the predictions, and there is an indication that the published errors may be conservatively overestimated by about 20%. Our analysis suggests that the parallaxes of 9 Cepheids brighter than G = 6 may be systematically underestimated".

Gould *et al.* (2016) use a similar approach and compare TGAS to photometrically determined parallaxes for 100 RRab stars using the K band PL relation, and find that the errors in TGAS are overestimated. The error in parallax quoted in GDR1 are inflated

Figure 1. This contribution is dedicated to the memory of Jan Cuypers (1956-2017) who died unexpectedly on the last day of February. Not only was he the head of the outreach department of the Royal Observatory of Belgium, and head of the Astronomy and Astrophysics department, Jan was heavily involved in *Gaia* in the context of DPAC Coordination Unit 7 on period determination and variable star classification. The picture was taken in 2010. It shows Jan fourth from the left with his colleagues from the Royal Observatory involved in *Gaia*.

compared to the formal parallax uncertainty (Eq. 4 and Appendix B in Lindegren *et al.* 2016), $\sigma_{\rm tgas}(\pi) = \sqrt{(A\sigma_{\rm int})^2 + \sigma_0^2}$, where $(A, \sigma_0) = (1.4, 0.2)$ is used in GDR1. Gould *et al.* propose that (1.1, 0.12) is more appropriate.

6. Outlook

The first data release of *Gaia* has shown the potential impact that this data can have on the calibration of the distance scale, and that the community seems ready for GDR2! The number of classical variables that can be expected is huge. From Table 20 in Robin *et al.* (2012) "Gaia Universe model snapshot" one can deduce that in the full catalog ($G < 20$), or at the bright end ($G < 12$), where additional abundance and detailed RV monitoring data will be available, one may expect 80 000 (400) RRab, 6500 (2200) classical Cepheids, and 40 000 (18 000) Mira variables. Eyer & Cuypers (2000) quote similar numbers.

In GDR2 one may already expect significant better precision in the parallaxes, as well as time series of the G band, and of the integrated BP and RP bands, providing colour information. There already may be an all-sky release and characterisation of RRL with sufficient epochs.

As became clear from GDR1, an important issue is the bright limit, that is currently near $G = 6$ and that has a huge impact on the availability of parallax data for the best known classical Cepheids with accurate *HST* parallaxes. Efforts are ongoing to bring this limit to $G = 3$ (Sahlmann *et al.* 2016), or even slightly brighter (Sahlmann *et al.*, this volume). An alternative route where *Gaia* could also contribute is to study Cepheids in clusters (Anderson *et al.* 2013; Chen *et al.* 2015). The well known Cepheids δ Cep and ζ Gem are located in clusters (Majaess *et al.* 2012a,b) that can provide alternative distances via main-sequence fitting.

Acknowledgements

This work has made use of data from the European Space Agency (ESA) mission *Gaia* (`www.cosmos.esa.int/gaia`), processed by the *Gaia* Data Processing and Analysis Consortium (DPAC, `www.cosmos.esa.int/web/gaia/dpac/consortium`). Funding for the DPAC has been provided by national institutions, in particular the institutions participating in the *Gaia* Multilateral Agreement.

References

Anderson, R. I., Ekström, S., Georgy, C., *et al.* 2014, *A&A*, 564, A100

Anderson, R. I., Eyer, L., & Mowlavi, N. 2013, *MNRAS*, 434, 2238

Anderson, R. I., Saio, H., Ekström, S., Georgy, C. & Meynet, G. 2016, *A&A*, 591, A8

Arenou, F. & Luri, X. 1999, *ASP-CS*, 167, 13

Beaton, R. L., Freedman, W. L., Madore, B. F., Bono, G., *et al.* 2016, *ApJ*, 832, 210

Benedict, G. F., McArthur, B. E., Feast, M. W., Barnes, T. G., *et al.* 2007, *AJ*, 133, 1810

Benedict, G. F., McArthur, B. E., Feast, M. W., Barnes, T. G., *et al.* 2011, *AJ*, 142, 187

Bhardwaj, A., Kanbur, S. M., Macri, L. M., *et al.* 2016a, *AJ*, 151, 88

Bhardwaj, A., Ngeow, C.-C., Kanbur, S. M., & Singh, H. P. 2016b, *MNRAS*, 458, 3705

Bhardwaj, A., Macri, L. M., Rejkuba, M., *et al.* 2017, *AJ*, 153, 154

Bono, G., Castellani, V., Marconi, M. 2000, *ApJ*, 529, 293

Bono, G., Pietrinferni, A., Marconi, M., *et al.* 2016, *Com. Konkoly*, 105, 149

Braga, V. F., Dall'Ora, M., Bono, G., *et al.* 2015, *ApJ*, 799, 165

Casertano, S., Riess, A. G., Anderson, J., Anderson, R. I., *et al.* 2016, *ApJ*, 825, 11

Casertano, S., Riess, A. G., Bucciarelli, B., Lattanzi, M. G., *et al.* 2017, *A&A*, 599, A67

Ceraski, W. 1905, *AN*, 168, 29

Chaboyer, B. 1999, in: A. Heck & F. Caputo (eds), *Post-Hipparcos Cosmic Candles* (Kluwer, Dordrecht), p. 111

Chen, X., de Grijs, R., & Deng, L. 2015, *MNRAS*, 446, 1268

Chiavassa, A., Pasquato, E., Jorissen, A., *et al.* 2011, *A&A*, 528, A120

Clementini, G., Ripepi, V., Leccia, S., *et al.* 2016, *A&A*, 595, A133

Coppola, G., Dall'Ora, M., Ripepi, V., *et al.* 2011, *MNRAS*, 416, 1056

Dyck, H. M., Benson, J. A., van Belle, G. T., & Ridgway, S. T. 1996, *AJ*, 111, 1705

Eyer, L. & Cuypers, J. 2000, *ASP-CS*, 203, 71

Feast, M. W. & Catchpole, R. M. 1997, *MNRAS*, 286, L1

Fiorentino, G. & Monelli, M. 2012, *A&A*, 540, A102

Fouqué, P., Arriagada, P., Storm, J., *et al.* 2007, *A&A*, 476, 73

Fraser, O. J., Hawley, S. L., & Cook, K. H. 2008, *AJ*, 136, 1242

Freedman, W. L., Madore, B. F., Gibson, B. K., Ferrarese, L., *et al.* 2001, *ApJ*, 553, 47

Gaia Collaboration, Brown, A. G. A., Vallenari, A., Prusti, T., *et al.* 2016a, *A&A*, 595, A2

Gaia Collaboration, Clementini, G. Eyer, L., *et al.* 2017, *A&A*, in press (arXiv: 1705.00688)

Gaia Collaboration, Prusti, T., de Bruijne, J. H. J., Brown, A. G. A., *et al.*, 2016b, *A&A*, 595, A1

Gingold, R. A. 1976, *ApJ*, 204, 116

Gingold, R. A. 1985, *MemSAIt*, 56, 169

Glass, I. S. & Lloyd Evans, T. 1981, *Nature*, 291, 303

Gould, A., Kollmeier, J. A., & Sesar, B. 2016, *arXiv*, 1609.06315

Gratton, R. G., Bragaglia, A., Clementini, G., *et al.* 2015, *A&A*, 421, 937

Groenewegen, M. A. T., 2000, *A&A*, 363, 901

Groenewegen, M. A. T., 2013, *A&A*, 550, A70

Groenewegen, M. A. T. & Jurkovic, M. 2017a, *A&A*, in press (arXiv: 1705.00886)

Groenewegen, M. A. T. & Jurkovic, M. 2017b, *A&A*, in press (arXiv: 1705.04487)

Groenewegen, M. A. T. & Salaris, M. 1999, *A&A*, 348, L33

Groenewegen, M. A. T., Udalski, A., & Bono, G. 2008, *A&A*, 481, 441

Hoffmeister, C. 1929, *AN*, 236, 233

Inno, L., Bono, G., Matsunaga, N. *et al.*, 2016, *ApJ*, 832, 176

Ireland, M. J., Tuthill, P. G., Bedding, T. R., *et al.* 2004, *MNRAS*, 350, 365
Ita Y., Tanabé, T., Matsunaga, N., *et al.*, 2004, *MNRAS*, 347, 720
Jacyszyn-Dobrzeniecka, A. M., Skowron, D. M., Mróz, P., *et al.* 2016, *AcA*, 66, 149
Jacyszyn-Dobrzeniecka, A. M., Skowron, D. M., Mróz, P., *et al.* 2017, *AcA*, 67, 1
Kamath, D., Wood, P. R., Van Winckel, H., & Nie, J. D. 2016, *A&A*, 586, L5
Karczmarek, P., Wiktorowicz, G., Iłkiewicz, K., *et al.* 2017, *MNRAS*, 466, 2842
Kervella, P., Montargès, M., Ridgway, S. T., *et al.* 2014, *A&A*, 564, A88
Kervella, P., Trahin, B., Bond, H. E., *et al.* 2017, *A&A*, 600, A127
Klein, C. R., Richards, J. W., Butler, N. R., & Bloom, J. S. 2014, *MNRAS*, 440, L96
Koen, C. 1992, *MNRAS*, 256, 65
Koen, C. & Laney, D. 1998, *MNRAS*, 301, 582
Lindegren, L., Lammers, U., Bastian, U., *et al.* 2016, *A&A*, 595, A4
Lutz, T. E. & Kelker, D. H. 1973, *PASP*, 85, 573
Macri, L. M., Ngeow, C.-C., Kanbur, S. M., Mahzooni, S., & Smitka, M. T. 2015, *AJ*, 149, 117
Majaess, D., Turner, D., & Gieren, W. 2012a, *ApJ*, 747, 145
Majaess, D., Turner, D., Gieren, W., Balam, D., & Lane, D. 2012b, *ApJ*, 748, L8
Marconi, M., Coppola, G., Bono, G., Braga, V., *et al.* 2015, *ApJ*, 808, 50
Matsunaga, N., Feast, M. W., & Menzies, J. W. 2009, *MNRAS*, 397, 933
Matsunaga, N., Feast, M. W., & Soszyński, I. 2011, *MNRAS*, 413, 223
Monson, A. J., Freedman, W. L., Madore, B. F., *et al.* 2012, *ApJ*, 759, 146
Muraveva, T., Palmer, M., Clementini, G., *et al.* 2015, *ApJ*, 807, 127
Navarrete, C., Catelan, M., Contreras Ramos, R., *et al.* 2017, *arXiv*, 1704.03031
Neeley, J. R., Marengo, M., Bono, G., *et al.* 2015, *ApJ*, 808, 11
Ngeow, C.-C., Kanbur, S. M., Bhardwaj, A., & Singh, H. P. 2015a, *ApJ*, 808, 67
Ngeow, C.-C., Sarkar, S., Bhardwaj, A., Kanbur, S. M., & Singh, H. P. 2015b, *ApJ*, 813, 57
Oudmaijer, R. D., Groenewegen, M. A. T., & Schrijver, H. 1998, *MNRAS*, 294, L41
Persson, S. E., Madore, B. F., Krzemiński, W., *et al.* 2004, *AJ*, 128, 2239
Pickering, E. C., Colson, H. R., Fleming, W. P., & Wells, L. D. 1901, *ApJ*, 13, 226
Planck Collaboration, Ade, P. A. R., Aghanim, N., Arnaud, M., *et al.* 2016, *A&A*, 594, A13
Riebel, D., Margaret M., Fraser, O., *et al.* 2012, *ApJ*, 723, 1195
Riess, A. G., Casertano, S., Anderson, J., MacKenty, J. & Filippenko, A.V. 2014, *ApJ*, 785, 161
Riess, A. G., Macri, L. M., Hoffmann, S. L., Scolnic, D., *et al.* 2016, *ApJ*, 826, 56
Ripepi, V., Marconi, M., Moretti, M. I., *et al.* 2014, *MNRAS*, 437, 2307
Ripepi, V., Moretti, M. I., Marconi, M., *et al.* 2012, *MNRAS*, 424, 1807
Ripepi, V., Moretti, M. I., Marconi, M., *et al.* 2015, *MNRAS*, 446, 3034
Robin, A. C., Luri, X., Reylé, C., *et al.* 2012, *A&A*, 543, A100
Sahlmann, J., Martín-Fleitas, J., Mora, A., *et al.* 2016, *SPIE*, 9904, E2E
Samus N. N., Durlevich O. V., Kazarovets E V., Kireeva N. N., Pastukhova E. N. 2017, *Astron. Rep.*, 61, 80
Sollima, A., Cacciari, C., & Valenti, E. 2006, *MNRAS*, 372, 1675
Soszyński, I., Udalski, A., Kubiak, M., *et al.* 2004, *AcA*, 54, 129
Soszyński, I., Udalski, A., Kubiak, M., *et al.* 2005, *AcA*, 55, 331
Soszyński, I. & Wood, P. R. 2013, *ApJ*, 763, 103
Soszyński, I., Wood, P. R., & Udalski, A. 2013, *ApJ*, 779, 167
Storm, J., Gieren, W., Fouqué, P., Barnes, T. G., *et al.* 2011a, *A&A*, 534, A94
Storm, J., Gieren, W., Fouqué, P., Barnes, T. G., *et al.* 2011b, *A&A*, 534, A95
van Leeuwen, F. 2007, *A&A*, 474, 653
van Leeuwen, F., Feast, M. W., Whitelock, P. A., & Laney, C. D. 2007, *MNRAS*, 379, 723
Wallerstein, G. 2002, *ARAA*, 114, 689
Whitelock, P. A. & Feast, M. 2000, *MNRAS*, 319, 759
Whitelock, P. A., Feast, M., & van Leeuwen, F. 2008, *MNRAS*, 386, 313
Wood P. R., 2000, *PASA*, 17, 18
Wood P. R., Alcock, C., Allsman, R. A., *et al.* 1999, in: T. Le Bertre, A. Lebre A. & C. Waelkens (eds), *IAU Symp. 191, Asymptotic Giant Branch Stars* (Kluwer, Dordrecht), p. 151

Astronomy and Astrophysics in the Gaia sky
Proceedings IAU Symposium No. 330, 2017
A. Recio-Blanco, P. de Laverny, A.G.A. Brown
& T. Prusti, eds.

doi:10.1017/S1743921317006469

Wide Binaries in TGAS: Search Method and First Results

Jeff J. Andrews[1,2], **Julio Chanamé**[3,4] **and Marcel A. Agüeros**[5]

[1]Foundation for Research and Technology - Hellas, IESL, Voutes, 71110 Heraklion, Greece
email: andrews@physics.uoc.gr

[2]Physics Department & Institute of Theoretical & Computational Physics,
University of Crete, 71003 Heraklion, Crete, Greece

[3]Instituto de Astrofísica, Pontificia Universidad Católica de Chile,
Av. Vicuña Mackenna 4860, 782-0436 Macul, Santiago, Chile

[4]Millennium Institute of Astrophysics, Santiago, Chile

[5]Department of Astronomy, Columbia University,
550 West 120th Street, New York, NY 10027, USA

Abstract. Half of all stars reside in binary systems, many of which have orbital separations in excess of 1000 AU. Such binaries are typically identified in astrometric catalogs by matching the proper motions vectors of close stellar pairs. We present a fully Bayesian method that properly takes into account positions, proper motions, parallaxes, and their correlated uncertainties to identify widely separated stellar binaries. After applying our method to the $>2 \times 10^6$ stars in the Tycho-Gaia astrometric solution from Gaia DR1, we identify over 6000 candidate wide binaries. For those pairs with separations less than 40,000 AU, we determine the contamination rate to be $\approx$5%. This sample has an orbital separation (a) distribution that is roughly flat in log space for separations less than $\sim$5000 AU and follows a power law of $a^{-1.6}$ at larger separations.

Keywords. binaries: visual, astrometry

---

1. Introduction

Binary systems exist with orbital separations extending from $\sim$10 $R_\odot$ to $\sim$pc. Unresolved binaries at the smallest separations are typically found by identifying radial velocity variations indicative of orbital motion, while binaries at intermediate separations can be identified by observing the stars' motions through their orbits over many years (e.g., Raghavan *et al.* 2010). At the largest separations, pairs of nearby stars are typically identified as being in a binary by matching their proper motions; these common proper motion pairs have separations from 10^2 to $\geqslant 10^4$ AU (Luyten 1971).

Beyond $\approx 10^3$ AU, the components of wide binaries are not expected to interact significantly over their lifetimes. These pairs can be considered mini-open clusters (Binney & Tremaine 1987) and are uniquely powerful for answering questions that cannot be addressed by other stellar systems. For example, wide binaries consisting of white dwarfs, subdwarfs, or M-dwarfs with FGK main sequence companions have been used to calibrate metallicity scales and age-rotation relations (e.g., Lépine *et al.* 2007). Other white dwarf-main sequence pairs and double white dwarfs have been used to determine the initial-to-final mass relation for white dwarfs (e.g., Zhao *et al.* 2012). At the widest separations, the orbits of stellar binaries are dynamically affected by interactions with passing stars, giant molecular clouds, and the Galactic tide, each of which acts to stochastically disrupt them (Weinberg *et al.* 1987). These binaries have been used as probes of the gravitational potential of their host, including the contribution of dark objects and of dark matter in general (e.g., Yoo *et al.* 2004).

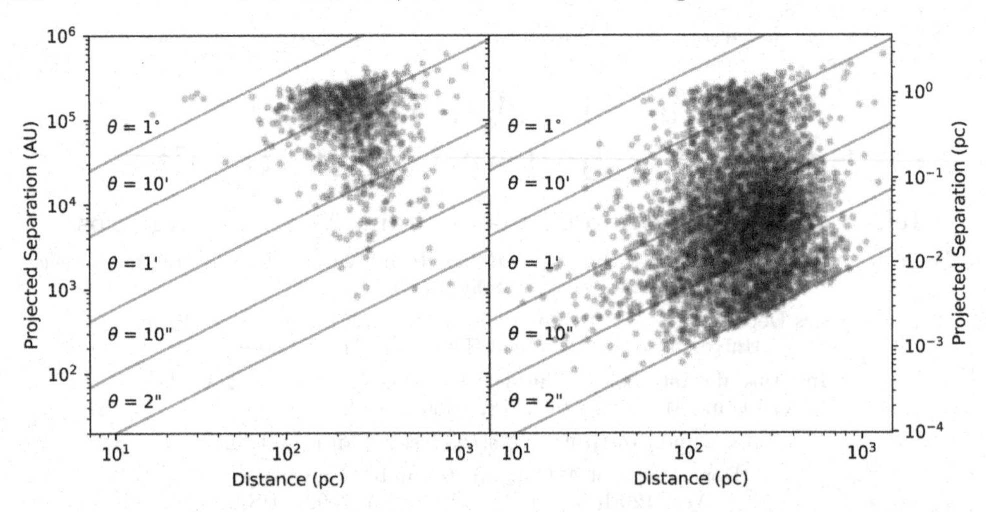

Figure 1. The distance and projected separation distribution of pairs from our random alignment catalog (left panel) and our catalog of >6000 candidate pairs (right panel). The locus of points at projected separations ∼1 pc are primarily random alignments since they exist in both panels. At separations less than 4×10^4 AU, there are few random alignments. Comparison between the two panels indicates that the contamination rate of random alignments in our sample of wide binaries is ≈5% for pairs with separations less than 4×10^4 AU.

Due to its extremely precise astrometry, the final *Gaia* catalog will produce accurate positions, proper motions, and parallaxes for more than a billion stars in the Galaxy. This data set offers a unique opportunity to robustly identify a sample of wide stellar binaries. Using the Tycho-*Gaia* Astrometric Solution (TGAS) from the first *Gaia* data release (Lindegren *et al.* 2016), we define a Bayesian statistical method that accounts for the correlated uncertainties in *Gaia*'s astrometry to identify wide binaries.

2. Method Overview

We identify stellar pairs using the five dimensions of phase space provided for each star in the TGAS catalog: right ascension (α), declination (δ), proper motion in right ascension and declination (μ_α, μ_δ) and parallax (ϖ).

We simplify the problem by reparameterizing the 10 observables (five for each star) into eight for each pair; instead of having a separate sky position and proper motion for each star, we use one position and proper motion for the stellar pair and include an angular separation (θ) and a proper motion difference ($\Delta\mu$) between the two stars. We then split the observables into one set containing the (α, δ) and (μ_α, μ_δ) position of the stellar pair, and another set containing the difference between the observed values for each of these parameters as well as the individual parallaxes (θ, $\Delta\mu$, ϖ_1, and ϖ_2).

Our method relies on a Bayesian formalism with two classes: every pair of stars is either a random alignment of unassociated stars or a genuine wide binary. For each stellar pair, we calculate two prior probabilities and two likelihoods. Since the number of stellar binaries in a sample scales with the size of the sample, N, while the number of randomly aligned pairs scales with $N(N-1)/2$, any arbitrary pair has a strong prior of being a random alignment. The exact value of this prior, which depends on the local density of a pair's position and proper motion, needs to be determined for every pair.

The probability of random alignments scales with the phase space volume of stars multiplied by the local stellar density in that phase space. Therefore, the random alignment

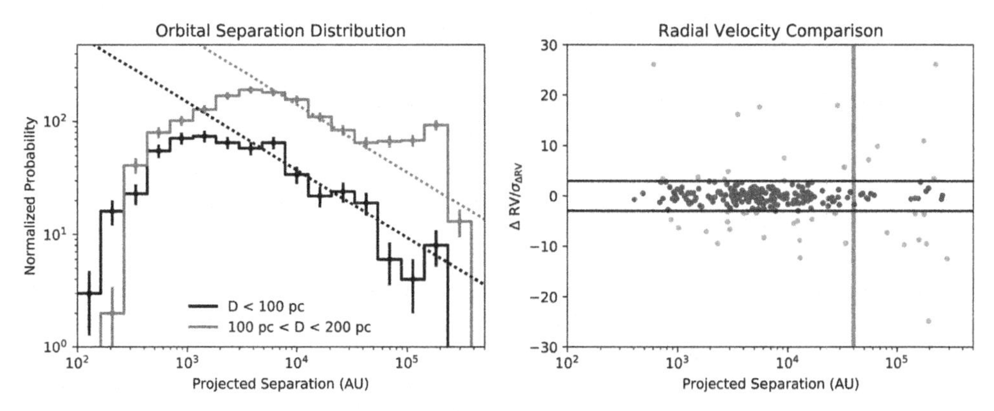

Figure 2. Left panel – The orbital separation distribution of our candidate pairs at distances less than 100 pc (black) and at distances between 100 and 200 pc (red). At large separations, the distribution scales with $a^{-1.6}$ (dotted lines), while at smaller separations, the distribution flattens and scales with a^{-1}. Right panel – Comparison between the radial velocities of the two stars in the subset of our candidate pairs observed by the RAVE survey. Most pairs have radial velocities consistent at the 3σ level (blue), bounded by the gray horizontal lines. Several pairs have somewhat discrepant radial velocities (yellow). Particularly for those pairs with projected separations less than 4×10^4 AU, these pairs could indicate the presence of unresolved binaries in our sample; some of our wide binaries may in fact be hierarchical triple systems.

likelihood increases linearly with θ and $\Delta\mu$, and scales with the local density in (α, δ) and (μ_α, μ_δ) space. Calculating the corresponding likelihood that a particular stellar pair is a genuine binary has traditionally been determined as the likelihood that the two stars have matching proper motions. TGAS astrometry is precise enough that we need to account for the fact that orbital motion may cause slight differences in the proper motions of the two stars. For details of how we account for these differences as well as the exact forms of our two prior probabilities and likelihoods see Andrews *et al.* (2017).

Random alignment catalog. To estimate the rate of contamination in our sample, we adapt the procedure of Lépine & Bongiorno (2007) and generate a catalog of randomly aligned wide binaries by shifting each star by $+2°$ in declination and + 3 mas yr^{-1} in both μ_α and μ_δ. We then apply our algorithm to identify matches between this shifted star and the original, unshifted catalog. Every resulting pair is a random alignment.

Our wide binary sample. Figure 1 shows the distance and projected separation of candidate pairs in the two catalogs of matched binaries produced using our Bayesian method. The left panel shows those pairs produced due to random alignments while the right panel shows our catalog of genuine binaries. The locus of points in both panels at projected separations of ~1 pc indicates that our sample is dominated by random alignments at these separations. At separations less than ≈4×10^4 AU, our sample is dominated by genuine pairs. By comparing the numbers of pairs with $\theta < 4 \times 10^4$ AU in both panels, we estimate the rate of contamination due to random alignments to be ≈5%.

3. First Results

Orbital separation distribution. The left panel of Figure 2 shows the orbital separation distribution for pairs within the nearest 100 pc (black) and for pairs at distances between 100 and 200 pc (red). These samples are subject to strong observational biases, but using a more sophisticated analysis in which we account for selection effects, we determine two things: first, the distribution of the sample at relatively larger separations is well characterized by a $P(a) \sim a^{-1.6}$ power law (indicated by the dotted lines). The increase in the sample of binaries at distances between 100 and 200 pc at $s \sim 10^5$ AU is due to contamination from random alignments. Were this uptick genuine, a corresponding increase would be observed in the sample at distances within 100 pc. Second, the turnover at $s \sim 3 \times 10^3$ AU is a genuine characteristic of the sample; at smaller separations, the distribution follows one closer to Öpik's law, i.e., $P(a) \sim a^{-1}$. These results are in general agreement with those observed in the wide binary samples of Chanamé *et al.* (2004) and Lépine & Bongiorno (2007).

Radial velocities. Large scale radial velocity surveys such as the Radial Velocity Experiment (RAVE; Kunder *et al.* 2017) provide a useful test of our method; the radial velocities of the components of genuine wide binaries should be consistent. The right panel of Figure 2 shows the radial velocity difference, scaled by the uncertainty in this difference, between the two stars in wide binaries for those pairs in which RAVE radial velocities are available. Horizontal lines bound the region in which the two radial velocities are consistent at the 3 σ level. Pairs at projected separations smaller than $4{\times}10^4$ AU (gray vertical line) have a contamination rate of $\approx$5%, and indeed, the stars overwhelmingly have consistent radial velocities. There are a number of pairs (yellow) which have inconsistent radial velocities despite their low contamination fraction. Some of the components of these wide binaries may contain unresolved binaries, and may therefore be hierarchical triple (or higher order) systems.

4. Future Expectations

The second data release from *Gaia* will contain $>10^9$ stars with five astrometrically measured dimensions of phase space. Identifying associated stellar pairs in such a large catalog presents a unique data analysis challenge. Nevertheless, our method is designed for scalability and use with a high performance computing cluster. Using the increase in catalog size as a guide, this data release may contain between 10^5 and 10^6 wide binaries.

Acknowledgements

This work has made use of data from the European Space Agency (ESA) mission *Gaia* (https://www.cosmos.esa.int/gaia), processed by the *Gaia* Data Processing and Analysis Consortium (DPAC, https://www.cosmos.esa.int/web/gaia/dpac/consortium). Funding for the DPAC has been provided by national institutions, in particular the institutions participating in the *Gaia* Multilateral Agreement.

References

Andrews, J. J., Chanamé, J., & Agüeros, M. A. 2017, *ArXiv preprint*:1704.07829
Binney, J. & Tremaine, S., 1987, *Galactic Dynamics*
Chanamé, J. & Gould, A., 2004, *ApJ*, 601, 289
Kunder, A., *et al.* 2017, *AJ*, 153, 75
Lépine, S. & Bongiorno, B., 2007, *AJ*, 133, 889
Lépine, S., Rich, R. M., & Shara, M. M., 2007, *ApJ*, 669, 1235
Lindegren, L., *et al.*, 2016, *A&A*, 595, A4
Luyten, W. J., 1971, *Ap&SS*, 11, 49
Raghavan, D., *et al.*, 2010, *ApJS*, 190, 1
Weinberg, M. D., Shapiro, S. L., & Wasserman, I., 1987, *ApJ*, 312, 367
Yoo, J., Chanamé, J., & Gould, A., 2004, *ApJ*, 601, 311
Zhao, J. K., Oswalt, T. D., Willson, L. A., Wang, Q., & Zhao, G., 2012, *ApJ*, 746, 144

Astronomy and Astrophysics in the Gaia sky
Proceedings IAU Symposium No. 330, 2017
A. Recio-Blanco, P. de Laverny, A.G.A. Brown & T. Prusti, eds.

doi:10.1017/S1743921317005373

The white dwarf mass-radius relation with Gaia, Hubble and FUSE

Simon R. G. Joyce[1], **Martin A. Barstow**[1], **Sarah L. Casewell**[1], **Jay B. Holberg**[2] **and Howard E. Bond**[3]

[1]Dept. of Physics & Astronomy, University of Leicester,
University Road, Leicester, LE1 7RH
email: srgj1@le.ac.uk mab@leicester.ac.uk slc25@le.ac.uk

[2]University of Arizona, LPL,
Tucson, AZ, USA
email: holberg@argus.lpl.arizona.edu

[3]Pennsylvania State University, University Park, PA, USA
email: bond@stsci.edu

Abstract. White dwarfs are becoming useful tools for many areas of astronomy. They can be used as accurate chronometers over Gyr timescales. They are also clues to the history of star formation in our galaxy. Many of these studies require accurate estimates of the mass of the white dwarf. The theoretical mass-radius relation is often invoked to provide these mass estimates. While the theoretical mass-radius relation is well developed, observational tests of this relation show a much larger scatter in the results than expected. High precision observational tests to confirm this relation are required. Gaia is providing distance measurements which will remove one of the main source of uncertainty affecting most previous observations. We combine Gaia distances with spectra from the Hubble and FUSE satelites to make precise tests of the white dwarf mass-radius relation.

Keywords. stars : distances, stars : white dwarfs, stars : fundamental parameters : masses : radii

---

1. Introduction

White dwarfs (WDs) are the remains of stars below ~ 8 $M_\odot$ which have ended their nuclear burning main sequence lifetime. They are no longer generating heat so they slowly cool down from temperatures of over 100,000 K to a few thousand Kelvin, and will eventually fade into the hypothetical black dwarfs. The study of WDs can lead to a greater understanding of the stellar population throughout the history of our galaxy. For example, the mass of the WD we see today is related to the mass of its progenitor star by the initial-final mass relation (Weidemann 2000). Also, the predictable way that WDs cool over billions of years means that if we can measure a WD's mass and temperature, we can derive its age, and by extension, the ages of associated stars (e.g. Fontaine *et al.* 2001, Oswalt 2012). In order to use WDs for such studies, we must have a reliable way of finding the mass and radius of the WD since these are the main factors that determine the rate at which it will cool down. The ammount of energy stored depends on the mass of the WD and the radius (as it relates to surface area) limits how quickly heat can be radiated away.

The key to obtaining accurate masses for WDs is the mass-radius relation (MRR) which allows us to derive the mass and radius of a WD from spectroscopic observations (Tremblay *et al.* 2017). Despite its well accepted theoretical basis, the white dwarf MRR has been difficult to test observationally due to the difficulty in making observations of sufficient precision. Studies (e.g. Holberg *et al.* 2012, Parsons *et al.* 2016) have shown that the few WDs we can measure reliably are consistent with the general form of the MRR. However, the results are not precise enough to distinguish between refinements to the MRR which take into account the finite temperature of the WD and the thickness of the surface hydrogen layer (Fontaine *et al.* 2001). A major problem has been the large uncertainty in the distances which are required input to calculations of the mass and radius. A further problem is that many targets do not have parallaxes available in the Hipparcos catalogue. Gaia DR1 has increased the number of targets for which both parallax and spectral data are available, making them suitable targets to include in this study.

We present the results of combining Gaia DR1 parallax data with optical spectroscopy from the Hubble Space Telescope (HST) and far-UV spectroscopy from FUSE which will allow us to make high precision tests of the white dwarf mass-radius relation.

2. Observational tests of the MRR

In order to test the MRR we measure the mass and radius using the spectroscopic method (Bergeron *et al.* 1992). Following this method we fit models to the Balmer and Lyman absorption lines in the WD spectrum which are caused by the thin layer of hydrogen on the surface of the WD. The extreme gravity of WDs causes pressure broadening of the absorption lines. The surface gravity ($log\ g$) and temperature (T_{eff}) are obtained from the model. The data are flux calibrated so the model normalisation can be combined with the distance from Gaia to calculate the radius of the WD (eqn 1).

$$R = \sqrt{D^2 \times norm} \div R_{\odot} \tag{2.1}$$

Once the radius is known, we combine it with log g to calculate the mass (eqn 2). This method is dependent on knowing the distance to the WD unlike the usual method of calculating the mass from log g and (eqn 2) which takes the ratio (M/R) from the MRR assuming it is correct.

$$M = \frac{gR^2}{G} \tag{2.2}$$

The Gaia TGAS catalogue (Brown *et al.* 2016) only contains parallaxes for ~6 WDs (Tremblay *et al.* 2017) due to the fact that most blue stars were not included. For this study, we make use of a sample of WDs which are in binaries with main sequence stars. These are known as Sirius-like systems. By taking the parallax for the main sequence star, we can infer the distance of the WD.

One disadvantage of Sirius-like systems is that the main sequence star is much brighter than the WD and it can be difficult to obtain a spectrum of the WD which is not contaminated by scattered light from the primary star. Many of these systems are too close to be resolved even with HST. In these situations, it is possible to obtain the spectrum of the WD in far-UV wavelengths, where it is much brighter than the main sequence star. Even for unresolved systems, we can obtain the Lyman line spectrum of the WD with little contamination from the companion star.

Our sample includes Balmer line spectra from HST and far UV (Lyman line) spectra from FUSE. In principal, the results from either set of lines should be consistent, although there is evidence that this might no be true for WDs above 50,000 K (Barstow *et al.* 2003). Several of our targets have data available for both wavelength ranges which allows us to test the validity of this assumption.

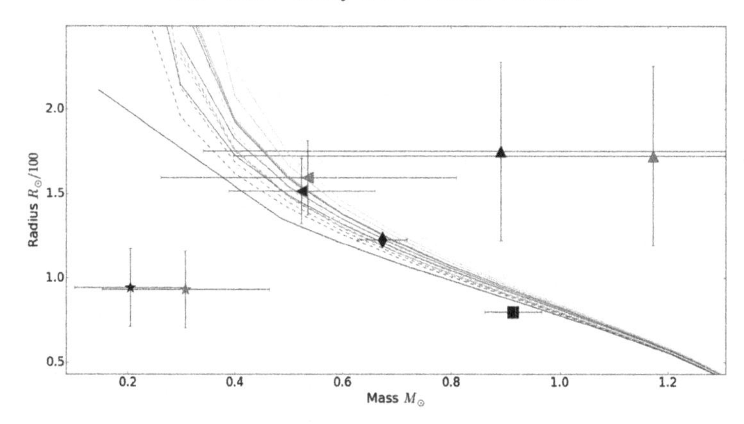

Figure 1. The mass-radius results from HST (black) and FUSE (grey) calculated using parallax data from Hipparcos compared to the theoretical mass radius relation (Fontaine *et al.*2001). The temperature ranges from zero temperature (solid black line) to 45,000 k (light grey). The symbols are Sirius B (square), HZ43 B (star), 14 Aur Cb (upright triangle), HD2133 B (left pointing triangle), HR1358 B (diamond).

3. Results

When the mass-radius values are calculated using the parallaxes avaiable from Hipparcos (van Leeuwen. 2007) there is no correlation with the mass radius relation and the error bars are too large to draw any meaningful conclusion (fig 1). In contrast, the results using the same spectra with parallaxes from Gaia show much better agreement with the MRR (fig 2).

The HST results show better agreement with the MRR than the FUSE data which appears to have a systematic offset to higher mass values. When comparing results for targets where we have both Balmer and Lyman line spectra, the mass is higher for the FUSE spectra by $\sim$0.2 $M_{\odot}$. Only HD2133 B has HST and FUSE mass values that are consistent with each other. In contrast to the mass estimates, the radius values are consistent for all targets when comparing Lyman and Balmer line results. This indicates that the cause of the offset is more likely due to the log g parameter found from the models rather than scatter due to the uncertainty in the parallaxes. It will require further investigation to properly understand the significance and cause of this offset.

The error bars in the mass axis have been reduced to the level where they no longer cover the full spread in theoretical models calculated for different temperatures. This raises the possibility that the data can now be used to distinguish between models with different temperature and H layer thickness.

4. Future work

The current set of results show the potential of this method and data set for testing the theoretical mass-radius relation for white dwarfs. The distance errors, which have been the main source of uncertainty in previous studies have been significantly reduced in the Gaia DR1 cataloge. Currently the average uncertainty in the distances in this sample is still 0.39 mas. We anticipate that in DR2 and beyond the parallax errors will be reduced to around 6.7 μas. In addition, distances for many more targets will be included for which we already have high quality spectra. This combination of high S/N spectra and highly accurate distance measurements will finally allow us to test the subtle variations of the MRR.

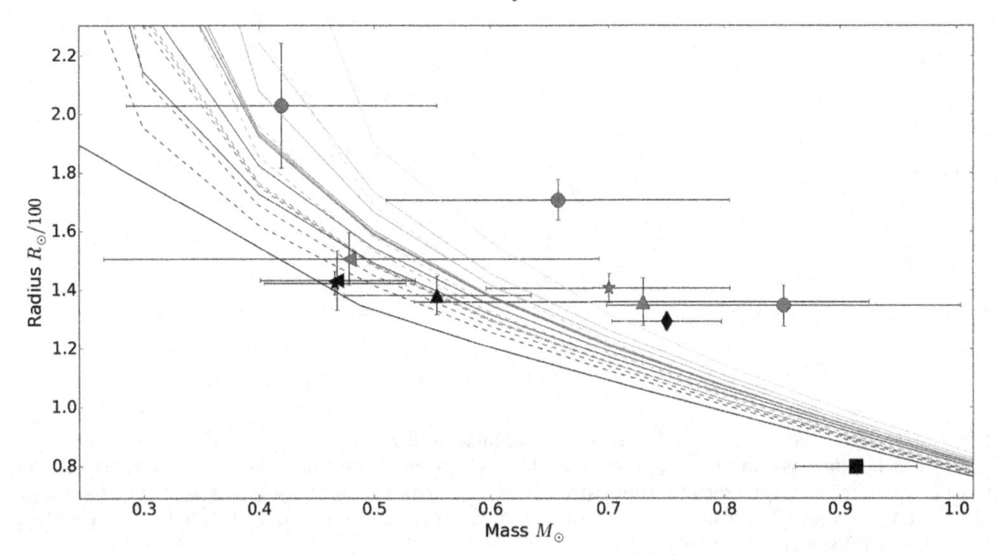

Figure 2. The mass-radius results from HST (black) and FUSE (grey) calculated using parallax data from Gaia TGAS. The symbols are the same as for fig 1. All other FUSE targets are circles.

The data are now approaching the stage where the distance will no longer be a source of significant uncertainty. This will start to uncover the effects of the other uncertainties in the data due to the stellar models used and the reliability of the fitting technique. We plan to use this set of high quality spectra to quantify these remaining sytematic effects so that they can be taken into account when future studies are undertaken which make use of spectra from thousands of white dwarfs.

5. Acknowledgements

SRGJ acknowledges support from the Science and Technology Facilities Council (STFC, UK). MAB acknowledges support from the Gaia post-launch support programme of the UK Space Agency. This work has made use of data from the European Space Agency (ESA) mission *Gaia*, processed by the *Gaia* Data Processing and Analysis Consortium (DPAC). Funding for the DPAC has been provided by national institutions, in particular the institutions participating in the *Gaia* Multilateral Agreement.

References

Barstow, M. A, Casewell, S. L., Catalan, S., Copperwheat, C. *et al.* 2014, *arXiv*, 1407.6163B
Barstow, M. A., Good, S. A., Burleigh, M., *et al.*, 2003, *MNRAS*, 344, 562
Bergeron, P., Saffer, R. A., & Liebert, J. 1992, *ApJ*, 394, 228
Chandrasekhar S., 1931, *ApJ*, 74, 81
Fontaine, G., Brassard, P., & Bergeron, P. 2001, *PASP*, 113, 409
Gaia Collaboration, Brown, A. G. A., Vallenari, A., Prusti, T., de Bruijne, J. H. J., & Mignard, F. *et al.* 2016, *A&A*, 595, 2G
Hamada, T. & Salpeter, E. E. 1961, *ApJ*, 134, 683
Holberg, J. B., Oswalt, T. D., & Barstow, M. A., 2012, *AJ*, 143, 68
van Leeuwen, F., 2007, *A&A*, 474, 653
Oswalt T. D. 2012, *JASS*, 29, 1750
Parsons, S. G., *et al.*, 2016, *MNRAS*, 458, 2793
Tremblay, P.-E., Gentile-Fusillo, N., Raddi, R., *et al.* 2017, *MNRAS*, 465, 2849T
Weidemann, V., 2000, *A&A*, 363, 647

Astronomy and Astrophysics in the Gaia sky
Proceedings IAU Symposium No. 330, 2017
A. Recio-Blanco, P. de Laverny, A.G.A. Brown & T. Prusti, eds.

doi:10.1017/S1743921317006792

Optical interferometry and Gaia parallaxes for a robust calibration of the Cepheid distance scale

Pierre Kervella[1,2], Antoine Mérand[3], Alexandre Gallenne[4], Boris Trahin[1], Simon Borgniet[1], Grzegorz Pietrzynski[5,7], Nicolas Nardetto[6] and Wolfgang Gieren[7]

[1]Unidad Mixta Internacional Franco-Chilena de Astronomía (CNRS UMI 3386), Departamento de Astronomía, Universidad de Chile, Santiago, Chile

[2]LESIA (UMR 8109), Observatoire de Paris, PSL Research University, CNRS, UPMC, Univ. Paris-Diderot, 92195 Meudon, France, email: `pierre.kervella@obspm.fr`

[3]European Southern Observatory, Karl-Schwarzschild-str. 2, D-85748 Garching, Germany.

[4]European Southern Observatory, Alonso de Córdova 3107, Casilla 19001, Santiago 19, Chile.

[5] Nicolaus Copernicus Astronomical Center, Polish Academy of Sciences, Warszawa, Poland

[6] Laboratoire Lagrange, UMR7293, Université de Nice Sophia-Antipolis, CNRS, Observatoire de la Côte d'Azur, 06000 Nice, France

[7] Universidad de Concepción, Departamento de Astronomía, Casilla 160-C, Concepción, Chile

Abstract. We present the modeling tool we developed to incorporate multi-technique observations of Cepheids in a single pulsation model: the Spectro-Photo-Interferometry of Pulsating Stars (SPIPS). The combination of angular diameters from optical interferometry, radial velocities and photometry with the coming Gaia DR2 parallaxes of nearby Galactic Cepheids will soon enable us to calibrate the projection factor of the classical Parallax-of-Pulsation method. This will extend its applicability to Cepheids too distant for accurate Gaia parallax measurements, and allow us to precisely calibrate the Leavitt law's zero point. As an example application, we present the SPIPS model of the long-period Cepheid RS Pup that provides a measurement of its projection factor, using the independent distance estimated from its light echoes.

Keywords. stars: distances, stars: individual (RS Pup), stars: oscillations, stars: variables: Cepheids, cosmology: distance scale

---

1. Introduction

One century after the discovery of their Period-Luminosity relation (the Leavitt law) by Leavitt & Pickering (1912), Cepheids are still the keystone of the empirical cosmic distance ladder. However, due to the large distances of Cepheids (particularly the long-period pulsators), only relatively imprecise parallax measurements are available (e.g., from Hipparcos; van Leeuwen *et al.* 2007). As a result, the calibration of the zero point of the Leavitt law using Galactic Cepheids is insufficient accurate. Gaia's DR2 high accuracy parallaxes will bring a tremendous improvement, with distances to hundreds of individual Cepheids measured to a few percent accuracy. Alternatively, a classical avenue to measure the distances to individual Galactic and LMC Cepheids is the Parallax-of-Pulsation (PoP) method, also known as the Baade-Wesselink (BW) technique. We here briefly present the SPIPS modeling tool that we developed to reproduce the classical observables of the pulsation of Cepheids (radial velocity, photometry, interferometry,...), and how we plan to use it in conjunction with the Gaia parallaxes to calibrate the PoP technique and eventually the Leavitt law.

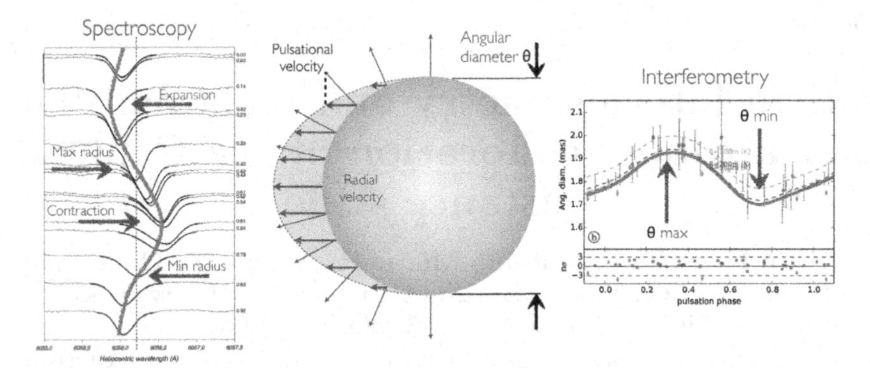

Figure 1. Principle of the classical Parallax-of-Pulsation distance determination on the Cepheid β Dor. *Left panel:* Doppler displacement of the Fe I line at 6056 Å obtained from high resolution HARPS spectra (Nardetto *et al.* 2006). *Right panel:* Interferometric angular diameter as a function of the pulsation phase (Breitfelder *et al.* 2016).

2. The Parallax-of-Pulsation technique and SPIPS

The PoP technique relies on the comparison of (1) the amplitude of the pulsation of the star from the integration of its radial velocity curve measured using spectroscopy and (2) the change in angular diameter estimated from its brightness and color, or measured using interferometry. The distance of the Cepheid is then obtained by fitting the linear and angular amplitudes (see, e.g., Storm *et al.* 2011). Although potentially precise at the percent level, the major weakness of the PoP technique is that it uses a numerical factor to convert disk-integrated radial velocities into photospheric velocities: the projection factor (p-factor; Nardetto *et al.* 2017). It is defined as the ratio of the pulsational velocity over the radial velocity projected on the line of sight. The absolute accuracy of the PoP technique is currently limited by our knowledge of the p-factor, that is degenerate with the distance d (the derived quantity is d/p). This parameter is particularly complex to model, as it represents simultaneously the spherical geometry, the limb darkening, and the gas dynamics in the Cepheid atmosphere.

The *Spectro Photo Interferometry of Pulsating Stars* (SPIPS; Mérand *et al.* 2015) model of a Cepheid is built assuming that the star is a radially pulsating sphere, for which the pulsational velocity $v_{\mathrm{puls}}(t)$ and the effective temperature $T_{\mathrm{eff}}(t)$ are the basic parameters. The cyclic variation of these parameters is represented using classical Fourier series or periodic splines functions. The photometry in all filters is computed using ATLAS9 atmosphere models, considering their bandpasses and zero points. The interstellar reddening is parametrized using the standard $E(B-V)$ color excess for Milky Way dust ($R_V = 3.1$) and computed for the phase-variable T_{eff}. Angular diameters including limb darkening are produced by the model to match the interferometric measurements, using SATLAS models (Neilson *et al.* 2013). Finally, circumstellar envelopes are included in the modeling (Gallenne *et al.* 2012; Gallenne *et al.* 2013), as well as the presence of stellar companions (Gallenne *et al.* 2015; Gallenne *et al.* 2017). SPIPS can be employed with only photometry and radial velocity, but interferometric angular diameters are independent of interstellar absorption, allowing to de-correlate the reddening and the effective temperature. Angular diameters thus bring a considerable added value to the quality and robustness of the derived parameters. Recent applications of SPIPS to Cepheids can be found in Mérand *et al.* (2015) (δ Cep, η Aql), Breitfelder *et al.* (2016) (Cepheids with HST/FGS parallaxes) and Breitfelder *et al.* (2015) (Type 2 Cepheid κ Pav).

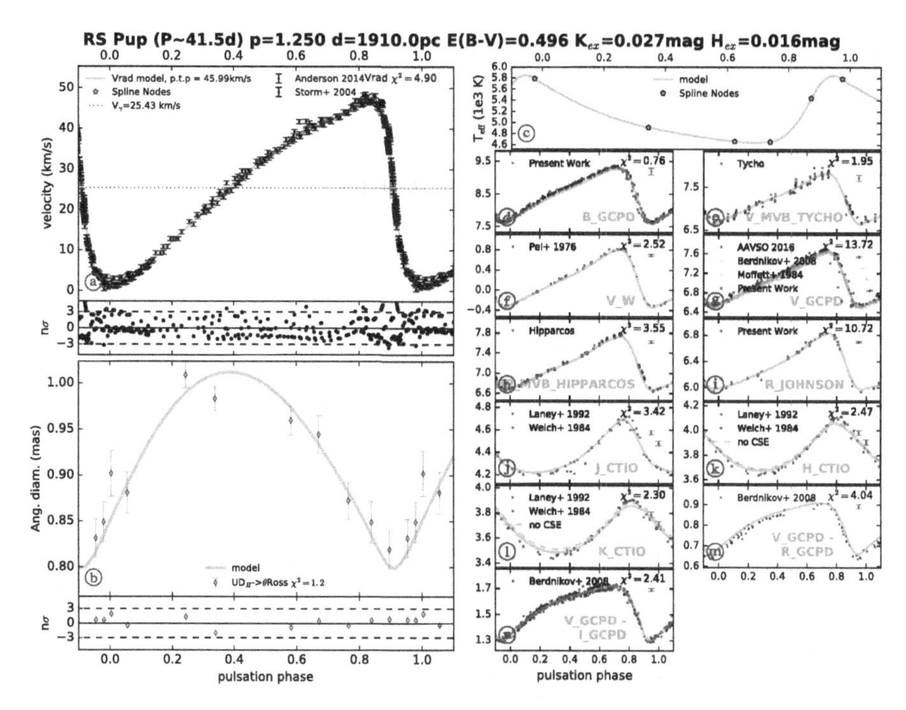

Figure 2. SPIPS combined fit of the observations of RS Puppis ($P = 41.5$ days). The radial velocity measurements are presented in the upper left panel, and the interferometric angular diameters in the lower left panel. The model's effective temperature is shown in the upper right panel, and the photometry in the different sub-panels of the right column (Kervella *et al.* 2017).

3. The projection factor of RS Puppis and the $p - P$ relation

RS Pup is one of the intrinsically brightest Cepheids in the Galaxy, and it is remarkable due to its large circumstellar nebula that reflects the light variations of the Cepheid. From the light echoes, Kervella *et al.* (2014) derived its distance of $d = 1910 \pm 80$ pc. Knowing the distance resolves the intrinsic distance/p-factor degeneracy of SPIPS and gives access to p. The pulsation model (Fig. 2) has been presented by Kervella *et al.* (2017), giving a best-fit p-factor of $p = 1.25 \pm 0.06$ ($\pm 5\%$). The addition of the p-factor of RS Pup to the limited set of existing p-factor measurements is particularly valuable, as it was up to now mostly limited to short and intermediate period Cepheids (Fig. 3). The other exception is ℓ Car ($P = 35.5$ days), whose parallax was measured by Benedict *et al.* (2007). The p-factors of ℓ Car and RS Pup are statistically identical within their uncertainties. The simple model of a constant $p = 1.29 \pm 0.04$ over the sampled range of Cepheid periods (3.0 to 41.5 days) is consistent with the observations ($\chi^2_{\rm red} = 0.9$; Kervella *et al.* 2017).

4. The contribution of Gaia

Gaia's DR2 in April 2018 will considerably improve the accuracy of Galactic Cepheid parallaxes compared to DR1 (Casertano *et al.* 2017; Gaia Collaboration 2017). Based on the foreseen accuracy, more than 400 Cepheid parallaxes will be measured with an accuracy $< 3\%$, out of which approximately one hundred to $< 1\%$. The observations of a sample of 18 long-period Cepheid using the spatial scanning mode of the HST/WFC3 by Casertano *et al.* (2016) will complement this data set. This unprecedented catalog will make it possible to obtain an extremely accurate calibration of the Leavitt law and other fundamental properties of Cepheids. However, to reach this accuracy implies that

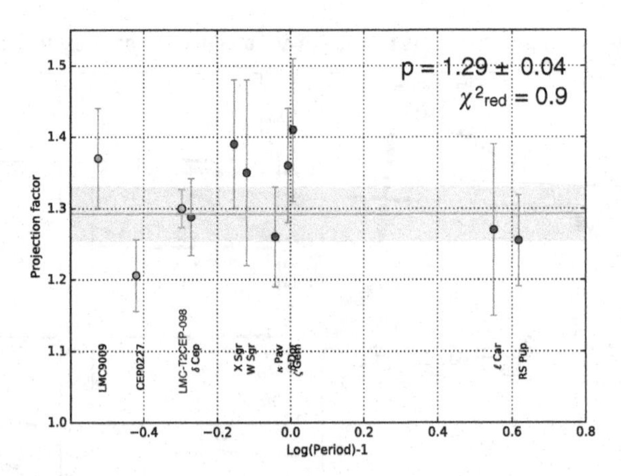

Figure 3. Measured p-factors of Cepheids (from Kervella *et al.* 2017). The green point is the measurement of the binary type 2 Cepheid OGLE-LMC-T2CEP-098 by Pilecki *et al.* (2017).

the pulsation models employed to interpret the observations must also reach this level of accuracy. The flexibility of the SPIPS modeling enables the integration of all types of new and archival observations in a consistent and robust approach. This will contribute to mitigate the influence of various systematics of observational origin, and bring the accuracy of the calibration of the Cepheid distance scale down to 1%.

Acknowledgements

The authors acknowledge the support of the French Agence Nationale de la Recherche (ANR-15-CE31-0012-01 UnlockCepheids). W.G. and G.P. gratefully acknowledge the BASAL Centro de Astrofisica y Tecnologias Afines (CATA) PFB-06/2007. W.G. acknowledges support from the Chilean Millenium Institute of Astrophysics (MAS) IC120009. This research has received funding from the European Research Council (Horizon 2020 grant No 695099).

References

Benedict, G. F., McArthur, B. E., Feast, M. W., *et al.* 2007, *AJ*, 133, 1810
Breitfelder, J., Kervella, P., Mérand, A., *et al.* 2015 *A&A*, 576, A64
Breitfelder, J., Mérand, A., Kervella, P., *et al.* 2016, *A&A*, 587, A117
Casertano, S., Riess, A. G., Anderson, J., *et al.* 2016, *ApJ*, 825, 11
Casertano, S., Riess, A. G., Bucciarelli, B., & Lattanzi, M. G. 2017, *A&A*, 599, A67
Gaia Collaboration, Clementini, G., Eyer, L., *et al.* 2017, *A&A*, in press, arXiv:1705.00688
Gallenne, A., Mérand, A., & Kervella, P. 2012, *A&A*, 538, A24
Gallenne, A., Mérand, A., Kervella, P., *et al.* 2013, *A&A*, 558, A140
Gallenne, A., Mérand, A., Kervella, P., *et al.* 2015, *A&A*, 579, A68
Gallenne, A., Kervella, P., Mérand, A., Evans, N. R., & Proffitt, C. 2017, these proceedings
Kervella, P., Bond, H. E., Cracraft, M., *et al.* 2014, *A&A*, 572, A7
Kervella, P., Trahin, B., Bond, H. E., *et al.* 2017, *A&A*, 600, A127
Leavitt, H. S. & Pickering, E. C. 1912, *Harvard College Observatory Circular*, 173, 1
Mérand, A., Kervella, P., Breitfelder, J., *et al.* 2015, *A&A*, 584, A80
Nardetto, N., Mourard, D., Kervella, P., *et al.* 2006, *A&A*, 453, 309
Nardetto, N., Poretti, E., Rainer, M., *et al.* 2017, *A&A*, 597, A73
Neilson, H. R. & Lester, J. B. 2013, *A&A*, 554, A98
Pilecki, B., Gieren, W., Smolec, R., *et al.* 2017, *ApJ*, in press (arXiv:1704.07782)
Storm, J., Gieren, W., Fouqué, P., *et al.* 2011, *A&A*, 534, A94

Astronomy and Astrophysics in the Gaia sky
Proceedings IAU Symposium No. 330, 2017
A. Recio-Blanco, P. de Laverny, A.G.A. Brown & T. Prusti, eds.

doi:10.1017/S1743921317005609

Gaia view of low-mass star formation

C. F. Manara[1], †, T. Prusti[1], J. Voirin[1] and E. Zari[2]

[1]Scientific Support Office, Directorate of Science, European Space Research and Technology Centre (ESA/ESTEC), Keplerlaan 1, 2201 AZ Noordwijk, The Netherlands
email: cmanara@cosmos.esa.int

[2]Leiden Observatory, Niels Bohrweg 2, 2333 CA Leiden, the Netherlands

Abstract. Understanding how young stars and their circumstellar disks form and evolve is key to explain how planets form. The evolution of the star and the disk is regulated by different processes, both internal to the system or related to their environment. The former include accretion of material onto the central star, wind emission, and photoevaporation of the disk due to high-energy radiation from the central star. These are best studied spectroscopically, and the distance to the star is a key parameter in all these studies. Here we present new estimates of the distance to a complex of nearby star-forming clouds obtained combining TGAS distances with measurement of extinction on the line of sight. Furthermore, we show how we plan to study the effects of the environment on the evolution of disks with Gaia, using a kinematic modelling code we have developed to model young star-forming regions.

Keywords. -

1. Introduction

Protoplanetary disks, made of dust and gas, are formed around young stars and are the birthplace of planets. These disks are evolving with time, and the evolution of their gas and dust content strongly impacts what kind of planets are formed, and their migration in the disk (e.g., Thommes *et al.* 2008; Morbidelli & Raymond 2016). Each of the processes known to impact disk evolution modifies the dust and gas content differently, and it is therefore mandatory to understand which processes are at place during the various disk evolution phases to be able to build a predictive planet formation theory.

The physical processes impacting disk evolution are either internal to the star-disk system, as in the cases of viscous evolution, internal photoevaporation, and disk wind driven evolution (e.g., Alexander *et al.* 2014; Bai 2016), or related to the stellar environment where disks form, as in the case of external photoevaporation or dynamical interactions (e.g., Clarke 2007; Pfalzner *et al.* 2005). While the former have been studied in details both from a theoretical and observational point of view, the latter are more elusive to be studied observationally, and their impact on disk evolution and planet formation is still to be understood. This is particularly true for dynamical interactions in young clusters and their impact on disk evolution. Pioneering studies have been carried out only in a limited number of environments, mainly in the Orion Nebula Cluster (e.g., de Juan Ovelar *et al.* 2012).

The ESA mission Gaia (Gaia Collaboration *et al.* 2016) will be of enormous impact in the study of the evolution of young stars and of their disks. In particular, it provides solid estimates of the distance to young stars, a key parameter to derive any stellar and disk property, and it allows one to establish kinematic membership of stars in clusters. Here we present our initial work on these aspects of young stars and disks evolution.

† Present address: European Southern Observatory, Karl-Schwarzschild-Strasse 2, 85748 Garching bei Munchen, Germany, e-mail: cmanara@eso.org.

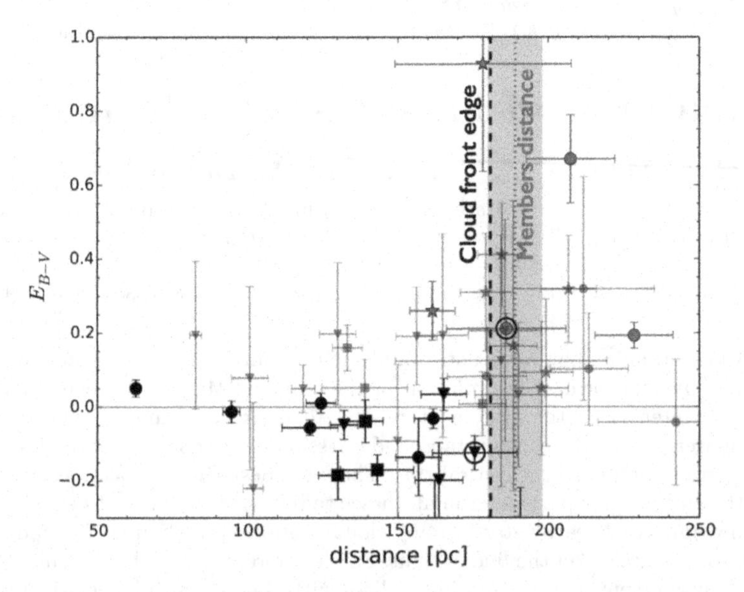

Figure 1. Color excess as a function of distance for stars located on the line-of-sight of the Chamaeleon I cloud. The black dashed line is the position of the front edge of the cloud determined by the reddening turn-on method. The red dotted line is the average distance of the members, shown here for comparison, and the red shaded regions is the uncertainty on this value. We highlight with black circles the stars used to compute the distance of the reddening turn-on, thus the front edge of the cloud. From Voirin *et al.*, subm.

2. Gaia DR1 distances to the Chameleon clouds

Distances to nearby star-forming regions are nowadays still uncertain. Even if their distance can be as small as ~50-300 pc, these nearby regions contain mainly highly extincted low-mass stars, which are thus not bright enough to be part of the Hipparcos catalog, the most extensive catalog of parallaxes prior to Gaia. However, the distance to these regions is key to constrain several crucial parameters to study young stars, such as the stellar luminosity and mass, the age, and the disk mass and size.

With the advent of the Gaia catalog, and in particular of the TGAS sub-sample, we have decided to investigate whether we could confirm or update the previous assumed distances to the clouds in one of the closest and most studied star-forming complex, the Chamaeleon-Musca complex (e.g., Luhman 2008). This region contains several hundreds of young stars, which are surrounded by disks in many cases (e.g., Luhman 2004), and are still accreting material from these disks (e.g., Manara *et al.* 2016, 2017). The distances to these clouds have been long debated in the literature and are assumed to be ~160, 178, and 140 pc, for the Chamaeleon I, II, and III clouds, respectively (Whittet *et al.* 1997), while the distance to the Musca cloud is still unknown.

We have queried the TGAS sub-sample in the Gaia archive for known members of any of these clouds, and we have found 8 members of the Chamaeleon I cloud and none in the other clouds. Of these 8 stars included in the TGAS catalog, 4 were also included in the Hipparcos catalog. We have compared the distances inferred by converting the parallaxes using an anisotropic prior (Astraatmadja & Bailer-Jones 2016), and we observe that the distances obtained from the TGAS parallaxes are systematically larger than the correspondent Hipparcos distances by ~5-25 pc, although the two values are compatible within 1σ. Furthermore, the TGAS distances to these objects suggest the presence of a north-south gradient in the Chamaeleon I cloud, with difference of ~20 pc in the distances to the stars, which are found more distant in the southern part of the cloud. When considering these objects, the mean distance to the cloud is 189±9±10 pc, where the quoted uncertainties are the statistical and systematic ones, respectively. However, this value is based on a limited number of objects, and the method of computing the distance to a region using the measured distances to its confirmed members cannot be used for the other clouds in this complex at this stage.

Following Whittet *et al.* (1997), we measure the distance to star-forming clouds by measuring the color-excess due to extinction for objects located on the line-of-sight of the cloud. The presence of the cloud results in an increase of the color-excess, and the distance at which this happens is representative of the distance of the front edge of the cloud. With respect to Whittet *et al.* (1997), the TGAS catalog gives us access to many more stars with directly measured parallaxes. We show in Fig. 1 the measured color-excess vs the distance to the stars on the line-of-sight of the Chamaeleon I cloud. The distance at which we observe a turn-on in the color-excess is 181±10±10 pc, which is compatible with the distance of the members, and is considered as the distance to the region. We similarly derive a distance of 181±10±10 pc for the Chamaeleon II cloud, of 199±15±13 pc to the Chamaeleon III cloud, and we constrain the Musca cloud to be closer than 600 pc (Voirin *et al.*, subm.). We confirm the distance to the Chamaeleon II cloud, while we find further away distances for the Chamaeleon I and III clouds, giving for the first time a firm estimate of the distance to the latter. Within the uncertainties, these values are compatible with the hypothesis that all these clouds are part of the same large-scale structure of clouds.

While we have applied this method only to the clouds in the Chamaeleon-Musca complex, this could be applied to any cloud with limited amount of foreground interstellar material. This method is particularly interesting for clouds which contain no members with measured distances, either because of the high extinction or because star formation has not started yet. The larger number of line-of-sight and members stars with measured parallaxes that will be available with DR2 will thus allow us to determine the distance to several other star-forming regions.

3. Gaia as a tool to study dynamical evolution of young stars

Young stars are usually located in clustered environment (e.g., Lada & Lada 2003) where dynamical interactions with other members of the region can affect the evolution of disks (e.g., Armitage & Clarke 1997; Pfalzner *et al.* 2015). Gaia can give us the possibility to probe observationally both the dynamical properties of young clusters (e.g., Allison 2012) and the effect of dynamical encounters on disk evolution. Members of star-forming regions are usually identified by the presence of an excess emission at near- and mid-infrared wavelengths due to the presence of a disk. Such method has two main limitations imposed by the small coverage of infrared satellites and by being very little sensitive to objects with small infrared excess, such as disk-less young stars. While the former limitation is now getting overcome by all-sky surveys such as WISE, the latter is possibly biasing our membership lists in young regions. Indeed, surveys of young stars done with other methods, such as measuring Hα- or UV-excess, or by obtaining spectra of large number of objects in a region, are showing that a large population of young stars is present in the surrounding of the main star-forming site(e.g., De Marchi *et al.* 2011; Comerón *et al.* 2013; Sanchez *et al.* 2014; Sacco *et al.* 2015). We think that this sparse population could be either the outcome of dynamical interactions in the main star-forming site, or due to the fact that star formation happens at all degrees of clustering. Both hypotheses are interesting and should be further investigated.

We have been developing a method to assign membership to stars in clusters purely based on their kinematic properties measured by the Gaia satellite. This method is based on the maximum likelihood estimate described by Lindegren *et al.* (2000) and de Bruijne (1999), where the membership of a star in a region is assigned by finding the stars sharing a common velocity pattern for which the likelihood is maximized. We have added the possibility to model two populations: a clustered population sharing a common velocity pattern, and a field population (Zari *et al.*, in prep.). With our method we can thus assign a probability to each star to be either member of the cluster or of the field, thus allowing us to identify possible members of a cluster with no assumptions on the presence of a disk or accretion.

We are now testing this method using N-body simulations of young clusters (e.g., Parker *et al.* 2014) and the GUMS simulation of the stars in the galaxy (Robin *et al.* 2012) using the current estimates of the uncertainty on the astrometric parameters in Gaia DR2. The first results show that we assign a probability of being members of the cluster larger than 50% to >95% of the real members of the cluster, with only <5% of field stars erroneously classified as members.

Acknowledgements

This work has made use of data from the European Space Agency (ESA) mission *Gaia* (`https://www.cosmos.esa.int/gaia`), processed by the *Gaia* Data Processing and Analysis Consortium (DPAC, `https://www.cosmos.esa.int/web/gaia/dpac/consortium`). Funding for the DPAC has been provided by national institutions, in particular the institutions participating in the *Gaia* Multilateral Agreement.

References

Alexander, R., Pascucci, I., Andrews, S., *et al.* 2014, Protostars and Planets VI, 475
Allison, R. J. 2012, *MNRAS*, 421, 3338
Armitage, P. J. & Clarke, C. J. 1997, *MNRAS*, 285, 540
Astraatmadja, T. L. & Bailer-Jones, C. A. L. 2016, *ApJ*, 833, 119
Bai, X.-N. 2016, *ApJ*, 821, 80
Clarke, C. J. 2007, *MNRAS*, 376, 1350
Comerón, F., Spezzi, L., López Martí, B., & Merín, B. 2013, *A&A*, 554, A86
de Bruijne, J. H. J. 1999, *MNRAS*, 310, 585
de Juan Ovelar, M., Kruijssen, J. M. D., Bressert, E., *et al.* 2012, *A&A*, 546, L1
De Marchi, G., Paresce, F., Panagia, N., *et al.* 2011, *ApJ*, 739, 27
Gaia Collaboration, Brown, A. G. A., Vallenari, A., *et al.* 2016, *A&A*, 595, A2
Gaia Collaboration, Prusti, T., de Bruijne, J. H. J., *et al.* 2016, *A&A*, 595, A1
Lada, C. J. & Lada, E. A. 2003, *ARA&A*, 41, 57
Lindegren, L., Madsen, S., & Dravins, D. 2000, *A&A*, 356, 1119
Luhman, K. L. 2004, *ApJ*, 602, 816
Luhman, K. L. 2008, Handbook of Star Forming Regions, Volume II, 169
Manara, C. F., Fedele, D., Herczeg, G. J., & Teixeira, P. S. 2016, *A&A*, 585, A136
Manara, C. F., Testi, L., Herczeg, G. J., *et al.* 2017, arXiv:1704.02842
Morbidelli, A. & Raymond, S. N. 2016, Journal of Geophysical Research (Planets), 121, 1962
Parker, R. J., Wright, N. J., Goodwin, S. P., & Meyer, M. R. 2014, *MNRAS*, 438, 620
Pfalzner, S., Umbreit, S., & Henning, T. 2005, *ApJ*, 629, 526
Pfalzner, S., Vincke, K., & Xiang, M. 2015, *A&A*, 576, A28
Robin, A. C., Luri, X., Reylé, C., *et al.* 2012, *A&A*, 543, A100
Sacco, G. G., Jeffries, R. D., Randich, S., *et al.* 2015, *A&A*, 574, L7
Sanchez, N., Inés Gómez de Castro, A., *et al.* 2014, *A&A*, 572, A89
Thommes, E. W., Matsumura, S., & Rasio, F. A. 2008, Science, 321, 814
Whittet, D. C. B., Prusti, T., Franco, G. A. P., *et al.* 1997, *A&A*, 327, 1194

Astronomy and Astrophysics in the Gaia sky
Proceedings IAU Symposium No. 330, 2017
A. Recio-Blanco, P. de Laverny, A.G.A. Brown & T. Prusti, eds.

doi:10.1017/S1743921317006147

Calibration and characterisation of the Gaia Red Clump

L. Ruiz-Dern, C. Babusiaux, F. Arenou, C. Danielski, C. Turon and P. Sartoretti

GEPI, Observatoire de Paris, PSL Research University, CNRS UMR 8111 - 5 Place Jules Janssen, 92190 Meudon, France
email: laura.ruiz-dern@obspm.fr

Abstract. We present new empirical Colour-Colour and Effective Temperature-Colour Gaia Red Clump calibrations. The selected sample takes into account high photometric quality, good spectrometric metallicity, homogeneous effective temperatures and low interstellar extinctions. From those calibrations we developed a method to derive the absolute magnitude, temperature and extinction of the Gaia RC. We tested our colour and extinction estimates on stars with measured spectroscopic effective temperatures and Diffuse Interstellar Band (DIB) constraints. Within the Gaia Validation team these calibrations are also being used, together with asteroseismic constraints, to check the parallax zero-point with Red Clump stars.

Keywords. Gaia Red Clump, photometry, stellar parameters: interstellar extinction, effective temperature

1. Introduction

The Gaia Data Release 1 (DR1) provided high precision parallaxes for 2 million stars. Trigonometric parallaxes are, indeed, the most direct method to measure distances. However, if we want to determine distances to further objects than what the Gaia parallax precision allows today, we need to take advantage of *standard candles*. That is, objects with a well-determined absolute magnitude and bright enough to be observed in distant regions. The difference between this known absolute magnitude and the apparent one gives the distance modulus of the source.

Red Clump (RC) stars are known to be good distance indicators (e.g. Udalski (2000), Alves (2000), Groenewegen (2008), Laney *et al.* (2012)). This is because they show weakly dependence of M_I on colour, age and chemical composition, and because they are very abundant in the solar neighbourhood, so their absolute magnitude may be accurately determined. The larger the number of sources the lower will be the statistical error in distance calculations.

The purpose of this work is to better describe and characterise the Red Clump for distance calculations. By analysing the photometric colour-colour relations of these stars, we may derive their interstellar extinctions, and consequently be able to obtain their absolute magnitudes and their distances. However, we have found that atmosphere models of red giant stars do not fit the observations: there is a *gap* in between no matter the photometric bands or the stellar evolution models used. Moreover, we remind that there is no Gaia calibrated filter model for the DR1. Both reasons led us to develop a purely empirical photometric calibration of RC stars.

We present here the colour vs $G\text{-}K_s$ and T_{eff} vs $G\text{-}K_s$ photometric calibrations derived for the RC, and some applications where they are being used to characterise these stars.

2. Empirical photometric calibration

The precision of empirical calibrations rely on the accuracy of the sample selection and the robustness of the method.

To select the data we have mainly considered six constraints (see Ruiz-Dern *et al.* (2017, submitted) for more details and a complete list of the sources used):

• *Photometric quality*: we selected visual and infrared wavelengths by including G B V H_P B_T V_T J K_s photometry (with uncertainties) from the Gaia DR1, Hipparcos, Tycho-2 and 2MASS catalogues

• *No binaries nor multiple systems*

• *Spectroscopic metallicities*

• *Giants subset*: to ensure no contamination from other spectral types, we applied the following colour and parallax criteria:

$$G - K_s > 1.6 \tag{2.1}$$

$$m_G + 5 + 5\ \log_{10}\left(\frac{\varpi + 2.32\ \sigma_\varpi}{1000}\right) < 2.5 \tag{2.2}$$

where the factor 2.32 on the parallax error corresponding to the 99th percentile of the parallax probability density function.

• *Interstellar extinction*: to avoid introducing possible bias due to derredening, we kept only stars with $A_0 < 0.03$, where A_0 is the interstellar extinction at $\lambda = 550$ nm (Gaia reference value). We used the most up-to-date 3D local extinction map of Capitanio *et al.* (2017) (a poster review can also be found in this proceedings)

• *Effective temperature*: only for the effective temperature - G-K_s calibration, we included the spectroscopic effective temperatures from the 13h release (DR13) of the APOGEE survey (Holtzman *et al.* (2015), García Pérez *et al.* (2016), SDSS Collaboration *et al.* (2016)), so we could have a larger homogeneous sample

We got a sample of 1329 stars for the colour vs G-K_s calibrations, and of 548 stars for the effective temperature vs G-K_s.

To derive these accurate photometric relations, we implemented a Monte Carlo Markov Chain (MCMC) method which allows us to account and deal with the uncertainties of both the predictor and response variables in a robust way. All calibrations were derived with respect to the G-K_s colour because of their actual and future broad use.

The general fitting formula adopted was:

$$Y = a_0 + a_1\ X + a_2\ X^2 + a_3\ [\mathrm{Fe/H}] + a_4\ [\mathrm{Fe/H}]^2 + a_5\ X\ [\mathrm{Fe/H}] \tag{2.3}$$

where X is G-K_s, Y is (for CC) a given colour or (for $T_{\rm eff}$ relations) the $T_{\rm eff}$, and a_i are the coefficients to be estimated. In order to provide the most accurate fit for each relation, the process penalises the complex terms by using the Deviance Information Criterion (DIC) (Plummer (2008)). We checked for outliers at 3σ from the fit. If outliers were found, the one was removed and the complete process was run again.

We obtained 23 colour vs G-K_s calibrations with a median dispersion in magnitude between 0.03 and 0.05, plus the $T_{\rm eff}$ vs G-K_s relation with a median dispersion of $\sim 57K$. To test our $T_{\rm eff}$ vs G-K_s fit we transformed the G-K_s colour to V-K_s through our V-K_s vs G-K_s calibration, and compared it to other $T_{\rm eff}$ vs V-K_s relations in the literature: Ramírez & Meléndez (2005) and González Hernández & Bonifacio (2009), both based on the infrared flux method technique, and Huang *et al.* (2015) based on interferometry. For solar metallicities we found discrepancies up to $\sim 90K$.

3. Red Clump characterisation

The Gaia DR1 HR diagram published in Gaia Collaboration *et al.* (2016) clearly shows the effect of the interstellar extinction on the RC region: the clump appears more elongated and tilted with respect to the Hipparcos RC, meaning that we need to account for the interstellar extinction if we want to use the G magnitude.

By subsetting the DR1 sample to the TGAS sample with 10% parallax precision and low extinction stars, we may also observationally detect other substructures of the giant branch such as the Red Giant Branch Bump and the Secondary Red Clump (Ruiz-Dern *et al.* (2017, submitted)). We have zoomed into the RC region and applied our T_{eff} vs G-K_s relationship to the theoretic T_{eff} of Padova isochrones (Bressan *et al.* (2012), Parsec 2.7) at different metallicites and at different ages. We find that the RC position fits properly the isochrones. The RGB bump, however, appears to be brighter in the isochrones (Ruiz-Dern *et al.* (2017, submitted)).

As we combine the calibrations with an extinction coefficient model, we can characterise some RC parameters, such as the effective temperature and the photometric interstellar extinction. To take into account the dependency of the extinction coefficients on the star Spectral Energy Distribution and on the extinction itself, we used an analytical model of those dependencies. We computed the extinction coefficient using the Fitzpatrick & Massa (2007) extinction law for extinctions A_0 from 0 to 5 and using Kurucz spectra for a logg of 2.5 with T_{eff} from 4000 to 6500 K. We modelled the dependency of the extinction coefficient k as a function of A_0, colour and effective temperature (i.e. $k_\lambda = f(A_0, T_{\mathrm{eff}})$ and $k_\lambda = f(A_0, colour)$, respectively). Regarding the coefficient of the Gaia G band, k_G, we used instead the empirical calibration of Danielski *et al.* (2017, in prep.), who actually makes use of the photometric relationships of this work to derive k_G.

We tested the method on the APOGEE DR13 data. The photometric effective temperatures obtained in this work agree with the spectrometric data of the survey with very low dispersion. Similarly, we obtained a precision of about 11% for the derived interstellar extinctions. We compared the DIB equivalent widths for the APOGEE stars in common with Zasowski *et al.* (2015) with the extinctions obtained here, and our A_0 with the A_K provided in the APOGEE catalogue. See Danielski *et al.* (2017, in prep.) for details.

4. Use of calibrations within the Gaia Data Validation

To test the quality of the Gaia astrometric data it is important to check the zero point of parallaxes and their precision. A way to do it is by using stars distant enough so that their estimated distance uncertainty is better than the Gaia parallax precision (for TGAS meaning $\sigma_{\varpi_{star}} < 0.1$ mas). For the Gaia Data Release 1, we used the APOKASC distances of Rodrigues *et al.* (2014). It contains 948 Tycho-2 sources which we used to check the Gaia parallax zero point (Arenou *et al.* (2017)).

For the Gaia Data Release 2 validation, we will use our photometric calibrations to derive the temperature of the asteroseismic giants, allowing to have a much larger sample of distance modulus to compare with Gaia data.

5. Conclusions

We presented a purely empirical, robust and complete first calibration of the Gaia RC, through colour-G-K_s and T_{eff}-G-K_s relations (see more details in Ruiz-Dern *et al.* (2017, submitted)). The work is being extended to other spectral types, such as dwarfs stars. These calibrations were also used to empirically derive the insterstellar extinction coefficient of the Gaia G band, k_G (Danielski *et al.* (2017, in prep.)).

We applied our photometric calibrations to the Padova isochrones to check the position of the RC on an HR diagram as well as the position of the Red Giant Branch Bump. We also implemented a method that combines the calibrations with an extinction coefficient model and the empirical k_G to derive photometric effective temperatures and interstellar extinctions. We successfully tested it on the APOGEE DR13 data. The photometric interstellar extinctions obtained are being used as input in the new 3D interstellar extinction map of Capitanio *et al.* (2017).

These calibrations have allowed us to determine the absolute magnitude of the RC (Ruiz-Dern *et al.* (2017, submitted)). Thus, more precise distances to large scale structures will be able to be derived.

Finally, their usage is extended to the Gaia astrometric and photometric data validation of DR2, allowing to obtain distance modulus for larger samples of stars and, thus, to improve the verification of, for instance, the Gaia parallax zero-point.

Acknowledgements

This work has made use of data from the European Space Agency (ESA) mission *Gaia* (`https://www.cosmos.esa.int/gaia`), processed by the *Gaia* Data Processing and Analysis Consortium (DPAC, `https://www.cosmos.esa.int/web/gaia/dpac/consortium`). Funding for the DPAC has been provided by national institutions, in particular the institutions participating in the *Gaia* Multilateral Agreement. We acknowledge financial support from the *Centre National d'Etudes Spatiales* (CNES) fellowship program, and from the *Agence Nationale de la Recherche* (ANR) through the STILISM project.

References

Alves, D. R. 2000, *ApJ*, 539, 732
Arenou, F., Luri, X., Babusiaux, C., Fabricius, C., *et al.* 2017, *A&A*, 599, A50
Bressan, A., Marigo, P., Girardi, L., *et al.* 2012, *MNRAS*, 427, 127
Capitanio, L., Lallement, R., J. L. Vergely, *et al.* 2017, *ArXiv e-prints*
Danielski, C., Babusiaux, C., L. Ruiz-Dern, *et al.* 2017 (in prep.)
Fitzpatrick, E. L. & Massa, D. 2007, *ApJ*, 663, 320
Gaia Collaboration, Brown, A. G. A., Vallenari, A., Prusti, T., *et al.* 2016, *A&A*, 595, A2
García Pérez, A. E., Allende Prieto, C., Holtzman, J. A., *et al.* 2016, *AJ*, 151, 144
González Hernández, J. I. & Bonifacio, P. 2009, *A&A*, 497, 497
Groenewegen, M. A. T. 2008, *A&A*, 488, 935
Holtzman, J. A., Shetrone, M., Johnson, J. A., *et al.* 2015, *AJ*, 150, 148
Huang, Y., Liu, X.-W., Yuan, H.-B., *et al.* 2015, *MNRAS*, 454, 2863
Laney, C. D., Joner, M. D., & Pietrzyński, G 2012, *MNRAS*, 419, 1637
Plummer, Martyn 2008, *Biostatistics*, 3, 523
Pourbaix, D., Tokovinin, A. A., Batten, A. H., *et al.* 2009, *VizieR Online Data Catalog*
Ramírez, I. & Meléndez, J. 2005, *ApJ*, 626, 446
Rodrigues, T. S., Girardi, L., Miglio, A., Bossini, D., *et al.* 2014, *MNRAS*, 445, 2758
Ruiz-Dern, L., Babusiaux, C., Arenou, F., Turon, C., & R. Lallement 2017 (submitted), *A&A*
SDSS Collaboration *et al.* 2016, *ArXiv e-prints*
Udalski, A. 2000, *ApJ Letters*, 531, L25
Zasowski, G., Ménard, B., Bizyaev, D., *et al.* 2010, *ApJ*, 798, 35

Astronomy and Astrophysics in the Gaia sky
Proceedings IAU Symposium No. 330, 2017
A. Recio-Blanco, P. de Laverny, A.G.A. Brown
& T. Prusti, eds.

doi:10.1017/S1743921317005427

White dwarfs in the Gaia era

P.-E. Tremblay[1], N. Gentile-Fusillo[1], J. Cummings[2], S. Jordan[3], B. T. Gänsicke[1] and J. S. Kalirai[2,4]

[1]Department of Physics, University of Warwick,
CV4 7AL, Coventry, UK
email: P-E.Tremblay@warwick.ac.uk

[2]Center for Astrophysical Sciences, Johns Hopkins University,
3400 North Charles Street, Baltimore, MD 21218, USA

[3]Astronomisches Rechen-Institut, Zentrum für Astronomie der Universität Heidelberg,
D-69120 Heidelberg, Germany

[4]Space Telescope Science Institute,
3700 San Martin Drive, Baltimore, MD 21218, USA

Abstract. The vast majority of stars will become white dwarfs at the end of the stellar life cycle. These remnants are precise cosmic clocks owing to their well constrained cooling rates. *Gaia* Data Release 2 is expected to discover hundreds of thousands of white dwarfs, which can then be observed spectroscopically with *WEAVE* and *4MOST*. By employing spectroscopically derived atmospheric parameters combined with *Gaia* parallaxes, white dwarfs can constrain the stellar formation history in the early developing phases of the Milky Way, the initial mass function in the 1.5 to 8 $M_\odot$ range, and the stellar mass loss as well as the state of planetary systems during the post main-sequence evolution.

Keywords. astrometry, white dwarfs, solar neighborhood, Galaxy: stellar content, Galaxy: evolution

---

1. Background

The European Space Agency (ESA) astrometric mission *Gaia* will determine positions, parallaxes, and proper motions for a full sky sample with $V \lesssim 20$ mag, corresponding to ~1% of the stars in the Galaxy (Gaia Collaboration *et al.* 2016a). In addition to astrometric measurements, the *Gaia* catalog will include G passband photometry, low-resolution spectrophotometry in the blue (BP, 330–680 nm) and red (RP, 640–1000 nm), and (for bright stars, $G \lesssim 15$) higher-resolution spectroscopy in the region around 860 nm with the Radial Velocity Spectrometer (Jordi *et al.* 2010, Carrasco *et al.* 2014). The final data release is expected to include between 250,000 and 500,000 white dwarfs (Torres *et al.* 2005, Carrasco et al. 2014).

Gaia DR1 was limited to G passband photometry and the astrometric solution for stars in common with the *Hipparcos* and Tycho-2 catalogs (Gaia Collaboration *et al.* 2016b). However, not all *Hipparcos* and Tycho-2 stars are found in *Gaia* DR1 owing to source filtering. Tremblay *et al.* (2017) recovered only 6 directly observed white dwarfs but they have also selected 46 members of wide binaries for which the companion has a precise *Gaia* DR1 parallax. They compared observed and predicted spectroscopic distances and found that the mean agreement is better than 1% between *Gaia* parallaxes, published atmospheric parameters, and theoretical mass-radius relations (Fontaine et al. 2001). In this work, we examine the prospects of studying degenerate stars in *Gaia* DR2 planned for April 2018.

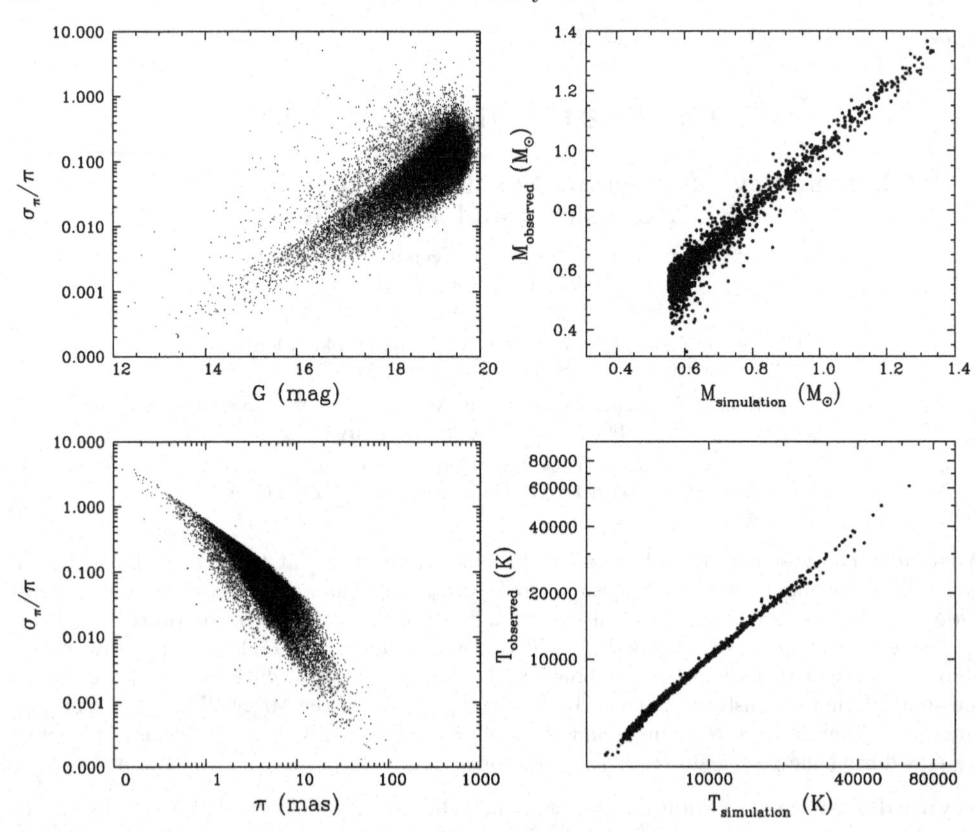

Figure 1. *(Left:) Gaia* DR2 parallax errors as a function of *G*-band magnitude (top left) and parallax (bottom left) for our simulated Galactic white dwarf population. *(Right:)* Photometric fit for all white dwarfs within 100 pc in our *Gaia* DR2 mock catalog. We rely on *Gaia* data alone and our simulation does not include binary evolution. *Gaia* colours are predicted assuming a ratio of 1/4 between He- and H-atmospheres.

2. White dwarfs in Gaia Data Release 2

Gaia DR2 is expected to have the astrometric solutions and integrated G, $G_{\rm BP}$ and $G_{\rm RP}$ photometric fluxes for all sources with acceptable formal standard errors. It will have significantly higher individual precision due to a longer measurement time (22 months instead of 11 months, which is already 36% of the total mission time). Systematic errors are also expected to decrease significantly resulting from a more sophisticated calibration, including a better definition of the line spread function, the application of a chromaticity correction, a more accurate calibration of the basic angle variation, and a calibration and correction of micro clanks.

We have simulated the *Gaia* DR2 white dwarf population using assumptions similar to those presented in Tremblay *et al.* (2016) for the magnitude-limited SDSS sample. The main difference is that *Gaia* is all-sky and it is currently expected that remnants of all colours will be released in DR2 given the discussion above. Our simulation is similar to those presented in Torres *et al.* (2005) and Carrasco *et al.* (2014), except that we use the most recent predicted DR2 errors on parallaxes and colours drawn from `PyGaia`†. The total number of white dwarfs in *Gaia* DR2 is expected to be between 250,000 and 500,000 depending on the Galactic model. In particular, by extrapolating the current

† https://github.com/agabrown/PyGaia

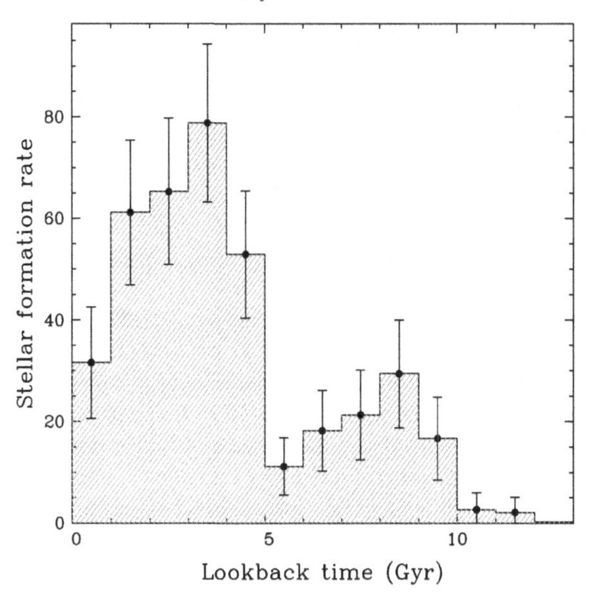

Figure 2. Derived total stellar formation rate (relative number) as a function of lookback time for the Galactic disk at 8 kpc from Galactic centre (see Tremblay *et al.* 2014). The source of the data is the volume complete 20 pc white dwarf sample (Giammichele *et al.* 2012) and the initial-to-final mass relation of Kalirai *et al.* (2008, 2009) to convert white dwarf parameters to initial stellar parameters. We accounted for biases due to the missing main-sequence stars and the velocity dispersion as a function of total stellar age in the direction perpendicular to the plane of the disk. The error bars take into account number statistics uncertainties and are derived from the uncorrected number of white dwarfs. We do not account for radial mixing and as such this represents the formation of stars that are at present day at 8 kpc Galactic radius.

remnant space density in the complete 13 pc sample (Holberg *et al.* 2016), we obtain ~2200 white dwarfs within 50 pc, the maximum distance for which we can hope for a volume complete DR2 sample (Carrasco *et al.* 2014).

Fig. 1 (left panel) illustrates the *Gaia* DR2 parallax precision as a function of G-band magnitude and parallax for our mock white dwarf catalog. It is noted in particular that all white dwarfs within 100 pc will have a distance precision better than ~4%. We performed *Gaia* data only photometric fits of all simulated remnants within 100 pc. *Gaia* colours were predicted assuming a ratio of 1/4 between He- and H-atmospheres. Fig. 1 (right panel) demonstrates that it is possible to extract white dwarf parameters within a few percent from *Gaia* data alone within 100 pc. It is observed that *Gaia* integrated photometric fluxes are not very sensitive to the atmospheric composition since the bandpasses are broad, yet the photometric precision is large enough for accurate $T_{\rm eff}$ determinations. We note that our mock catalog does not include binary evolution, and as such the actual *Gaia* catalog is expected to have a significant number of outliers when using single white dwarf models.

3. Future outlook

Starting with DR2 it will be feasible to compare predicted spectroscopic distances (Holberg *et al.* 2008) for large samples of pure-hydrogen (DA) and pure-helium atmosphere remnants (DB) with Gaia parallaxes. This will effectively test white dwarf atmosphere and structure models at the 1% level over a wide range of cooling ages and masses. It will also be possible to derive precise mass distributions for other white dwarf subtypes, such as cool metal rich objects (DZ, DQ) and magnetic remnants, providing crucial information on the internal structure and progenitors of these spectral types. Furthermore, *Gaia* DR2 will be invaluable in characterising the binary white dwarf population (Toonen *et al.* 2017) and identifying a halo white dwarf sample from proper motions and calculate its luminosity function (Cojocaru *et al.* 2015). Most of these applications can only be performed for a small part of the *Gaia* DR2 sample with currently available independent spectroscopic observations. *WEAVE* (Dalton *et al.* 2016) and *4MOST* (de Jong *et al.* 2014) spectroscopic follow-up surveys will thus have a major impact in enlarging the science output from *Gaia* white dwarfs.

Gaia DR2 will make it possible to update the local stellar formation history shown in Fig. 2 as derived from white dwarfs within 20 pc (Tremblay *et al.* 2014). The Gaia sample within 100 pc will be 10 times larger, significantly reducing the error bars from number statistics. Once again, *WEAVE* and *4MOST* spectroscopic follow-ups will be essential to improve the precision of the conversion of white dwarf parameters to initial stellar parameters from the initial-to-final mass relation (Cummings *et al.* 2016). *Gaia* will evidently allow for a strong synergy between white dwarf research and Galactic archeology.

Acknowledgement

This work has made use of data from the European Space Agency (ESA) mission *Gaia* (https://www.cosmos.esa.int/gaia), processed by the *Gaia* Data Processing and Analysis Consortium (DPAC, https://www.cosmos.esa.int/web/gaia/dpac/consortium). Funding for the DPAC has been provided by national institutions, in particular the institutions participating in the *Gaia* Multilateral Agreement.

References

Bergeron, P., Saffer, R. A., & Liebert, J. 1992, *ApJ*, 394, 228

Carrasco, J. M., Catalán, S., Jordi, C., *et al.* 2014, *A&A*, 565, A11

Cojocaru, R., Torres, S., Althaus, L. G., Isern, J., & García-Berro, E. 2015, *A&A*, 581, A108

Cummings, J. D., Kalirai, J. S., Tremblay, P.-E., Ramirez-Ruiz, E., & Bergeron, P. 2016, *ApJL*, 820, L18

Dalton, G., Trager, S., Abrams, D. C., *et al.* 2016, *Proceedings of the SPIE*, 9908, 99081G

Fontaine, G., Brassard, P., & Bergeron, P. 2001, *PASP*, 113, 409

Gaia Collaboration, Brown, A. G. A., Vallenari, A., *et al.* 2016, *A&A*, 595, A2

Gaia Collaboration, Prusti, T., de Bruijne, J. H. J., *et al.* 2016, *A&A*, 595, A1

Giammichele, N., Bergeron, P., & Dufour, P. 2012, *ApJS*, 199, 29

Holberg, J. B., Bergeron, P., & Gianninas, A. 2008, *AJ*, 135, 1239

Holberg, J. B., Oswalt, T. D., Sion, E. M., & McCook, G. P. 2016, *MNRAS*, 462, 2295

de Jong, R. S., Barden, S., Bellido-Tirado, O., *et al.* 2014, *Proceedings of the SPIE*, 9147, 91470M

Jordi, C., Gebran, M., Carrasco, J. M., *et al.* 2010, *A&A*, 523, A48

Kalirai, J. S., Hansen, B. M. S., Kelson, D. D., *et al.* 2008, *ApJ*, 676, 594

Kalirai, J. S., Saul Davis, D., Richer, H. B., *et al.* 2009, *ApJ*, 705, 408

Toonen, S., Hollands, M., Gaensicke, B. T., & Boekholt, T. 2017, *A&A*, in press, arXiv:1703.06893

Torres, S., García-Berro, E., Isern, J., & Figueras, F. 2005, *MNRAS*, 360, 1381

Tremblay, P.-E., Kalirai, J. S., Soderblom, D. R., Cignoni, M., & Cummings, J. 2014, *ApJ*, 791, 92

Tremblay, P.-E., Cummings, J., Kalirai, J. S., *et al.* 2016, *MNRAS*, 461, 2100

Tremblay, P.-E., Gentile-Fusillo, N., Raddi, R., *et al.* 2017, *MNRAS*, 465, 2849

Astronomy and Astrophysics in the Gaia sky
Proceedings IAU Symposium No. 330, 2017
A. Recio-Blanco, P. de Laverny, A.G.A. Brown & T. Prusti, eds.

doi:10.1017/S1743921317005518

Runaway companions of supernova remnants with Gaia

Douglas Boubert[1], Morgan Fraser[2] and N. Wyn Evans[1]

[1]Institute of Astronomy, University of Cambridge,
Madingley Rise, Cambridge, CB3 0HA, United Kingdom
email: d.boubert@ast.cam.ac.uk, nwe@ast.cam.ac.uk

[2]School of Physics, O'Brien Centre for Science North, University College Dublin,
Belfield, Dublin 4, Ireland
email: morgan.fraser@ucd.ie

Abstract. It is expected that most massive stars have companions and thus that some core-collapse supernovae should have a runaway companion. The precise astrometry and photometry provided by Gaia allows for the systematic discovery of these runaway companions. We combine a prior on the properties of runaway stars from binary evolution with data from TGAS and APASS to search for runaway stars within ten nearby supernova remnants. We strongly confirm the existing candidate HD 37424 in S147, propose the Be star BD+50 3188 to be associated with HB 21, and suggest tentative candidates for the Cygnus and Monoceros Loops.

Keywords. supernovae: general – binaries: close – ISM: supernova remnants

---

1. Introduction

Most massive stars in the range $8 < M/M_{\odot} < 40$ are born in binaries (e.g. Sana *et al.* 2012) and explode in core-collapse supernovae (SNe, e.g. Heger *et al.* 2003). In the SN explosion the envelope of the primary is ejected at more than 5000 km s^{-1} and forms a SN remnant (SNR) shell around the system (e.g. Reynolds 2008). This mass loss can unbind the binary and result in the ejection of the secondary as a runaway star (Blaauw 1961). The ejection velocity is primarily determined by the orbital velocity prior to the SN. Processes such as mass transfer or a common envelope phase can shrink the separation and increase the orbital velocity. Discovering the runaway former companion of a SNR progenitor can thus place tight constraints on the evolutionary history of the progenitor binary. We conduct a targeted search for runaway companions of ten SNRs within 2 kpc using astrometry and photometry in the first data release of the Gaia satellite.

2. Method

We consider all stars within the central quarter radius of each SNR present in a cross-match of the Tycho-Gaia Astrometric Solution (TGAS, Gaia Collaboration *et al.* 2016) with the AAVSO Photometric All-Sky Survey (APASS). We construct predictive models for the hypotheses that a given star is the runaway companion or is a background star. Our prior incorporates our best knowledge of the distance and age of each SNR, a prior on the properties of runaway stars (calculated using the binary evolution software *binary_c*, Izzard *et al.* 2006), and a distribution for the reddening along the line-of-sight to each star (generated based on samples from the Green *et al.* 2015 dust-map). The likelihood function translates these model parameters into predicted observables for each star, which are the parallax, proper motions and G band magnitude from TGAS and $B-V$ colour from APASS. The background model is a kernel density estimate generated

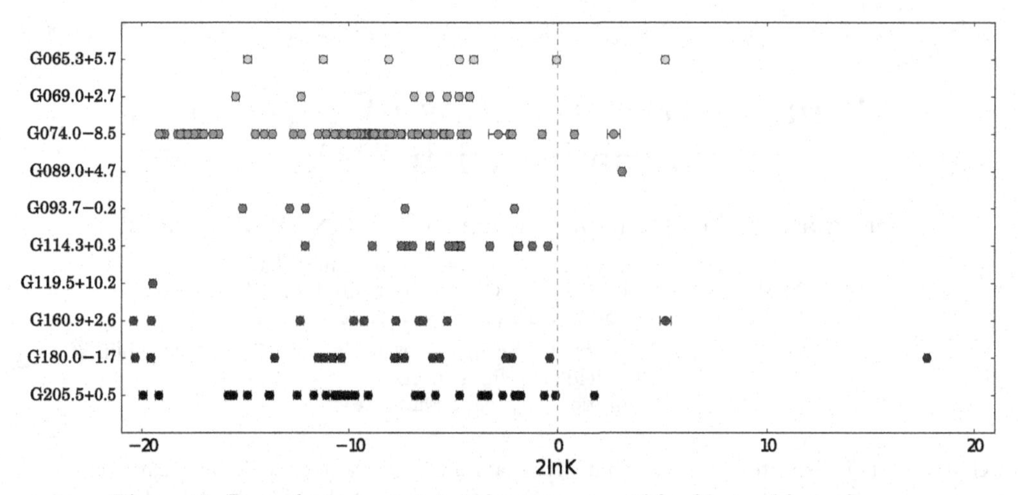

Figure 1. Bayes factor comparing the runaway and background hypotheses.

from stars drawn from an annulus of width 10 deg around the SNR. Further details on the implementation of this method can be found in Boubert *et al.* (2017).

3. Discussion

In Figure 1 we show the Bayes factor K for every star in each SNR. Seven of our candidates have a Bayes factor which favours the runaway hypothesis. Three can be dismissed as contaminants due either to anomalously large errors in APASS or them being foreground stars which are not well described by the background model. The most favoured of the remaining stars is HD 37424 in S147 which Dinçel *et al.* (2015) had suggested as the runaway companion based on a past spatial coincidence with the associated pulsar. Our best novel candidate is the Be star BD+50 3188 in HB 21. The Be star phenomenon has been linked to mass transfer in a binary preceding a SN explosion (Pols *et al.* 1991) and thus that BD+50 3188 is a Be star supports its runaway candidacy. The other candidates are TYC 2688-1556-1 in the Cygnus Loop and HD 261393 in the Monoceros Loop. The release of Gaia DR2 in April 2018 will allow us to both rule out the existence of a runaway companion for nearby SNRs and to extend our search to more distant SNRs.

This work has made use of data from the European Space Agency (ESA) mission Gaia (`https://www.cosmos.esa.int/gaia`), *processed by the Gaia Data Processing and Analysis Consortium (DPAC,* `https://www.cosmos.esa.int/web/gaia/dpac/consortium`*). Funding for the DPAC has been provided by national institutions, in particular the institutions participating in the Gaia Multilateral Agreement.*

References

Blaauw, A. 1961, *Bulletin of the Astronomical Institutes of the Netherlands*, 15, 265
Boubert, D., Fraser, M., Evans, N. W., Green, D., & Izzard, R. G. 2017, arXiv:1704.05900
Dinçel, B., Neuhäuser, R., Yerli, S. K., *et al.* 2015, *MNRAS*, 448, 3196
Gaia Collaboration, Brown, A. G. A., Vallenari, A., *et al.* 2016, *A&A*, 595, A2
Heger, A., Fryer, C. L., Woosley, S. E., Langer, N., & Hartmann, D. H. 2003, *ApJ*, 591, 288
Green, G. M., Schlafly, E. F., Finkbeiner, D. P., *et al.* 2015, *ApJ*, 810, 25
Izzard, R. G., Dray, L. M., Karakas, A. I., Lugaro, M., & Tout, C. A. 2006, *A&A*, 460, 565
Pols, O. R., Cote, J., Waters, L. B. F. M., & Heise, J. 1991, *A&A*, 241, 419
Reynolds, S. P. 2008, *ARA&A*, 46, 89
Sana, H., de Mink, S. E., de Koter, A., *et al.* 2012, *Science*, 337, 444

Astronomy and Astrophysics in the Gaia sky
Proceedings IAU Symposium No. 330, 2017
A. Recio-Blanco, P. de Laverny, A.G.A. Brown
& T. Prusti, eds.

doi:10.1017/S1743921317005464

The TGAS HR diagram of barium stars

A. Escorza[1,2] †, H. M. J. Boffin[3], A. Jorissen[2], L. Siess[2,4], S. Van Eck[2], S. Shetye[2,1], D. Pourbaix[2,4] and H. Van Winckel[1]

[1]Institute of Astronomy, KU Leuven, Celestijnenlaan 200D, B-3001 Leuven, Belgium
email: ana.escorza@kuleuven.be

[2]Institut d'Astronomie et d'Astrophysique, Université Libre de Bruxelles, Campus Plaine C.P. 226, Boulevard du Triomphe, B-1050 Bruxelles, Belgium

[3]ESO, Garching bei München, Germany

[4]F.R.S.-FNRS, Belgium

Abstract. Barium stars are formed via binary interaction with a former AGB companion. Observations are needed to constrain theoretical models and better understand their evolution and surface composition. We present the HR diagram of Ba and related stars, using the recently released TGAS parallaxes, and the mass distribution of the Ba giants that we derived from it.

Keywords. stars: binaries, stars: evolution, Hertzsprung-Russell diagram

---

1. Introduction

Barium (Ba) stars (Bidelman & Keenan 1951), are chemically peculiar giants which show overabundances of elements produced by the slow neutron-capture (s-) process of nucleosynthesis. They got polluted by a former AGB companion in a low- or intermediate-mass binary system (e.g., McClure 1984; Boffin & Jorissen 1988). Binary evolution models fail to reproduce their observed chemical and orbital properties. Observations are essential to constrain interaction physics in these binary systems, which will lead to a better understanding of their formation.

We present a Hertzsprung-Russell diagram (HRD) of Ba and CH (low metallicity equivalents) stars making use of the Tycho-Gaia Astrometic Solution (TGAS, Lindegren *et al.* 2016). We also show the mass distribution of the Ba giants. This work is part of our research to combine the results of our dedicated observational programme with our state-of-the-art binary evolution models produced with BINSTAR (Siess *et al.* 2013) to explore all aspects of the evolution of Ba stars.

2. Methods

Atmospheric parameters of a sample of 437 Ba and CH stars were derived by modelling their spectral energy distribution (SED). We used photometry available in the literature and looked for the best-fitting MARCS model atmosphere (Gustafsson *et al.* 2008) in a parameter-grid search. The temperature ($T_{\rm eff}$) was assigned from the best-fitting model, and the luminosity (L) was obtained by integrating the SED over all wavelengths and

† This work has made use of data from the European Space Agency (ESA) mission *Gaia* (https://www.cosmos.esa.int/gaia), processed by the *Gaia* Data Processing and Analysis Consortium (DPAC, https://www.cosmos.esa.int/web/gaia/dpac/consortium). Funding for the DPAC has been provided by national institutions, in particular the institutions participating in the *Gaia* Multilateral Agreement.

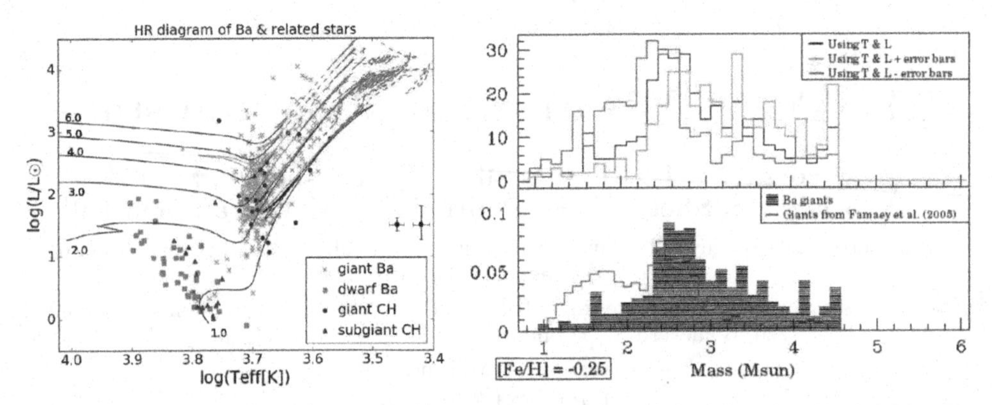

Figure 1. HRD of Ba and CH stars, and mass distribution of Ba and normal K and M giants.

applying the distance modulus derived from the TGAS parallax. Hipparcos parallaxes (ESA 1997) were used when the source was too bright to be part of TGAS. We fixed the metallicity of the atmospheric models to [Fe/H] = -0.25 because the distribution of spectroscopically determined metallicities for Ba stars peaks at this value.

The location of stars on the HRD provides information about their masses and evolutionary status. We derived the mass distribution of the Ba giants interpolating between our STAREVOL (Siess 2006) evolutionary tracks computed also with [Fe/H] = -0.25.

3. Results and discussion

Figure 1 shows the HRD of Ba and CH stars, including the average and maximum error bars, and the mass distribution of Ba giants, which peaks at about 2.6 $M_{\odot}$. The latter is compared with a sample of K and M giants (Famaey *et al.* 2005). The Ba giants concentrate on the red clump, which suggests that they behave like other red giants.

Metallicity influences our result. It affects the stellar lifetime as well as the location of the stars on the HRD through the atmospheric model used to compute $T_{\rm eff}$ and L. This effect will be studied by Escorza *et al.* 2017 (in prep.). Spectroscopically determined metallicities are a key ingredient to obtain information about individual masses.

The HRD presented in this publication will be used by Escorza *et al.* 2017 (in prep.) to compare the evolutionary stage of Ba stars with their orbital parameters. Any correlation found could give us clues about the evolution and interaction history of these binaries.

References

Bidelman, W. P. & Keenan, P. C. 1951, *ApJ*, 114, 473
Boffin, H. M. J. & Jorissen, A. 1988, *A&A*, 205, 155-163
ESA 1997, *ESA Special Publication*, 1200
Famaey, *et al.* 2005, *A&A*, 430, 165-186
Gustafsson, *et al.* 2008, *A&A*, 486, 951-970
Lindegren, L., *et al.* 2016, *A&A*, 595, A4
McClure, R. D. 1984, *PASP*, 96, 117-127
Siess, L. 2006, *A&A*, 448, 717-729
Siess, L. *et al.* 2013, *A&A*, 550, A100

Astronomy and Astrophysics in the Gaia sky
Proceedings IAU Symposium No. 330, 2017
A. Recio-Blanco, P. de Laverny, A.G.A. Brown & T. Prusti, eds.

doi:10.1017/S1743921317006433

Dynamical masses of Cepheids from the GAIA parallaxes

A. Gallenne[1], P. Kervella[2], A. Mérand[3], N. R. Evans[4] and C. Proffitt[5]

[1]European Southern Observatory, Alonso de Córdova 3107, Casilla 19001, Santiago 19, Chile

[2]LESIA, Obs. de Paris, CNRS UMR 8109, UPMC,Univ. Paris 7, 5 Pl. Jules Janssen, 92195 Meudon, France

[3]European Southern Observatory, Karl-Schwarzschild-Straße 2, 85748 Garching bei München, Germany

[4]Smithsonian Astrophysical Observatory, MS 4, 60 Garden Street, Cambridge, MA 02138, USA

[5]Space Telescope Science Institute, 3700 San Martin Drive, Baltimore, MD 21218, USA

Abstract. The mass of a Cepheid is a fundamental parameter for studying the pulsation and evolution of intermediate-mass stars. But determining this variable has been a long-standing problem for decades. Detecting the companions (by spectroscopy or imaging) is a difficult task because of the brightness of the Cepheids and the close orbit of the components. So most of the Cepheid masses are derived using stellar evolution or pulsation modeling, but they differ by 10-20 %. Measurements of dynamical masses offer the unique opportunity to make progress in resolving this mass discrepancy.

The first problem in studying binary Cepheids is the high contrast between the components for wavelengths longer than 0.5 μm, which make them single-line spectroscopic binaries. In addition, the close orbit of the companions ($<$ 40 mas) prevents us from spatially resolving the systems with a single-dish 8m-class telescope. A technique able to reach high spatial resolution and high-dynamic range is long-baseline interferometry. We have started a long-term program that aims at detecting, monitoring and characterizing physical parameters of the Cepheid companions. The GAIA parallaxes will enable us to combine interferometry with single-line velocities to provide unique dynamical mass measurements of Cepheids.

Keywords. techniques: interferometric, stars: Cepheids, binaries

1. Introduction

Cepheids are powerful astrophysical laboratories providing fundamental clues for studying the pulsation and evolution of intermediate-mass stars. However, one of the most critical parameters, the mass, is a long-standing problem because of the 10-20 % discrepancy between masses predicted from stellar evolution and pulsation models. Cepheids in binary systems are the only tool to constrain models and make progress on this discrepancy. Studying Cepheid companions can also provide insight on the impact of binarity on the calibration of the period-luminosity relation from the Baade-Wesselink technique and the IR surface-brightness method. Detectable companions are mostly hot main-sequence stars, therefore the flux contribution in the near-IR might is often negligible ($<$ 1-2%), but can be as large as 10 % in V.

Most of known companions are too close to the Cepheid ($<$ 40 mas) to be resolved with single-dish 8m class telescopes. In addition, the high-contrast between the components makes the detection even more difficult. But long-baseline interferometry (LBI) is able to reach high-spatial resolution and high-dynamic range, which allow us to accurately determine the astrometric position of some companions.

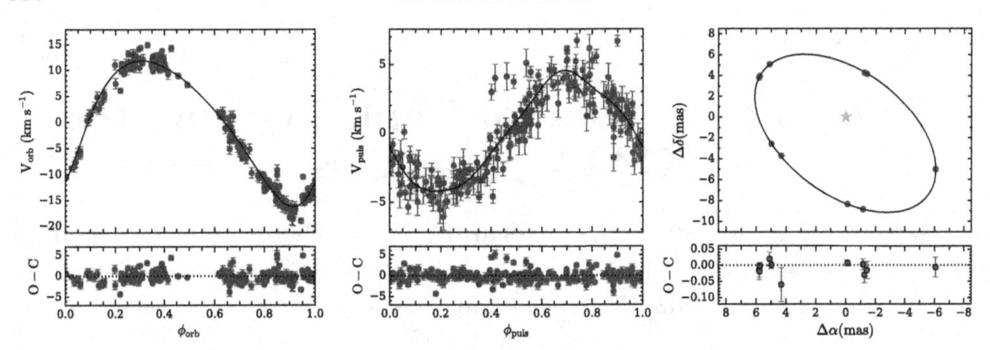

Figure 1. Combined fit of single-line velocities with astrometry for the Cepheid V1334 Cyg. Left: orbital velocity. Middle: pulsation velocity. Right: astrometric motion of the companion.

2. Combining spectroscopy, interferometry and GAIA

Because of the brightness of the Cepheid, it is challenging to spatially/spectrally detect companions using ground-based telescopes. With spectroscopy in the V band, the companion's lines are difficult to disentangle from the Cepheid's, mainly because of the high-contrast. However, the more massive companion causes the Cepheid to wobble (as the two components orbit their center of mass). This is more easily detectable from radial velocities (RVs) of the Cepheids, which makes them single-line spectroscopic binaries. From space, ultraviolet observations from HST/STIS can provide the spectra of the companions. However, broad features (some of them are fast rotators) and blended lines complicate the analysis, and can prevent the determination of accurate RVs. Combining astrometry and RVs of both components will yield estimates on the masses and distances.

With interferometry, we have observed some binary Cepheids in both the northern and southern hemisphere. We detected companions separated by less than 50 mas from the Cepheids with contrast as high as 6.5 mag in H. For this purpose, we have created a dedicated tool, named `CANDID`, which searches for high-contrast companions from interferometric data (Gallenne *et al.* 2015). The tool delivers, among other things, the flux ratio f, the relative astrometric separation $(\Delta\alpha, \Delta\delta)$, and the (non-)detection level of the companion based on χ^2 statistics. So far, we have detections for six binary Cepheids (V1334 Cyg, AW Per, RT Aur, BP Cir, AX Cir, S Mus), with projected separations ranging from 1.5 to 40 mas and flux ratios (in H) from 0.8 to 4 %. The astrometric positions can be combined with the RVs of the primary, as shown in Fig. 1 for V1334 Cyg, to provide the full set of orbital elements, including a, i and Ω previously unknown (Gallenne *et al.* 2013, 2014). However, the distance and masses are still degenerate, and unfortunately no accurate parallaxes exist for these binary Cepheids. The coming GAIA parallaxes will allow us to break this degeneracy, and we will then be able to combine interferometry with single-line velocities to provide dynamical mass measurements of Cepheids.

References

Gallenne, A., Mérand, A., Kervella, P., *et al.* 2014, *A&A*, 561, L3
Gallenne, A., Mérand, A., Kervella, P., *et al.* 2015, *A&A*, 579, A68
Gallenne, A., Monnier, J. D., Mérand, A., *et al.* 2013, *A&A*, 552, A21

Astronomy and Astrophysics in the Gaia sky
Proceedings IAU Symposium No. 330, 2017
A. Recio-Blanco, P. de Laverny, A.G.A. Brown & T. Prusti, eds.

doi:10.1017/S1743921317005683

Confronting the Gaia and NLTE spectroscopic parallaxes for the FGK stars

Tatyana Sitnova, Lyudmila Mashonkina and Yury Pakhomov

Institute of Astronomy, Russian Academy of Sciences,
Pyatnitskaya 48, 119017, Moscow, Russia
email: sitnova@inasan.ru

Abstract. The understanding of the chemical evolution of the Galaxy relies on the stellar chemical composition. Accurate atmospheric parameters is a prerequisite of determination of accurate chemical abundances. For late type stars with known distance, surface gravity (log g) can be calculated from well-known relation between stellar mass, $T_{\rm eff}$, and absolute bolometric magnitude. This method weakly depends on model atmospheres, and provides reliable log g. However, accurate distances are available for limited number of stars. Another way to determine log g for cool stars is based on ionisation equilibrium, i.e. consistent abundances from lines of neutral and ionised species. In this study we determine atmospheric parameters moving step-by-step from well-studied nearby dwarfs to ultra-metal poor (UMP) giants. In each sample, we select stars with the most reliable $T_{\rm eff}$ based on photometry and the distance-based log g, and compare with spectroscopic gravity calculated taking into account deviations from local thermodinamic equilibrium (LTE). After that, we apply spectroscopic method of log g determination to other stars of the sample with unknown distances.

Keywords. stars: fundamental parameters, abundances.

Nearby dwarfs, -2.6 ⩽ [Fe/H] ⩽ 0.2. For 20 stars, with accurate non-spectroscopic atmospheric parameters, namely, $T_{\rm eff}$ from the infrared flux method (IRFM, Alonso *et al.* 1996, Casagrande *et al.* 2011) and log g from the Hipparcos parallax (van Leeuwen2007), which was measured with an error of no more than 10 % we found consistent within 0.06 dex NLTE abundances from lines of Fe I and Fe II. For other 30 stars of the sample with larger parallax errors we determined spectroscopic log g. NLTE leads to up to 0.5 dex higher log g compared to LTE. Our whole stellar sample has 22 stars in common with the Gaia DR1 (Gaia Collaboration *et al.* 2016). Employing the Gaia DR1 and Hipparcos parallaxes leads to consistent surface gravities for all our stars within 80 pc from the Sun and, thus, consistent spectroscopic and distance based surface gravities. For the more distant stars, up to d = 250 pc, our spectroscopic determinations were found to agree well with log g_{Gaia}, i.e. Δ log g (Gaia - Sp) = -0.01 $\pm$ 0.07, while systematically lower log g are obtained, when using the Hipparcos parallaxes, with Δlog g(Hipparcos - Sp) = -0.15 $\pm$ 0.12 for the stars more distant than 80 pc. This proves that the NLTE spectroscopic method can provide reliable surface gravity for distant objects.

Giants, -4.0 ⩽ [Fe/H] ⩽ -1.5. Our giant sample contains 36 stars from dwarf spheroidal galaxies (dSphs) Sculptor, Ursa Minor, Sextans, Fornax, Boötes I, Ursa Major II, and Leo IV, and 24 stars from the MW halo. $T_{\rm eff}$ were obtained with calibration from Ramírez & Meléndez 2005 for the V-IJK colours. For giants with distance based log g and [Fe/H] ⩾ -3.7 we found consistent abundances from Fe I and Fe II and Ti I and Ti II in NLTE. NLTE leads to positive abundance difference from neutral and ionised species in giants with [Fe/H] ⩽ -3.7. For MW halo stars we determined spectroscopic log g, the six stars among them have Gaia parallax measurements. For the four stars, our log g_{sp} are consistent with log g_{Gaia} within the error bars, and the difference log g (Sp - Gaia) does not exceed 0.11 dex. For example, for CD-38 245, log g_{Gaia} = 1.91, and we adopted parameters 4900/2.0/-3.9, which lead to Fe I – Fe II = 0.11 dex, and Ca I – Ca II = -0.05 ± 0.07 dex in NLTE. An outlier is HD 8724, with log g (Gaia-Sp) = -0.76 dex despite of small statistical, σ (log g_{Gaia}) = 0.08 dex, σ (log g_{Sp}) = 0.24 dex, and systematic errors, Δ log g (Sp - Gaia) = -0.07 dex. To reconcile the Fe I and Fe II abundances with log g_{Gaia} = 2.05, $T_{\rm eff}$ of HD 8724 should be increased by $\simeq$ 400 K. This seems very unlikely, because all estimates, based on the IRFM are close to $T_{\rm eff}$ = 4560 K derived in this study.

For *ultra MP halo stars, [Fe/H] ⩽ -4.0* distances are not available, and reliable Fe II lines cannot be detected. We selected 11 UMP stars, where Ca I and Ca II lines can be detected in their observed spectra. We adopted photometric $T_{\rm eff}$ and log g from evolutionary tracks (Yi *et al.* 2004), and checked Ca I-II NLTE ionisation equilibrium. For all stars of the sample, Ca I and Ca II agree in NLTE within error bars. An abundance difference Ca I – Ca II does not exceed 0.16 dex in absolute value in NLTE, while it is negative, and reaches -0.90 dex in LTE. Ca I/Ca II NLTE ionization equilibrium can be used for determination of atmospheric parameters of UMP stars.

Conclusions. We found that, in wide range of luminosity and metallicity, different methods of atmospheric parameter determination based on photometry, distances, NLTE ionisation equilibrium, evolutionary tracks can provide consistent atmospheric parameters for given star. We also note that $T_{\rm eff}$-colour calibrations are not available for [Fe/H] ⩽ -4 stars, but desired. For UMP stars, determination of $T_{\rm eff}$ by extrapolating the calibration to lower [Fe/H] leads to large uncertainty. From independent spectroscopic method we confirm that Gaia DR1 parallaxes are accurate.

Acknowledgments

This work has made use of data from the European Space Agency (ESA) mission *Gaia* (https://www.cosmos.esa.int/gaia), processed by the *Gaia* Data Processing and Analysis Consortium (DPAC, https://www.cosmos.esa.int/web/gaia/dpac/consortium). Funding for the DPAC has been provided by national institutions, in particular the institutions participating in the *Gaia* Multilateral Agreement. The authors are grateful to IAU and the grant on Leading Scientific Schools 9951.2016.2. A part of this study is carried out within a project 'The Formation and Evolution of the Galactic Halo', International Space Science Institute, Bern, Switzerland. We made use the ESO Science Archive Facility, SIMBAD, MARCS, 2MASS, SDSS, and VALD databases.

References

Alonso, A., Arribas, S., & Martinez-Roger, C. 1996, *A&AS*, 117, 227

Casagrande, L., Schönrich, R., Asplund, M., Cassisi, S., Ramírez, I., Meléndez, J., Bensby, T., & Feltzing, S. 2011, *A&A*, 530, A138

Gaia Collaboration, Brown, A. G. A., Vallenari, A., Prusti, T., de Bruijne, J. H. J., Mignard, F. *et al.* 2016, *A&A*, 595, A2

Ramírez, I. & Meléndez, J. 2005, *ApJ*, 626, 465

van Leeuwen, F. 2007, *A&A*, 474, 653

Yi, S. K., Demarque, P., & Kim, Y.-C. 2004, *Ap&SS*, 291, 261

Astronomy and Astrophysics in the Gaia sky
Proceedings IAU Symposium No. 330, 2017
A. Recio-Blanco, P. de Laverny, A.G.A. Brown
& T. Prusti, eds.

doi:10.1017/S1743921317005567

Double, triple and quadruple-line spectroscopic binary candidates within the Gaia-ESO Survey

Thibault Merle[1], Sophie Van Eck[1], Alain Jorissen[1]
Mathieu Van der Swaelmen[1], Gregor Traven[2] and Tomaz Zwitter[2]

[1]Institut d'Astronomie et d'Astrophysique, Université Libre de Bruxelles
CP 226, boulevard du Triomphe, B-1050 Brussels, Belgium
email: thibault.merle@ulb.ac.be

[2]Faculty of Mathematics and Physics, University of Ljubljana,
Jadranska 19, 1000, Ljubljana, Slovenia

Abstract. The Gaia-ESO Survey (GES, Gilmore *et al.* 2012) provides a unique opportunity to detect spectroscopically multiplicity among different populations of the Galaxy using the cross-correlation functions (CCFs). We present here the GES internal Data Release 4 (iDR4) results of the detection of double, triple and quadruple-line spectroscopic binary candidates (SBs) and discuss some peculiar systems.

Keywords. (stars:) binaries: spectroscopic, techniques: radial velocities, techniques: spectroscopic, methods: data analysis, surveys, catalogs

---

We developed a method based on the computation of the CCF successive derivatives to detect multiple peaks and determine their radial velocities, even when the peaks are strongly blended. The Detection Of Extrema (DOE) code therefore allows to automatically detect multiple line spectroscopic binaries (SBn, $n \geqslant 2$). We reveal about 354 SBn candidates in iDR4 (342 SB2, 11 SB3 and even one SB4), including only nine SBn already known in the literature. About 98% of these SBn candidates are new because of their faint visual magnitude ($V < 19$), as clearly illustrated by the comparison with the Geneva-Copenhagen Survey (GCS, Holmberg *et al.* 2009; see left panel of Fig. 1). In particular:

• An orbital solution could be computed for two new SB2: 06404608+0949173 in NGC 2264 (known as V642 Mon) and 19013257-0027338 in Berkeley 81.

• Three giant SB2 candidates (right panel of Fig. 1) are surprising, since they should have a mass ratio very close to 1. Interestingly, the one with the lowest gravity (a CoRoT target) shows a very short estimated period of about 7.5 d compared to minimum periods of 25 d (resp. 200 d) for samples of K (resp. M) giants (Mermilliod *et al.* 2007, Jorissen *et al.* 2009). This system will provide important insight into the evolution of close binaries composed of stars with similar (low) masses.

• Eleven SB3 have been discovered (left and middle panels of Fig. 2) and given the stability issues raised by the dynamics of such systems, it is important to increase the statistics of their detection and to constrain the hierarchy of their orbits. A monitoring of the most probable SB3 candidates have been proposed with UVES/VLT.

• HD 74438 (A2V, V = 7.5) is a candidate quadruple system (SB4, right panel of Fig. 2), member of the very young open cluster IC 2391 (50 Myr). Indeed, the velocities of the A and B components (corresponding to the highest CCF peaks on the top right panel of Fig. 2) vary slowly and oppositely to each other, with similar variation amplitudes

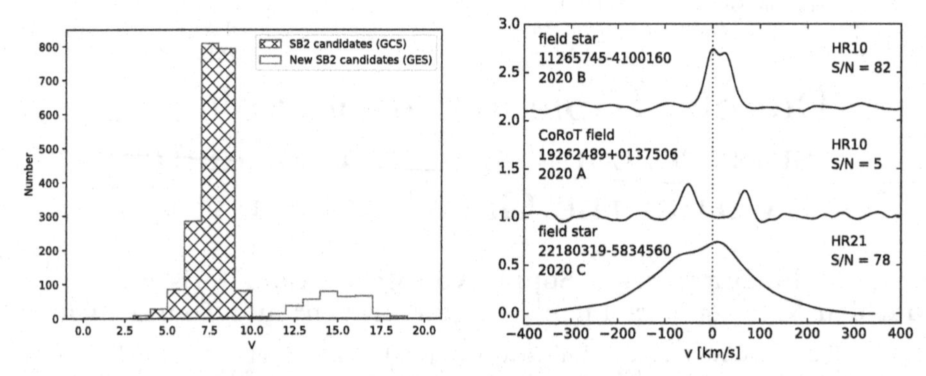

Figure 1. Left: Histogram of the SB2 candidates within the GCS and within the GES. Right: The CCFs of the three SB2 giant candidates.

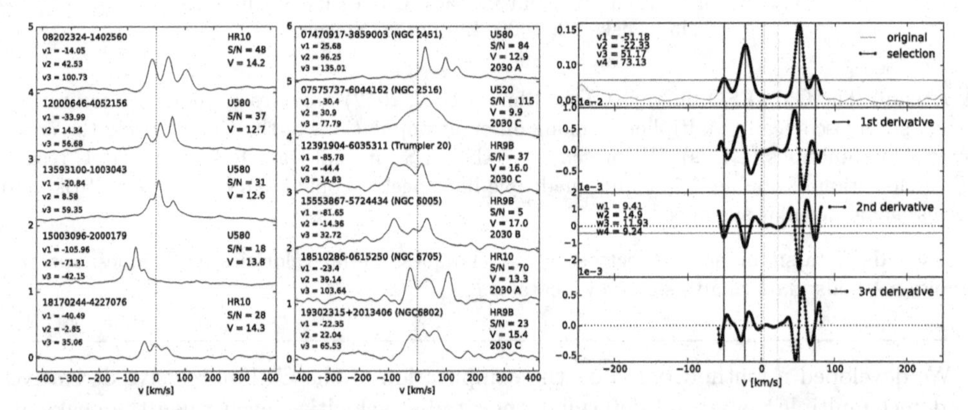

Figure 2. Left: five SB3 candidates in the field. Middle: six SB3 candidates in clusters of various ages. Right: CCF and the three successive derivatives of the SB4 candidate.

with respect to the cluster velocity (14.8 ± 1 km s^{-1}), as expected in a binary system ($\Delta v_r(B)/\Delta v_r(A) = M_A/M_B$). The same prevails for the CD pair. From a preliminary analysis, we found an upper limit on the period of each pair: $P_{AB} < 155$ d and $P_{CD} < 1.2$ d. Such an SB4 candidate would clearly deserve a follow-up.

The frequency of SBn ($n \geqslant 2$) found in the GES iDR4 sample is 0.7%, comparable to the RAVE detection rate (Matijevič *et al.* 2010). Our automated method has allowed an efficient discovery of many new multiple systems (Merle *et al.*, submitted). New CCFs calculated with adapted masks are under investigation to increase the rate detection (see M. Van der Swaelmen's contribution, this issue). We warn against the use of atmospheric parameters for these system components not derived by SB-specific pipelines. Our method can easily be applied to other spectroscopic surveys like the ESA Gaia mission.

References

Gilmore, G., Randich, S. Asplund *et al.* 2012 *The Messenger*, 147, 25
Holmberg, J., Nordström, B., & Andersen, J. 2009 *A&A* 501, 941
Jorissen, A., Frankowski, A., Famaey, B., & Van Eck, S. 2009 *A&A* 498, 489
Matijevič, G., Zwitter, T., Munari, U. *et al.* 2010 *AJ*, 140, 184
Merle, T., Van Eck, Jorissen, A., Van der Swaelmen, M. *et al.* submitted to *A&A*
Mermilliod, J.-C., Andersen, J., Latham, D. W., & Mayor M. 2007 *A&A*, 473, 829

Astronomy and Astrophysics in the Gaia sky
Proceedings IAU Symposium No. 330, 2017
A. Recio-Blanco, P. de Laverny, A.G.A. Brown & T. Prusti, eds.

doi:10.1017/S1743921317005439

Stellar Parameters, Chemical composition and Models of chemical evolution

T. Mishenina[1], M. Pignatari[2,3,6], B. Côté[3,4,5,6] F.-K. Thielemann[7], C. Soubiran[8], N. Basak[1], T. Gorbaneva[1], S. A. Korotin[1,9], V. V. Kovtyukh[1], B. Wehmeyer[7], S. Bisterzo[3,10,11], C. Travaglio[3,10,11], B. K. Gibson[2,6], C. Jordan[2,6], A. Paul[4], C. Ritter[3,6] and F. Herwig[3,6]

[1]Astron. Obs., Odessa Natl. Univ., Shevchenko Park, 65014, Odessa, Ukraine
email: tmishenina@ukr.net
[2]E.A. Milne Centre for Astrophys., Univ. of Hull, HU6 7RX, United Kingdom
[3]The NuGrid Collaboration, http://www.nugridstars.org
[4]Dep. of Phys. and Astron., Univ. of Victoria, Victoria, BC, V8W 2Y2, Canada
[5]Natl. Superconducting Cyclotron Lab., Michigan State Univ., East Lansing, MI, 48824, USA
[6]Joint Inst. for Nuclear Astrophys. Center for the Evolution of the Elements, USA
[7]Dep. of Phys., Univ. of Basel, Klingelbergstrabe 82, 4056 Basel, Switzerland
[8]Lab. d'Astrophys. de Bordeaux, Univ. Bordeaux - CNRS, B18N, 33615 Pessac, France
[9]Crimean Astrophys. Obs., Nauchny, 298409, Crimea
[10]INAF, Astrophys. Obs. Turin, Strada Oss. 20, I-10025 Pino Torinese (Turin), Italy
[11]B2FH Association, Turin, Italy

Abstract. We present an in-depth study of metal-poor stars, based high resolution spectra combined with newly released astrometric data from Gaia, with special attention to observational uncertainties. The results are compared to those of other studies, including Gaia benchmark stars. Chemical evolution models are discussed, highlighting few puzzles that are still affecting our understanding of stellar nucleosynthesis and of the evolution of our Galaxy.

Keywords. stars: abundances, Galaxy: abundances, Galaxy: evolution.

---

1. Observations and abundance determination

The stellar spectra considered in this work have been obtained with the echelle spectrograph SOPHIE on the 1.93m telescope of OHP (France) which has a resolving power of R = 75 000 and covers the wavelengths range 4400 – 6800 ÅÅ. The atmospheric parameter determinations are from Mishenina *et al.* (2017). The abundances of the investigated elements Li, O, Na, Mg, Al, Si, Ca, Ni, Co, Mn, Y, Zr, Ba, La, Ce, Nd, Sm, Eu and Gd were determined for our target stars under LTE and NLTE approximations.

2. Chemical evolution

The theoretical galactic chemical evolution (GCE) calculations compared with these observations have a number of uncertanties and approximations to take into account, possibly leading to different results.

In the Fig. 1, we present a comparison between our results, and a number of GCE models produced using different codes. The black lines presented code OMEGA, a one-zone model (solid and dashed lines correspond to the massive star yields and the no-cutoff prescriptions for the stellar remnant masses, respectively. The black dotted lines represent NuGrid Set 1 extension massive star yields Côté *et al.* (2016). The GCE model predictions by Bisterzo *et al.* (2014) are shown with red lines (solid line - thin disk, dashed line - thick disk, dashed-dotted line - halo). The green solid line the solar neighbourhood

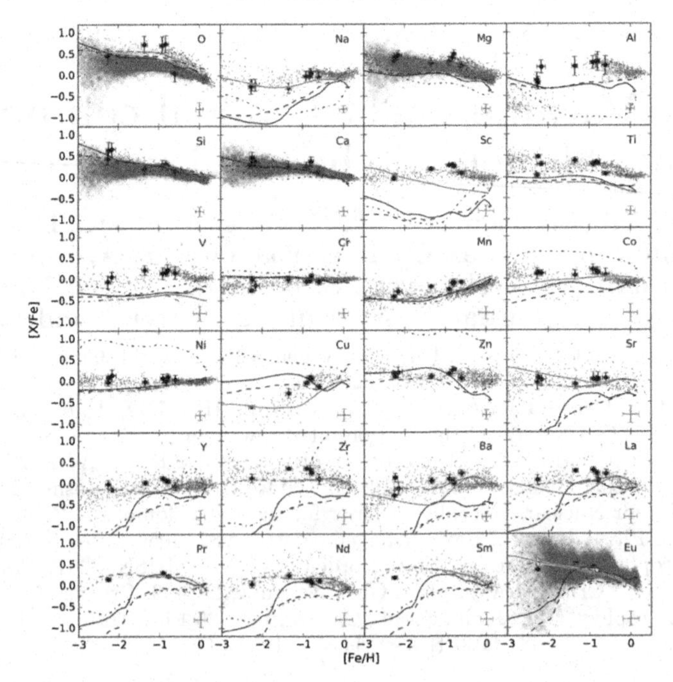

Figure 1. The trends of [El/Fe] vs. [Fe/H] for our stellar sample are marked as full symbols, for other studies as blue and magenta symbols.

chemical evolution model described by Hughes *et al.* (2008), realised with the GEtool software package. Results from the inhomogeneous GCE model by ICE code Wehmeyer, Pignatari & Thielemann (2015) are shown with magenta crosses. Details for the different codes and setup of the GCE models are given in Mishenina *et al.* (2017).

3. Results and Conclusions

– The abundances for 14 to 27 elements were derived using both LTE and NLTE approaches for 10 stars.

– The main sources of GCE uncertainty are from stellar yields and from different assumptions in GCE simulations, e.g., the stellar mass range on which stellar yields are applied, the interpolation scheme between stellar models, the stellar initial mass function, the star formation history, the star formation efficiency (related to the gas fraction), the treatment of SNe Ia, the astrophysical sites for heavy elements, and the galaxy framework (single- or multi-zone).

– Predictions from different GCE models produce a scatter larger than observational errors for many elements. Finally, we confirm the well-known difficulties in reproducing the evolution of [Sc/Fe], [Ti/Fe], and [V/Fe].

References

Bisterzo S., Travaglio C., Gallino R., Wiescher M. & Käppeler F. 2014, *ApJ*, 787, 10
Côté B., West C., Heger A. *et al.* 2016, *MNRAS*, 463, 3755
Hughes G. L., Gibson, B. K., Carigi L. *et al.* 2008, *MNRAS*, 390, 1710
Mishenina T., Pignatari M., B. Côté *et al.* 2017, *eprint arXiv:1705.03642*
Wehmeyer B., Pignatari M. & Thielemann F.-K. 2015, *MNRAS*, 452, 1970

Astronomy and Astrophysics in the Gaia sky
Proceedings IAU Symposium No. 330, 2017
A. Recio-Blanco, P. de Laverny, A.G.A. Brown
& T. Prusti, eds.

doi:10.1017/S1743921317006196

Long term near infrared observation of very bright stars at Kagoshima University

Takairo Naagaya

Graduate School of Science and Engineering, Kagoshima University, 1-21-35 Korimoto, Kagoshima 890-0065, Japan
email: nagayama@sci.kagoshima-u.ac.jp

Abstract. We are monitoring nearby long period variable stars (LPVs) in the near infrared K band to establish their precise Period-Luminosity relation. However, they are very bright in the near-infrared and it is difficult to observe them because they are easily saturated on the modern near-infrared camera. We developed a special ND filter, named Local Attenuation Filter (LAF), to observe very bright stars. Using LAF, we can observe not only the very bright targets without saturation but also reference stars in the same image. We can perform the accurate relative photometry for the bright stars. We present this new method to observe bright stars as well as the status of our monitoring of nearby LPVs.

1. Introduction

We, Kagoshima university, are monitoring about 1000 long period variable stars (LPVs) selected from the IRAS 2-colors diagram in the near-infrared K band to reveal the structure of Milky Way using the NIR period luminosity relation (PLR) of LPVs. The PLR of LPVs was found and has been mainly calibrated in the Large Magellanic Clouds (Glass and Evans 1981, Feast 1989), but the calibration in Milky Way is not performed well (Nakagawa *et al.* 2016). The improvement of PLR is still needed for the precise distance determination of galactic LPVs. In order to establish the high precision PLR of LPVs in Milky Way, the excellent distance measurements for the nearby and less foreground extinction stars are necessary. Hence, we are also monitoring the nearby LPVs with the water maser emission whose distances are expected to be determined by VERA (Japanese VLBI network). However, such nearby LPVs are very bright in the NIR wavelength, typically 1-5 mag in the K band. The modern astronomical instruments are too sensitive, and detector is easily saturated. How can we observe it ?

2. The status of our nearby LPV monitoring

We had observed such bright stars with large defocus, but reference stars in the same image were also defocused and not detected. Therefore, we needed to observe photometric standard stars separately. In order to observe very bright near-infrared stars without saturation, we developed a special ND filter in which only small portion works as the ND filter with a transparency of 1/5000 but the other part does not attenuate the flux at all. We named this filter Local Attenuation Filter (hereafter LAF5000). Fig.1 is a picture of LAF5000 in the filter cassette. The detail of LAF5000 is described in Nagayama(2016). Since only the flux through this patch is attenuated and the fluxes passing outside the patch are not attenuated, the attenuated region is generated on the part of detector array if we install LAF5000 near the telescope focal plane. We can therefore observe the attenuated bright star, together with the not attenuated field stars, simultaneously. The field stars can be used as the reference stars for the relative photometry.

Figure 1. (left) LAF5000 in the filter cassette. (right) An image obtained with LAF5000. The star indicated by the arrow is actually very bright but attenuated by the local attenuated patch in this image.

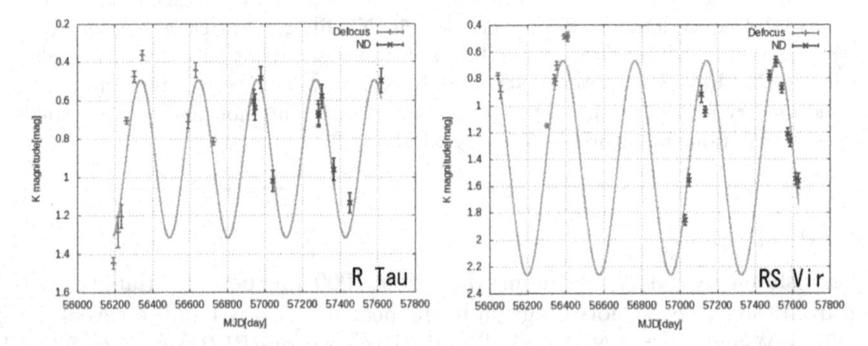

Figure 2. K band light curves of R Tau (left) and RS Vir (right). The data points with MJD<56800 are obtained by the large defocus but >56800 are obtained with LAF5000.

3. Local Attenuation Filter

We are monitoring about 50 bright LPVs for the PLR calibration. Fig. 2 shows the K band light curves of two nearby LPVs, R Tau and RS Vir. These light curves are combination of two observation methods, the large defocus (MJD $<$ 56800) and LAF5000 (MJD $>$ 56800), but we can see that they are connected very smoothly. The number of targets are now limited by the association of bright water maser because the only method to determine the accurate distance of them before Gaia is the VLBI observation for the maser sources. However, we are considering to extend our target to all nearby LPVs. Nearby LPVs are very bright in NIR, but not so bright in the Gaia wavelength. Therefore, they are not saturated in the Gaia photometry and we expect that Gaia determine their distance accurately.

This research is supported by the Optical and Near-infrared Astronomy Inter-University Cooperation Program of Japan and JSPS KAKENHI Grant Number JP25103509.

References

Feast, M. W., Glass, I. S., Whitelock, P. A. & Catchpole, R. M., Monthly Notices of the Royal Astronomical Society (ISSN 0035-8711), vol. 241, Nov. 15 , 1989, p. 375-392 (1989)

Glass, I. S. & Evans, T. L., *Nature*, vol. 291, May 28-June 4, 1981, p. 303, 304. (1981)

Nakagawa, A. *et al.* Publications of the Astronomical Society of Japan, Volume 68, Issue 5, id.78 14 pp., (2016)

Nagayama, T. Proceedings of the SPIE, Volume 9912, id. 991237 6 pp. (2016)

Astronomy and Astrophysics in the Gaia sky
Proceedings IAU Symposium No. 330, 2017
A. Recio-Blanco, P. de Laverny, A.G.A. Brown & T. Prusti, eds.

doi:10.1017/S1743921317005543

The Baade-Wesselink p-factor of Cepheids in the Gaia area

Nicolas Nardetto

Laboratoire Lagrange, UMR 7293, Université de Nice Sophia-Antipolis, CNRS, Observatoire de la Côte d'Azur, France
email: Nicolas.Nardetto@oca.eu

Abstract. With the next *Gaia* release (expected in April 2018), the distance of about 300 Galactic Cepheids will be derived with a precision of better than 3%. These distances will be used first to constrain the Cepheid period-luminosity relation, but they will also bring strong constrains on the physics of Cepheids, through the projection factor, a physical quantity used in the inverse Baade-Wesselink (BW) method.

Keywords. Cepheids, projection factor, Gaia, distances

---

1. The astrophysics behind the projection factor

The BW method is used to determine the distance of Cepheids in the Milky Way and beyond, in the Magellanic Clouds, and consists in combining the angular size variations of the star with its linear size variation. The angular size variation can be determined using infrared surface-brightness relations (Storm *et al.* 2011, Storm *et al.* 2011b), interferometry (Kervella *et al.* 2004a) or even a full set of photometric and interferometric data (SPIPS approach; Merand *et al.* 2015). The linear size variation is deduced from spectroscopy. The radial velocity curve is first derived from a spectral line profile or a set of spectral line profiles (cross-correlation). The radial velocity curve is then multiplied by a projection factor, which is used to derive the true pulsation velocity curve of the star. Finally this pulsation velocity curve is time-integrated in order to derive the radius variation of the star. In this approach, the projection factor and the distance of the star are fully degenerate. Thus, if the distance is known, the p-factor can be derived. This has been done for several Cepheids already (Merand *et al.* 2005, Breitfelder *et al.* 2016, Kervella *et al.* 2017), but with *Gaia* parallaxes, it should be possible to derive the projection factor of about 300 Cepheids with a 3% precision. In this context, the p-factor decomposition into three sub-concepts proposed by Nardetto *et al.* 2007 will be useful in order to interpret the *Gaia* p-factors. For a Cepheid described simply by a uniform disk pulsating, the value of the projection factor is 1.5 (whatever the pulsation phase). But actually, the radial velocity of each surface element of the star is projected along the light of sight and weighted by the intensity distribution of the Cepheid. The limb-darkening of δ Cep reduces the p-factor significantly, and the so-called geometric projection factor (p_0, **step 1** in Fig. 1) is between 1.36 to 1.39, depending on the wavelength in the visible range. The time variation of the p-factor, due mainly to limb-darkening variation, is neglected as it has no impact on the distance (Nardetto *et al.* 2006b). However, a Cepheid is not simply a limb-darkened pulsating photosphere, it has also an extended atmosphere with various spectral lines (in absorption) forming at different levels from which we derive the radial velocity curve used in the BW method. Moreover, there is a velocity gradient in the atmosphere of the Cepheid, which can be measured from spectroscopic observations (**step 2**). Then, depending on the line considered, the amplitude of the radial velocity curve will not be the same and the resulting projection factor will

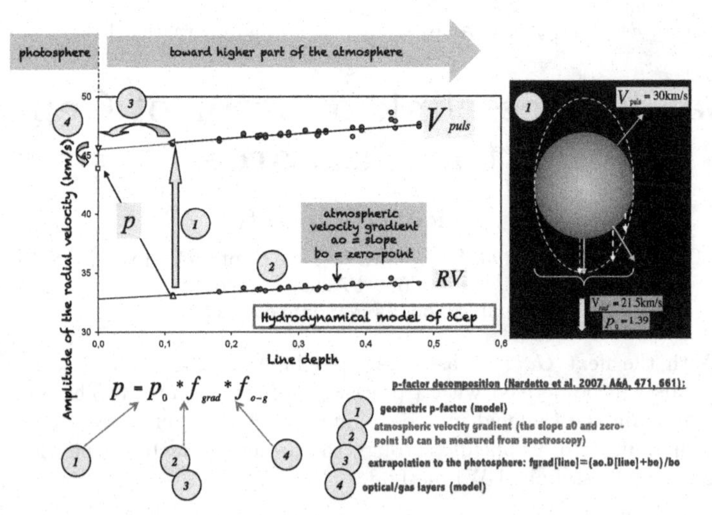

Figure 1. The p-factor decomposition is illustrated based on the model of δ Cep.

be different. In Fig. 1 ($f_{\rm grad}$, **step 3**), we show the impact of the atmospheric velocity gradient on the p-factor for a line forming rather close to the photosphere (line depth of about 0.1). The higher is the line forming region in the atmosphere, the lower is the projection factor (up to 3% compared to p_0 in the case of δ Cep). The last correction on the projection factor ($f_{\rm o-g}$, **step 4**) is more subtle. In spectroscopy, the radial velocity is actually a velocity associated with the moving *gas* in the line forming region, while in photometry or interferometry, we probe an *optical* layer corresponding to the black body continuum (i.e. the layer from which escape the photons). A correction on the projection factor of several percents (independent of the wavelength or the line considered) has to be considered. A relation between the period of Cepheids and the p-factor has been established using this approach for a specific line (Nardetto *et al.* 2007) or using the cross-correlation method (Nardetto *et al.* 2009). This decomposition of the projection into physical concepts has been recently validated in the case of δ Cep thanks to HARPS-N spectroscopic data (Nardetto *et al.* 2017). We thus have all the conceptual tools in hands in order to interpret the future Gaia p-factors.

Acknowledgements

The author acknowledges all his collaborators, the support of the French Agence Nationale de la Recherche (ANR) and the financial support from "Programme National de Physique Stellaire" (PNPS) of CNRS/INSU, France.

References

Breitfelder, J., Mérand, A., Kervella, P.,*et al.* 2016, *A&A*, 587, 117
Kervella, P., Nardetto, N., Bersier, D., *et al.* 2004, *A&A*, 416, 953
Kervella, P., Trahin, B., Bond, H. E., *et al.* 2017, *A&A*, 600, 127
Mérand, A., Kervella, P., Coudé du Foresto, V., *et al.* 2005, *A&A*, 438, 9
Mérand, A., Kervella, P., Breitfelder, J., *et al.* 2015, *A&A*, 584, 80
Nardetto, N., Fokin, A., Mourard, D., *et al.* 2004, *A&A*, 428, 137
Nardetto, N., Fokin, A., Mourard, D., *et al.* 2006b, *A&A*, 454, 332
Nardetto, N., Mourard, D., & Mathias, P., *et al.* 2007, *A&A*, 471, 661
Nardetto, N., Gieren, W., Kervella, P., *et al.* 2009, *A&A*, 502, 956
Nardetto, N., Poretti, E., Rainer, M., *et al.* 2017, *A&A*, 597, 73
Storm, J., Gieren, W., Fouqué, P., *et al.* 2011, *A&A*, 534, 94
Storm, J., Gieren, W., Fouqué, P., *et al.* 2011b,*A&A*, 534, 95

Astronomy and Astrophysics in the Gaia sky
Proceedings IAU Symposium No. 330, 2017
A. Recio-Blanco, P. de Laverny, A.G.A. Brown & T. Prusti, eds.

doi:10.1017/S1743921317006305

G-Band Period-Luminosity Relation For Galactic Cepheids Based on Gaia DR1 Measurements

Chow-Choong Ngeow[1], Anupam Bhardwaj[2,3] and Shashi M. Kanbur[4]

[1]Graduate Institute of Astronomy, National Central University, Jhongli 32001, Taiwan
email: cngeow@astro.ncu.edu.tw

[2]Department of Physics & Astrophysics, University of Delhi, Delhi 110007, India
[3]European Southern Observatory, Karl-Schwarzschild-Straße 2, Garching 85748, Germany
[4]Department of Physics, SUNY Oswego, Oswego, NY 13126, USA

Abstract. Classical Cepheids (hereafter Cepheids) are important standard candle as they obey the famous period-luminosity (PL) relation. Parallax measurements from Gaia offer a unique opportunity to derive or calibrate the PL relations for Galactic Cepheids, as traditionally their distances were measured via different methods. In this work, we attempted to derive the Gaia G-band PL relation based on the Gaia Data Release 1 (DR1) measurements. We adopted the inferred distances provided by Astraatmadja & Bailer-Jones (2016), calculated using two priors in a Bayesian analysis, and cross-matched to known Galactic Cepheids. The resulting G-band PL relation, however, exhibits a much larger scatter than expected. Hence the inferred distances based on the Gaia DR1 parallaxes are not suitable for calibrating the Galactic PL relation, and future Data Releases with improved parallax measurements are desirable.

Keywords. Cepheids, distance scale, stars: distances

---

1. Introduction and Motivation

Period-luminosity (PL) relation for classical Cepheids is an important rung on the extra-galactic distance ladder for the measurement of distances to nearby galaxies and hence the determination of the Hubble constant. Therefore it is desirable to calibrate the PL relation using Gaia's parallaxes (Casertano *et al.* 2016). The Gaia's Data Released 1 (DR1) includes the Gaia's G-band mean magnitudes and parallaxes (p) based on the Tycho-Gaia Astrometric Solution (TGAS, hereafter DR1 parallax) for ~ 2 million stars brighter than ~ 12 mag. Lindegren *et al.* (2016) and Casertano *et al.* (2016) compare the DR1 parallaxes to parallaxes calculated from an adopted PL relation for ~ 141 and ~ 212 Galactic Cepheids, respectively. They found good global agreements between the two sets of parallaxes. Hence, the goal of this work is attempted to derive the Gaia's G-band PL relation for Galactic Cepheids based on the Gaia DR1 data, in order to evaluate the performance of DR1 parallaxes in deriving the Cepheid PL relations.

2. Data, Method and Results

Astraatmadja & Bailer-Jones (2016a, hereafter ABJ-II) demonstrated that in presence of measurement errors, reciprocal of measured parallaxes ($1/p$) is not a good estimator for distances r. Instead, Bayesian approach with proper choice of prior need to be used to infer distance r from the measured parallaxes p together with it uncertainty σ_p. ABJ-II recommended two priors: a exponential decreasing space density prior and a prior based on Milky Way model. Later, these priors were applied to DR1 parallaxes by Astraatmadja & Bailer-Jones (2016b, hereafter ABJ-III), who derived distances r to the stars in TGAS.

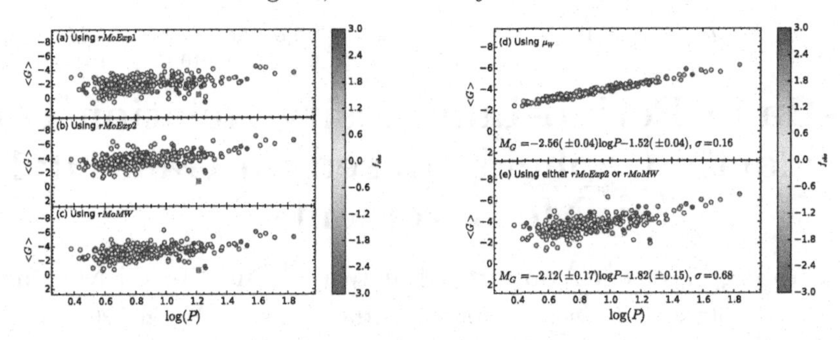

Figure 1. The G-band PL relation for Galactic Cepheids with difference adopted distances. The magenta square in the left panel is for RW Cam, the outlier in the comparison of parallaxes as shown in Casertano *et al.* (2016). Color bar represents the value of $f_{obs} = \sigma_p/p$, restricted to $[-3, 3]$ for display purpose. There are some Cepheids with $f_{obs} > |3|$, shown as darkest blue/red colors. The average f_{obs} for this sample is 0.22. Error bars are omitted for clarity.

We cross-matched catalog in ABJ-III to $\sim$ 400 Galactic Cepheids listed in Ngeow (2012, hereafter N2012), and 246 Cepheids were found matched in both catalogs.

The absolute magnitudes in Gaia G-band for the matched Galactic Cepheids were derived using four available distances or distance modulus: (1) distance modulus based on a calibrated period-Wesenheit relation given in N2012, μ_W; (2) inferred distance based on the exponentially decreasing space density prior with a scale height $L = 0.11$kpc, $rMoExp1$; (3) same as (2) but with $L = 1.35$kpc, $rMoExp2$; and (4) inferred distance based on the Milky Way prior, $rMoMW$. The mode distances were adopted for (2)-(4). The G-band extinctions were estimated via $A_G/A_V = 0.695$ (ABJ-II), where $A_V = 3.23E(B-V)$ and $E(B-V)$ is taken from N2012. Fig. 1(a)-(c) displays the resulted G-band PL relation, which shows that the PL relation based on the inferred distances from DR1 parallaxes, given in ABJ-III catalog, displays large scatter when compared to the PL relation constructed using the μ_W distance modulus from N2012 (Fig. 1[d]) . These PL relations are also shallower, and in general fainter (especially the one based on the $rMoExp1$ distance), than the PL relation shown in the Fig. 1(d). ABJ-III found that when $r < 2000$pc, $rMoMW$ give a better agreement to distances based on external method; while for $r > 2000$pc, $rMoExp2$ shows a better result. Assume μ_W gives the true distance, we adopted either $rMoMW$ or $rMoExp2$ based on the true distance. The revised PL relation is shown in Fig. 1(e) with reduced scatter. In conclusion, distances inferred from DR1 parallaxes are not suitable to calibrate the PL relation for Galactic Cepheids, and future Data Releases from Gaia are needed to improve such calibration.

This work has made use of data from the European Space Agency (ESA) mission *Gaia* (`https://www.cosmos.esa.int/gaia`), processed by the *Gaia* Data Processing and Analysis Consortium (DPAC, `https://www.cosmos.esa.int/web/gaia/dpac/consortium`). Funding for the DPAC has been provided by national institutions, in particular the institutions participating in the *Gaia* Multilateral Agreement.

References

Astraatmadja, T. L. & Bailer-Jones, C. A. L. 2016a, *The Astrophysical Journal*, 832, 137
Astraatmadja, T. L. & Bailer-Jones, C. A. L. 2016, *The Astrophysical Journal*, 833, 119
Casertano, S., Riess, A. G., Bucciarelli, B., & Lattanzi, M. G. 2016, *Astronomy & Astrophysics*, 599, 67
Lindegren, L., Lammers, U., Bastian, U., *et al.* 2016, *Astronomy & Astrophysics*, 595, A4
Ngeow, C.-C. 2012, *The Astrophysical Journal*, 747, 50

Astronomy and Astrophysics in the Gaia sky
Proceedings IAU Symposium No. 330, 2017
A. Recio-Blanco, P. de Laverny, A.G.A. Brown & T. Prusti, eds.

doi:10.1017/S1743921317005415

The mass-ratio distribution of spectroscopic binaries along the main-sequence

Henri M. J. Boffin[1] **and Dimitri Pourbaix**[2]

[1]ESO, Garching, Germany
email: hboffin@eso.org

[2]Université Libre de Bruxelles and FRS-FNRS, Belgium
email: pourbaix@astro.ulb.ac.be

Abstract. Binarity is now a well-established quality affecting a large fraction of stars, and recent studies have shown that the fraction of binaries is a function of the spectral type of the primary star, with most massive stars being member of a close binary system. By cross-matching TGAS with S_{B^9}, we went one step further and derived the mass ratio distribution of binary systems as a function of the spectral type of the primary star. This, combined with the binary fraction, provides very strong constraints on star formation and critical input for stellar population models.

Keywords. stars: binaries, stars: evolution, Hertzsprung-Russell diagram

---

The distribution of the masses of the two components of a binary system, M_1 and M_2 are clues to critical questions related to the formation and evolution of binaries. We aim here to derive the distribution of the mass ratio, $q = M_2/M_1$, as a function of the primary mass, M_1, thanks to the Tycho-Gaia Astrometic Solution (TGAS; Lindegren *et al.* 2016).We use the Oct. 2016 version of the S_{B^9} catalogue (Pourbaix *et al.* 2004) that contains a large set of spectroscopic binaries gathered from the literature. Our systems are divided into single-lined spectroscopic binaries (SB1), for which we only have the spectroscopic mass function, and double-lined spectroscopic binaries (SB2), for which we already have the mass ratio.

TGAS and S_{B^9} are cross-matched to select all binary systems containing a main sequence primary, for which the relative error on the parallax was below 16%. This provided us with a catalogue of 142 K, 340 G, 421 F, 369 A, and 153 B stars, i.e. 1425 stars in total. Using their B- and V-magnitudes, combined to an estimate of the extinction, A_V, enabled us to put all our objects in the colour-magnitude diagram, $B - V$ vs. M_V.

For the SB1 systems, their position in the Hertzsprung–Russell diagram was used to determine their mass from a comparison with the PARSEC stellar evolutionary tracks (Bressan *et al.* 2012). This was done with a weighting scheme that takes into account the error bars as well as the time a star of a given mass spends at a certain location of the colour-magnitude diagram. The mode's mass of our samples is, resp., 0.85 $M_\odot$ (K stars), 1.05 $M_\odot$ (G), 1.2 $M_\odot$ (F), 1.8 $M_\odot$ (A), and 4.0 $M_\odot$ (B). Using a different metallicity would imply a different mass, but this has negligible impact on the mass ratios we determine.

To determine the mass-ratio distribution (MRD) of our SB1, we make use of their spectroscopic mass function and the primary mass derived above and apply the Richardson-Lucy deconvolution technique (Boffin 2010). This can be done for our 5 samples and the result is shown in Fig. 1. The final MRD is then obtained by summing up the MRD coming from the SB1 with that of the SB2 (obtained directly). The main issue is whether we are suffering from observational biases, i.e. if the S_{B^9} catalogue contains relatively

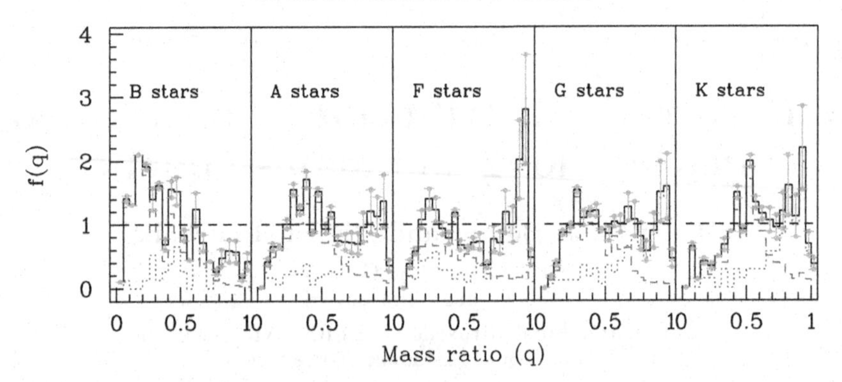

Figure 1. Mass-ratio distributions for the stars in the S_{B^9} catalogue per their spectral type. The purple (dashed) histogram is the MRD obtained for SB1 by deconvolution, while the red, dotted line shows the MRD for the SB2. The final, heavy black line, MRD is the sum of both, with the green dots showing the variations if we count only half of the SB2 or twice the SB2. For illustration purpose, the dashed line shows the uniform distribution.

too few or too many SB2 compared to SB1. Our large samples and the fact that the data come from various, independent sources, should ensure that any bias, if present, is quite small. For illustration purpose, we show in Fig. 1 what we would have if we added only half (resp. twice) the distribution of SB2.

Our results clearly indicate that the MRD is a function of the spectral type of the primary, and thus of its mass, although (except for B stars) the general trend is to have a MRD that is relatively flat, with the *K*-stars lacking smaller companions and the *F*-stars showing an excess of twins.

Using our primary mass distribution as well as our MRD, we can further obtain the distribution of the companion's mass as a function of the primary's spectral type. It appears that up to a given value, around 0.4 $M_\odot$, the distributions are peaked towards lower masses, but in a much less steep way than would be expected from the Salpeter or Chabrier IMF.

Acknowledgement

This work has made use of data from the European Space Agency (ESA) mission *Gaia* (`https://www.cosmos.esa.int/gaia`), processed by the *Gaia* Data Processing and Analysis Consortium (DPAC, `https://www.cosmos.esa.int/web/gaia/dpac/consortium`). Funding for the DPAC has been provided by national institutions, in particular the institutions participating in the *Gaia* Multilateral Agreement.

References

Boffin, H. M. J. 2010, *A&A* 524, A14
Bressan, A. *et al.* 2012, *MNRAS* 427, 127
Lindegren, L. *et al.* 2016a *A&A*, 595, A4
Pourbaix, D. *et al.* 2004, *A&A* 424, 727

Astronomy and Astrophysics in the Gaia sky
Proceedings IAU Symposium No. 330, 2017
A. Recio-Blanco, P. de Laverny, A.G.A. Brown & T. Prusti, eds.

doi:10.1017/S1743921317005385

OB stars towards NGC 6357 and NGC 6334

Delphine Russeil
Aix Marseille Univ, CNRS, LAM, Laboratoire d'Astrophysique de Marseille, Marseille, France
email: delphine.russeil@lam.fr

Abstract. The star forming regions NGC6334 and NGC6357 are amid the most active star-forming complexes of our Galaxy where massive star formation is occuring. Both complexes gather several HII regions but they exhibit different aspects: NGC6334 is characterised by a dense molecular ridge where recent massive star formation is obvious while NGC6357 is dominated by the action of the stellar cluster Pismis 24 which have shaped a large cavity. To understand and compare the formation of massive stars in these two regions requires to precise the distance and characterise the proper motions of the O to B3 stellar population in these regions.

Keywords. stars: distances, stars: early-type, ISM: HII regions

1. Overview and GAIA perspectives

The NGC6334 and NGC6357 (Fig. 1a) are two active star-forming complexes for which it is possible to study the formation and the evolution of massive stars from their earliest phase to their impact on the environment due to their HII regions. From the OB stars distribution, mainly distributed between 1 and 3 kpc (Russeil *et al.* 2012), a distance of 1.75 kpc for both regions is usualy adopted. From the GAIA-DR1 catalogue (e.g. Arenou *et al.* 2017, Lindegren *et al.* 2016) one can underline (Fig. 1b and c) a stellar layer around 1.12 kpc (main peak of the histogram Fig. 1c at $\sim$0.89 mas) while few O-B3 stars towards NGC 6334 and NGC 6357 have a measured parallactic distance (see Table 1) in the GAIA DR1 catalogue (Astraatmadja *et al.* 2016). Despite our small sample the mean distance obtained is 2.23 kpc or 1.23 kpc depending on the prior. The Milky-way prior gives better agreement with Wu *et al.* 2014 and Chibueze *et al.* 2014 who gives for NGC 6334 a maser parallax distance between 1.25 and 1.35 kpc.

In parallel, WISE-22μm and HERSCHEL-70μm data were used to identify (Russeil *et al.* 2016) bow shocks features (Fig. 1d) which are usualy related to run-away stars (e.g. Gvaramadze *et al.* 2011). The identified features are either seen in Hα or radio continuum suggesting that run-away stars can be massive.

In addition, an age estimate of O-B3 stars (Russeil *et al.*, in preparation) suggest that young (log(age) $<$ 7) stars are found at the periphery of NGC 6357. They can have been formed in Pismis 24 but quickly expelled.

Table 1. O-B3 stars with parallax distance in NGC 6334 and NGC 6357.

HII region	Star Ident.	Spectral type	Dist.[1] kpc	GAIA dist.[2] kpc	GAIA dist.[3] kpc
NGC 6334 - GUM64C	CD-3511482	B0.5e	1.41	2.71±2.20	1.21±0.43
NGC 6334 - GUM64b	HD319702	O8III	2.11	1.66±1.82	1.17±0.36
NGC 6334 - GUM64b	CD-3511484	B1V	1.10	2.10±1.93	1.25±0.38
NGC 6334 - G351.2+0.5	HD156738	O6.6III(f)	1.58	2.67±2.06	1.31±0.40
NGC 6357	HD157504	WN5b	–	2.75±2.13	1.26±0.39
NGC 6357	HD319881	O6Vn	1.66	1.49±1.73	1.19±0.50

Notes: [1] Spectro-photometric distance (Russeil *et al.* 2012). [2] Distances from Gaia DR1 catalogue parallaxes computed by Astraatmadja *et al.* 2016 with the exponential decreasing space density prior with L = 1.35 kpc (systematics uncertainties of 0.3mas included). [3] Same as [2] but using the Milky-Way prior.

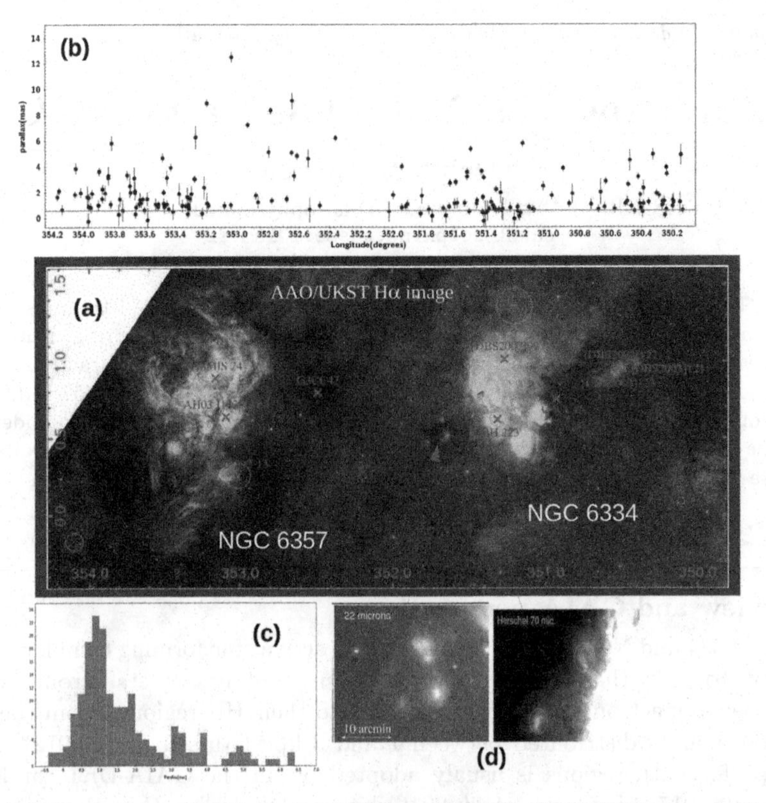

Figure 1. (a) Hα image of NGC 6334 and NGC 6357. (b) GAIA DR1 sources ($0.5° < b < 1.5°$) parallaxes (histogram in (c)) versus longitude (the horizontal line underlines the 1.75 kpc parallax). (d) Candidate run-away stars identified by Russeil *et al.* 2016.

To summarise, the next GAIA release will be used to 1) determine/precise statistically the distance of the OB stars and the different stellar layers present along the line of sight of NGC 6334 and NGC 6357, 2) identify and precise the proper motion of the run-away stars in order to clarify their origin and their possible star-formation impact.

Acknowledgements

This work has made use of data from the European Space Agency (ESA) mission *Gaia* (`https://www.cosmos.esa.int/gaia`), processed by the *Gaia* Data Processing and Analysis Consortium (DPAC, `https://www.cosmos.esa.int/web/gaia/dpac/consortium`). Funding for the DPAC has been provided by national institutions, in particular the institutions participating in the *Gaia* Multilateral Agreement.

References

Arenou, *et al.* 2017, *A&A*, 599, A50
Astraatmadja, *et al.* 2016, *ApJ*, 833, 119
Chibueze, *et al.* 2014, *ApJ*, 784, 114
Gvaramadze, *et al.* 2011, *A&A*, 535, A29
Lindegren, *et al.* 2016, *A&A*, 595, A4
Russeil, *et al.* 2012, *A&A*, 538, A142
Russeil, *et al.* 2016, *A&A*, 587, A135
Wu, *et al.* 2014, *A&A*, 566, A17

Astronomy and Astrophysics in the Gaia sky
Proceedings IAU Symposium No. 330, 2017
A. Recio-Blanco, P. de Laverny, A.G.A. Brown & T. Prusti, eds.

doi:10.1017/S1743921317005592

Gaia observations of naked-eye stars: status update

J. Sahlmann[1,2], A. Mora[2,3], J. M. Martín-Fleitas[2,3], A. Abreu[2,4], C. Crowley[2,5] and M. Fink[2]

[1]Space Telescope Science Institute, 3700 San Martin Drive, Baltimore, MD 21218, USA
email: jsahlmann@stsci.edu

[2]European Space Agency, ESAC, P.O. Box 78, Villanueva de la Cañada, 28691 Madrid, Spain

[3]Aurora Technology, Crown Business Centre, Heereweg 345, 2161 CA Lisse, The Netherlands

[4]Elecnor Deimos Space, R. de Poniente 19, Ed. Fiteni VI, 28760 Tres Cantos, Madrid, Spain

[5]HE Space Operations BV, Huygensstraat 44, 2201 DK Noordwijk, The Netherlands

Abstract. ESA's Gaia space astrometry mission is performing an all-sky survey of stellar objects. At the beginning of the nominal mission in July 2014, an operation scheme was adopted that enabled Gaia to routinely acquire observations of all stars brighter than the original limit of G$\sim$6, i.e. the naked-eye stars. We present the current status and extent of those observations.

Keywords. space vehicles: instruments, catalogs, astrometry, stars: distances

---

1. Gaia's bright limit

In the Gaia focal plane, the SkyMapper CCDs (SM1 and SM2) identify the the star-like sources that Gaia will observe. Data of stars not identified in the SkyMapper are not downlinked, thus are lost. The original Gaia bright limit of $G \simeq 6$ was improved to $G \simeq 3$ by tuning the onboard parameters of the SkyMapper star detection algorithm (Sahlmann *et al.* 2016b; Martín-Fleitas *et al.* 2014). For the 230 stars brighter than $G=3$, we are pursuing two solutions in order to observe them as well:

Forced SkyMapper Imaging: This consists of forcing the acquisition of full-frame SkyMapper images and is in operation since the beginning of Gaia's nominal mission.

Virtual Object Synchronisation: This uses Virtual Objects whose associated CCD windows are placed at defined locations. They usually fall on 'empty' regions of the sky and, for instance, serve to estimate the sky background. The idea of Virtual Object synchronised observations is to predict the focal plane crossing of a very bright star and to place a Virtual Object window on top of it. The method has been successfully tested and its implementation for the brightest 50 stars (G<1.75) is underway.

2. Very bright star science cases

Very bright stars with magnitudes $G < 6$, i.e. the $\sim$6000 stars observable with the naked eye, are among the best studied astronomical objects. Securing Gaia data for those stars is a unique science opportunity, in particular in what concerns astrometry because no other current or planned observatory can obtain global astrometry at sub-milliarcsecond level of this stellar sample. Science cases include but are not limited to:

- Parallaxes and proper motions about 10 times more precise than from Hipparcos, e.g. of bright massive stars that are fundamental anchor points for stellar astrophysics.
- Orbit constraints for very bright binary stars (at least 25% of the sample).

- Discover new exoplanets, in particular around very bright A and F stars.
- Accurate masses of known exoplanets discovered by radial-velocity monitoring. This will for instance include GJ 676A whose astrometric orbit caused by planet b was already determined from the combination of ground-based astrometry and radial velocities (Sahlmann *et al.* 2016a). Gaia naked-eye star observations can make similar work possible for tens of known exoplanets.

3. Forced SkyMapper Imaging

At every predicted passage of a very bright star, Gaia records 5 seconds of SkyMapper full-frame data. The PSF core saturates and a nominal model does not reproduce the high spatial frequencies, but the images measuring ∼5'×6' contain plenty of astrometric information. The data are non-nominal and treated with an off-line pipeline (Sahlmann *et al.* 2016b; Gaia Collaboration *et al.* 2016b,a). This method has been in operation since October 2014. Its disadvantages are that only SkyMapper data are collected, which have a fixed integration time (CCD gating) and they are undersampled by a factor of two (∼0.1" × 0.3" effective pixel size) compared to the astrometric field CCD data. The more powerful solution of virtual object synchronisation can mitigate this.

The technique of forced SkyMapper imaging is also applied to capture images of extremely dense fields (to mitigate effects of crowding) and of events when stars are observed close to Jupiter's limb (for scene reconnaissance).

4. Virtual Object Synchronisation

Because Gaia is spinning and precessing, this method relies on accurate temporal (∼9 ms) and spatial (∼1") predictions of very bright star passages in the Gaia focal plane. These prediction capabilities were demonstrated in several tests reaching ∼70% success rates (defined as capture of the stellar core). Using improved prediction models we aim at >90% success rate. The critical advantage of this method is that it gives access to SkyMapper, astrometric field, and spectro-photometric data of extremely bright stars.

In summary, there is no bright limit for Gaia astrometric observations, however core saturation poses challenges both for naturally detected stars ($G < 6$) and in the forced SkyMapper images. Virtual object synchronisation may mitigate some of those problems for the 50 stars brighter than G=1.75.

Acknowledgements

This work has made use of data from the European Space Agency (ESA) mission *Gaia* (`https://www.cosmos.esa.int/gaia`), processed by the *Gaia* Data Processing and Analysis Consortium (DPAC, `https://www.cosmos.esa.int/web/gaia/dpac/consortium`). Funding for the DPAC has been provided by national institutions, in particular the institutions participating in the *Gaia* Multilateral Agreement.

References

Gaia Collaboration, Brown, A. G. A., Vallenari, A., *et al.* 2016a, *A&A*, 595, A2
Gaia Collaboration, Prusti, T., de Bruijne, J. H. J., *et al.* 2016b, *A&A*, 595, A1
Martín-Fleitas, J., Sahlmann, J., Mora, A., *et al.* 2014, in SPIE, Vol. 9143
Sahlmann, J., Lazorenko, P. F., Ségransan, D., *et al.* 2016a, *A&A*, 595, A77
Sahlmann, J., Martín-Fleitas, J., Mora, A., *et al.* 2016b, in SPIE, Vol. 9904

Astronomy and Astrophysics in the Gaia sky
Proceedings IAU Symposium No. 330, 2017
A. Recio-Blanco, P. de Laverny, A.G.A. Brown & T. Prusti, eds.

doi:10.1017/S1743921317005610

The TGAS HR diagram of S-type stars

Shreeya Shetye[1,2] †, Sophie Van Eck[1], Alain Jorissen[1], Hans Van Winckel[2] and Lionel Siess[1]

[1]Institut d'Astronomie et d'Astrophysique, Université Libre de Bruxelles, CP 226, Boulevard du Triomphe, B-1050 Bruxelles,Belgium
email: Shreeya.Shetye@ulb.ac.be

[2]Instituut voor Sterrenkunde (IvS), KU Leuven, Celestijnenlaan 200D, B-3001 Leuven, Belgium

Abstract. S-type stars are late-type giants enhanced with s-process elements originating either from nucleosynthesis during the Asymptotic Giant Branch (AGB) or from a pollution by a binary companion. The former are called intrinsic S stars, and the latter extrinsic S stars. The atmospheric parameters of S stars are more numerous than those of M-type giants (C/O ratio and s-process abundances affect the thermal structure and spectral synthesis), and hence they are more difficult to derive. Nevertheless, high-resolution spectroscopic data of S stars combined with the TGAS (Tycho-Gaia Astrometric solution) parallaxes were used to derive effective temperatures, surface gravities, and luminosities. These parameters allow to locate the intrinsic and extrinsic S stars in the Hertzsprung-Russell diagram.

Keywords. S stars, AGB stars, TGAS, HR diagram

1. Introduction

S stars are late-type giants showing ZrO molecular bands along with TiO bands as the most characteristic distinctive spectral features (Merrill 1922). The C/O ratio of S stars ranges from 0.5 to 1 suggesting that they are transition objects between M-type giants (C/O $\sim$ 0.5) and carbon stars (C/O $>$ 1) on the Asymptotic Giant Branch (AGB) (Iben and Renzini 1983). Their spectra show signatures of overabundances in s-process elements (Smith and Lambert 1990).

The evolutionary status of S stars as AGB stars was challenged when Tc lines (an s-process element with no stable long-lived isotope) were reported as missing in some S stars (Merrill 1952; Smith and Lambert 1986; Jorissen *et al.* 1993). This puzzle regarding the evolutionary status of S stars was solved when it was perceived that the Tc-poor S stars belong to binary systems (Smith and Lambert 1986; Jorissen *et al.* 1993). S stars may therefore be classified into two different classes: Tc-rich as intrinsic S stars that are genuine thermally-pulsing AGB (TP-AGB) stars and Tc-poor as extrinsic S stars that owe their s-process element overabundances to a mass transfer from a former AGB companion which is now a white dwarf. They are the cooler analogues of barium stars.

The thermal structure of the atmospheres of S stars depends on effective temperature (T_{eff}), surface gravity (log g), [Fe/H], C/O as well as [s/Fe] (s-process element abundances). The abundance analysis of S stars requires a reliable determination of all these stellar atmosphere parameters.

† This work has made use of data from the European Space Agency (ESA) mission *Gaia* (https://www.cosmos.esa.int/gaia), processed by the *Gaia* Data Processing and Analysis Consortium (DPAC, https://www.cosmos.esa.int/web/gaia/dpac/consortium). Funding for the DPAC has been provided by national institutions, in particular the institutions participating in the *Gaia* Multilateral Agreement.

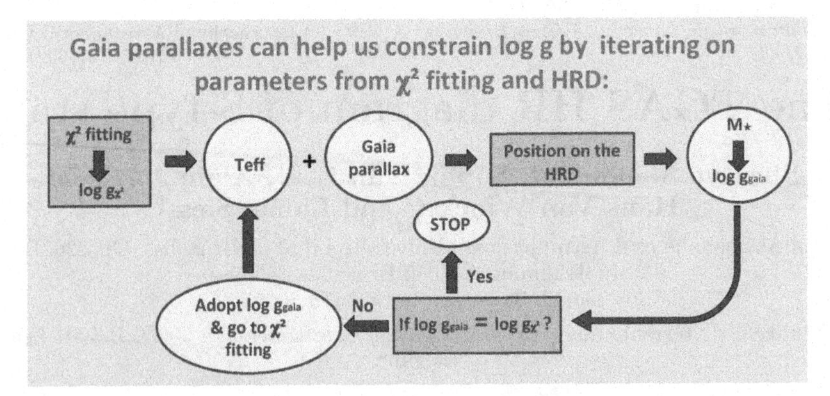

Figure 1. Algorithm adopted to constrain log g, comparing the location of the stars in the HR diagram with the evolutionary tracks from the STAREVOL code.

2. Stellar sample and parameter determination

Our sample consists of S stars from the General Catalog of S stars (Stephenson 1984) with $V \leqslant 11$ and $\delta \geqslant -30°$, thus observable with HERMES (High Efficiency and Resolution Mercator Echelle Spectrograph, mounted on the 1.2m Mercator Telescope at the Roque de Los Muchachos Observatory, La Palma; Raskin *et al.* 2011). Furthermore, a condition is imposed on the TGAS parallaxes (Gaia Collaboration 2016), considering only those stars with a small error on the parallax ($\sigma_{\varpi} \leqslant 0.3\varpi$). With these conditions, the sample amounts to 18 S stars.

The stellar parameters are derived using the MARCS grid of atmospheric models for S stars (Van Eck *et al.* 2017) containing more than 3500 models covering the parameter space in $T_{\rm eff}$, log g, [Fe/H], C/O and [s/Fe] ratios. The comparison between observed and synthetic spectra is then performed by a χ^2-fitting procedure, summing over all spectral pixels in spectral bands approximately 200 Å wide. The model with the lowest χ^2 value is chosen as the best fitting model.

The distance and luminosity of these stars are derived from the TGAS parallaxes. Comparison between the positions of the stars in the HR diagram constructed from TGAS parallaxes with the evolutionary tracks from the STAREVOL code (Siess and Arnould 2008) yields the mass, hence the surface gravity of our stars ($\log g_{\rm Gaia}$). Because $\log g_{\rm Gaia}$ and log g derived from the χ^2 fitting do not always agree, we derive a new surface gravity estimate as explained in Figure 1. This iteration on the stellar parameters ensures that the adopted log g is consistent with the TGAS parallaxes.

3. HR diagram of S stars

The temperatures obtained after constraining log g with the TGAS parallaxes lead to the HR diagram of S stars presented in Figure 2. Intrinsic S stars are cool and luminous objects likely on the TP-AGB in the HR diagram. On the other hand, the extrinsic S stars are hotter and intrinsically fainter on the early AGB or red giant branch (RGB) except for HD 150922 and HD 191226 which seem more evolved in the 4-6 $M_\odot$ range. Also, the intrinsic S star on the 1 $M_\odot$ track is intriguing because the third dredge-up is expected to occur only for masses larger than 1.3 $M_\odot$ according to stellar evolution predictions (Karakas and Lugaro 2016). Nevertheless, the occurrence of the third dredge-up for low mass stars (<1.3 $M_\odot$) was also found in low-luminosity s-process-rich post-AGB stars (De Smedt *et al.* 2015). The evolutionary tracks have a strong metallicity dependence, impacting on the mass determination. Spectral analysis to constrain [Fe/H] is ongoing.

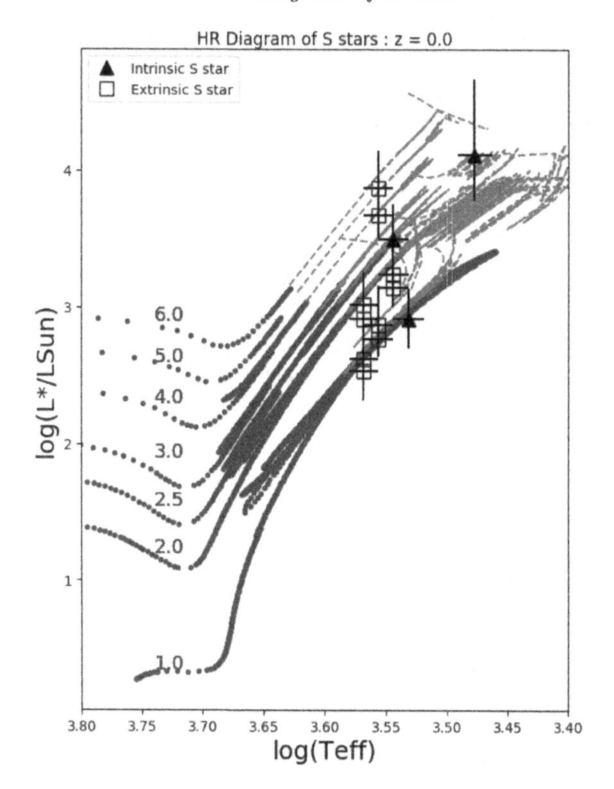

Figure 2. HR diagram of S stars with STAREVOL evolutionary tracks for [Fe/H] = 0.00. The dark blue dotted tracks represent pre-AGB phases and the pink dashed tracks represent the AGB phase.

4. Conclusion

The S stars set strong constraints on mixing and nucleosynthesis processes in AGB stars. However, it is difficult to probe the onset of third dredge-up with TGAS data since there are only 3 intrinsic S stars (identified from the analysis of their Tc lines) as yet. This is due to the limitations of Gaia DR1 on red sources which put a bias against the very evolved and red intrinsic S stars.

References

De Smedt, K., *et al.* 2015, *A&A*, 583, A56
Gaia Collaboration 2016, *A&A*, 595, A2
Iben, Jr., I. & Renzini, A. 1983, *ARA&A*, 21, 271
Jorissen, A., *et al.* 1993, *A&A*, 271, 463
Karakas, A. I. & Lugaro, M. 2016, *ApJ*, 825, 26
Merrill, P. W. 1922, *ApJ*, 56, 457M
Merrill, P. W. 1952, *ApJ*, 116, 21
Raskin, G., *et al.* 2011, *A&A*, 526, A69
Siess, L. & Arnould, M. 2008, *A&A*, 489, 395
Smith, V. V. & Lambert, D. L. 1986, *ApJ*, 311, 843
Smith, V. V. & Lambert, D. L. 1990, *ApJS*, 72, 387
Stephenson, C. B. 1984, *Publications of the Warner & Swasey Observatory*, 3, 1
Van Eck, S. *et al.* 2017, *A&A*, 601, A10

Astronomy and Astrophysics in the Gaia sky
Proceedings IAU Symposium No. 330, 2017
A. Recio-Blanco, P. de Laverny, A.G.A. Brown
& T. Prusti, eds.

doi:10.1017/S1743921317005804

Understanding Li enhancement in K giants and role of accurate parallaxes

Raghubar singh and B. E. Reddy

Indian Institute of Astrophysics,
Bangalore 560034, India
email: raghubar.singh@iiap.res.in

Indian Institute of Astrophysics,
Bangalore 56003, India
email: ereddy@iiap.res.in

Abstract. Our recent studies based on a large sample of K giants with Hipparcos parallaxes and spectroscopic analysis resulted more than a dozen new Li-rich K giants including few super Li-rich ones. Most of the Li-rich K giants including the new ones appear to occur at the luminosity bump in the HR diagram. However, one can't rule out the possibility of overlap with the clump region where core He-burning K giants reside post He-flash at the tip of RGB. It is important to distinguish field K giants of clump from the bump region in the HR diagram to understand clues for Li production in K giants. In this poster, we explore whether GAIA parallaxes improve to disentangle clump from bump region, more precisely.

Keywords. Lithium, GAIA, RGB, planet engulfment, parallax.

1. Introduction

Lithium is one of the primordial elements produced along with H and He during Big Bang nucleosynthesis (BBN). The observed Li abundance of $\log A(\mathrm{Li}) = 2.2$ dex in metal-poor stars is considered as primordial abundance. The observed value is significantly less than the value, $\log A(\mathrm{Li}) = 2.72$ dex (Cyburt *et al.* 2008), predicted by BBN models based on measured baryon density using WMAP. However, observations show that Li abundance ($\log A(\mathrm{Li}) = 3.3$ dex) in ISM and young stellar objects is about a magnitude more than the primordial value. It is not clear which are the sources of excess Li, and level of their contribution to the current Li values in the Galaxy. Recent observations suggest Red giant branch (RGB) stars may be significant source of Li to the Galaxy. This is contrary to prediction of standard stellar models which predict a maximum of $\log A(\mathrm{Li}) = 1.5$ dex in low mass (1-2.5$M_{\odot}$) RGB or K giants (Iben *et al.* 1967a). As per standard models, Li gets depleted as star evolves from main sequence to RGB due to deep convection and 1st dredge-up. However, a small group of K giants are found to have large amounts of Li in their photosphere (see Kumar *et al.* 2011), some times exceeding their natal clouds, $\log A(\mathrm{Li}) = 3.3$ dex Exact location, of Li-rich K giants in the H-R diagram is not well established. Location has important implications for our understanding of Li origin in K giants. In this context, improved accuracies in astrometry from GAIA will help.

2. Results & discussion

Origin for Li excess in giants has been a subject of study for over three decades, since its discovery by Wallerstein *et al.* 1982. There are three main hypotheses for excess Li in

K giants: a) Retaining of main sequence Li abundance due to inefficient mixing process., b) Internal nucleosynthesis and dredge-up process, and c) External source such as planet or brown dwarf engulfment. First of the three, an inefficient mixing, is obviously not the likely scenario as few K giants have been shown to have Li abundance which exceeds ISM values (Kumar *et al.* 2011). As per other two scenarios (b and c), current observational results do not rule out either of the two possibilities. Recent studies suggest occurrence of Li-rich K giants all along the RGB indicating some kind of external origin such as planet engulfment which can happen anywhere on the RGB. It is suggested that Li in photospheres of stars with respect to hydrogen can be enhanced by engulfment of planets (for example Earth) which have Li values similar to ISM but devoid of hydrogen

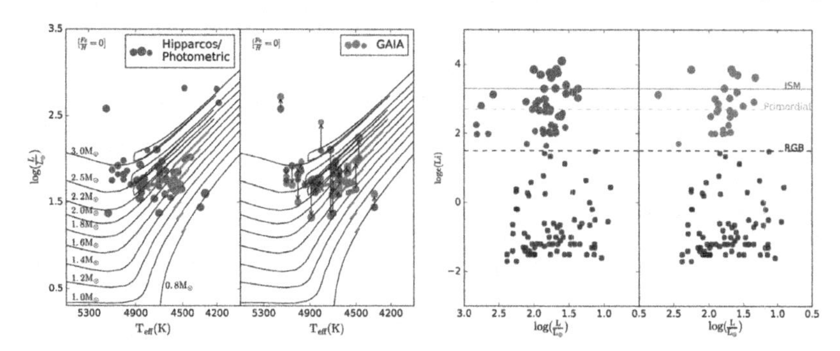

However, as shown in figures, most of the Li-rich K giants seems to be confined to a narrow range of luminosity in H-R diagram overlapping with luminosity bump and red clump regions. It is not clear whether Li enhancement is linked to internal mixing process during luminosity bump evolution or He-flash at the tip of RGB. Surveys for Li-rich K giants such as Kumar *et al.* 2011 suggest Li enhancement occur either at bump or clump. Their results do not show Li-rich K giants before or after the bump region. In this poster, we show Li abundances of known Li-rich K giants and Li abundance measurements of K giants along the RGB. It appears that scatter in luminosity reduced in the case of GAIA parallaxes compared to earlier estimates. But as our preliminary results indicate it is difficult to make conclusion though, in many cases GAIA results shows significantly different luminosities.

3. Conclusion

Preliminary analysis shows that GAIA results may help to reduce scatter in luminosities of Li rich K giants which in turn provide clues for Li enhancement excess in K giants.

This work has made use of data from the European Space Agency (ESA) mission *Gaia* (`https://www.cosmos.esa.int/gaia`), processed by the *Gaia* Data Processing and Analysis Consortium (DPAC, `https://www.cosmos.esa.int/web/gaia/dpac/consortium`). Funding for the DPAC has been provided by national institutions, in particular the institutions participating in the *Gaia* Multilateral Agreement.

References

Cyburt, R. H., Fields, B. D., & Olive, K. A. 2008, *JCAP*, 12, 11
Iben, Jr. I. 1967, *ApJ*, 147, 650
Kumar, Y. B., Reddy, B. E., & Lambert, D. L. 2011, *ApJ*, L12, 730
Wallerstein, G., & Sneden, C. 1982, *ApJ*, 577-584, 255

Astronomy and Astrophysics in the Gaia sky
Proceedings IAU Symposium No. 330, 2017
A. Recio-Blanco, P. de Laverny, A.G.A. Brown & T. Prusti, eds.

doi:10.1017/S1743921317006767

Detection of spectroscopic binaries: lessons from the Gaia-ESO survey

Mathieu Van der Swaelmen[1], Thibault Merle[1], Sophie Van Eck[1], Alain Jorissen[1] and Tomaž Zwitter[2]

[1]Institut d'Astronomie et d'Astrophysique, Université Libre de Bruxelles, Belgium
email: mathieu.van.der.swaelmen@ulb.ac.be

[2]Faculty of Mathematics and Physics, University of Ljubljana, Slovenia

Abstract. The Gaia-ESO survey (GES; Gilmore *et al.* (2012), Randich *et al.* (2013)) is a spectroscopic survey complementing the Gaia mission to bring accurate radial velocities and chemical abundances for 10^5 stars. Merle *et al.* (submitted to A&A; see also this volume) developped a tool (DOE) to detect multiple peaks in the cross-correlation functions (CCFs) of GES spectra. Using the GIRAFFE HR10 and HR21 settings, we were able to compare the efficiency of our SB detection tool depending on the wavelength range and resolution. We show that a careful design of CCF masks can improve the detection rate in the HR21 settings. HR21 spectra are similar to the ones produced by the RVS spectrograph of the Gaia mission, though the lower resolution of RVS spectra may result in a lower detection efficiency than the case of HR21. Analysis of RVS spectra in the context of spectroscopic binaries can take advantage of the lessons learnt from the GES to maximize the detection rate.

Keywords. binaries: spectroscopic, techniques: radial velocities, techniques: spectroscopic, surveys, catalogs

---

1. Methods & Results

We run Monte-Carlo simulations to generate HR10 and HR21 spectra ($R \sim 21500$ and $R \sim 18000$, resp.) of a pair of twin (non-rotating) stars for various levels of S/N and radial velocity separations $\Delta v_{\rm rad}$. We then apply our DOE pipeline on the simulated spectra. Fig. 1 shows the SB2 detection efficiency in HR10 and HR21. The green dots (respectively the red triangles) indicate ($\Delta v_{\rm rad}$, S/N) conditions when DOE is able to detect the two expected peaks in more than 95% of cases (resp., conditions when DOE failed at detecting two expected peaks in more than 95% of cases). Blue plusses represent intermediate cases making detection efficiency dependent of the noise: (i) due to the noise, spurious peaks may appear or (ii) thanks to the noise, the two peaks have different height (despite being a pair of twins) and become discernible to DOE for small $\Delta v_{\rm rad}$. Our simulations show that HR10 allows a more efficient detection, with a good detection rate as soon as S/N $\geqslant 2$ and $\Delta v_{\rm rad} \geqslant 25\,{\rm km\,s^{-1}}$. On the other hand, HR21 allows the detection of SB2 with $\Delta v_{\rm rad} \geqslant 35\,{\rm km\,s^{-1}}$ with S/N $\gtrsim 10$. Merle *et al.* performed a full analysis of GES DR4 spectra looking for stellar multiplicity and their smallest detected $\Delta v_{\rm rad}$ is $25\,{\rm km\,s^{-1}}$ for HR10 and $60\,{\rm km\,s^{-1}}$ for HR21, well in line with our predictions.

In order to improve the HR10 and HR21 CCFs, we selected a set of weakly-blended lines in the range [5330 Å, 5610 Å] and [8430 Å, 8990 Å] and used them to compute synthetic spectra. Fig. 2 compares the GES/CASU CCF (old) to our new CCF for the object 07272578-0310066. While the binary nature was already obvious in HR10 with the GES CCF, it was not detectable in HR21 with the GES CCF. As expected, our new CCFs give the same results for HR10, showing that the method is robust and that our new

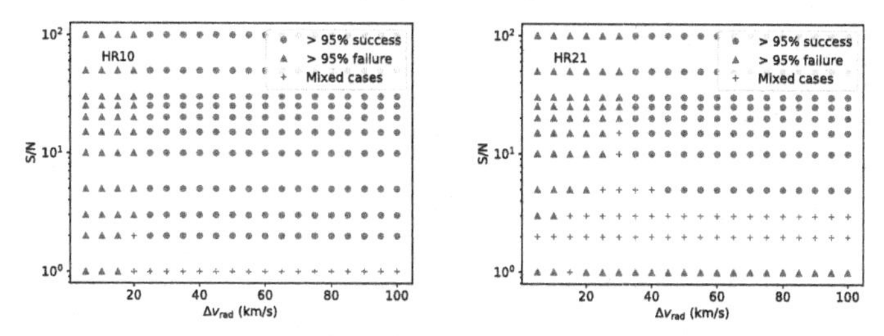

Figure 1. Detection efficiency for HR10 and HR21

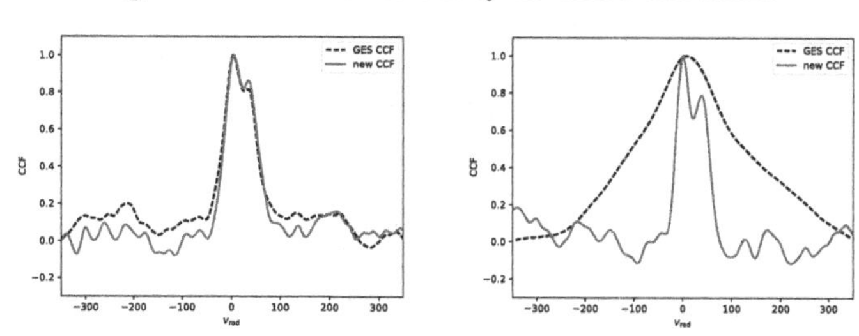

Figure 2. HR10 (left) and HR21 (right) CCF of 07272578-0310066 at MJD 57032.153726 and 57032.247332 (resp.). Black dashed curve is the GES/CASU CCF, red solid curve is the newly computed CCF.

mask allows to retrieve already known SB2s. However, unlike the GES CCF, the new HR21 CCF has narrower peaks and now also reveals the SB2 nature. The broad profile of the HR21 GES CCF is due to the presence of strong lines (Ca II triplet, strong Mg line, Paschen lines) in the range [8430 Å, 8990 Å]. Since our masks do not include such lines, we get narrower CCFs. We performed preliminary tests on a subset of SB2s identified by Merle *et al.* for which the SB2 nature is detected in HR10 GES CCFs but not in HR21 ones. Our new CCFs now allow to detect the SB2 nature of $\sim 35\%$ objects based on their HR21 CCFs (26 objects out of 72).

2. Conclusion

After applying this analysis to the whole GES survey, we expect to improve our SB catalogue: 1/ by completing the time series of velocity measurements for already known objects (when HR21 did not previously lead to SB detection); 2/ by detecting new SBs among the objects observed only with the HR21 setup; 3/ by detecting new SBs among the objects for which HR10 CCFs did not lead to SB detection (very low S/N, phase not favourable to see all components, etc.). Our work shows that very low S/N spectra (> 2 for HR10; > 5 for HR21) are still usable in the context of radial velocity measurement and thus, for SB detection. Since the HR21 spectral domain resembles that of the Gaia RVS, RVS CCFs may suffer from similar broadening issues and may benefit from a careful design of correlating masks.

References

Gilmore, G., Randich, S. *et al.* 2012, *The Messenger*, 25, 147

Merle, T., Van Eck, S., Jorissen, A., Van der Swaelmen, M. *et al.* 2017, *A&A*, submitted

Randich, S., Gilmore, G., & Gaia-ESO Consortium 2013, *The Messenger*, 47, 154

Astronomy and Astrophysics in the Gaia sky
Proceedings IAU Symposium No. 330, 2017
A. Recio-Blanco, P. de Laverny, A.G.A. Brown
& T. Prusti, eds.

doi:10.1017/S1743921317005634

S stars in the Gaia era: stellar parameters and nucleosynthesis

Sophie Van Eck[1], Drisya Karinkuzhi[1], Shreeya Shetye[1,2], Alain Jorissen[1], Stéphane Goriely[1], Lionel Siess[1], Thibault Merle[1] and Bertrand Plez[3]

[1]Institut d'Astronomie et d'Astrophysique, Université Libre de Bruxelles, CP 226, Boulevard du Triomphe, B-1050 Bruxelles, Belgium
email: svaneck@ulb.ac.be

[2]Instituut voor Sterrenkunde (IvS), KU Leuven, Celestijnenlaan 200D, B-3001 Leuven, Belgium

[3]LUPM, Université Montpellier 2, CNRS, F-34095 Montpellier, France

Abstract. S stars are s-process and C-enriched (0.5<C/O<1) red giants. Their abundances can be determined thanks to a new grid of MARCS model atmospheres covering their whole parameter range. Detailed abundance determinations in intrinsic S stars (TP-AGB) and extrinsic S stars (binary masqueraders) can provide strong constraints on the s-process nucleosynthesis: in particular, the s-process temperature can be determined using zirconium and niobium abundances, independently of stellar evolution models. Synthetic spectra of dwarf S stars have been computed and will be sought for in spectroscopic survey data, constraining their luminosity thanks to Gaia parallaxes.

Keywords. Nuclear reactions, nucleosynthesis, abundances; stars: AGB and post-AGB; stars: atmospheres; stars: abundances.

---

1. Introduction

S-type stars have effective temperatures similar to those of M-type stars, but show prominent ZrO molecular bands besides the TiO bands typical of M-type stars. Roughly 50% of S stars are *intrinsic*, i.e. thermally-pulsing asymptotic giant branch (TP-AGB) stars experiencing thermal pulses and third dredge-up episodes, while the remaining 50% are *extrinsic* and owe their overabundances to a pollution from a binary companion, formerly a TP-AGB star, now an extinct WD. Intrinsic S stars are enriched in technetium, while Tc has completely decayed in extrinsic S stars ($\tau_{1/2}(^{99}\mathrm{Tc}) = 0.21 \times 10^6$ yrs).

2. S star model atmospheres

A grid of MARCS model atmospheres (Gustafsson *et al.*, 2008) has been computed for S and SC stars (Van Eck *et al.* 2017) covering a five-dimensional parameter space: 2700K< Teff <4000K (step 100K); $\log g = 0 - 5$ (step of 1), C/O (from 0.5 to 0.99); [s/Fe] (from 0 to +2 dex), and [Fe/H] (0 or -0.5). Parameters are derived from a comparison between synthetic and observed spectra (Mercator HERMES, Raskin *et al.* 2011) and photometric colors (Geneva, SAAO). Synthesis of dwarf S star spectra has been attempted (Fig. 1). Given the lack of a currently available LaO linelist, the only prominent differences between dwarf M and S stars are (i) the ZrO bands at 6400Å and (ii) the ZrO band at 9300Å ($\Delta\nu$=0). The latter, unfortunately, coincides with the H_2O (201-000) band at 9360±150Å, and separating the strong telluric component from the weak stellar absorption is by no means easy. Therefore, distinguishing S and M dwarfs is not trivial.

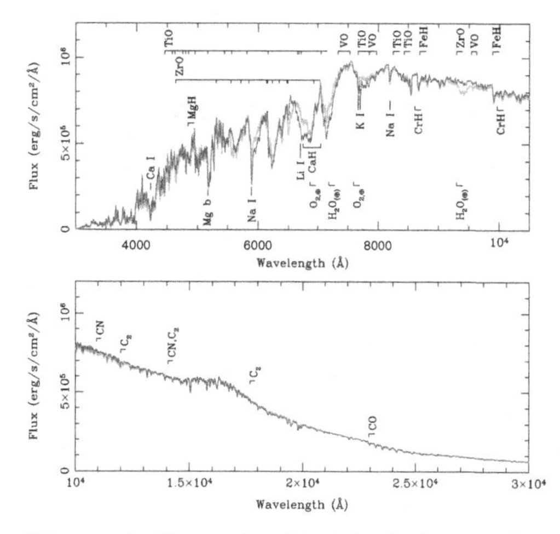

Figure 1. Example of synthetic low-resolution spectra: thick green line: dwarf S star (Teff=3800 K, log g=4, C/O=0.75, [s/Fe]=1); thin blue line: dwarf M star (Teff=3800 K, log g=4, C/O=0.50, [s/Fe]=0).

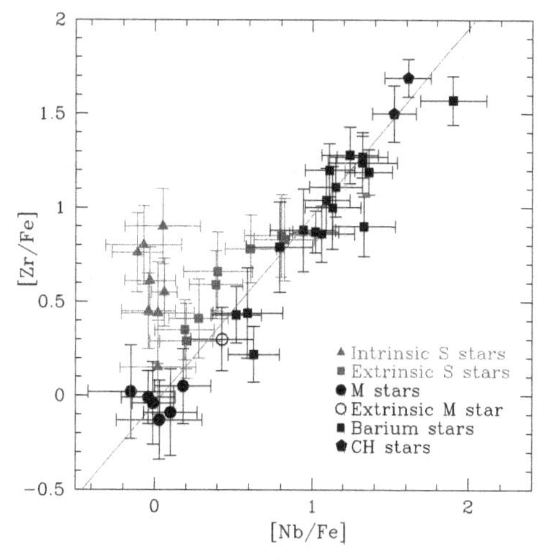

Figure 2. [Zr/Fe] and [Nb/Fe] for intrinsic and extrinsic S stars, barium stars, non-enriched M giants (used as reference), and a slightly enriched M giant labelled as 'extrinsic M star'.

3. Niobium and zirconium abundances: a thermometer

Since ^{93}Zr decays ($\tau_{1/2} = 1.53$ Myr) into mono-isotopic Nb, the Nb abundance represents a new powerful diagnostic to separate the families of intrinsic and extrinsic stars, besides the original method based on Tc detection only. Indeed, Fig. 2 is an update of Neyskens *et al.* (2015) including a new, extended sample of extrinsic stars. It shows that Tc-rich stars (intrinsic S stars) are Nb-poor, while Tc-poor stars (extrinsic S stars and barium stars) are Nb-rich.

For extrinsic stars it can be demonstrated (Neyskens *et al.* 2015) that, when the abundance of the s-processed material dominates over the initial composition:

$$\left[\frac{\mathrm{Zr}}{\mathrm{Fe}}\right] = \left[\frac{\mathrm{Nb}}{\mathrm{Fe}}\right] + \log\frac{N_s(\mathrm{Zr})}{N_s(\mathrm{Nb})} - \log\frac{N_\odot(\mathrm{Zr})}{N_\odot(\mathrm{Nb})}. \tag{3.1}$$

From Fig. 2 it can be seen that, for extrinsic stars, [Zr/Fe] as a function of [Nb/Fe] follows nicely a straight line of slope 1 (red line on Fig. 2). The y-intercept of this straight line allows to determine $\log \frac{N_s(\mathrm{Zr})}{N_s(\mathrm{Nb})}$, which is a function of the s-process operation temperature. Therefore Zr and Nb abundances in extrinsic stars provide a sensitive s-process thermometer.

References

Gustafsson, B., Edvardsson, B., Eriksson, K., Jørgensen, U. G., Nordlund, Å & Plez, B., 2008, *A&A*, 486, 951–970

Van Eck, S., Neyskens, P., Jorissen, A., Plez, B., Edvardsson, B., Eriksson, K., Gustafsson, B., Jørgensen, U. G., & Nordlund, Å, 2017, *A&A*, 601, A10

Raskin, G., Van Winckel, H., Hensberge, H., Jorissen, A., Lehmann, H. *et al.*, 2011, *A&A*, 526, A69

Neyskens, P., Van Eck, S., Jorissen, A., Goriely, S., Siess, L., & Plez, B., 2015, *Nature*, 517, 174

Astronomy and Astrophysics in the Gaia sky
Proceedings IAU Symposium No. 330, 2017
A. Recio-Blanco, P. de Laverny, A.G.A. Brown
& T. Prusti, eds.

doi:10.1017/S1743921317005397

Observational Facilities of Sternberg Astronomical Institute for Ground-Based Photometric Study of Newly Identified GAIA Objects, — CV-candidates.

Irina Voloshina and Valerian Sementsov

Sternberg Astronomical Institute, Lomonosov Moscow State University,
Moscow, 119991, Russia, Universitetsky prospekt, 13
email: voloshina.ira@gmail.com

Abstract. The extended observational program for study of cataclysmic variables is realized in Sternberg Astronomical Institute during the last years. A few telescopes of Crimean Observational Station equipped with a different devices, — UBV photometer and two CCD camera, are used for observations. Among the close binary systems (CBS), cataclysmic variables are the most interesting objects because of the outburst activity and variety of their observational features. They could serve a good laboratory for study of physical processes in CBS. GAIA provides astronomers with a new ample opportunity for investigation of cataclysmic variables. Though the relative faintness of detected objects it is still possible to carry out a high accuracy ground-based observations with our equipment. Obtained ground-based data permit us to confirm classification of detected CV-candidates, to determine the physical characteristics with a sample of new cods and improve the current understanding of their nature.

Keywords. binaries: close, cataclysmic variables, techniques: photometric

---

Cataclysmic variables are low-mass CBS in the late evolutionary stages, where the primary is a WD and the secondary is a late-type star. It fills its Roche Lobe and transfers mass to the WD. The gas stream flows from the secondary to the WD forming an accretion disk around it. Both components, along with the accretion disk and the gaseous flow from the optical star contribute to the total system brightness. Most of CVs have orbital periods from 80^m to 15^h. The geometry of the satellite and the scanning law determine the sets of allocated frequencies in GAIA data (with periods in the neighborhood of 106.5^m, 6^h and 63^d). Continuous ground-based observations will provide an opportunity to seriously clarify the data on the variability of irregular objects such as CV.

We will carry out the long-term observations of the objects discovered recently in GAIA project using CCD photometers installed at telescopes of the Sternberg Astronomical Institute in Crimea. The light detector on the 50-cm telescope is an Apogee Alta U8300 (3326×2504 pel, 1 pel = 5.4 μm) and an Apogee 47 (528×512 pel, 1 pel = 12 μm) on the 60-cm telescope. We performed observations mostly in the R or R_c bands, because the sensitivity of these CCD detectors is highest in the red (5800-6600 A). The duration of observational sets is about 5–6 hours, on average. The uncertainty of a single measurement depended on the star brightness, and is approximately the same for both telescopes, $\sigma \sim 0.01 - 0.05^m$. The new 2.5-m telescope near Kislovodsk (KGO) with CCD photometer could be used for particularly faint objects from our program.

Acknowledgments. This study has been supported via grant of the RFBR N 17-52-53200. We acknowledge ESA Gaia, DPAC and the Photometric Science Alerts Team (http://gsaweb.ast.cam.ac.uk/alerts).

Table 1. List of CV candidates in the Northern Hemisphere from GAIA releases ($< 18^m$)

#Name	α_{J2000}	δ_{J2000}	G	SIMBAD identification
Gaia17ayu	$18^h 57^m 36.90^s$	$+32°08'35.95''$	16.5^m	
Gaia17ayl	19 13 34.4472	+37 23 22.272	16.9	
Gaia17ayi	20 33 35.9952	+24 27 05.472	17.6	
Gaia17asx	20 49 26.4480	+19 45 05.544	17.69	
Gaia17asw	02 36 27.6072	+50 44 34.224	17.63	
Gaia17asc	20 20 51.2424	+22 10 20.532	15.41	
Gaia17arj	20 52 27.2280	+31 50 26.880	17.97	
Gaia17aqx	18 04 19.3416	+13 21 37.800	17.29	
Gaia17aqj	19 41 45.4992	+33 54 01.296	17.5	
Gaia17apr	19 25 00.0768	+43 00 07.956	16.47	
Gaia17aop	22 10 18.0600	+53 08 35.484	17.86	
Gaia17ahf	05 03 32.8392	+69 09 47.160	16.56	
Gaia17agc	03 50 35.00	+35 32 47.0	16.48	(1.7″) CRTS J035034.9+353247 CV?
Gaia17afs	17 35 17.2776	+01 32 49.488	16.42	
Gaia17afq	22 57 39.6816	+50 43 04.080	17.91	
Gaia17afp	21 17 21.0120	+45 58 47.568	17.83	
Gaia17ade	22 29 00.3	+26 37 07	16.1	(0.65″) GALEX J222900.3+263707
Gaia17abv	18 02 31.3488	+30 58 29.1	16.84	
Gaia17aaz	18 35 12.82	+38 20 04.4	13.62	(0.07″) V* LL Lyr
Gaia16cfx	10 29 37.749	+41 40 46.35	17.67	(0.08″) SDSS J102937.74+414046.3 CV*
Gaia16cft	21 12 04.5408	+36 35 29.508	16.34	
Gaia16cba	01 39 26.7264	+49 40 53.580	17.69	
Gaia16caf	00 43 04.2024	+53 17 15.936	17.4	
Gaia16bzy	00 46 25.8720	+38 20 23.820	17.86	
Gaia16bzo	23 33 21.6000	+55 03 41.976	17.53	
Gaia16bzc	22 24 52.0968	+52 05 38.112	17.91	
Gaia16bxc	20 12 40.0296	+25 10 26.472	17.93	
Gaia16bww	16 36 33.1200	+39 33 13.140	15.31	
Gaia16buu	18 32 05.0448	+09 22 21.684	16.55	
Gaia16bos	22 24 52.9560	+53 30 01.512	17.79	
Gaia16bno	21 01 40.49	+21 57 30.9	17.62	(1.10″) MASTER OT J210140.49+215730.9 ev
Gaia16blu	22 28 33.4392	+50 40 16.068	17.72	
Gaia16bln	21 46 39.94	+09 21 19.3	17.79	(0.62″) CRTS CSS110613 J214640+092119 CV?
Gaia16ble	21 26 24.16	+25 38 27.2	16.52	(0.57″) MASTER OT J212624.16+253827.2 CV*
Gaia16bis	07 49 28.013	+19 04 52.10	16.53	(0.15″) SDSS J074928.01+190452.1 DN*
Gaia16bhk	05 05 15.0000	+06 17 07.800	16.75	
Gaia16bfb	05 43 29.0760	+77 20 24.576	16.8	
Gaia16bbz	19 16 39.4344	+46 21 07.236	17.69	
Gaia16bbf	16 47 05.087	+19 33 34.98	17.98	(0.16″) SDSS J164705.08+193334.9 CV?
Gaia16baw	19 49 11.6736	+29 06 41.256	17.88	
Gaia16azk	05 32 20.9592	+03 57 31.536	15.35	
Gaia16azd	20 51 59.2392	+34 49 46.128	16.71	
Gaia16ayw	04 59 55.09	+20 48 51.1	17.82	(0.13″) CRTS MLS101225 J045955+204851 CV?
Gaia16awq	14 47 47.6592	+66 08 47.616	14.85	
Gaia16aue	13 05 25.5816	+05 43 24.996	17.68	
Gaia16asd	19 31 49.5024	+09 49 39.216	17	
Gaia16art	20 31 36.8472	+08 48 30.204	14.97	
Gaia16apa	00 15 38.213	+26 36 57.45	13.89	(1.15″) 2MASS J00153821+2636574 CV*
Gaia16amd	20 35 29.77	+06 36 53.3	16.49	(0.56″) MASTER OT J203529.80+063652.8 DN*
Gaia16aid	08 10 57.4536	+27 15 12.492	14.07	
Gaia16aht	21 07 21.5352	+27 31 11.604	17.9	
Gaia16ahk	23 00 25.1160	+41 31 18.732	16.92	
Gaia16ahi	23 27 16.2360	+41 31 48.612	16.78	
Gaia16agx	20 36 48.8040	+29 31 27.192	17.51	
Gaia16agl	17 34 29.6568	+14 34 38.532	17.59	
Gaia16afh	18 58 38.706	+46 02 07.83	17.16	(0.68″) GALEX J185838.7+460207 CV*
Gaia16aeb	20 49 12.7896	+65 01 08.472	17.65	
Gaia16adh	00 38 27.0480	+25 09 25.020	16.43	
Gaia16adb	12 09 30.2592	+76 09 11.916	16.94	
Gaia16acz	01 46 44.6544	+48 26 49.668	15.9	
Gaia16abv	06 04 24.7584	+54 07 28.740	17.62	
Gaia16abj	21 44 08.5464	+82 12 34.092	17.75	
Gaia15adf	08 19 36.06	+19 15 40.1	16.2	(0.57″) CRTS J081936.1+191540 CV?
Gaia15abx	23 19 09.18	+33 15 39.8	17.88	(0.20″) 1RXS J231909.9+331544 CV*
Gaia15abh	22 18 29.5584	+39 48 37.476	16.32	
Gaia15abg	00 22 53.2	+13 40 41	16.88	(0.43″) GALEX J002253.2+134041 UV
Gaia15aan	16 05 47.996	+24 05 31.06	13.03	(0.04″) NAME 400d J160547.5+240524 CV*
Gaia14adn	09 59 08.7336	+81 53 35.592	15.95	
Gaia14ade	23 50 52.01	+28 58 59.5	17.78	

Astronomy and Astrophysics in the Gaia sky
Proceedings IAU Symposium No. 330, 2017
A. Recio-Blanco, P. de Laverny, A.G.A. Brown
& T. Prusti, eds.

doi:10.1017/S1743921317005452

The nearby triple star HIP 101955

Fang, Xia

Purple Mountain Observatory, Chinese Academy of Sciences, Nanjing 210008, China
email: xf@pmo.ac.cn

Abstract. The nearby triple star HIP 101955 with strongly inclined orbit still remains. Thus the long-term dynamical stability deserves to be discussed based on the new dynamical state parameters (component masses and kinematic parameters) derived from fitting the accurate three-body model to the radial velocity, the Hipparcos Intermediate Astrometric Data (HIAD), and the accumulated speckle and visual data. It is found that the three-body system remains integrated and most likely undergoes Kozai cycles. With the already accumulated high-precision data, the three-body effects cannot always be neglected in the determination of the dynamical state. And it is expected that this will be the general case under the available Gaia data.

Keywords. binaries: close, stars: kinematics and dynamics, stars: individual (HIP 101955)

1. Introduction

According to the Multiple Star Catalog Tokovinin (1997), the nearby triple star HIP 101955 with low orbit hierarchy and strongly inclined orbit still remains. Thus the stability of this system deserves to be studied with improved dynamical state parameters. The three components were optically resolved for the first time by Balega *et al.* (2002) and Malogolovets *et al.* (2007). Detailed study of this triple system were presented in these papers and Tokovinin & Latham (2017). However, these results are not satisfactory.

In fact, the main factors that decide the quality of dynamical state determination are the time span, the precision of observations, and the accuracy of the dynamical model. All these factors are taken into consideration in our redetermination. First, in the previous orbital determinations, only relative position data (RPD) from speckle interferometric observations made in 1998 $\sim$ 2004 are used. In our redetermination, apart from spectroscopic and visual RPD accumulated since 1934, the HIAD, and radial velocities are used jointly. Therefore, both the time span and precision of the observational data are significantly increased. Second, the accuracy is increased by using the three-body model, which turns out to be necessary for us to obtain a better result.

2. Observational Data and fitting results

With three kinds of observational data, the maximum likelihood estimate of model parameters is usually obtained by minimizing the following objective function

$$\chi^2 = \sum_{i=1}^{N} \left(\frac{y_i - y(x_i; a_1 \cdots a_M)}{\sigma_i}\right)^2, \tag{2.1}$$

where y_i is the observed value with standard error σ_i, and $y(x_i; a_1 \cdots a_M)$ is the calculated value depending on the model parameters $a_1 \cdots a_M$.

Taking into consideration the two-body effect on the motion of A_m (the center of mass of A_a, A_b), which is previously ignored when the inner and outer orbits are determined separately, we simultaneously fit the double two-body model to all observational data. There are 21 model parameters, including the 7 inner orbital elements, 7 outer orbital elements and the mass ratio q. The other parameters are 5 motion state parameters $(\alpha_c, \delta_c, \varpi_c, \dot{\alpha}_c, \dot{\delta}_c)$ describes the constant motion of Cm_3 (the mass center of the 3-body

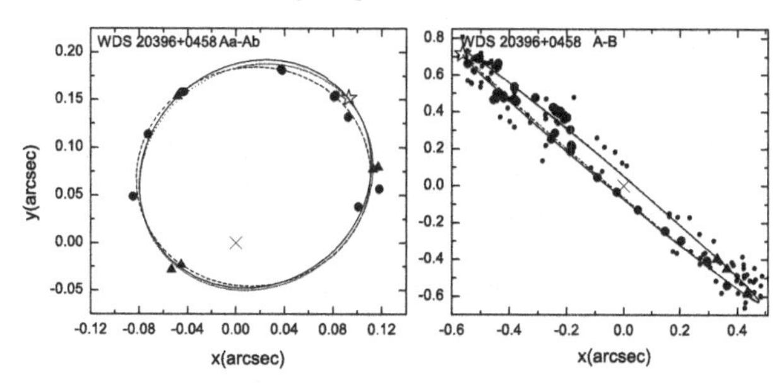

Figure 1. Inner and outer orbital motions of HIP 101955. The filled circles are the RPD used in our fitting and triangle points are RPD from Tokovinin & Latham (2017).

system) in the ICRS , and ρ which is related to the magnitude difference (Δm) and the mass ratio (q) by $\rho = \frac{r}{1+r} - q$ ($r = 10^{-0.4\Delta m}, q = M_{Ab}/(M_{Aa} + M_{Ab})$). Here, as well as in the following, the Bounded Variable Least Squares (BVLS) algorithm Lawson & Hanson (1995) is used to give the least-square solution.

The fitted inner and outer orbits are plotted in Figure 1 as dotted curves, while the orbits provided by Tokovinin & Latham (2017) are plotted as dashed curves.

Using the three-body model, we have again 21 parameters to fit. The parameter set $(\alpha_c, \delta_c, \varpi_c, \dot{\alpha}_c, \dot{\delta}_c)$, and ρ are the same as in the double two-body model. The other parameters are component masses (M_{Aa}, M_{Ab}, M_B), the initial (1991.25) position and velocity $(r_x, r_y, r_z, v_x, v_y, v_z)_{Ab}$ of Ab relative to Aa, the initial position and velocity $(r_x, r_y, r_z, v_x, v_y, v_z)_B$ of B relative to Aa.

The preliminary fitting results are plotted in Figure 1. The open stars represent the best-fit position at the epoch 1991.25 and the solid curves are the trajectories. The χ^2 of this result is much reduced in comparison with the result using double two-body model.

3. Summary

The dynamical state parameters of HIP 101955 modeled as a three-body system are determined. Based on this result, the long-term dynamical stability is explored for 10^3 of the outer period. The instantaneous two-body orbital elements, e.g., a_{in}, a_{out}, e_{out}, are found to be constant. Other parameters vary significantly. In comparison with the result by our simultaneous fit using the double two-body model, the maximum positional deviation reaches ~50 mas within the time interval [1891.25, 2091.25]. Therefore, with the already accumulated high-precision data, the three-body effects cannot always be neglected in the determination of the dynamical state of a triple star. This should be the general case as the Gaia data are available.

Acknowledgements

This research has made use of the Washington Double Star Catalog maintained at the U.S. Naval Observatory (WDS, http://ad.usno.navy.mil/wds/dsl.html). This research is supported by the National Natural Science Foundation of China under Grant Nos. 11673071 and 11203086.

References

Balega, I. I., Balega, Y. Y., Hofmann, K.-H. *et al.* 2002, *A&A*, 385,87
Lawson, C. & Hanson, R. (ed.) 1995, *in Solving Least Squares Problems*, 279
Malogolovets, E. V., Balega, Yu. Yu., & Rastegaev, D. A. 2007, *AstBu*, 62, 111
Tokovinin, A. A. 1997, *A&AS*, 124, 75
Tokovinin , A. A. & Latham D. W., 2017, *ApJ*, 838, 54

Astronomy and Astrophysics in the Gaia sky
Proceedings IAU Symposium No. 330, 2017
A. Recio-Blanco, P. de Laverny, A.G.A. Brown & T. Prusti, eds.

doi:10.1017/S1743921317005440

A new method for orbit determination on the Gaia SB1s

Wang Xiaoli

Yunnan Observatories, CAS, Kunming, 650011, China
email: wangxl@ynao.ac.cn

Abstract. An iterative method to determine the self-consistent orbital solutions of single-lined spectroscopic binaries (SB1s) along with compatible physical properties of component stars via a simultaneous fit including both the Hipparcos Intermediate Astrometric Data (HIAD) and radial velocity data is introduced in this work. For the method, a stellar evolutionary model is used to distribute the total mass and luminosity to the primary and the secondary and update the ratio of the semimajor axes of the photocenter to the primary orbits. Once the Gaia Intermediate Astrometric Data (GIAD) are released, the method can be applied to study the Gaia SB1s and give self-consistent orbital solutions and compatible physical properties of component stars.

Keywords. single-lined spectroscopic binaries, orbital determination, fundamental parameters.

---

1. Introduction

Since the Hipparcos catalogue has been published in 1997 (ESA (ed.) (1997)), the photocentric orbital solution of some spectroscopic binaries have already been determined by HIAD (Van Leeuwen& Evans(1998); Jancart *et al.* (2005); Ren & Fu(2010), Ren & Fu(2013)). In these work, they treated the semimajor axis of the photocenter orbit as a free fitting parameter which couldn't make full use of spectroscopic parameters or considered the the primary orbit is consistent with the photocenter orbits which neglected the influence of the secondary. Therefore, we have developed an appropriate method that makes full use of the HIAD and the radial velocity data, and gives a self-consistent orbit as well as the compatible physical properties of the component stars (Wang *et al.* (2015)).

Comparing with Hipparcos mission, the precision of Gaia is much higher at order of magnitude (Lindegren *et al.* (2016)). Then, our new method can be used to determine the orbits of the SB1s in combination of the GIAD and radial velocity data in the future, and the preliminary physical properties of the known SB1s can be derived.

Although the orbital solutions of SB1s mentioned above rely on stellar evolutionary model, it is helpful for developing binary statistical study. Meanwhile, these results can guide follow-up observation and derive physical parameters of component stars independent of stellar evolutionary model. Then, these physical parameters can be used to test stellar evolutionary model.

2. The new method

For an SB1, the semimajor axis (a_0) of the photocenter orbit can be expressed as

$$a_0 = \lambda a_1 = \lambda \varpi \frac{K_1 P \sqrt{1-e^2}}{2\pi \sin i}, \tag{2.1}$$

where a_1, ϖ, K_1, P, e, and i are the semi-major axis of the primary orbit, the system parallax, the semi-amplitude of the radial velocity curve of the primary, the orbital period, the orbital eccentricity, and the orbital inclination, respectively. The parameter λ can be expressed as

$$\lambda = [1 - \frac{10^{0.4H_p}(1+q^{-1})}{1+10^{0.4H_{p2}}}]. \tag{2.2}$$

The value of λ is between 0 and 1. The parameters q, H_p, H_{p2} in Equation (2.2) represent the mass ratio $\frac{M_2}{M_1}$, and the Hipparcos magnitude of a binary system and the secondary, respectively. H_{p2} can be calculated from the mass-absolute Hipparcos magnitude (Arenou *et al.* (2000)).

$$H_{p2} = -13.5 \log M_2 + 5.07. \tag{2.3}$$

A simultaneous fit to the revised HIAD(Van Leeuwen(2007)) and Radial velocity data is used to derive the orbital solution. The iterative fitting process begins with $\lambda = 1$. The modified grid method developed by (Ren & Fu(2010)) is used to search for the global optimization solution. Based on this solution, the value of λ is updated in combination with the stellar evolutionary model, more information refer to Wang *et al.* (2015).

With the method mentioned above, we have study eight SB1s and found that HIP 7143 and HIP 45333 have relative large luminosity ratio of the secondary and primary which may be resolved by the latest observation, and this have already been proved by Halbwachs *et al.* (2014) and Fekel *et al.* (2015).

3. Conclusions

In combination of spectroscopic orbital parameters and Hipparcos parallax, we found that the minimum angular semi-major axes of more than 1200 SB1s in the 9th Catalogue of Spectroscopic Binary Orbits (Pourbaix *et al.* (2004)) is bigger than 0.3mas. This means that the method mentioned above can be used to study these systems, and give the self-consistent orbital solutions along with compatible physical properties of component stars. And the secondaries of these systems which can be observed by the latest observation can be picked out for further observation and analysis.

Acknowledgements

This work is supported by the NSFC under Grant No.11603072.

References

Arenou, F., Halbwachs, J., Mayor, M., Palasi, J., & Udry, S. 2000, in: B. Reipurth & H. Zinnecker (eds.), *The Formation of Binary Stars*, Proc. IAU Symposium No. 200 (Cambridge: Cambridge Univ. Press), p. 135

ESA (ed.) 1997, in *The Hipparcos and Tycho Catalogues* (Paris: ESA) ESASP-1200

Fekel, F. C., Williamson, M. H., Muterspaugh, M. W. *et al.* 2015, *AJ*, 149, 63

Jancart, S., Jorissen, A., Babusiaux, C., & Pourbaix, D. 2005, *A&A*, 442, 365

Halbwachs, J.-L., Arenou, F., Pourbaix, D. *et al.* 2014, *MNRAS*, 422, 14

Lindegren L, Lammers U, Bastian U *et al. arXiv preprint arXiv:*1609.04303, 2016

Pourbaix, D. & Boffin, H. M. J. 2003, *A&A*, 398, 1163

Pourbaix, D., Tokovinin, A. A., Batten, A. H. *et al.* 2004, *A&A*, 424, 727

Ren, S. L. & Fu, Y. N. 2013, *AJ*, 145, 81

Ren, S. L. & Fu, Y. N. 2010, *AJ*, 139, 1075

Van Leeuwen, F., & Evans, D. W. 1998, *A&AS*, 130, 157

Van Leeuwen, F. 2007, Hipparcos: The New Reduction of the Raw Data, Vol. 350 (Berlin: Springer)

Wang, X. L., Ren, S. L., & Fu, Y. N. 2015, *AJ*, 150, 4

Astronomy and Astrophysics in the Gaia sky
Proceedings IAU Symposium No. 330, 2017
A. Recio-Blanco, P. de Laverny, A.G.A. Brown
& T. Prusti, eds.

doi:10.1017/S1743921317005476

Clarification of the formation process of the super massive black hole by Infrared astrometric satellite, Small-JASMINE

Taihei Yano, JASMINE-WG

National Astronomical Observatory of Japan, 2-21-1, Osawa, Mitaka, Tokyo 181-8588, Japan
email: yano.t@nao.ac.jp

Abstract. Small-JASMINE (hearafter SJ), infrared astrometric satellite, will measure the positions and the proper motions which are located around the Galactic center, by operating at near infrared wave-lengths. SJ will clarify the formation process of the super massive black hole (hearafter SMBH) at the Galactic center. In particular, SJ will determine whether the SMBH was formed by a sequential merging of multiple black holes. The clarification of this formation process of the SMBH will contribute to a better understanding of merging process of satellite galaxies into the Galaxy, which is suggested by the standard galaxy formation scenario. A numerical simulation (Tanikawa and Umemura, 2014) suggests that if the SMBH was formed by the merging process, then the dynamical friction caused by the black holes have influenced the phase space distribution of stars. The phase space distribution measured by SJ will make it possible to determine the occurrences of the merging process.

Keywords. Galaxy: bulge, Galaxy: kinematics and dynamics, Galaxy: structure

---

1. Introduction

A super massive black hole (SMBH) resides in the central region of a galaxy. Its possible physical formation process is divided into two; a gas accretion onto a seed BH or a merger of BHs. From a viewpoint of the standard scenario for galaxy formation, it is natural to regard a SMBH as an end result of the merger of BHs. However, it was considered that there exist fatal problems in forming a BH merger. Recently, N-body simulations with a high resolution have overcome such problems and have shown that a sequential merger of multiple BHs is possible. Accordingly we aim at observationally investigating whether a SMBH in the Milky Way resulted from BH mergers through the analysis of stellar kinematics. A numerical simulations (Tanikawa and Umemura 2014) suggests that the merger process of BHs influenced the distributions of density and velocity of bulge stars within a distance of 10∼100pc from the Galactic center. Our method to verify the presence/absence of BH mergers is to assess the distribution function corresponding this region.

2. Method

First, we model the distribution functions form the results of numerical simulations of the Galactic bulge. Secondly, we deduce the distribution function P from the mock catalogue which we make with the assumed observational errors for stars within the circular region with the radius of 0.7deg centering around the Galactic center (the region that SJ will observe). Third, we make a few different models Q_i (i is the suffix for each model) of the distribution functions. Finally, we judge which model Q_i fits with P the best using the test of Kullback-Leibler (KL) divergence. Here the KL divergence means

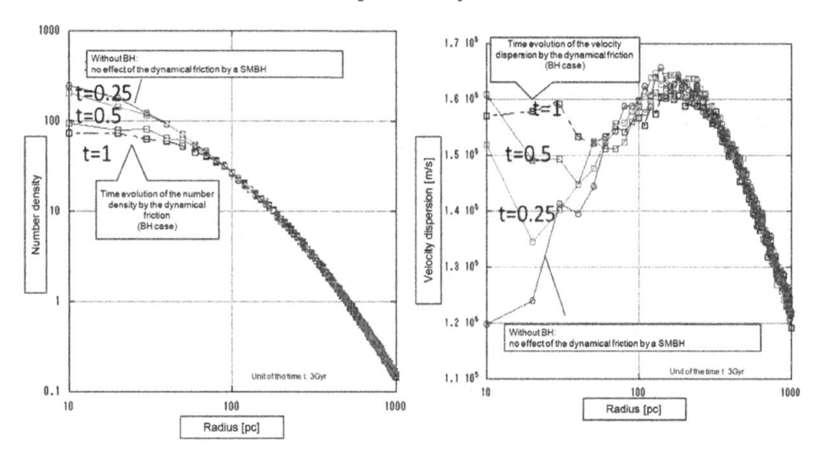

Figure 1. Evolution of the number density (left) and the velocity dispersion (right) of the central region of the Galactic bulge.

the expectation of the logarithmic difference between P and Q_i. Note that the divergence gives a smaller value for a more agreement case.

3. Preparation of models and mock catalogue:

We prepare two models of central region of the Galactic bulge with/without SMBH. We also prepare a mock catalogue using the above model with SMBH including the observational error and the effect of extinction by the dust (Gonzalez *et al.*, 2012). Bulge model parameters are assumed as follows: Bulge mass $M_{bulge} = 9 \times 10^9 M_{sun}$, BH total mass $M_{BH} = 4 \times 10^6 M_{sun}$, Virial velocity $v = 120$km/s, and Calcuratied time $T = 3$Gyr.

4. Evolution of the Number Density and Velocity Dispersion — Dynamical Friction by the Black Holes —

Evolutions of the number density and the velocity dispersion of the central region of the Galactic bulge are shown in Fig. 1. When the merger process of black holes has occurred, dynamical friction by black holes make the distribution of stars and velocity dispersion within the radius of 100pc change.

5. Results

We check whether we correctly judje using KL divergence. We have investigated 64000 trials. We obtain the numbers of stars, error of parallaxes, error of proper motion in order to judge with a confidence level of 99.7%. Criterion for judging with high confidence level is as follows: Number of stars we need is more than 3500 stars. Error of parallaxes and proper motions are less than 20μas and 200μas/yr, respectively.

References

Tanikawa, A. & Umemura, 2014, *MNRAS*, 440, 652

Gonzalez, O. A., Rejkuba, M., Zoccali, M., Valenti, E., Minniti, D., Schultheis, M., Tobar, R., & Chen, B. 2012, *A&A*, 543, 13

Astronomy and Astrophysics in the Gaia sky
Proceedings IAU Symposium No. 330, 2017
A. Recio-Blanco, P. de Laverny, A.G.A. Brown & T. Prusti, eds.

doi:10.1017/S1743921317005531

Mathematical Assessment of Physical and Chemical Processes from the Middle B to the Early F Type Main Sequence Stars

Kutluay Yüce[1] and Saul J. Adelman[2,3]

[1]Department of Astronomy and Space Sciences, Faculty of Science, University of Ankara, TR-06100 Anadolu, Ankara, Turkey,
email: Kutluay.Yuce@ankara.edu.tr
[2]Department of Physics, The Citadel, 171 Moultrie Street, Charleston, SC 29409, United States of America,
email: adelmans@citadel.edu
[3]Guest Investigator, Dominion Astrophysical Observatory, Victoria, Canada

Abstract. The middle B to the early F Main Sequence stars have some of the most quiet stellar atmospheres. In this part of the HR diagram we find stars with atmospheres in radiative equilibrium. They lack the convective circulations of the middle F and cooler stars and the massive stellar winds of hotter stars. When stars of different mass evolve off the Main Sequence in this part of the HR Diagram their evolutionary paths do not cross initially. Thus stars with the same effective temperature and surface gravity have the same luminosity and mass. By comparing their elemental abundances, we might be able to identify physical processes which cause any differences in their abundances. Here we begin with stars whose effective temperatures and surface gravities are similar, and which have been analyzed by us using spectra obtained from the Dominion Astrophysical Observatory (DAO). Improvements in our knowledge of the energy distributions of stars (for example via GAIA measurements) should lead to improved estimates of stellar effective temperatures and surface gravities.

---

1. Introduction

Preston (1974) formulated the designation chemically peculiar (CP) stars on the Upper Main Sequence (MS). Adelman *et al.* (2003) showed that two of his classes the Mercury-Manganese (HgMn) and the metallic-line (A*m*) stars were a single class. Evolutionary tracks indicated that the hot A*m* stars (spectral type A0-A2) have evolved from the stars at the cool end of the HgMn class on the MS. Later Adelman & Unsure (2007) discovered there is no clear demarcation between the surface abundances of those stars classified as normal and as A*m* stars.

Yüce & Adelman (2014) describe the detailed chemical composition characteristics of non-magnetic B, A, and early F type stars for 32 elements using graphical techniques to understand better the relationship of normal and non-magnetic CP stars. The spread of the abundance anomalies for a given element tends to be smaller among the Am stars than among the HgMn stars. Adelman *et al.* (2015) found the abundances of the SB2 component stars HR 104 A, B and θ Aql A follow the solar pattern while those of θ Aql B follow those of a weak non-magnetic CP star. We progress in our correlation study of the physical parameters and surface chemical composition of middle Main Squence band stars. The study has the potential for some interesting surprises to describe how we separate the A*m* from the normal A stars using our DAO series analyses coauthored with Dr. Austin F. Gulliver. Now we propose to study non-magnetic normal and chemically peculiar B, A, and F stars by creating an index which measures the degree to which such stars have values between those of solar abundance A stars and the classical Am/Fm stars.

2. Findings

We investigate whether metallic line Am and normal A stars with similar values of effective temperature and surface gravity have the same abundances obtained from high resolution fine analyses. The values of microturbulent velocity for the cooler Am stars (T < 8500 K) are greater than those of both the hotter Am and all normal stars. The microturbulent velocity increases from 2 km s^{-1} for early Am type stars with T $\geqslant$ 8500 K to 5 km s^{-1} for the late Am stars.

For Am stars our index is based on log g, and systematic chemical peculiarities in iron group and heavy elements. A key parameter for the Am phenomena is the titanium abundance. There are clearly differences between light elements (MgAlSiS) and iron group elements (ScTiVCrMnFeCoNi) for the stars in the region of log g $\geqslant$ 3.90 and for the stars evolving from the MS (log g < 3.90). The richness relative to the Sun in the iron group elements of the MS stars is more than those of both the light elements and the evolving from MS stars. Our computations show that Ca and/or Sc are underabundant in many MS Am stars. The CaSc abundances relative to the Sun have an average value of -0.6 dex for 10 Am Main Squence stars whereas 12 evolved Am stars show solar values within our error limits.

For Normal A stars our index is based obtained during this study is based on log g, and systematic peculiarities in only light elements with the evolution. The relative element abundances in the light (MgAlSiS), iron group (ScTiVCrMnFeCoNi), heavy elements (BaLaCeNdEu) are nearly solar/normal for the normal A stars with log g close to 4 while these stars in the light elements are more underabundant than the stars evolving from the MS. The CaSc abundance relative to the Sun is solar in the averaged values for the 6 Main Squence stars and 15 evolved normal stars.

3. Conclusion

In this investigation we notice that the effects due to surface gravity and other physical parameters on the physical properties, and chemical structure distinction. Chemical peculiarities are clearly evident in iron group and heavy elements of the MS Am stars and in light elements of the evolved normal stars within our DAO series. Another particularly surprising result, which is an indicator of the Am phenomenon, is that the observed abundances of all of elements with Z > 22 are overabundant for such stars.

Acknowledgements

Dr. Kutluay YÜCE thanks Ankara University. Prof. Dr. Saul J. ADELMAN thanks Dr. James E. Hesser, Director of the DAO for the observing time. SJA's contribution to this paper was supported in part by grants from The Citadel Foundation.

References

Adelman, S. J., Adelman, A. S., & Pintado, O. I. 2003, *A&A*, 397, 267
Adelman, S. J. & Unsuree, N. 2007, *BaltA*, 16, 183
Adelman, S. J., Yüce, K., & Gulliver, A. F. 2015, *PASP*, 127, 509
Preston, G. W. 1974, *ARAA*, 12, 257
Yüce, K. & Adelman, S. J. 2014, *PASP*, 126, 345

Astronomy and Astrophysics in the Gaia sky
Proceedings IAU Symposium No. 330, 2017
A. Recio-Blanco, P. de Laverny, A.G.A. Brown
& T. Prusti, eds.

doi:10.1017/S174392131700549X

Massive companions of binary systems

D. Jableka[1], S. Zola[1,2], B. Zakrzewski[2], J. M. Kreiner[2] and W. Ogloza[2]

[1]Astronomical Observatory of the Jagiellonian University,
Orla 171, 30-244 Krakow, Poland
email: jableka@oa.uj.edu.pl
[2]Mt. Suhora Observatory, Pedagogical University,
ul. Podchorazych 2, 30-084 Krakow, Poland
email: szola@oa.uj.edu.pl

Abstract. We examined the O-C diagrams of eclipsing binary systems and selected these exhibiting cyclic shape, either sinusoidal or quasi sinusoidal. Assuming these variations being due to the Light Time Travel effect (LTE), we estimated the parameters of companions with the Monte Carlo method. As a result, we identified nearly two dozen of eclipsing systems that might have companions with a minimum mass larger than that of a neutron star. Their masses fall into the range between 1.7 and 34 solar masses. This sample of triples with high mass companions can be confirmed with the help of observations gathered by Gaia: parallaxes and astrometric measurements.

Keywords. stars: distances, binaries: eclipsing

---

1. Sample selection

This work is aimed at discovering massive companions, and we assumed the threshold at the mass of a neutron star. For this purpose, we examined the O-C diagrams created from minima timings collected by Kreiner (2004), and based on their shape, exhibiting significant changes and possible cyclic variations as well as large amplitude, we selected 79 binaries out of more than a thousand. For these, assuming that the changes are being due to companions, we made preliminary computations and derived the orbits of third bodies. The sample of interest of this work was limited to these binaries which meet the minimal mass of 1.7 solar masses (for details see: Jableka *et al.*, 2013).

2. Monte Carlo computations

The inverse problem requires up to nine parameters, some of them are correlated. We applied the Monte Carlo search method to find the best fit in the nine-dimension space and allows to fit all nine, non orthogonal parameters simultaneously. We did the errors estimation by the jackknife re-sampling technique. For all systems we adjusted the linear ephemerides and a periodical LTE. In most cases we also fitted the quadratic term in the ephemerides.

No quadratic terms in the ephemerides of Y Cam, SZ Cam, RX Gem, T LMi and RW Per were needed. For these systems a periodical term is clearly noticeable. For U Cep, TU Her, CC Her, SW Oph and Y Psc mixed quadratic and periodical terms solution were required. Three systems: V602 Aql, V442 Cas and BO Gem, have their O-C diagrams with poor coverage, as for those, only one minimum in the minima timings is visible. V602 Aql and BO Gem have visible inflection points on both sides of relatively well covered minima, what suggests a possibility of periodical solutions. The same concerns

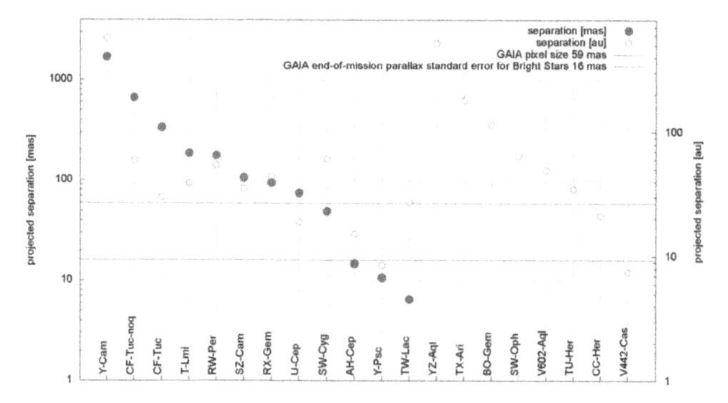

Figure 1. Projected separations of tertiaries to the center of mass of binary systems

YZ Aql and RX Gem in which the scatter in timings is comparable with the amplitude of the periodical solution. We obtained a very good fit for TW Lac and the residuals between observations and the periodic fit, may indicate an existence of another body, making this system to be a quadruple one.

3. Summary

Taking into account the total mass of the binary systems considered (taken from the literature), we calculated the third bodies mass functions, and, from these, their lower mass limits. For systems to which the distance is known, we also calculated the projected distances of the companions to the center of mass of the companion and the binary, treated as a single mass. The results are shown in Fig. 1. All hypothetical companions have wide or very wide orbits, orbiting the binaries in a distance from 6 to more than 160 AU. For the systems with known distances this translates to the projected distance (for inclination of 90 degrees) in the range between about 7 and 1100 mas.

For 18 systems we derived the minimum masses larger than than the threshold limit of 1.7 $M_{\odot}$. The most massive tertiary (about 34 $M_{\odot}$) was obtained for YZ Aql. Having such large masses, the companions should be detected in photometric and/or spectroscopic observations. In the case of these undetected, despite of their high mass, e.q. CF Tuc (Dogru *et al.*, 2009), an improved distance determination and astrometric measurements made by Gaia can confirm their presence, and if not, a more detail investigation of magnetic cycles theories will be required.

The Applegate mechanism, often considered as an explanation of cyclic or pseudo-cyclic trends in the O-C diagrams of eclipsing binaries, could be applied to 4 systems in our sample (SZ Cam, V442 Cas, U Cep and Y Psc). The amplitude of variations is too large for remaining 14 binaries, and the Applegate mechanism can not be responsible for such large O-C variations.

Aknowledgements

This study was partly funded by the NCN 2012/05/E/ST9/03915 grant.

References

Dogru, D., Erdem, A., Dogru, S. S., & Zola, S. 2009, *MNRAS*, 397, 1647
Jableka, D., Zola, S., Kreiner, J. M., & Zakrzewski, B. 2013, *CEAB*, 37, 195
Kreiner, J. M. 2004, *AcA*, 54, 207

[illegible]

[illegible]

8. Summary

[illegible]

[illegible] has been confirmed [illegible]

[illegible]

[illegible]

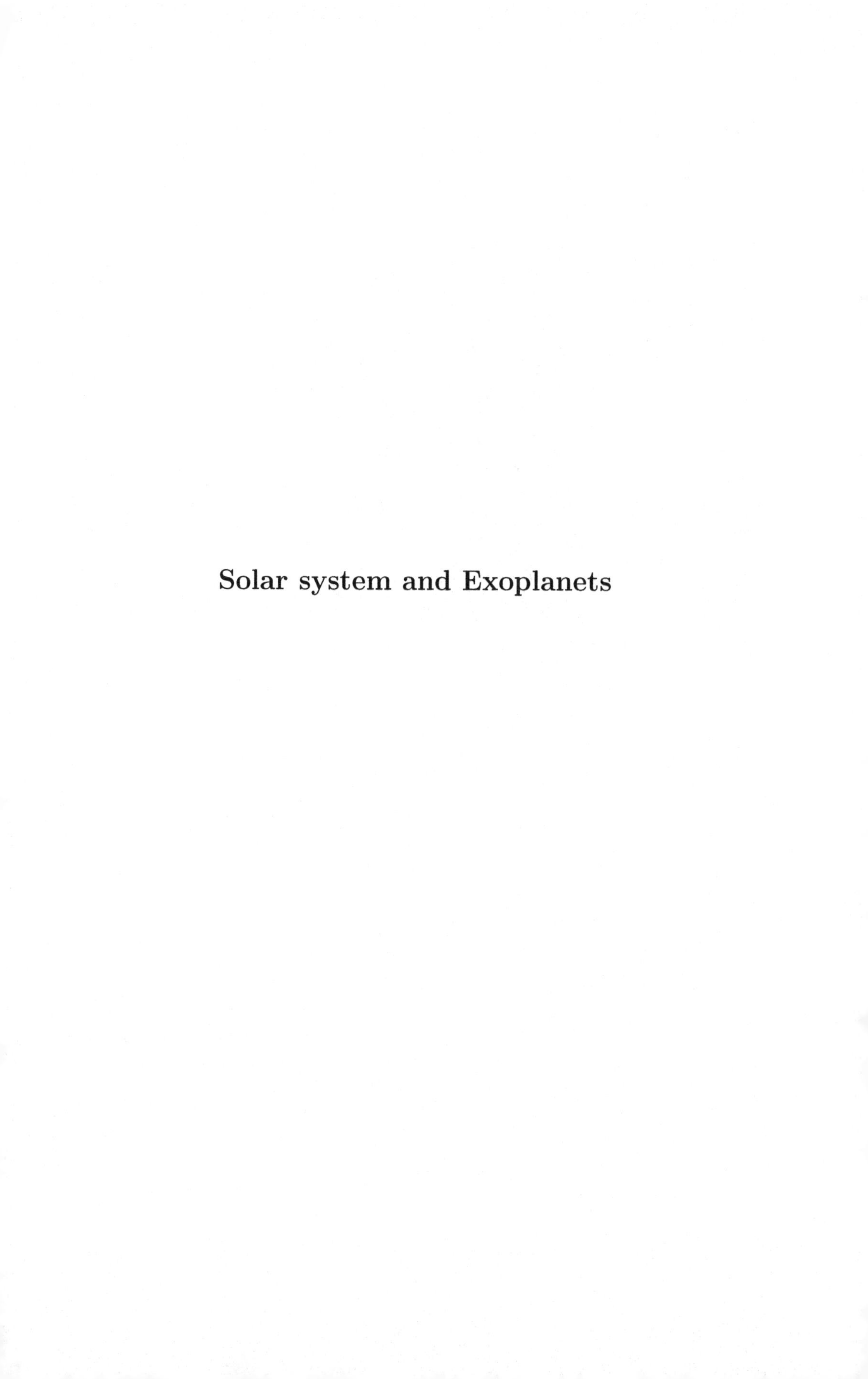

Solar system and Exoplanets

François Mignard and Alejandra Rrecio-Blanco during the conference dinner.

Astronomy and Astrophysics in the Gaia sky
Proceedings IAU Symposium No. 330, 2017
A. Recio-Blanco, P. de Laverny, A.G.A. Brown & T. Prusti, eds.

doi:10.1017/S1743921317006421

Characterisation of exoplanet host stars: A window into planet formation

Nuno C. Santos[1,2]

[1]Instituto de Astrofísica e Ciências do Espaço, Universidade do Porto, CAUP, Rua das Estrelas, 4150-762 Porto, Portugal
email: nuno.santos@astro.up.pt

[2]Departamento de Física e Astronomia, Faculdade de Ciências, Universidade do Porto, Rua do Campo Alegre, 4169-007 Porto, Portugal

Abstract. The detection of thousands of planets orbiting stars other than the Sun has shown that planets are common throughout the Galaxy. However, the diversity of systems found has also raised many questions regarding the process of planet formation and evolution. Interestingly, but perhaps not unexpectedly, crucial information to constraint the planet formation models comes from the analysis of the planet-host stars. In this talk I will review why it is so important to study and understand the stars when finding and characterising exoplanets. I will then present some of the most relevant star-planet relations found to date, and how they are helping us to understand planet formation and evolution. I will end with a presentation of the future steps in this field, including what Gaia will bring to help constrain the properties of planet-host stars, as well as to the star-planet connection.

Keywords. planetary systems, stars: abundances, planetary systems: formation

1. Introduction

Since the detection of a giant planet orbiting the solar-type star 51 Peg Mayor & Queloz (1995) more than 3500 extrasolar planets have been published†. The impact of these discoveries is considerable, both scientifically and socially. They represent the first firm steps of humankind towards the detection and characterisation of other planets similar to our Earth. In this process, this multi-disciplinary domain is opening new bridges between different fields in Astrophysics (e.g. stellar astrophysics, solar system research) and other areas of knowledge such as geophysics (Valencia *et al.* 2006) and biology (Kaltenegger *et al.* 2010). Together these bring new hopes of finding an Earth-like planet where life may have evolved.

The diversity of discovered planets is raising new questions and opening new pathways. One point is already clear, however: even though the precise frequency of the different kinds of planets in the Galaxy is a matter of debate, the community presently agrees that planets, in particular rocky planets like our Earth, are very common around solar type FGK and M stars (see e.g. Udry & Santos 2007, Howard *et al.* 2012, Bonfils *et al.* 2013). This conclusion is fully supported by state-of-the-art planet formation models based on the core-accretion paradigm, that further predict low mass/radius planets to largely surmount the number of their jovian or neptune-like counterparts (e.g. Mordasini *et al.* 2012).

The strong progress in this field is well illustrated by Fig. 1, where we plot, in the left panel, a mass-period diagram of the discovered exoplanets. The plot illustrates not only the large diversity of discovered planets, but also the clustering of planets around three

† For an up-to-date list see http://exoplanet.eu

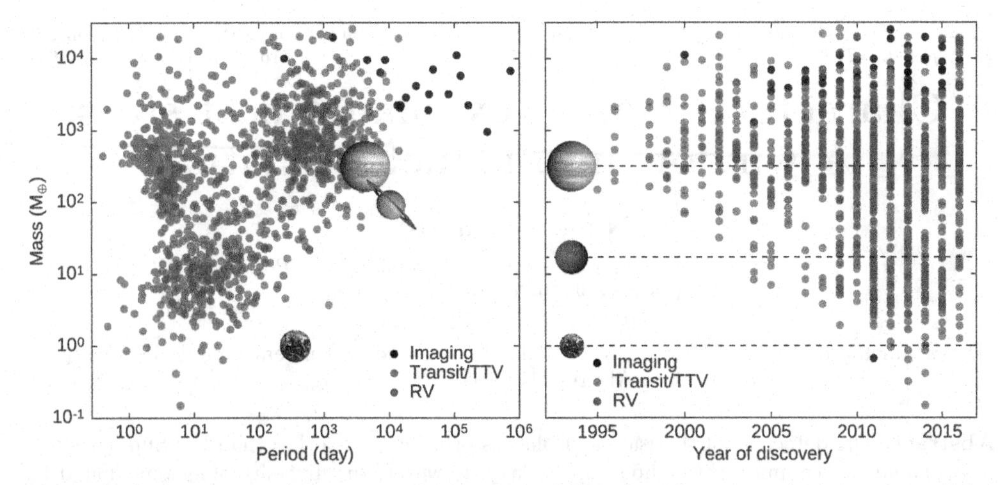

Figure 1. *Left*: Mass-period diagram for known exoplanets with a mass determination. Solar system planets Jupiter, Saturn, and the Earth are illustrated for reference. *Right*: Planet mass as a function of year of discovery. Solar system planets Jupiter, Neptune, and the Earth are illustrated for reference. From Adibekyan (2017).

main regions: the short period giant planets, also dubbed hot jupiters, occupy the upper-left part of the plot. Cool and temperate giants (more similar to Jupiter and Saturn) occupy the upper right part of the plot. Finally, the lower part of the diagram is populated by an increasing population of rocky and neptune-like planets. This latter population, though not the most represented in the plot (because its also the most difficult to detect), is likely the most common. In the right panel of this figure we also show the exoplanet mass as a function of the year of discovery, a plot that illustrates the progress in exoplanet discoveries and in the techniques used to detect them.

While the number and variety of discovered planets is still an important asset for exoplanet research (e.g. with an impact for the models), the focus of extrasolar planet researchers is now moving towards two main lines: i) the detection of lower and lower mass planets, with the goal of finding an Earth sibling, and ii) the detailed characterisation of known exoplanets, including their interior structures and atmospheres. Both lines of research have already seen their own success. Though it is out of the scope of the present paper to review in detail these results, we would like to point the reader to the recent reviews by Mayor *et al.* (2014), Fortney *et al.* (2014), and Burrows (2014).

Inherent to the detection and characterisation of exoplanets is the study of their host stars. In most cases, planet search and characterisation methods† only observe the star, not the planet itself. Phenomena related to stellar activity, stellar granulation, and oscillations (that act in different timescales) are particularly nasty for exoplanet detection and characterisation efforts using the radial velocity method (see e.g. Dumusque *et al.* 2011). They can prevent us from finding planets, if the perturbation is larger than the planet induced signal, or even give us false candidates, if they produce a periodic and stable signal over a few rotational periods (e.g. Figueira *et al.* 2010). Stellar activity is also particularly relevant when dealing with transit searches. Not only it induces strong photometric modulations (that need to be filtered), but also they induce in-transit fluctuations that prevent us from having precise values for the transit depth, and hence the planet radius (e.g. Oshagh *et al.* 2013). In brief, different sources of noise are a strong

† Namely the radial velocity and transit methods, responsible for the discovery of the huge majority of the known exoplanets.

challenge in planet detection and characterisation efforts. The understanding of the different physical phenomena as well as ways to model or subtract them is now one of the most important avenues to guarantee the success of future ground- and space-based exoplanet projects.

A precise derivation of the physical characteristics of exoplanets is also intimately connected to our ability to derive the stellar properties. For example, when a planet is found transiting, the measurement precision on the planetary radius depends directly on the precise knowledge of the stellar radius (e.g. Torres *et al.* 2008, Mortier *et al.* 2013a). The stellar mass is also a key ingredient for the derivation of planet masses using the radial velocity method. The age of a planet can also only be known through the derivation of the stellar age. All these ingredients, together with the stellar irradiation (that depends also in the orbital distance) are fundamental to understand the potential for habitability.

Finally, it has been shown that the chemical composition of the stars is intimately connected to the frequency, architecture, and chemical composition of the discovered planets. This is likely not unexpected, since planets are formed in proto-planetary disks, being thus one of the outcomes of the star formation process. This review will concentrate on this aspect, namely by presenting and discussing some of the most relevant aspects of the star-planet connection. In Sect. 2 we will discuss the relation between stellar abundances and planet frequency. In Sect. 3 we will then see how stellar properties relate with the architecture of the discovered exoplanet systems and the chemical composition of the planets themselves. Finally, in Sect.4 we will present the future prospects in this research, not forgetting the role of Gaia.

2. Stellar abundances and planet frequency

A number of different studies pointed the existence of a strong relation between the properties and frequency of the newfound planets and those of their host stars. In this respect, the well known correlation between the stellar metallicity and the frequency of giant planets is a good example. Large spectroscopic studies (e.g. Santos *et al.* 2001, Santos *et al.* 2004, Fischer & Valenti 2005, Sousa *et al.* 2011) confirmed the initial suspicions of a positive correlation between the probability of finding a giant planet and the metal content of the stars (See Fig. 2). Curiously, this strong metallicity-giant planet correlation was not found for the lowest mass planets (Sousa *et al.* 2011, Buchhave *et al.* 2012). Both results, however, are in full agreement with the expectations from the most recent models of planet formation based on the core-accretion paradigm (e.g. e.g. Mordasini *et al.* 2012, and discussion therein). We should add, however, that recent results seem to suggest that the overall metal content of the stars may still be relevant for the formation of the lower mass planets (Zhu *et al.* 2016, Wang & Fischer 2015). The higher abundances of alpha elements in metal-poor planet hosts also points in that direction (Adibekyan *et al.* 2013).

Interestingly, recent results also suggest that on the other mass limit, planet formation may follow a different path. Santos *et al.* (2017) have shown that stars hosting planets with mass above $\sim 4\,M_{Jup}$ are metal poor when compared with stars hosting lower mass, giant planets (Fig. 3), a result that was shown to be statistically significant. This result suggests that above $\sim 4\,M_{Jup}$ giant planets may be mainly formed via a different physical process, likely a disk instability mechanism (e.g. Cai *et al.* 2006).

Its important to add that these results have only been possible thanks to the increase in the number of discovered planets, but also due to the existence of precise and uniform

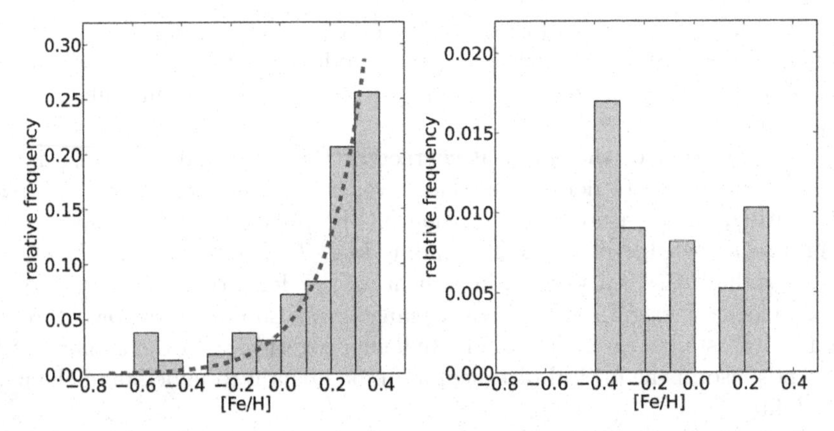

Figure 2. Frequency of giant (left panel) and neptune-mass (right panel) planets detected by radial velocity surveys as function of stellar metallicity (Sousa *et al.* 2011). From Mayor *et al.* (2014).

values for the stellar parameters and metallicities for planet host stars, such as the ones compiled in the SWEET-Cat database (Santos *et al.* 2013, Andreassen *et al.* 2017).

Although the general metallicity-giant planet correlation is reasonably well established, many details are still missing that may hold the clue to new and important details concerning planet formation. For example, the exact shape of the metallicity-planet correlation is still debated (Udry & Santos 2007, Johnson *et al.* 2010, Mortier *et al.* 2013b). The understanding of this issue may be critical to point out the mechanisms responsible for the formation of giant planets across the whole metallicity range (Matsuo *et al.* 2007), or to the understanding of the frequency of planets in the MilkyWay. The role of the abundances of other elements is also being discussed, with some curious trends being a strong matter of debate concerning e.g. the abundances of the light element lithium (e.g. Israelian *et al.* 2009, Baumann *et al.* 2010, Figueira *et al.* 2014).

It is worth adding that a role in the formation of giant planets has also been assigned to stellar mass. It is now widely accepted that the frequency of giant planets orbiting M dwarfs is considerably lower than that found for FGK dwarfs (e.g. Bonfils *et al.* 2013), at least regarding the short-period domain. This result is expected from the models of planetary formation following the core-accretion paradigm (e.g. Mordasini *et al.* 2012, Ida & Lin 2005).

3. Stellar abundances, planet architecture, and planet composition

The role of stellar properties on the formation of different architectures of planetary systems has also been addressed. Among these, initial suspicions have been raised concerning the metallicity-orbital period relation (e.g. Queloz *et al.* 2000, Sozzetti 2004). Hot-Jupiters have often been identified as orbiting particularly metal-rich stars, even if this trend had not been confirmed from a statistical point of view. Interestingly, recent results do support the existence of a period-metallicity correlation. Beaugé & Nesvorny (2013) have shown that among Kepler small planets there is a lack of short period objects orbiting metal-poor stars. A similar trend has also been found by Adibekyan *et al.* (2013), who have shown that among planets discovered by radial velocity surveys, in all mass domains metal-rich stars have longer period planets than their metal-poor counterparts. It is also interesting to add that preliminary results from an ESO Large Program to search for planets orbiting metal-poor stars (Santos *et al.* 2017) have also failed to detect

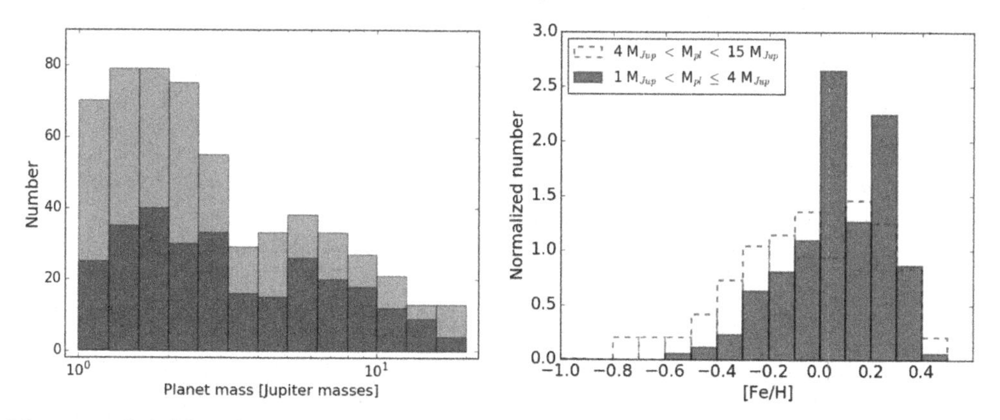

Figure 3. *Left*: Mass distribution of giant planets orbiting solar-type stars. The green histogram includes all hosts, while the blue histogram includes only a selected subsample. A "valley" around $4\,M_{Jup}$ is seen. Right: Metallicity distributions for planets with mass above and below $4\,M_{Jup}$. See Santos *et al.* (2017) for details.

short period planets, even if some intermediate period cases were found (e.g. Mortier *et al.* 2016, Faria *et al.* 2016).

Results from Dawson & Murray-Clay (2013) have also raised the possibility that stellar metallicity may be related to the orbital eccentricity. Analyzing the eccentricity-period diagram for giant planets orbiting solar-type stars they have shown that for intermediate period planets the eccentricity values are higher if the stars are metal-rich. This result, as well as the period-metallicity correlation mentioned above, can likely be explained if higher metallicity stars were able to produce more planets: this would lead to stronger planet-planet scattering, a process that is able to migrate planets into shorter period orbits and lead to the formation of higher eccentricity systems. We note that evidence exists that short period giant planets may result from the outcome of violent migration processes (e.g. Winn *et al.* 2010, Sotiriadis *et al.* 2017). Alternatively, planet-disk interactions could also play a role if the process is metallicity dependent (Tsang *et al.* 2014).

The chemical composition of the stars also seems to be reflected on the structure of the planets that were formed. For instance, the presence of a core (or at least the heavy element content of a giant planet) has been suggested to be related to the metallicity of the star (Guillot *et al.* 2006). Furthermore, even though most studies dealing with the star-planet connection have focused on the global metallicity as a proxy for the metal-content of the star (and likely of the proto-planetary disk), specific chemical abundance ratios may also have an impact on the planets themselves. Different chemical abundances in the disk may result in the formation of planets having different composition and structure (e.g. Carter-Bond *et al.* 2012, Delgado Mena *et al.* 2010, Dorn *et al.* 2015, Thiabaud *et al.* 2015, Santos *et al.* 2015, Dorn *et al.* 2017), a fact that may even change their habitability potential (Noack *et al.* 2014).

Understanding if the chemical abundances we see on the host star are related to the chemical composition we observe on the orbiting planet may provide valuable clues for the modelling of the observations. For instance, it is known that abundance ratios such as Fe/Mg and Fe/Si are very similar on the Sun, Earth, Venus, and Mars – see discussion in Dorn *et al.* (2015). In a recent paper, Dressing *et al.* (2015) has shown that 5 known rocky planets (Kepler-10b, Kepler-36b, Kepler-78b, and Kepler-93b, as well as CoRoT-7b) having precise measurements of the mass and radius seem to follow the same line

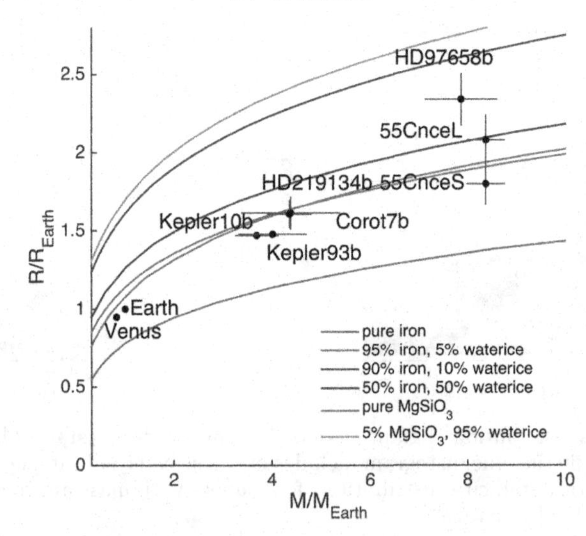

Figure 4. Mass-radius diagram including very low mass (potentially) rocky planets found by different surveys. The lines represent models of planets having different compositions. This plot illustrates how the chemical composition of a planet can be inferred from its position in this plot. From Dorn *et al.* (2017).

in the mass-radius diagram: a line corresponding to the expected position of Earth-like planets (see also Fig. 4). A detailed chemical analysis of 3 of these stars together with a simple stoichiometric model has also shown that the expected planet composition in indeed compatible with Earth-like (Santos *et al.* 2015). If confirmed, these results may be used to constrain the models of planet composition, as long as the abundances of some key elements are known in the stellar atmosphere. Furthermore, this would imply that stars from different galactic populations may even be able to form planets with different composition (Adibekyan *et al.* 2016).

4. Exoplanet research in the Gaia era

The results presented in this paper illustrate how the study of stars with planets and their link with the properties of the planets they host is providing important clues about the processes of planet formation and evolution. Of course, several of the correlations discussed above were only possible thanks to the continuous increase in the number of planets found. As usual, when a new parameter space is explored, unexpected results are obtained. We can thus expect that new star-planet relations will appear in the future as a whole new generation of planet search and characterisation instruments becomes reality. These include ground-based optical spectrographs such as ESPRESSO (ESO-VLT), capable of achieving down to sub m/s precision in RV. A whole new generation of infrared (IR) high resolution spectrographs is also on its way, including instruments such Carmenes@Calar Alto, Spirou@CFHT, and NIRPS (for ESO's 3.6-m). To these we should add the dawn of a new generation of ground- and space-based projects that will search and characterize transiting low mass/radius planets. These include the ESA mission CHEOPS, NASA's TESS (both expected to fly in 2018/2019) and further ahead the ESA PLATO2.0 mission. Finally, a new set of high resolution spectrographs is being planed for a new generation of ELT telescopes.

In this context Gaia will also play a key role. On the one hand, the precise astrometric data gathered is expected to allow the detection of thousands of giant planets orbiting solar-type FGK and M stars, as well as to confirm (and derive precise orbital parameters and masses) for dozens of giant planets discovered by radial-velocity surveys (e.g. Perryman 2014, Sozzetti *et al.* 2014). Given that higher amplitude astrometric signals are expected for higher-mass and longer-period planets, Gaia will focus its planet harvest in this domain, thus probing a mass and period regime that has not yet been fully explored. Finally, the photometry of Gaia will also allow to detect several transiting candidates (Dzigan & Zucker 2012).

But the role of Gaia in the exoplanet field does not limit to planet detections. The exquisite stellar distances will bring new valuable constraints for the derivation of stellar properties (e.g. Stassun *et al.* 2017). These will allow to characterise in detail the properties of planet hosts. Such is fundamental for a precise derivation of planet masses and radii (and thus the planet mean densities): both quantities depend on the measurement of the stellar mass and radius. Furthermore, stellar ages are the only way we can possibly derive the age of the system, an important parameter to understand the planet formation and evolution process. These inputs are all fundamental for the full success of future planet characterisation projects.

I would like to thank the whole exoplanet team at the Instituto de Astrofísica e Ciências do Espaço for the work that led to many of the results presented in this review. This work was supported by Fundação para a Ciência e a Tecnologia (FCT, Portugal) through the research grant through national funds and by FEDER through COMPETE2020 by grants UID/FIS/04434/2013&POCI-01-0145-FEDER-007672 and PTDC-/FIS-AST/1526/2014&POCI-01-0145-FEDER-016886, as well as through Investigador FCT contract nr. IF/00169/2012/CP0150/CT0002.

References

Adibekyan, V. Zh., Delgado Mena, E., Sousa, S. G., *et al.* 2012, *A&A*, 547, A36
Adibekyan, V. Zh., Figueira, P., Santos, N., *et al.* 2013, *A&A*, 560, A51
Adibekyan, V. Zh., Figueira, P., & Santos, N. 2016, *OLEB*, 46, 351
Adibekyan, V. Zh. 2017, *ASP Conference Series*, in press (arXiv:1701.01661)
Andreasen, D. T., Sousa, S. G., & Tsantaki, M. 2017, *A&A*, 600, A69
Beaugé, C. & Nesvorný, D. 2013, *ApJ*, 763, 12
Baumann, P., Ramírez, I., Meléndez, J., Asplund, M., & Lind, K. 2010, *A&A*, 519, 87
Carter-Bond, J. C., O?Brien, D. P., & Raymond, S. N. 2012, *ApJ*, 760, 44
Bonfils, X., Delfosse, X., Udry, S., *et al.* 2013, *A&A*, 549, A109
Burrows, A. S. 2014, *Nature*, 513, 345
Buchhave, Lars A., Latham, David W., Johansen, Anders, *et al.* 2012, *Nature*, 486, 375
Cai, K., Durisen, R. H., & Michael, S., *et al.* 2006, *ApJ*, 636, L149
Dawson, Rebekah I. & Murray-Clay, Ruth A. 2013, *ApJ*, 676, L24
Delgado Mena, E., Israelian, G., González Hernández, J. I., *et al.* 2010, *ApJ*, 725, 2349
Dorn, C., Khan, A., Heng, K., *et al.* 2015, *A&A*, 577, A83
Dorn, C., Hinkel, N. R., & Venturini, J. 2017, *A&A*, 597, A38
Dressing, C. D., Charbonneau, D., Dumusque, X., *et al.* 2015, *ApJ*, 800, 135
Dumusque, X., Lovis, C., Ségransan, D., *et al.* 2011, *A&A*, 535, A55
Dzigan, Y. & Zucker, S. 2012, *ApJ*, 753, L1
Figueira, P., Marmier, M., Bonfils, X., *et al.* 2010, *A&A*, 513, L8
Faria, J. P., Santos, N. C., Figueira, P., *et al.* 2016, *A&A*, 589, A25
Figueira, P., Faria, J. P., Delgado-Mena, E., *et al.* 2014, *A&A*, 570, A21
Fischer, Debra A. & Valenti, Jeff 2005, *ApJ*, 622, 1102
Guillot, T., Santos, N. C., Pont, F., Iro, N., Melo, C., & Ribas, I. 2006, *A&A*, 453, L21

Howard, A. W., Marcy, G. W., Bryson, S. T., *et al.* 2012, *ApJS* 201, 15
Ida, S. & Lin, D. N. C. 2005, *ApJ* 626, 1045
Israelian, Garik, Delgado Mena, Elisa, & Santos, Nuno C. 2009, *Nature*, 462, 189
Johnson, J. A., Aller, K. M., Howard, A. W., & Crepp, J. R. 2010, *PASP* 122, 905
Kaltenegger, L. & Sasselov, D. 2010, *Astrobiology*, 10, 89
Lissauer, Jack J., Dawson, Rebekah I., & Tremaine, Scott 2014, *Nature*, 513, 336
Matsuo, T., Shibai, H., Ootsubo, T., & Tamura, M. 2007, *ApJ*, 662, 1282
Mayor, M. & Queloz, D. 1995, *Nature*, 355, 278
Mayor, M., Lovis, C., & Santos, N. C. 2014, *Nature*, 513, 328
Mordasini, C., Alibert, Y., Benz, W., Klahr, H., & Henning, T. 2012, *A&A*, 541, A97
Mortier, A., Santos, N. C., Sousa, S. G., *et al.* 2013a, *A&A*, 558, A106
Mortier, A., Santos, N. C., Sousa, S. G., *et al.* 2013b, *A&A*, 557, A70
Mortier, A., Faria, J. P., Santos, N. C., *et al.* 2016, *A&A*, 585, A135
Noack, L., Godolt, M., von Paris, P., *et al.* 2014, *P&SS*, 98, 14
Oshagh, M., Santos, N. C., Boisse, I., *et al.* 2013, *A&A*, 556, A19
Perryman, M., Hartman, J., Bakos, Gáspár A., & Lindegren, Lennart 2014, *ApJ*, 797, 14
Queloz, D., Mayor, M., Weber, L., *et al.* 2000, *A&A*, 354, 99
Santos, N. C., Israelian, G., & Mayor, M. 2001, *A&A*, 373, 1019
Santos, N. C., Israelian, G., & Mayor, M. 2004, *A&A*, 415, 1153
Santos, N. C., Sousa, S. G., Mortier, A., *et al.* 2013, *A&A*, 556, A150
Santos, N. C., Adibekyan, V., Mordasini, C., *et al.* 2015, *A&A*, 580, L13
Santos, N. C., Mortier, A., Faria, J. P., *et al.* 2016, *A&A*, 566, A35
Santos, N. C., Adibekyan, V., Figueira, P., *et al.* 2017, *A&A*, 415, 1153
Sotiriadis, Sotiris, Libert, Anne-Sophie, Bitsch, Bertram, Crida, Aurélien 2017, *A&A*, 598, A70
Sousa, S. G., Santos, N. C., Israelian, G., Mayor, M., & Udry, S. 2011, *A&A*, 533, A141
Sozzetti, A. 2004, *MNRAS*, 354, 1194
Sozzetti, A., Giacobbe, P., Lattanzi, M. G., *et al.* 2014, *MNRAS*, 437, 497
Stassun, K. G., Collins, K. A., & Gaudi, B. S. 2017, *AJ*, 153, 136
Torres, Guillermo, Winn, Joshua N. & Holman, Matthew J. 2008, *ApJ*, 677, 1324
Thiabaud, A., Marboeuf, U., Alibert, Y., *et al.* 2014, *A&A*, 562, A27
Tsang, David, Turner, Neal J. & Cumming, Andrew 2014, *ApJ*, 782, 113
Udry, S. & Santos, N. C. 2007, *ARAA*, 45, 397
Valencia, D., O'Connell, R. J., & Sasselov, D. 2006, *Icarus*, 181, 545
Wang, Ji, Fischer, Debra A. 2015, *AJ*, 149, 14
Winn, Joshua N. & Fabrycky, Daniel, Albrecht, Simon, Johnson, John Asher 2010, *ApJ*, 718, L145
Zhu, Wei, Wang, Ji, Huang, Chelsea 2016, *ApJ*, 832, 196

Astronomy and Astrophysics in the Gaia sky
Proceedings IAU Symposium No. 330, 2017
A. Recio-Blanco, P. de Laverny, A.G.A. Brown
& T. Prusti, eds.

doi:10.1017/S1743921317005828

Exploring the Solar System using stellar occultations

Bruno Sicardy

LESIA, Observatoire de Paris, PSL Research University, CNRS, Sorbonne Universités, UPMC Univ. Paris 06, Univ. Paris Diderot, Sorbonne Paris Cité
emails: bruno.sicardy@obspm.fr

Abstract. Stellar occultations by solar system objects allow kilometric accuracy, permit the detection of tenuous atmospheres (at nbar level), and the discovery of rings. The main limitation was the prediction accuracy, typically 40 mas, corresponding to about 1,000 km projected at the body. This lead to large time dedicated to astrometry, tedious logistical issues, and more often than not, mere miss of the event. The Gaia catalog, with sub-mas accuracy, hugely improves both the star positions, resulting in achievable accuracies of about 1 mas for the shadow track on Earth. This permits much more carefully planned campaigns, with success rate approaching 100%, weather permitting. Scientific perspectives are presented, e.g. central flashes caused by Plutos atmosphere revealing hazes and winds near its surface, grazing occultations showing topographic features, occultations by Chariklos rings unveiling dynamical features such as proper mode "breathing".

Keywords. occultations, astrometry, solar system: general, planets: rings, Kuiper Belt

---

1. Introduction

Stellar occultations represent a very poweful tool that allow us to study a great variety of planetary bodies in the solar system. They occur when these bodies pass in front of a star. The photometric monitoring of those events at a given site permit to measure the length of the occultation phenomenon, thus providing one "occultation chord" per site. Combining those chords, one may obtain the shape and size of the object at kilometric accuracy. In fact, that accuracy is linked to the acquisition rate during the event. As the typical speed of the occultation shadow on Earth's surface is 20 km s^{-1}, an acquisition rate of, say, ten frames per second (fps) results in a resolution of 2 km per data point on the object. Moreover, the occultation may be not sudden, but gradual, due to the presence of a tenuous atmosphere. The sensitivity of ground-based occultations is such that pressure levels as small as a few nbar can detected. Finally, material surrounding the body, like rings, cometary jets or dust shells, can be revealed, while remaining out of reach of any direct imaging techniques.

Ultimately, the resolution obtained during an occultation is set by the Fresnel diffraction, which has a scale of $\lambda_F = \sqrt{\lambda D/2}$, where λ is the wavelength of observation and D is the geocentric distance of the object. For events involving remote objects in the solar systems like Trans-Neptunian Objects (TNOs), with $D = 20 - 50$ au, this results in $\lambda_F <\sim 1$ km. Moreover, depending on the type and magnitude of the star, the stellar diameter is also of the order of a fraction of km to a few km when projected at the object. This resolution power is far larger, by almost 3 orders of magnitude, than any classical direct imaging: for instance it is at best 500 km using the Hubble Space Telescope to observe Pluto.

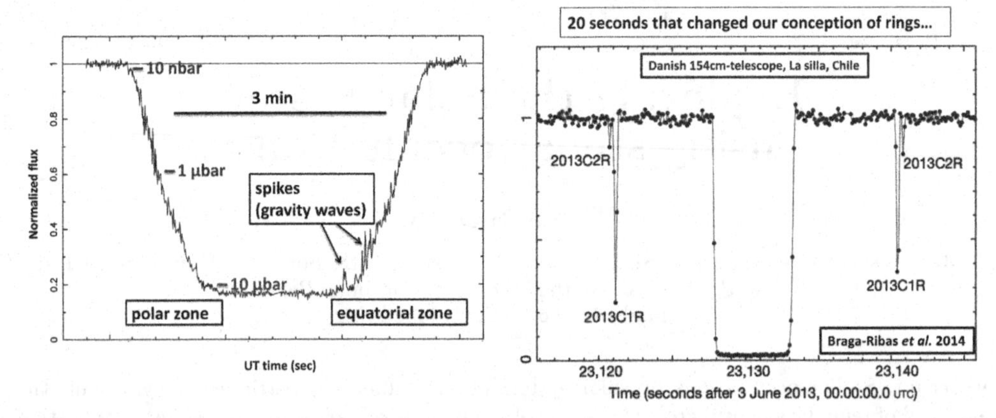

Figure 1. Left - An occultation by Pluto's atmosphere observed on July 18, 2012 with the adaptive optics camera NACO attached to the Very Large Telescope of the European Southern Observatory. The black horizontal bar indicates the time scale (3 minutes) and the red labelling-shows the typical atmospheric levels reached along the occultation light curve(obtained here at a rate of 5 fps). Right - One of the discovery light curves of Chariklos rings, obtained on June 3, 2013 from the Danish telescope at La Silla (Chile). The light curve was obtained at a rate of 10 fps and shows the occultation by the main body at the center, with the symmetrical detections of two rings (C1R and C2R) that surround that small object with diameter of about 250 km.

2. Results obtained by stellar occultations

Stellar ocultations have an impressive record of discoveries, published in high impact journals as *Nature* or *Science*. For instance, they allow to discover Uranus' narrow and dense rings (Bhattacharyya & Bappu 1977, Elliot *et al.* 1977), Neptune's ring arcs (Hubbard *et al.* 1986, Sicardy *et al.* 1986), measure Neptune's shape (Sicardy *et al.* 1986), probe Titan's atmosphere (Sicardy *et al.* 1990) and Triton's atmosphere (Elliot *et al.* 1997), discover material around the Centaur object Chiron (Elliot *et al.* 1995), reveal a spectacular expansion of Pluto's atmosphere between 1988 and 2002 (Elliot *et al.* 2003, Sicardy *et al.* 2003), measure the size of its satellite Charon at kilometric accuracy (Gulbis *et al.* 2006, Sicardy *et al.* 2006a), as well as the sizes of the very remote dwarf planets Eris (Sicardy *et al.* 2011) and Makemake (Ortiz *et al.* 2012), and more recently, discover dense and narrow rings around the Centaur object Chariklo, the first body ever known to possess ring besides the giant planets (Braga-Ribas *et al.* 2014).

Fig. 1 shows two examples of results obtained using stellar occultations, one illustrating the probing of Pluto's atmosphere, and the other one showing the discovery of rings around Chariklo.

3. The Gaia era

The main limitation of occultation have been so far the prediction accuracy. In fact, the typical errors induced by the pre-Gaia catalog were typically 20-40 milliarcsec (mas), depending on the star and the catalog used. This corresponds to typically 500-900 km at Pluto. As the latter has a radius of 1,190 km, this means that more than often, the event was missed just because of star catalog errors, not talking of course about smaller bodies for which that problem worsens.

This is illustrated in Fig. 2, where we compare a pre-Gaia prediction with a prediction using the Gaia Data Release 1 (DR1), combined with an improved Pluto's ephemeris generated by the Jet Propulsion Laboratory (JPL) before the flyby of the dwarf planet by the NASA spacecraft *New Horizons*. Note the drastic improvement brought by the

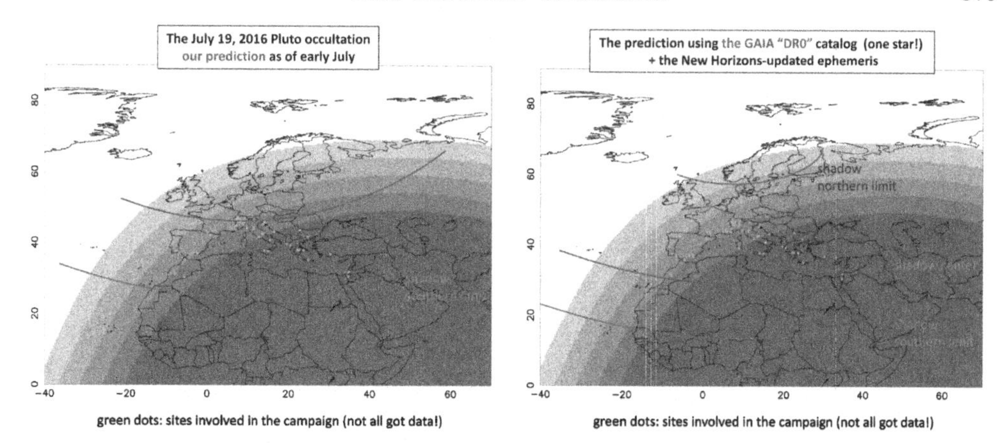

Figure 2. Left - Predicted of Pluto's shadow track on Earth (July 19, 2016 occultation) based on pre-Gaia astrometric catalogs. The lower curved line is the southern limit of the atmospheric shadow, while the middle curved line is the centrality of the event, where a central flash was planned. The dots indicate the locations of the sites involved in the campaign (some of them eventuelly clouded out). The darkest zone corresponds to astronomical night (Sun 18 deg below local horizon), while the white part of the map corresponds to daylight. Right - The same using the GAIA DR1 star position, showing a large drift to the south. The actual event was very close to that prediction, with an offset of about 5 mas with respect to the prediction (correspnding to less tha 100 km for the shadow track on Earth).

Gaia DR1 release. In this particular example, there was of a large error on the star position, which reached 100 mas, probably due to a local catalog problem in the region of the sky where the star was found.

The Gaia catalogs are free of such local problems, and permit accurate predictions well in advance. This is illustrated in Fig. 3, where we compare the pre-Gaia types of predictions with the post-DR1 predictions. The improvement is impressive, with a factor almost four in angular accuracy (typically 10 mas, vs. some 40 mas previously).

Note that because the DR1 catalog lacks of the star proper motions, its accuracy is degraded by several mas per year as the star moves with respect to its DR1 (2015.0) position. Morover, some stars had not enough visits in the DR1 catalog, resulting in cumulated errors of as much as 20 mas for their positions as of 2017.

4. The forthcoming Gaia releases and their implications

The main improvement brought by DR2 will be the access to the star proper motion, with an eventual accuracy on the star position of a fraction of mas. As an example, this corresponds to a few kilometers at the level of Chariklo and its rings, and typically 10 km for the most remote TNOs. This is a tremendous leap forward, and will represent a new era in the domain of stellar occultations by solar system objects. At this stage, the main limitation will be the internal accuracy of the body ephemeris. The ephemeris will be improved either by the bootstrapping approach illustrated in Fig. 3, or by looking at close angular approches (appulses) between the object and Gaia DR2 star positions. In a few cases, Gaia itself will release positions of bodies that can be used for ephemeris improvement.

This said, among the future applications of DR2, we may quote a few illustrative examples:

• **Topographic features** at the surface of small bodies may reach 10 km in height or depth. Thus, carefully planned occultations will permit to organize grazing events,

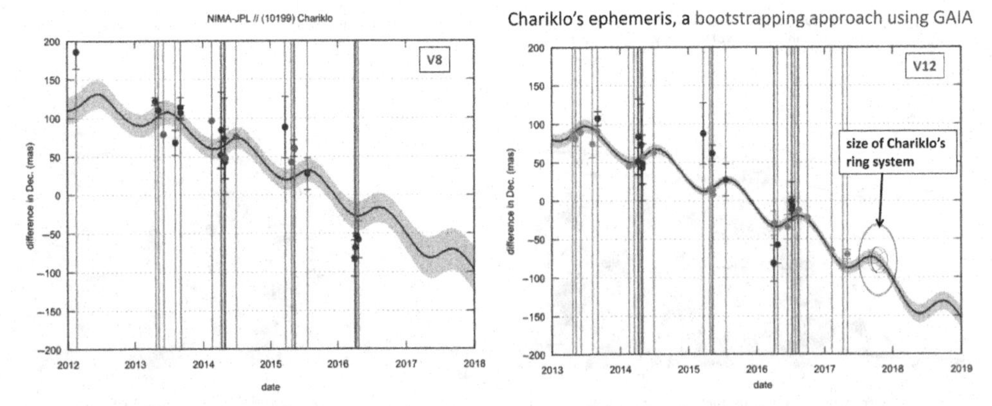

Figure 3. An illustration of the improvements of occultation predictions provided by Gaia. Left - The difference in declination between the NIMA v8 Chariklo's ephemeris and a reference ephemeris from the JPL plotted vs. time (taken from http://josselin.desmars.free.fr/tno/, see also Desmars *et al.* 2015). It was obtained using classical astrometric measurements (blue points) and previous positive stellar occultations (red points). The gray zone is the uncertainty associated with the ephemeris. Right - The same using positive Chariklo stellar occultations between June 2013 and April 2017, and star positions from the DR1 Gaia catalog (Gaia Collaboration 2016a, Gaia Collaboration 2016a). Note the drastic improvement of Chariklo's ephemeris quality, which is now sufficiently accurate to plan occultations by the rings and the main body This is illustrated by the insert of Chariklo's system in the right panel.

where various stations separated by a few kilometers will detect the star "flickering" along the limb, thus measuring these features, as was done in the 1970's to map the Moon topography near its poles. In other words, we will be in the position of not only monitor the occultations, but also use them for "geological" purposes. A particularly interesting object is Chariklo, as km-sized features and a possible elongation of the body may give rise to important mean motion resonances with the rings. Also, an initial debris disk surrounding that Centaur (and from which the rings emerged) may have left an equatorial ridge that could be detected during grazing occultations.

• Concerning objects with **atmosphere** (e.g. Titan, Triton or Pluto), there is a small region near the shadow center where a "central flash" can be observed. It stems from the focusing of light toward the observer, due to the refraction by the deepest atmospheric layers reached during the occultation. Such flashes provide accurate information on the distortions of the atmosphere, and therefore, the zonal wind regimes that maintain those distortions. Moreover, by observing at various wavelengths, we may detect a differential extinction that constrains the particle size distribution of the hazes responsible the dimming of stellar rays (see e.g. Sicardy *et al.* 2006b)

• **Chariklo's rings** may be not perfectly circular, as normal modes can distort their orbits, as observed for the Uranian rings (French *et al.* 1991). Again, the Gaia releases, together with updated Chariklo's ephemeris, will allow a good coverage in longitude of the rings, and then a possible detection of proper modes with various azimuthal numbers. This in turn will permit to estimate Chariklo's mass, its dynamical oblateness J_2 and the ring surface densities.

• Needless to say, the greatly enhanced accuracy of occultation predictions will allow us to observe many more events, thus **triggering the findings of still unknown remarkable features**. Among those potential discoveries, we may quote new ring systems, atmospheres around dwarf planets, contact binaries, cometary material, and seasonal effects in Titan, Pluto or Triton's atmosphere as time passes by.

Acknowledgement

The work leading to this results has received funding from the European Research Council under the European Community's H2020 2014-2020 ERC grant Agreement n° 669416 "Lucky Star". This work has made use of data from the European Space Agency (ESA) mission *Gaia* (https://www.cosmos.esa.int/gaia), processed by the *Gaia* Data Processing and Analysis Consortium (DPAC, https://www.cosmos.esa.int/web/gaia/dpac/consortium). Funding for the DPAC has been provided by national institutions, in particular the institutions participating in the *Gaia* Multilateral Agreement.

References

Braga-Ribas, F. *et al.* 2014, *Nature*, 508, 72
Bhattacharyya, J. C. & Bappu M. K. V. 1977, *Nature*, 270, 503
Elliot, J. L. *et al.* 2015, *Astron. & Astroph.*, 584, A96
Elliot, J. L. *et al.* 1977, *Nature*, 267, 328
Elliot, J. L. *et al.* 1995, *Nature*, 373, 46
Elliot, J. L. *et al.* 1997, *Science*, 278, 436
Elliot, J. L. *et al.* 2003, *Nature*, 424, 165
French, R. G. *et al.* 1991, in: J.T. Bergstralh, E.D. Miner & M.S. Matthews (eds.) *Uranus* (Univ. of Arizona Press, Tucson), p. 327
Gaia Collaboration, Brown, A. G. A., Prusti, T. *et al.* 2016a, *Astron. & Astroph.*, 595, A1
Gaia Collaboration, Lindegren, L., Lammers, U. *et al.* 2016a, *Astron. & Astroph.*, 595, A2
Hubbard, W. B. *et al.* 1988, *Nature*, 336, 462
Gulbis, A. A. *et al.* 2006, *Nature*, 439, 48
Hubbard, W. B. *et al.* 1986, *Nature*, 319, 636
Lellouch, E. *et al.* 1986, *Nature*, 324, 227
Ortiz, B. *et al.* 2012, *Nature*, 320, 729
Sicardy, B. *et al.* 1986, *Nature*, 320, 729
Sicardy, B. *et al.* 1990, *Nature*, 343, 350
Sicardy, B. *et al.* 2003, *Nature*, 424, 168
Sicardy, B. *et al.* 2006a, *Nature*, 439, 52
Sicardy, B. *et al.* 2006b, *J. Geophys. Res.*, 111, E11S91
Sicardy, B. *et al.* 2011, *Nature*, 478, 493

Astronomy and Astrophysics in the Gaia sky
Proceedings IAU Symposium No. 330, 2017
A. Recio-Blanco, P. de Laverny, A.G.A. Brown
& T. Prusti, eds.

doi:10.1017/S1743921317006226

Prediction of stellar occultations by distant solar system bodies in the Gaia era

Josselin Desmars[1], Julio Camargo[2,5], Bruno Sicardy[1], Felipe Braga-Ribas[3,2,5], Roberto Vieira-Martins[2,5], Marcelo Assafin[4], Diane Bérard[1] and Gustavo Benedetti-Rossi[5]

[1]LESIA, Observatoire de Paris, PSL Research University, CNRS, Sorbonne Universités, UPMC Univ. Paris 06, Univ. Paris Diderot, Sorbonne Paris Cité, 5 place Jules Janssen, 92195 Meudon, France, email: `josselin.desmars@obspm.fr`

[2]Observatório Nacional/MCT, R. General José Cristino 77, RJ 20921-400 Rio de Janeiro, Brazil

[3]Federal University of Technology-Paraná (UTFPR/DAFIS), Rua Sete de Setembro, 3165, CEP 80230-901, Curitiba, PR, Brazil

[4] Observatório do Valongo/UFRJ, Ladeira Pedro Antonio 43, RJ 20.080-090 Rio de Janeiro, Brazil

[5]Laboratório Interinstitucional de e-Astronomia - LIneA, Rua Gal. José Cristino 77, Rio de Janeiro- RJ 20921-400, Brazil

Abstract. Stellar occultations are a unique technique to access physical characteristics of distant solar system objects from the ground. They allow the measure of the size and the shape at kilometric level, the detection of tenuous atmospheres (few nanobars), and the investigation of close vicinity (satellites, rings) of Transneptunian objects and Centaurs. This technique is made successful thanks to accurate predictions of occultations. Accuracy of the predictions depends on the uncertainty in the position of the occulted star and the object's orbit. The Gaia stellar catalogue (Gaia Collaboration (2017)) now allows to get accurate astrometric stellar positions (to the mas level). The main uncertainty remains on the orbit. In this context, we now take advantage of the NIMA method (Desmars *et al.*(2015)) for the orbit determination and of the Gaia DR1 catalogue for the astrometry. In this document, we show how the orbit determination is improved by reducing current and some past observations with Gaia DR1. Moreover, we also use more than 45 past positive occultations observed in the 2009-2017 period to derive very accurate astrometric positions only depending on the position of the occulted stars (about few mas with Gaia DR1). We use the case of (10199) Chariklo as an illustration. The main limitation lies in the imprecision of the proper motions which is going to be solved by the Gaia DR2 release.

Keywords. Kuiper Belt, catalogs, astrometry, ephemerides, occultations.

---

1. Introduction

Stellar occultations are a unique technique to access physical characteristics of distant solar system objects from the ground: size and shape at the kilometric level, tenuous atmospheres at few nanobars, and investigation of close vicinity (satellites, rings). This technique is successful thanks to accurate predictions of occultations depending on the precision of star's position and the object's orbit.

Predictions of occultations by distant bodies is a difficult challenge. Indeed, good predictions require both accurate positions of the star and the body. As the objects are distant (from 15 to 90 au) and small in size (100-2000 km), the apparent size of the body is about 10 to 50 mas. A precision to the same level or less is required for good predictions. In comparison, 30 mas is equivalent to a coin of 1 euro seen at 200 km.

The uncertainty of the prediction comes from the uncertainty on the star and the object positions. For the star position, the source of uncertainties are the zonal errors in stellar catalogs (which is not the case anymore since Gaia DR1) and uncertainty in the proper motion. For the object's position, the source of uncertainty comes from the astrometric positions used for orbit determination. Since the Gaia DR1 (Gaia Collaboration (2017)), the main source of uncertainty in the predictions now comes from the ephemeris.

2. Methods of predictions

Assafin *et al.*(2012) and Camargo *et al.*(2014) propose method of prediction based on the JPL ephemeris and an offset deduced from observations. The method was successfull when the observations were made only few days before the occultation. Since mid-2013, we use the NIMA method for the predictions (Desmars *et al.*(2015)) using our own ephemeris NIMA. The main advantage is that we can used more observations (from the Minor Planet Center, from the Rio team and unpublished observations) and we have the control of the weighting process. Also, we can use astrometric positions derived from positive occultations.

3. Astrometry from previous occultations with Gaia

During an occultation, the star and the object have the same direction, meaning that at the time of the occultation, an astrometric position of the body can be derived from the position of the star. In particular, since the publication of the Gaia DR1, the accuracy of the star's positions is about 1-10 mas which is much more accurate than a classical astrometric position reduced from a CCD frame (about 300 mas). On the 2009-2017 period, 45 predicted occultations for 18 TNOs/Centaurs (Chariklo, Makemake, Eris, ...) have been successfully detected. We are able to derive an astrometric position for all of them. In particular, for Chariklo, we have observed 13 occultations. We used these positions derived from the Gaia DR1 catalog to refine the orbit. Figure 1 shows the difference between NIMA and JPL ephemerides in right ascension (left) and declination (right) by using observations reduced with UCAC4 in blue points (top) and by using also astrometric positions from occultations derived with the Gaia DR1 catalog in red points (bottom). The gray area around the line represents the 1-σ precision of the NIMA ephemeris. The orbit is clearly improved with the use of the Gaia DR1 catalog and the astrometric positions. With the previous version of the ephemeris, we have a precision of 30-40 mas in declination whereas with the new version the precision is less than 10 mas. The apparent size of Chariklo (25 mas) and its rings (80 mas) are represented for comparison in Fig. 1. In particular, now the precision is smaller than the apparent size of Chariklo. Thanks to the Gaia DR1 catalog and the refinement of the orbit, we are able to predict occultations to 10-mas level allowing to gather the observing stations on the ground.

4. Proper motion issue

Despite accurate astrometric positions, Gaia DR1 does not provide proper motions for all the stars (only stars in TGAS). Since the beginning of 2017, several publications provide proper motions for fainter stars.

Altmann *et al.*(2017) propose HSOY, a stellar catalog with proper motion for 583 million stars. Proper motions are derived from the Gaia DR1 and the PPMXL catalogs. More recently, Zacharias *et al.*(2017) derive proper motions from the Gaia DR1 and a

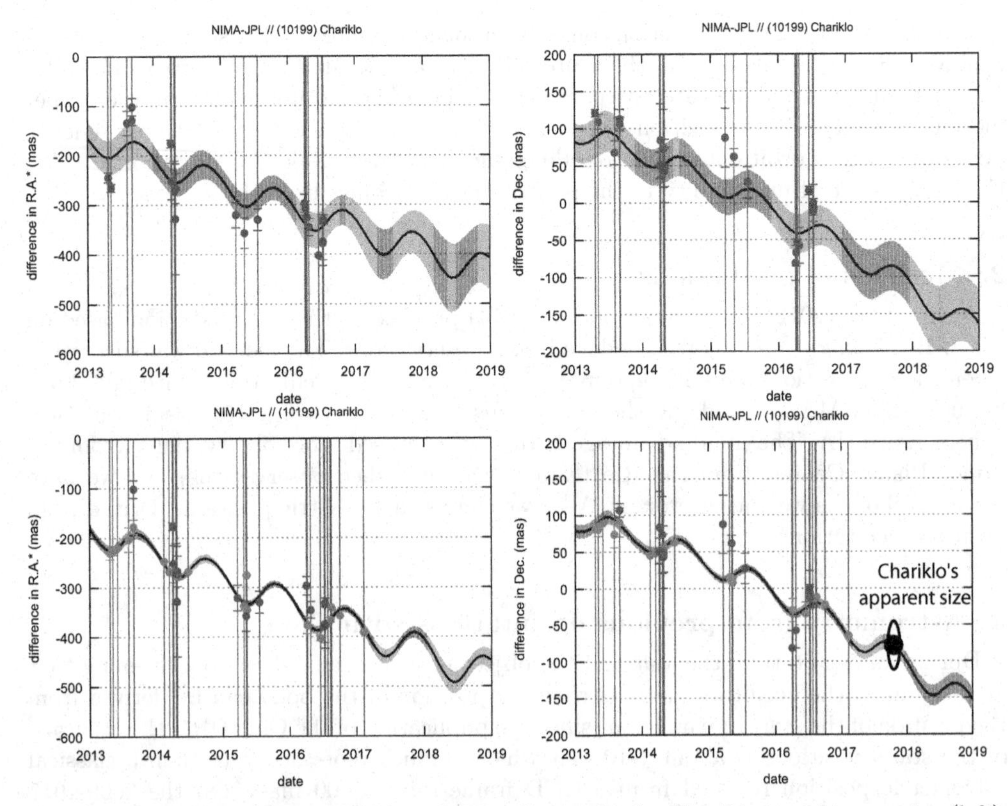

Figure 1. Difference between NIMA and JPL ephemerides for Chariklo in right ascension (left) and declination (right) by using only astrometric observations reduced with UCAC4 (top) and by using also astrometric positions from occultations (bottom). The blue points represent the observations reduced with UCAC4 and red points represent positions from occultations. The gray area represents the 1-σ precision of the NIMA ephemeris.

new reduction of US Naval Observatory CCD Astrograph Catalog. The UCAC5 catalog contains proper motions for 107 million stars. Finally, Tian *et al.*(2017) also propose the Gaia-PS1-SDSS (GPS1) proper motion catalog based on a combination of Gaia DR1, PS1, SDSS and 2MASS astrometry for 350 million sources.

Before the publication of these catalogs, we used a method from Dave Herald (personal communication) to derive proper motions from the Gaia DR1 and the UCAC4 catalogs. Basically, this method can be summarized as:

- Derive from UCAC4 positions and proper motion, the UCAC4 position at UCAC4 epoch E_U: (α_U, δ_U)
- Gaia position at Gaia epoch E_G : (α_G, δ_G)
- Proper motions computed as:

$$\mu_\alpha = \frac{\alpha_U - \alpha_G}{E_U - E_G}; \mu_\delta = \frac{\delta_U - \delta_G}{E_U - E_G}$$

In order to test the different sources, we compare proper motions for the 14 stars that have been occulted by Chariklo in the last few years. Fig. 2 shows the proper motion in declination for three different sources: HSOY, UCAC5 and Herald's method. Proper motions in UCAC5 and Herald's method are usually in a good agreement, which is expected as the Herald's method uses stars from UCAC. Nevertheless, proper motion for some stars still remains different. HSOY provides a different proper motion for most

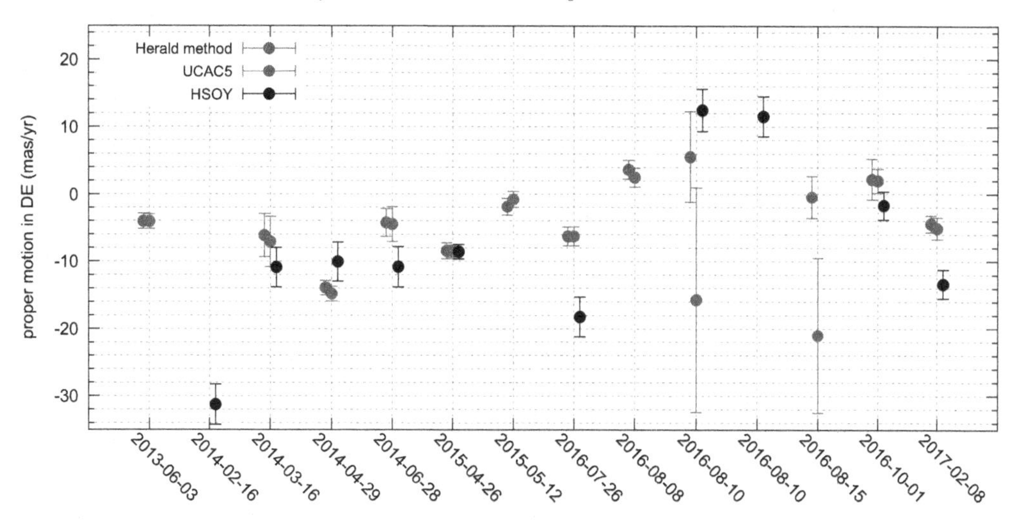

Figure 2. Comparison of proper motion in declination between Herald's method, HSOY and UCAC5 catalogs for the stars involved in Chariklo's occultations (dates of occultations are indicated in abscissa).

stars. Until the publication of the Gaia Data Release 2, the proper motion will remain an issue for accurate predictions of occultations.

5. Conclusion

Gaia-DR1 greatly improves the predictions of occultations. Before Gaia, the precision of prediction was about 30-40 mas whereas after Gaia DR1, the precision is about 10 mas. With the Gaia DR2, we expect to have a precision smaller than 1 mas. Moreover, accurate astrometric positions from positive occultations help to refine orbits without direct observations of the body with Gaia. A precision of few mas is now reachable for some objects (Chariklo, Pluto) leading to accurately predict grazing occultations or central flash and allowing to gather observers on the shadow's path. Nevertheless, until the publication of Gaia DR2, the proper motions will remain the main source of uncertainty for the predictions. The Gaia DR2 will also provide direct observations of some TNOs/Centaurs and help to greatly refine their orbit. Finally, we also plan to make use of surveys (LSST, Dark Energy Survey, ...) to derive astrometric positions.

This work has made use of data from the European Space Agency (ESA) mission *Gaia* (https://www.cosmos.esa.int/gaia), processed by the *Gaia* Data Processing and Analysis Consortium (DPAC, https://www.cosmos.esa.int/web/gaia/dpac/consortium). Funding for the DPAC has been provided by national institutions, in particular the institutions participating in the *Gaia* Multilateral Agreement.

References

Altmann, M., Roeser, S., Demleitner, M., Bastian, U., & Schilbach, E. 2017, *A&A*, 600, L4
Assafin, M., Camargo, J. I. B., Vieira Martins, R., *et al.* 2012, *A&A*, 541, A142
Camargo, J. I. B., Vieira-Martins, R., Assafin, M., *et al.* 2014, *A&A*, 561, A37
Desmars, J., Camargo, J. I. B., Braga-Ribas, F., *et al.* 2015, *A&A*, 584, A96
Tian, H.-J., Gupta, P., Sesar, B., *et al.* 2017, arXiv:1703.06278
Zacharias, N., Finch, C., & Frouard, J. 2017, *AJ*, 153, 166
Gaia Collaboration, Brown A. G. A., Vallenari A., *et al.*, 2016, *A&A*, 595, A2

Astronomy and Astrophysics in the Gaia sky
Proceedings IAU Symposium No. 330, 2017
A. Recio-Blanco, P. de Laverny, A.G.A. Brown
& T. Prusti, eds.

doi:10.1017/S1743921317005713

Prospects for asteroid mass determination from close encounters between asteroids: ESA's Gaia space mission and beyond

Anatoliy Ivantsov[1], **Daniel Hestroffer**[2] **and Siegfried Eggl**[2]

[1]Fen Fakültesi, Akdeniz Üniversitesi,
Dumlupınar Bulvarı Kampüs, TR-07058, Antalya, Turkey
email: ivantsov@akdeniz.edu.tr

[2]Institut de Mécanique Céleste et de Calcul des Éphémérides, Observatoire de Paris,
77 avenue Denfert-Rochereau, F-75014 Paris, France

Abstract. We present a catalog of potential candidates for asteroid mass determination based on mutual close encounters of numbered asteroids with massive perturbers (D>20 km). Using a novel geometric approach tuned to optimize observability, we predict optimal epochs for mass determination observations. In contrast to previous studies that often used simplified dynamical models, we have numerically propagated the trajectories of all numbered asteroids over the time interval from 2013 to 2023 using relativistic equations of motion including planetary perturbations, J2 of the Sun, the 16 major asteroid perturbers and the perturbations due to non-sphericities of the planets. We compiled a catalog of close encounters between asteroids where the observable perturbation of the sky plane trajectory is greater than 0.5 mas so that astrometric measurements of the perturbed asteroids in the Gaia data can be leveraged. The catalog v1.0 is available at ftp://dosya.akdeniz.edu.tr/ivantsov.

Keywords. Minor planets, asteroids, ephemerides, methods: numerical.

1. Introduction

Compared to the influence of the main Solar system bodies such as the Sun and the planets the gravitational interaction between asteroids tends to be small even during mutual close encounters. If the perturber is massive enough, close approach events can be used to constrain the mass of the perturbing asteroid through astrometric observation and analysis of the pre- and post-encounter trajectories. Given the large numbers of asteroids and potential mutual events the following questions become key to a successful asteroid mass determination campaign:

(*a*) Which asteroids should be observed?

(*b*) When should they be observed?

(*c*) What are the requirements to the accuracy of astrometric measurements allowing for reasonable constraints on a perturbing asteroid's mass?

In this paper we address the first two questions, as they can be answered to a satisfactory degree by precomputing asteroid-asteroid close encounters based on current trajectory information.

The 'classic' approach in modeling the close encounter between two asteroids has two evident assumptions: the gravitational interaction between the asteroids is instantaneous, so we can neglect interactions with other bodies during the close encounter, and the perturbed trajectory part of the small body can be considered a hyperbola with the centre of mass located at the nearest focus. The deflection of the velocity vector of the smaller asteroid due to the interaction with the more massive body can be used as a measure of the resulting perturbation.

The classic approach yields a seemingly simple way to assess the perturbation under certain assumptions. However, it is difficult to time observations in order to derive the precise asymptotes of the deflection hyperbola directly, given the number and magnitude of other gravitational interactions occurring at the same time. Close encounters between asteroids in the main belt are rather frequent, more than 1900 day^{-1}, with respect to the orbital revolution periods, see Sect. 3. Also, the perturbing bodies are not necessarily located in the orbital plane of the perturbed asteroid, so the superposition of the perturbations will change the flat second-order curve to a more complex three-dimensional curve, in particular when encounters are slow. Hence, a more sophisticated model than the simple two body scattering approach is needed in order to determine when to best observe close encounters so as to maximise the encounter signal and minimize the noise due to other gravitational perturbations.

2. Geometric approach

Assuming the three-dimensional perturbed trajectory of an asteroid to be a regular curve, it is possible to introduce the *Frenet-Serret* frame fixed to each point of the trajectory. Here the orthonormal basis spanning the Euclidean space $\mathbb{R}^3$ is given by tangent, normal and binormal vectors. Using this frame any regular three-dimensional trajectory can be decomposed uniquely into two functions: curvature and torsion.

The curvature and torsion have local extrema with a clear meaning: minima are related to the most 'straight' or 'flat' elements of the trajectory with respect to the local neighbourhood while maxima give 'turning' points of the trajectory, a vertex, where the tangent or normal vectors experience the highest rate of change. In other words the minima of curvature and torsion correspond to the less perturbed elements of the trajectory while the maxima give the opposite case. The second order curves that are solutions of a two-body problem have the curvature $\kappa = \frac{|1+e\cos v|^3}{p(1+2e\cos v+e^2)^{3/2}}$ and torsion $\tau = 0$ where p is the semi-latus rectum, e is the eccentricity, and v is the true anomaly of the perturbed orbit. As an example the curvature functions of (1) Ceres are presented in Fig. 1.

The trajectory of (1) Ceres resulting from a numerical integration of the equations of motion - the dynamical model is described in the following section - yields a large number of torsion and curvature extrema, Fig. 2. Those are a consequence of the perturbations by the planets and other asteroid perturbers, which are not in the same plane and change with time. All curvature and torsion functions of solar system objects will contain such contributions. We suggest to observe the perturbed asteroids at those epochs which coincide with the extrema of curvature and torsion closest to the encounter with the perturbing asteroid. The consecutive minima yield measurements of the greatest deflection angle or the 'accumulated effect' of perturbation while the intermediate maximum restricts the plane of the perturbation arc where the deflection angle is measured. An observational strategy based on this approach is bound to measure the *greatest* effect the perturber has on the perturbed body, before other perturbations change the orbit of the latter significantly.

3. Algorithm and dynamical model

In order to find the optimum epochs for observing the trajectory of perturbed asteroids we performed an exhaustive search of close encounters between

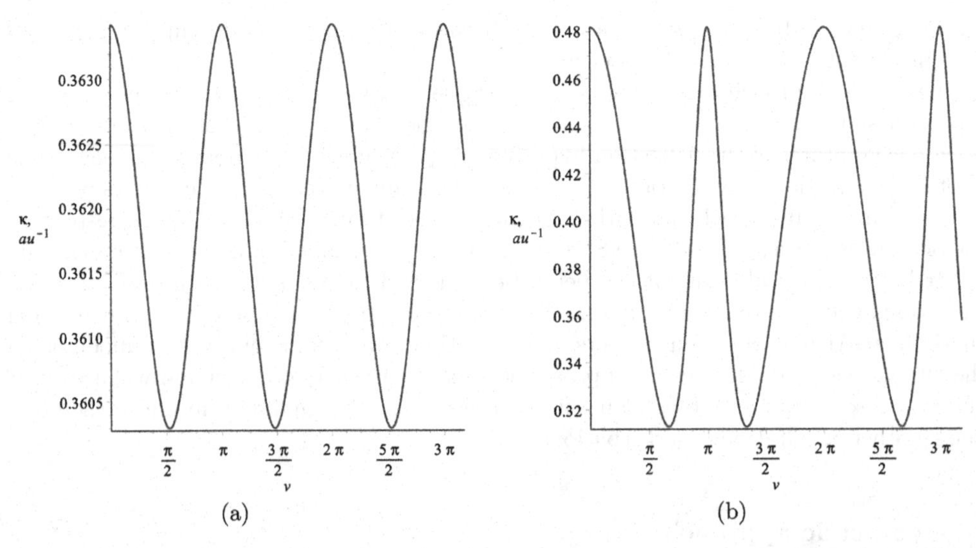

Figure 1. Curvature $\kappa(v)$ of the ideal orbit of (1) Ceres resulted from the two-body problem, $e = 0.0757$ (a), and the same orbit with exaggerated eccentricity, $e = 0.5$ (b).

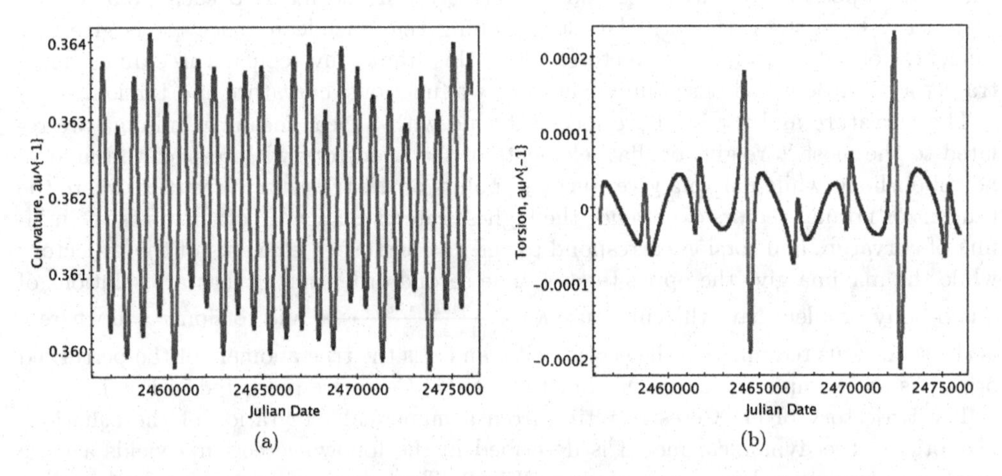

Figure 2. Curvature (a) and torsion (b) as functions of time resulting from an orbit-propagation of (1) Ceres.

2191† massive asteroids and 488,449 numbered asteroids (and TNOs) within 2013-2023. To this end, we numerically propagated the equations of motion for each asteroid including all planetary perturbations and expanded the trajectories in series of *Chebyshev* polynomials. We then determined all local minima in the pairwise distances between asteroids and retained only those encounters which occurred at distances less than 0.1 au (6,941,633 encounters), see Table 1. After a curvature and torsion analysis, we selected only those encounters that yield skyplane deviations in the motion of perturbed asteroids greater than 0.5 mas (535,182 encounters). The complete dynamical model used to find the optimum observation epochs consists of Parametrized Post Newtonian equations of motion including the planetary ephemeris JPL DE431, perturbations from the

† Masses of the largest asteroids were found either from the publications or were calculated assuming equivalent spheres using the diameters measured by the IRAS (Tedesco *et al.* 2004) and the NEOWISE teams (Mainzer *et al.* 2011) and a density 3 g/cm^3.

Table 1. Statistics of extreme values for the close encounters between asteroids at distances less 0.1 au between 2191 massive asteroids ($D > 20$ km) and 488,449 numbered asteroids from 2013 till 2023 (total number of such encounters found is 6,941,633).

Parameter	Minimum value	Maximum value
Minimal distance, au	2.34×10^{-5}	0.1
Relative speed, au/day	5.57×10^{-6}	0.0187
Minimal distance in Hill's radii	0.198	4397
Deflection angle, arcsec	0+	249

'BIG-16' asteroids (JPL HORIZONS), J_2 form factor of the Sun and non-sphericities of the planets to J_4. The propagator is based on an Adams-Moulton algorithm of variable order (1-12) and variable step; local relative error control is monitored to guarantee a relative error below 10^{-15} per step. Osculating elements of the processed asteroid orbits are based on data from JPL HORIZONS. Table 1 contains statistical information on the close encounter analysis.

4. Discussion

Curvature and torsion are invariant with respect to translations or rotations of the reference frame, however, both functions depend on the velocity and higher derivatives. The choice of the heliocentric or geocentric reference systems centered at the Sun or the Earth will introduce undesirable additional kinematic effects of the corresponding reference frames and should, thus, be avoided. This is the reason why the barycentric reference frame was used for the present calculations. The gravitational interaction between asteroids can be of the same order as Yarkovsky effect. The list of close encounters can furthermore be used for checking the adequacy of different ensembles of asteroids used in dynamical modeling.

5. Conclusions

We have presented a novel approach to determining optimum observation epochs for asteroid mass determination using asteroid-asteroid close encounters. The suggested approach for choosing observational moments corresponding to extrema in curvature and torsion of the perturbed asteroid trajectory can be useful for studying *any* particular orbit altering event, even if the target is subject to numerous other perturbations. Although the astrometric accuracy of the observations by the Gaia satellite is expected to be superior to the astrometry of the ground-based facilities, Gaia is not able to observe asteroids at specific epochs. Using the improved orbits resulting from Gaia data as well as the Gaia stellar catalog, close encounters between asteroids not to be observed by Gaia can be monitored by ground-based facilities. We have prepared a catalog listing excellent occasions and targets for astrometric observations that can greatly contribute to the effort of asteroid mass determination. The database of the close encounters and the observational moments is open to the public access. It is currently available at ftp://dosya.akdeniz.edu.tr/ivantsov. The interested reader is kindly requested to contact the corresponding author for further details.

References

Mainzer, A., Bauer, J., Grav, T., Masiero, J., Cutri, R. M., Dailey, J., Eisenhardt, P., McMillan, R. S., Wright, E., Walker, R., Jedicke, R., Spahr, T., Tholen, D., Alles, R., Beck, R.,

Brandenburg, H., Conrow, T., Evans, T., Fowler, J., Jarrett, T., Marsh, K., Masci, F., McCallon, H., Wheelock, S., Wittman, M., Wyatt, P., DeBaun, E., Elliott, G., Elsbury, D., Gautier, IV, T., Gomillion, S., Leisawitz, D., Maleszewski, C., Micheli, M., & Wilkins, A. 2011, *ApJ*, 731, 53

Tedesco, E. F., Noah, P. V., Noah, M., & Price, S. D. 2004, *NASA Planetary Data System*, IRAS Minor Planet Survey, IRAS-A-FPA-3-RDR-IMPS-V6.0

Astronomy and Astrophysics in the Gaia sky
Proceedings IAU Symposium No. 330, 2017
A. Recio-Blanco, P. de Laverny, A.G.A. Brown & T. Prusti, eds.

doi:10.1017/S174392131700535X

T_c-trend and terrestrial planet formation: The case of Zeta Reticuli

V. Adibekyano[1], E. Delgado-Mena[1], N. C. Santos[1,2], S. G. Sousa[1] and P. Figueira[1]

[1]Instituto de Astrofísica e Ciências do Espaço, Universidade do Porto, CAUP, Rua das Estrelas, 4150-762 Porto, Portugal
email: Vardan.Adibekyan@astro.up.pt

[2]Departamento de Física e Astronomia, Faculdade de Ciências, Universidade do Porto, Rua do Campo Alegre, 4169-007 Porto, Portugal

Abstract. Some studies suggested that the chemical abundance trend with the condensation temperature, T_c, is a signature of rocky planet formation. Very recently, a strong T_c trend was reported in ζ^2 Ret relative to its companion (ζ^1 Ret) and was explained by the presence of a debris disk around ζ^2 Ret. We re-evaluated the presence and variability of the T_c trend in this system with a goal to understand the impact of the presence of the debris disk on a star. Our results confirm the reported abundance difference between ζ^2 Ret and ζ^1 Ret and its dependence on the T_c. However, we also found that the T_c trends depend on the individual spectrum used. We conclude that for the ζ Reticuli system, for example, nonphysical factors can be at the root of the T_c trends for the case of individual spectra. For more details see Adibekyan *et al.* (2016b).

Keywords. Planetary systems, stars: abundances, stars: binaries

1. Abundance trends with condensation temperature

During the last decade astronomers have been trying to search for chemical signatures of terrestrial planet formation in the atmospheres of the hosting stars. Several studies explored a possible trend between the abundances of chemical elements and the condensation temperature (T_c) of the elements. This trend is usually called "T_c-trend". A T_c-trend was reported for several binary and field stars (including our Sun - Meléndez *et al.* 2009). While terrestrial planet formation, Galactic chemical evolution and stellar formation/evolution were proposed to explain the observed T_c-trend in Sun-like stars, its real nature is still debated (see Adibekyan *et al.* 2017 for a recent review).

Recently, Saffe *et al.* (2016) reported a positive T_c trend in the binary system, ζ^1 Ret – ζ^2 Ret. The authors explained the deficit of the refractory elements relative to volatiles in ζ^2 Ret as caused by the depletion of about $\sim$3 $M_\oplus$ rocky material. Here we re-evaluated the presence and variability of the T_c trend in this interesting system.

2. The ζ Reticuli system: stellar parameters and abundances

The ζ Reticuli binary system consists of two solar analogs where one of the stars (ζ^2 Ret) hosts a debris disk. Stellar parameters and chemical abundances of the stars are derived (from individual and combined high quality spectra) as described in the following works (Sousa *et al.* 2015, Adibekyan *et al.* 2015, Adibekyan *et al.* 2016a, Delgado-Mena *et al.* 2017).

3. ζ^1 Ret vs. ζ^2 Ret

In Fig. 1 (a) we compare the abundances of the two stars in the ζ Reticuli system against the T_c. There is a clear deficit of the refractory elements elative to volatiles in ζ^2 Ret, that can be related to the presence of the debris material.

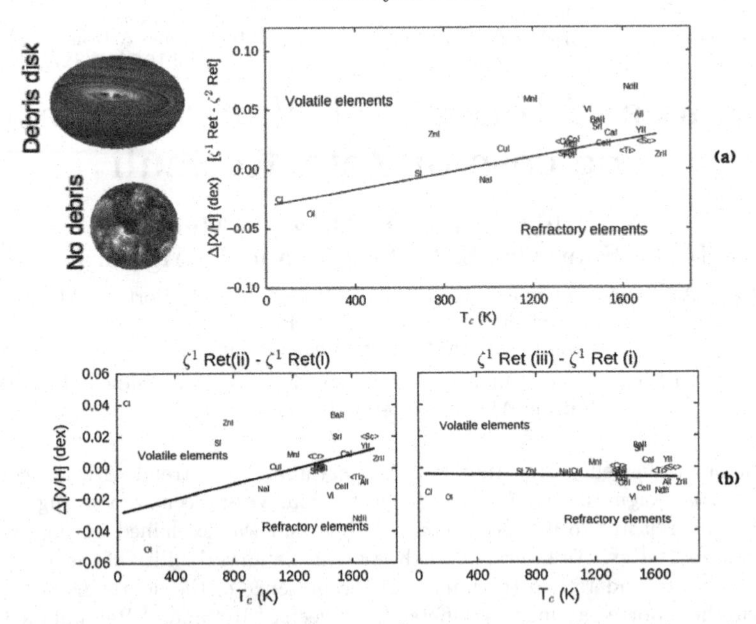

Figure 1. (a) Differential abundances (ζ^2 Ret − ζ^1 Ret) against T_c. The abundances are derived from very high S/N (> 1000) HARPS spectra. (b) Differential abundances against condensation temperature for ζ^1 Ret, derived from three highest S/N (> 350) individual spectra. The black lines are the results of the linear regression.

4. ζ^1 Ret vs. ζ^1 Ret (different spectra, different epochs)

In Fig. 1 (b) we compare the abundances of ζ^1 Ret derived from three individual spectra observed at different epochs. One can see significant but varying differences in the abundances of the same star from different individual high-quality spectra i.e. T_c-trend depends on the individual spectrum used (even if always of very high quality).

Acknowledgements

This work was supported by Fundação para a Ciência e Tecnologia (FCT) through national funds (project ref. PTDC/FIS-AST/7073/2014) and by FEDER through COMPETE2020 (ref. POCI-01-0145-FEDER-016880). V.A., E.D.M, P.F., N.C.S., and S.G.S. also acknowledge the support from FCT through Investigador FCT contracts of reference IF/00650/2015/CP1273/CT0001, IF/00849/2015, IF/01037/2013, IF/00169/2012, and IF/00028/2014, respectively, and POPH/FSE (EC) by FEDER funding through the program “Programa Operacional de Factores de Competitividade - COMPETE”. V.A. and E.D.M. acknowledge the support from IAU S330.

References

Adibekyan, V., Delgado-Mena, E., Feltzing, S., *et al.* 2017, *AN*, 338, 442
Adibekyan, V., Delgado-Mena, E., Figueira, P., *et al.* 2016b, *A&A*, 591, A34
Adibekyan, V., Delgado-Mena, E., Figueira, P., *et al.* 2016a, *A&A*, 592, A87
Adibekyan, V., Benamati, L., Santos, N. C., *et al.* 2015, *MNRAS*, 450, 1900
Delgado Mena, E., Tsantaki, M., Adibekyan, V., *et al.* 2017, [arXiv:1705.04349]
Meléndez, J., Asplund, M., Gustafsson, B., & Yong, D. 2009, *ApJ*, 704, L66
Saffe, C., Flores, M., Jaque Arancibia, M., Buccino, A., & Jofre, E. 2016, *A&A*, 588, A81
Sousa, S. G., Santos, N. C., Mortier, A., *et al.* 2015, *A&A*, 576, A94

Astronomy and Astrophysics in the Gaia sky
Proceedings IAU Symposium No. 330, 2017
A. Recio-Blanco, P. de Laverny, A.G.A. Brown
& T. Prusti, eds.

doi:10.1017/S1743921317006391

Shape and spin of asteroid 967 Helionape

G. Apostolovska[1], A. Kostov[2], Z. Donchev[2], E. Vchkova Bebekovska[1] and O. Kuzmanovska[1]

[1]Institute of Physics, Faculty of Science, Ss. Cyril and Methodius University, Skopje, Republic of Macedonia, Arhimedova 3, 1000 Skopje, Republic of Macedonia
email: gordanaaspostolovska@gmail.com

[2]Institute of Astronomy and National Astronomical Observatory, Bulgarian Academy of Sciences, Tsarigradsko Chaussee Blvd. 72, BG-1784, Sofia, Bulgaria

Abstract. Knowledge of the spin and shape parameters of the asteroids is very important for understanding of the conditions during the creation of our planetary system and formation of asteroid populations. The main belt asteroid and Flora family member 967 Helionape was observed during five apparitions. The observations were made at the Bulgarian National Astronomical Observatory (BNAO) Rozhen, since March 2006 to March 2016. Lihtcurve inversion method (Kaasalainen *et al.* (2001)), applied on 12 relative lightcurves obtained at various geometric conditions of the asteroid, reveals the spin vector, the sense of rotation and the preliminary shape model of the asteroid. Our aim is to contribute in increasing the set of asteroids with known spin and shape parameters. This could be done with dense lightcurves, obtained during small number of apparitions, in combination with sparse data produced by photometric asteroid surveys such as the Gaia satellite (Hanush (2011)).

Keywords. Minor planets, asteroids, photometric-Asteroids: individual: 967 Helionape

1. Observations and data Reduction

The observations were made by 50/70 cm Schmidt telescope equipped with FLI PL16803 CCD camera and 60 cm Cassegrain telescope with FLI PL9000 CCD camera. Aperture photometry of the asteroids and the comparison stars was performed using the software program CCDPHOT (Buie (1998)). For lightcurve analysis, we used the software package MPO Canopus v10.4 (Warner (2011)). The first observations of 967 Helionape were part of studies of the interrelations among Flora family asteroids (Apostolovska *et al.* (2009), Kryszczyńska *et al.* (2012)). According to NEOWISE, Helionape has a diameter of 10.216 km and albedo of 0.178 (Mainzer *et al.* (2016)).

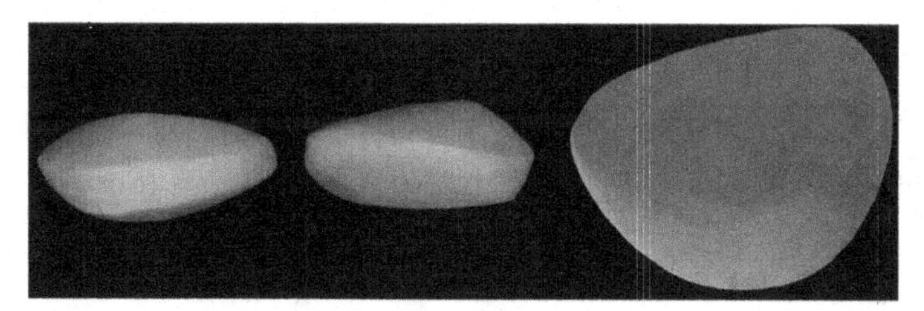

Figure 1. Shape model of 967 Helionape, shown at equatorial viewing illumination geometry, with rotational phases 90 apart (two pictures on the left) and the pole-on view on the right.

The table 1 gives the sidereal rotational period, sense of rotation, ecliptic coordinates λ and β of the pole solution and rough relative shape dimensions. Pole 2 is corresponding mirror solution to Pole 1.

Table 1. Parameters of the model

Asteroid 967 Helionape	Sidereal period (h)	Sence of rotation	Pole $\lambda(°)$	Pole $\beta(°)$	a/b	b/c
Pole 1	3.2339400	P	180.4	43.0	1.07	1.77
Pole 2	3.2339403	P	359.5	30.0	1.05	1.71
Pole 3	3.2339362	P	162.8	35.0	1.08	1.78

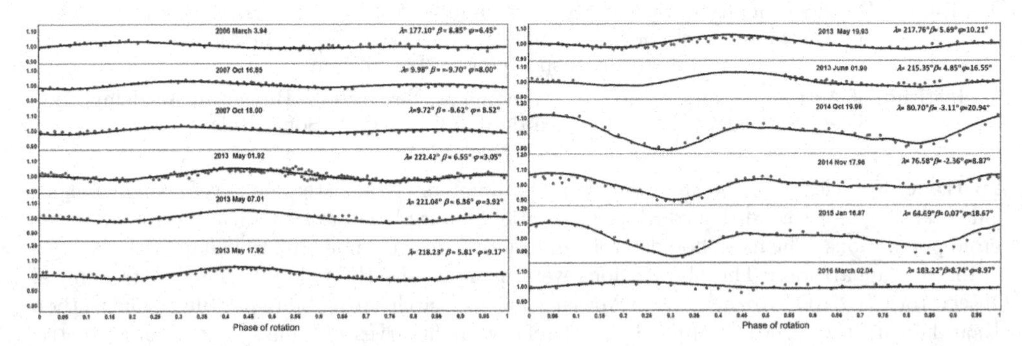

Figure 2. Lightcurves (points) obtained from observations for 967 Helionape superimposed on the lightcurves created by a model (solid line). In the graph are given, the phase angle (φ), the ecliptic longitude (λ) and latitude (β) of the asteroid referred to the date of the observation referring to the midtime of the lightcurve observed.

2. Conclusions

Using convex lightcurve inversion method we obtained three possible spin axes solutions with prograde sense of rotation and preliminary shape of 967 Helionape. One of the solutions, Pole 2 is corresponding mirror solution to Pole 1. Next opportunities for observing this asteroid through Schmidt telescope at BNAO Rozhen will be since August to December 2017 which could improve the calculated spin axes solutions and the obtained shape model. A search of the Asteroid Lightcurve Database (Warner *et al.* (2009)) and DAMIT (Ďurech *et al.* (2010)) has been shown that prior to the present paper there have been no reported results for the pole and shape of 967 Helionape.

Acknowledgments

Authors gratefully acknowledge observing grant support from the Institute of Astronomy and Rozhen National Astronomical Observatory, Bulgarian Academy of Sciences.

References

Apostolovska, G., Ivanova, V., & Kostov, A., 2009, *MPB*, 36, 27
Buie, M., 1998, http://www.boulder.swri.edu/ buie/idl/ccdphot.html
Ďurech, J., Sidorin, V., & Kaasalainen, M., 2010, *A&A*, 513, 46
Hanush, J., 2011, *Astron. Astrophys.* 530, 134
Kaasalainen, M., Torppa, J., & Muinonen, K., 2001, *Icarus*, 153, 37
Kryszczyńska, A., Colas, F., Polińska, M., *et al.*, 2012, *A&A*, 546, 72
Mainzer, A., Bauer, J., Cutri, R., *et al.*, 2016, *PDSS*, 247
Warner, B., Harris, A., & Pravec, P., 2009, *Icarus*, 202, 134, Updated 2017 April. http://www.minorplanet.info/lightcurvedatabase.html
Warner, B., 2011, MPO Canopus Software. http://www.MinorPlanetObserver.com

Astronomy and Astrophysics in the Gaia sky
Proceedings IAU Symposium No. 330, 2017
A. Recio-Blanco, P. de Laverny, A.G.A. Brown & T. Prusti, eds.

doi:10.1017/S174392131700566X

Preliminary Results of Low Dispersion Asteroid Spectroscopy Survey at NAO Rozhen†

Elena Vchkova Bebekovska[1], Galin Borisov[2,3], Zahary Donchev[3] and Gordana Apostolovska[1]

[1]Institute of Physics, Faculty of Natural Sciences and Mathematics, University Ss. Cyril and Methodius, Skopje, Republic of Macedonia
email: elenavckova@gmail.com

[2]Armagh Observatory, College Hill, BT61 9DG Armagh, Northern Ireland, UK
email: gbb@arm.ac.uk

[3]Institute of Astronomy and National Astronomical Observatory, Bulgarian Academy of Sciences

Abstract. We are presenting the first results of low dispersion spectroscopic observation of asteroids at Bulgarian National Astronomical Observatory Rozhen. Asteroids with unclassified spectra and brighter than 15 magnitude have been chosen. Besides just presenting the asteroid reflectance, classification according to Bus S. J. *et al.* (2012) has been done. The asteroid spectra of 590 Tomyris, 703 Noemi, 1596 Itzigsohn and 1826 Miller are presented together with standard spectra corresponding to the three best matches given by the public software tool M4AST (Popescu M. *et al.* (2012)). Our aim is to participate in the coordinated program of asteroids spectroscopy complementary to the observations of Gaia.

Keywords. minor planets, asteroids, techniques: spectroscopic

1. Introduction

Bulgarian National Astronomical Observatory (BNAO) - Rozhen (071 Rozhen) with astrometric observations has been already involved in GAIA Follow-Up Network for Solar System Objects since the end of 2011. One of our aims is to develop a coordinated program of asteroid spectroscopy complementary to Gaia's observations. In this paper the first results of asteroid spectroscopy at BNAO Rozhen are presented. Spectroscopic observations were made for four asteroids with unknown spectra: 590 Tomyris, 703 Noemi, 1596 Itzigsohn and 1826 Miller.

We obtained optical spectra of asteroids using the 2-m RCC telescope equipped with CCD VarsArray 1300B (pixel size 20 μm or 0.736 arcsec/px) in spectroscopic mode of FoReRo2 in its red channel (Jockers K. *et al.* (2000)). The spectroscopic characteristics are: low-dispersion grism Bausch & Lomb, working in the parallel beam of FoReRo2, with 300 lines/mm which gives 4.3 Å/px and 200 μm width slit which corresponds to 2.6 arcsec. We determined spectral types of the asteroids (Bus S. J. *et al.* (2012)) by the overall shapes of the spectra between 450 nm and 700 nm. For spectral analysis in our work we use public software tool M4AST (Popescu M. *et al.* (2012)). It covers aspects related to taxonomy, curve matching with laboratory spectra, space weathering models, and mineralogical diagnosis.

† Based on data collected with 2-m RCC telescope at Rozhen National Astronomical Observatory.

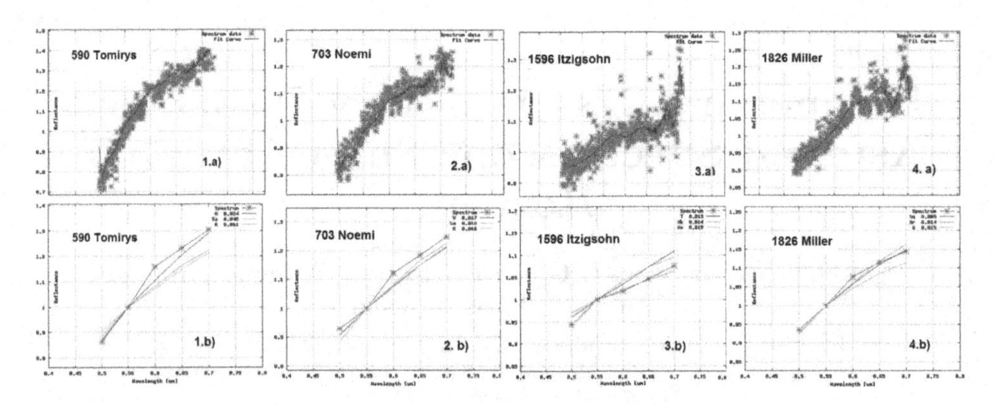

Figure 1. The asteroids observed spectrum is plotted in panels **a** and the standard spectra corresponding to the three best matches in panels **b**.

2. Observations and data reduction

Low dispersion spectroscopic observations of asteroids were taken on 29 December 2013. We have chosen asteroids with unclassified spectra and with apparent magnitudes not greater than 15. For each asteroid several (usually five) exposures were taken with approximate exposure time of 300 sec. In order to remove the solar contribution we obtained several images with exposure time of 300 sec. of a spectral solar analog star (SSAS) HD 28099. The wavelength interval of our observations is from 0.45 to 0.90 μm, but because of the strong fringing we had to cut at 0.7 μm.

For the preliminary data reduction of asteroid and SSAS spectra we did bias subtraction and flat field correction. The wavelength calibration of asteroid spectra was made by measuring the pixel position of the Ballmer lines in the spectrum of HD28099 and fitted them with their laboratory wavelengths. The background sky was obtained from the spectra along the slit, apart from object spectrum and was subtracted from the SSAS and asteroids spectra. The extinction correction was applied to each extracted (one dimension) spectrum (for the asteroid and SSAS). After these calibrations we divided asteroid spectrum by that one of the SSAS. To find the relative reflectance of the asteroid and to facilitate the comparison of spectra for different objects we normalised asteroids reflectivity to unity at 0.55 μm.

3. Results

Most of observed asteroids belong to some families and there are no published spectra. Besides plotting the spectra (the reflectance as a function of wavelength) classification according to Bus S. J. *et al.* (2012) has been done. The asteroid spectrum is plotted together with standard spectra corresponding to the three best matches (see figure 1).

590 Tomirys is a asteroid from the EOS family. We determined its spectral class as A type or probably Sa. **703 Noemi** is member of FLORA family and its reflectance corresponds to V type spectral class. **1596 Itzigsohn** is a outer main-belt asteroid and its reflectance spectrum fits well Xk type asteroid class spectra. **1826 Miller** is a asteroid from the EOS family and we determined its spectral class as Sq type.

References

Bus, S. J., Vilas, F. & Barucci, M. A. 2012, *Asteroids III*, Univ. of Arizona Press, 169-182.
Jockers, K. *et al.* 2000, *Kinematika i Fizika Neb. Tel. Suppl.*, 3, 13
Popescu, M. *et al.* 2012, *Astronomy & Astrophysics*, 544, A130

Astronomy and Astrophysics in the Gaia sky
Proceedings IAU Symposium No. 330, 2017
A. Recio-Blanco, P. de Laverny, A.G.A. Brown & T. Prusti, eds.

doi:10.1017/S1743921317005488

Solar system astrometry, Gaia, and the large surveys – a huge step ahead to stellar occultations by distant small solar system bodies

J. I. B. Camargo[1,2,3], M. V. Banda-Huarca[1,2,3], R. L. Ogando[1,2,3], J. Desmars[4,2], F. Braga-Ribas[5,2,3], R. Vieira-Martins[1,2,3], M. Assafin[6,2,3], B. Sicardy[4], D. Bérard[4], G. Benedetti-Rossi[1,2,3], L. A. N. da Costa[1,2,3], M. A. G. Maia[1,2,3], M. Carrasco-Kind[7] and A. Drlica-Wagner[8]

[1]Observatório Nacional/MCTIC, [2]Laboratório Interinstitucional de e-Astronomia (LIneA), [3]INCT do e-Universo,
Rua Gal. José Cristino, 77 20921-400 Rio de Janeiro RJ, Brazil
email: camargo@linea.gov.br

[4]Observatoire de Paris - Site de Meudon – CNRS : UMR8109,
Observatoire de Paris - 5, place Jules Janssen 92195 Meudon cedex, France

[5]Universidade Tecnológica Federal do Paraná,
Av. Sete de Setembro, 3165 80230-901 Rebouças, Curitiba - PR, Brazil

[6]Observatório do Valongo - Universidade Federal do Rio de Janeiro,
Ladeira do Pedro Antônio, 43 - Centro, Rio de Janeiro - RJ, 20080-090, Brazil

[7]University of Illinois, Department of Astronomy (NCSA - EUA) – Department of Astronomy, MC-221,
1002 W. Green Street, Urbana, IL 61801, United States

[8]Center for Particle Astrophysics - Fermi National Accelerator Laboratory (FERMILAB),
Pine Street Kirk Road, Batavia, IL 60510, United States

Abstract. The stellar occultation technique is a powerful tool to study distant small solar system bodies. Currently, around 2 500 trans-neptunian objects (TNOs) and Centaurs are known. With the astrometry from Gaia and large surveys like the Large Synoptic Survey Telescope (LSST), accurate predictions of occultation events will be available to tens of thousands of TNOs and Centaurs and boost the knowledge of the outer solar system.

Keywords. astrometry, ephemerides, occultations, Kuiper Belt

---

1. Introduction

Stellar occultation is a powerful technique to study distant small solar system bodies. It allows high angular resolution of the occulting (solar system) body from the analysis of a light curve acquired with high temporal resolution. If, on the one hand, stellar occultations present difficulties – they are transient events and require accurate predictions – on the other hand they can be considered magnitude independent, in the sense that it is not necessary to detect light from the occulting body during the occultation. With the results from the Gaia space mission, stellar and bright TNO positions will be known with unprecedented accuracy. Also, using Gaia stars as references, ground based astrometry of solar system bodies will be known to the milliarcsecond level. In this context, deep sky surveys also play an important role. The LSST, for instance, will map the visible sky from Cerro Pachón about twice a week and observe objects as faint as r∼24.5 in single images,

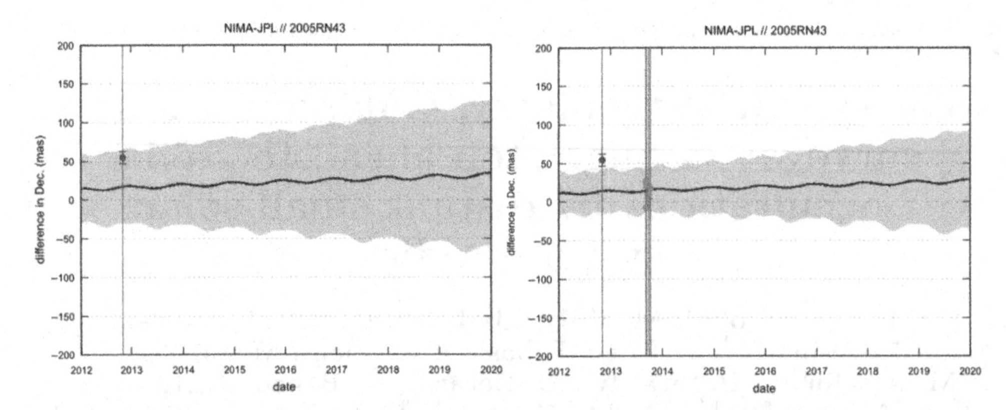

Figure 1. Black thick lines: differences in declination between the ephemeris determined by NIMA (Desmars *et al.* (2015)) and that from the JPL for the TNO 2005 RN43. Grey zones around thick lines: 1–σ uncertainty of NIMA's orbit. Left: orbit determination by NIMA using MPC data and averaged observations made at La Silla. Right: same as left panel plus observations from the DES. These latter observations were reduced with the Gaia DR1 (Lindegren *et al.* (2016)). Note that the uncertainties (grey zone) close to the dates of the most recent observations decrease considerably.

so that now we can have accurate predictions of stellar occultations for a large number of bodies, faint ones in particular. In other words, the prediction difficulty is overcome and the stellar occultation technique will now definitely be able to shed light on a so far "hidden" world of faint objects, telling us about their sizes and shapes, presence of atmosphere, and also about their immediate neighbourhoods.

2. Overview

We used public images taken by the Dark Energy Survey (DES) that were available at the NOAO archive (early 2015) to check the improvement in the orbit accuracy (Fig. 1). With accurate positions well distributed over few years, orbital accuracies better than 10 mas up to 1-3 years after the latest observation can be expected.

The LSST will increase the number of known TNOs by a factor ~16 (or ~40 000 objects) with new discoveries. In addition to a vibrating future that Gaia and the large surveys will bring to the study of the outer solar system through stellar occultations, a context of big data must also be considered.

3. Acknowledgement

This work has made use of data from the European Space Agency (ESA) mission *Gaia* (`https://www.cosmos.esa.int/gaia`), processed by the *Gaia* Data Processing and Analysis Consortium (DPAC, `https://www.cosmos.esa.int/web/gaia/dpac/consortium`). Funding for the DPAC has been provided by national institutions, in particular the institutions participating in the *Gaia* Multilateral Agreement.This research uses services or data provided by the NOAO Science Archive. NOAO is operated by the Association of Universities for Research in Astronomy (AURA), Inc. under a cooperative agreement with the National Science Foundation.

References

Desmars, J., Camargo, J. I. B., Braga-Ribas, F., *et al.* 2015, *A&A*, 584, 96
Lindegren, L., Lammers, U., Bastian, U., *et al.* 2016, *A&A*, 595, 4

Astronomy and Astrophysics in the Gaia sky
Proceedings IAU Symposium No. 330, 2017
A. Recio-Blanco, P. de Laverny, A.G.A. Brown & T. Prusti, eds.

doi:10.1017/S1743921317005403

Using Gaia spectrophotometric data for the purposes of asteroid taxonomy

Alberto Cellino[1], Paolo Tanga[2], Marco Delbo[2], Laurent Galluccio[2], Philippe Bendjoya[2] and Francesca De Angeli[3]

[1]INAF-Osservatorio Astrofisico di Torino, strada Osservatorio 20, I-10025, Pino Torinese, Italy
email: cellino@oato.inaf.it

[2]Université Côte d'Azur, Observatoire de la Côte d'Azur, CNRS, Laboratoire Lagrange, Bd de l'Observatoire, CS 34229, 06304 Nice Cedex 4, France
email: paolo.tanga@oca.eu, marco.delbo@oca.eu, laurent.galluccio@oca.eu, philippe.bendjoya@oca.eu

[3]Institute of Astronomy, University of Cambridge, Madingley Road, Cambridge CB3 0HA, UK
email: fda@ast.cam.ac.uk

Abstract. A new asteroid taxonomy will be an important result of Gaia observations of Solar System objects. Since Gaia observes asteroids in observing conditions and in an interval of wavelength which are slightly different with respect to normal ground-based observations, a dedicated observing campaign has been carried out at the Telescopio Nazionale Galileo in La Palma (Canary Islands, Spain). The obtained spectra have been used to generate a large number of synthetic clones, each one having slight changes with respect to its parent spectrum. These synthetic spectra are then used to feed the algorithm of taxonomic classification developed to reduce Gaia asteroid spectra. Processing of these data is in progress.

Keywords. minor planets, asteroids, methods: data analysis

1. Introduction

Using its BP/RP detector, Gaia observes asteroid spectra extending from blue to red for a number of targets of the order of 10^5. This is a much larger number with respect to currently available reflectance spectra obtained from the ground. Moreover, Gaia spectra are not obtained close to solar opposition, but at larger phase angles, between about 12 and 25 degrees†. Some reddening effects can therefore be expected. In order to use such big data-base of asteroid spectra to obtain a new taxonomic classification, an algorithm has been developed and will be used to process Gaia data. In order to test the algorithm, a spectroscopic survey aimed at obtaining spectra of asteroids of different taxonomic classes, preferentially observed at large phase angles, and in the same interval of wavelengths covered by Gaia, has been done using the 3.6-m Telescopio Nazionale Galileo (TNG) in La Palma (Spain). Some examples of the obtained spectra are shown in Fig. 1 (left panel). The TNG reflectance spectra have then been used to create a large number of synthetic spectra, each one exhibiting some realistic random variations with respect to its parent spectrum. These clones can be be used to feed to the algorithm of Gaia taxonomic classification.

† The phase angle is the angle between the directions to the Sun and to the observer, as seen from the asteroid.

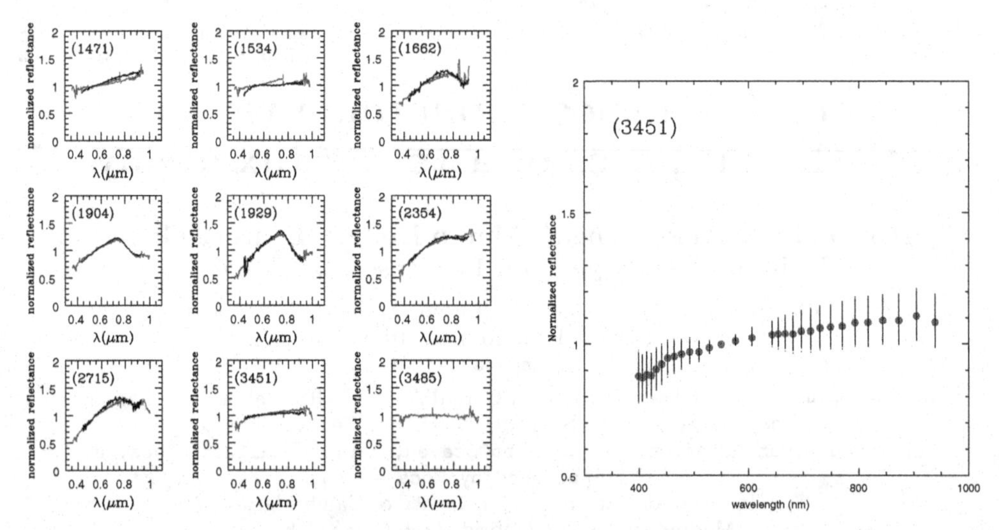

Figure 1. Left panel: a small subset of the asteroid reflectance spectra obtained during our observing campaign at TNG. The blue and red parts of the spectra have been measured separately for each object by using two different grisms. Superimposed black curves represent, when available, spectra previously obtained by the SMASS survey (Bus and Binzel (2002). Right panel shows, in the case of the reflectance spectrum of the Jupiter Trojan asteroid (3451) Mentor, the total variation exhibited by the full set of its clones at each of the discrete values of wavelength used in our analysis. Note that all spectra are normalized at the wavelength of 550 nm.

2. Spectral clones

The continuous TNG spectra have been discretized and used to generate a synthetic population of 100,000 "clone" spectra. The number of clones of asteroids belonging to different classes has been chosen according to the currently known relative abundances of different classes. Each clone represents a realistic random variation of its parent TNG spectrum. Fig. 1 (right panel) shows the total range of variation at different wavelengths of the generated clones of asteroid (3451) Mentor, a Jupiter Trojan. All the spectra are normalized at the wavelength of 550 nm. We are currently working to use the generated spectra to optimize the algorithm of taxonomic classification developed for Gaia.

3. Work in progress

Asteroid taxonomic classes are generally interpreted in terms of differences in surface composition. A great advantage of a Gaia-based taxonomy will be that of being based on a wider interval of wavelengths with respect to the most recent ground-bases classifications. As an example, we expect to be able to recover the old F class, which has been lost in modern taxonomies based on data not covering the blue spectral region, but are known to be well distinct in terms of polarimetric properties. Polarimetric data also suggest for the F class a possible cometary origin. A paper is currently in preparation.

Reference

Bus, S. J. & Binzel, R. P. 2002, *Icarus*, 158, 146

Astronomy and Astrophysics in the Gaia sky
Proceedings IAU Symposium No. 330, 2017
A. Recio-Blanco, P. de Laverny, A.G.A. Brown
& T. Prusti, eds.

doi:10.1017/S1743921317005877

Follow-up studies of Gaia transients at the Terskol Observatory

Vira Godunova[1], Volodymyr Reshetnyk[2], Andrii Simon[2], Sergii Velichko[1], Oleksandr Sergeev[1] and Volodymyr Taradii[1]

[1]ICAMER Observatory, NAS of Ukraine,
27 Acad. Zabolotnoho Str., Kyiv, Ukraine
email: godunova@mao.kiev.ua

[2]Faculty of Physics, National Taras Shevchenko University,
4 Acad. Glushkov Ave., Kyiv, Ukraine
email: reshetnykv@gmail.com

Abstract. Since 2015, scientific activities at the Terskol Observatory have been aimed at optical follow-up of stellar objects and asteroids detected within the framework of the Gaia mission. Two years of successful research have yielded new data and findings in this field. Photometric observations of Gaia transients allowed us to reveal physical characteristics of a good few of them. Moreover, we detected positions of dozens of asteroids which were reported by the GBOT group. In this paper, some results obtained from observations of transients Gaia16bkf, Gaia16bkn, Gaia17asz and newly detected asteroids are presented.

Keywords. techniques: photometric, asteroids, stars: variables: other, Gaia transients

---

1. Introduction

Facilities of the Terskol Observatory (the Northern Caucasus, 3100 m asl) have been heavily used for observations within the framework of the Gaia mission since 2015. The available telescopes Zeiss-2000 and Zeiss-600 provide good enough opportunities for long-term astrometric and photometric monitoring of stellar objects (SNe, CVs, YSOs, etc.) and asteroids. It should be noted that many advances in this field came from the development and use of specific instruments and techniques (Tarady *et al.* 2010).

2. Astrometry of newly detected asteroids

Scientific programmes on studies of asteroids have been run at the Terskol Observatory (IAU code B18) since the early 2000s (Godunova *et al.* 2014). In May 2015, we started to observe asteroids discovered within the Gaia project. Objects have been selected from the list of recently discovered asteroids prepared by the GBOT group (gbot.obspm.fr/index.php?page=asteroids) and the Gaia-FUN-SSO team. Asteroids have been observed down to V $\sim$ 21.5 mag, with individual exposure times of 60–180 s. For astrometric measures, an accuracy of about 0.15 arcsec was achieved. As of today, we could detect the following asteroids: G01366, G01378, G01831, G01893, G01899. G01900, G01764, G01773, G05150, G05164, G05168, G05165, G05089, G05154, G05117, G05120, G05096, G05829, G05865, G06018, G06028, G06029, G06030. In addition to that, we detected about 30 known main-belt asteroids and two unidentified objects, which appeared in the main target fields. All the positions were submitted to the IAU Minor Planet Center (www.minorplanetcenter.net).

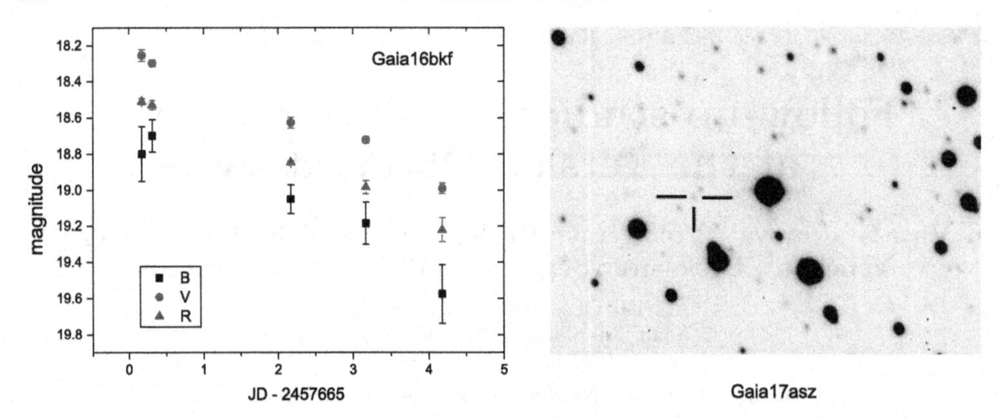

Figure 1. Follow-up of Gaia transients at Terskol Observatory. *Left:* The combined light curve of Gaia16bkf. *Right:* Gaia17asz, as seen with the 2-m telescope on 2017-03-24 (R filter).

3. Photometry of transients

In 2016–2017, a variety of Gaia transients discovered by ESA Gaia, DPAC and the Photometric Science Alerts Team (http://gsaweb.ast.cam.ac.uk/alerts) was observed at Terskol: Gaia16bkf, Gaia16bkn, Gaia16blg, Gaia16bnz, Gaia16bvs, Gaia16bvt, Gaia17agr, Gaia17agj, Gaia17aqm, Gaia17asz, Gaia17akp, etc. Some findings are listed below.

Gaia16bkf was detected at magnitude G=18.94 on 2016-09-24. This star near the Galactic plane was found 0.6 mag brighter than during its previous observations by Gaia which had demonstrated a minor variability (about 0.06 mag) in the last 1.5 years. Follow-up of this object started at Terskol on 2017-10-03, just after the alert was published. The BVRI photometry was performed with the 2-m telescope; CCD images were calibrated using NOMAD field stars. Figure 1 *(left panel)* depicts a rapid decay of Gaia16bkf with changing its color within four days.

Gaia16bkn was discovered at magnitude G=18.34 on 2017-10-01; it was previously classified as SN Ia by the GS-TEC (Blagorodnova *et al.* 2014). There was no previous detection of this object by Gaia; moreover, there is no progenitor object on archival images. Observations of Gaia16bkf showed that this object had dimmed from B=18.49±0.05, V=18.66±0.02, R=18.75±0.06 (on 2016-10-04) to B=19.05±0.05, V=19.05±0.14, R = 19.29±0.13 (on 2016-10-07). The magnitudes were calibrated against NOMAD field stars; they are not corrected for the Galactic foreground extinction.

Gaia17asz was discovered at magnitude G = 18.82 on 2017-03-17. There is no progenitor object on archival images. We confirmed Gaia17asz with images taken with the 2-m telescope on 2017-03-24 (see Fig. 1, *right panel*). The following magnitudes of the object were derived: B=19.4±0.1, V=19.9±0.1, R=20.1±0.1 (calibrated against Gaia DR1 field stars; not corrected for the interstellar extinction).

These and other results obtained at Terskol demonstrate that ground-based small and medium-sized telescopes remain a valuable tool for monitoring and investigation of newly detected objects. It is obvious that more systematic, integrated use of these instruments could lead to better information about transient events in the Universe.

References

Blagorodnova, N., *et al.* 2014, *MNRAS*, 442, 327
Godunova, V., *et al.* 2014, *Proc. of the 3rd Gaia-FUN-SSO workshop* (Paris Observatory), p. 131
Tarady, V., *et al.* 2010, *arxiv.1003.4875*

Astronomy and Astrophysics in the Gaia sky
Proceedings IAU Symposium No. 330, 2017
A. Recio-Blanco, P. de Laverny, A.G.A. Brown
& T. Prusti, eds.

doi:10.1017/S1743921317005361

Alerting observations of asteroids at the SBG telescope of the Kourovka Astronomical Observatory in the Gaia-FUN-SSO Network

Eduard Kuznetsov, Dmitry Glamazda, Galina Kaiser and Yulia Wiebe

Kourovka Astronomical Observatory,
Institute of Natural Sciences and Mathematics, Ural Federal University,
Lenin ave., 51, Yekaterinburg, Russia, 620000
email: eduard.kuznetsov@urfu.ru

Abstract. Regular astrometric observations of small bodies of the Solar System are conducted using a SBG telescope of the Kourovka Astronomical Observatory of the Ural Federal University. The first results of participation in Gaia-FUN-SSO network are presented.

Keywords. minor planets, asteroids

1. Introduction

Regular astrometric observations of small bodies of the Solar System are conducted using a SBG telescope of the Kourovka Astronomical Observatory of the Ural Federal University (AO UrFU). The four-axis SBG telescope with a 798.9 mm focal length is equipped with a Schmidt optical system and a 500 mm diameter main mirror. The aperture diameter is 420 mm. An Apogee Alta U32 CCD camera with a KAF-3200ME-1 CCD matrix containing 2184×1472 elements, each of size 6.8×6.8 μm is mounted at the main focus of the telescope. The scale of the CCD image is 1.76 arcsec/pixel. The field of view of the system is 61.2×42.5 arcmin. Limiting magnitude is 19 mag. The precision timing system uses a 12-channel GPS receiver Acutime 2000 GPS Smart Antenna. The SBG telescope and the CCD system are operated by the SBGControl software (Glamazda 2012) developed at AO UrFU.

2. Results

We participated in three campaigns of observation which were organized by the Gaia-FUN-SSO Network. The asteroid 2013 TV135 was observed in October 2014. The asteroid 2007 HB15 was not detected because it was very faint object for the SBG telescope. The near-Earth object 2014 HQ124 was not observed because the sky was very light in a nautical twilight near a day of summer solstice.

Analysis of quality of astrometric observations of small Solar System bodies with the SBG telescope is given in (Kaiser & Wiebe 2017). Fig. 1 and 2 show the root-mean-square (RMS) residuals (O–C) in a right ascending and declination depending on magnitudes of asteroids. We use follow notes: Main Belt asteroids are (•), near-Earth objects are (+) and potentially hazardous objects are (▲). Velocities of an apparent motion of some asteroids are given. The RMS residuals (O–C) in coordinates are from $0.07''$ to $0.15''$ for

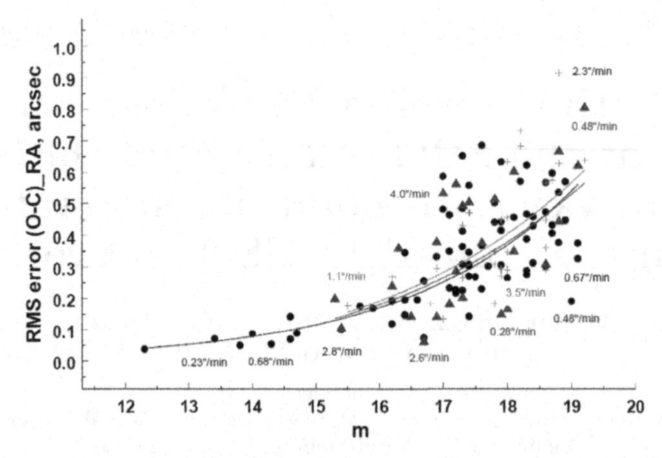

Figure 1. The RMS residuals (O–C) in a right ascending vs magnitudes m of asteroids.

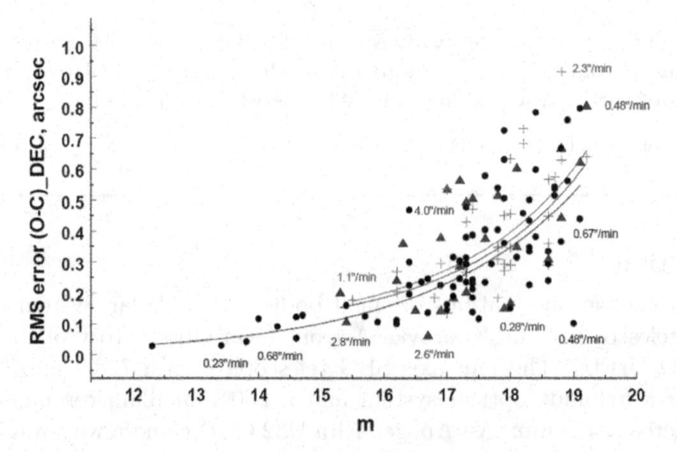

Figure 2. The RMS residuals (O–C) in a declination vs magnitudes m of asteroids.

objects with magnitude less than 14.5^m. If magnitude is risen to 19^m, the RMS residuals (O–C) grow to $0.8''$ for Main Belt asteroids and $1.0''$ for potentially hazardous objects.

There are not positive results of alerting observations of asteroids at the SBG telescope yet. Current the Gaia alerts asteroids are very faint to be observed with the SBG telescope. We are going to continue alerting observations in the Gaia-FUN-SSO Network.

3. Acknowledgments

G. K. and Yu. W. acknowledge funding from the Ministry of Education and Science of the Russian Federation (the basic part of the State assignment, RK no. AAAA-A17-117030310283-7). E. K. and D. G. acknowledge funding from the Government of the Russian Federation (Act no. 211, agreement no. 02.A03.21.0006).

References

Glamazda, D. 2012, *Astrophys. Bulletin*, 67, 237
Kaiser, G. & Wiebe, Yu. 2017, *Solar System Research*, 51, 233

Astronomy and Astrophysics in the Gaia sky
Proceedings IAU Symposium No. 330, 2017
A. Recio-Blanco, P. de Laverny, A.G.A. Brown & T. Prusti, eds.

doi:10.1017/S1743921317005658

Searching for planets around eclipsing binary stars using timing method: NSVS 14256825

Ilham Nasiroglu[1], Krzysztof Goździewski[2], Aga Słowikowska[3], Krzysztof Krzeszowski[3], Michal Żejmo[3], Staszek Zola[4] and Huseyin Er[1]

[1]Ataturk University, Department of Astronomy and Astrophysics, Erzurum, Turkey
email: inasir@atauni.edu.tr

[2]Faculty of Physics, Astronomy and Applied Informatics, N. Copernicus Univ., Toruń, Poland

[3]Janusz Gil Institute of Astronomy, University of Zielona Gora, Zielona Gora, Poland

[4]Astronomical Observatory, Jagiellonian University, Krakow, Poland

Abstract. We present four new mid eclipse times and an updated O-C diagram of the short period eclipsing binary NSVS14256825. The new data follow the (O-C) trend and its model proposed in Nasiroglu *et al.* (2017). The (O-C) diagram shows quasi-periodic variations that can be explained with the presence of a brown-dwarf in a quasi-circular circumbinary orbit.

Keywords. binaries: eclipsing, ephemerides, planetary systems, NSVS14256825

---

1. Introduction

Eclipse timing observations have provided evidence of third bodies orbiting binary systems through the Light Travel Time (LTT) effect. These deviations can be measured with a high accuracy and used to infer the presence of low-mass compaions (Irwin 1952, Goździewski *et al.* 2012, 2015).

The post-common envelope binary NSVS 14256825 is a member of the HW Vir family with a period of 2.65 hrs, consisting of a OB sub-dwarf and a M dwarf companion (Almeida *et al.* 2012). The cyclic behaviour of the (O-C) eclipse timings in NSVS 14256825 was attributed to one or two Jovian-type circumbinary planets (Beuermann *et al.* 2012, Almeida *et al.* 2013). However, the orbital stability of a 2-planet system was found strongly unlikely by Wittenmyer *et al.* (2013). Moreover, Hinse *et al.* (2014) also performed a detailed analysis and they concluded that the time span of timing measurements is not long enough to constrain the proposed planets. In this study, we present four new eclipse times (Tab. 1) that follow the trend of the (O-C) diagram shown recently by Nasiroglu *et al.* (2017). The observations were performed at the TUBITAK National Observatory (TUG, T100) and at the Adiyaman University Observatory (ADYU60) in Turkey.

2. Data and LTT Model

Nasiroglu *et al.* (2017) presented 83 new eclipse timings of NSVS 14256825 which extended the time span of previous (O-C) diagram by three years. It is marked as grey shaded region in Fig. 1. We note that four new points beyond the last epoch in Nasiroglu *et al.* (2017), i.e. November 3rd, 2016, fit the earlier third-body model very well. The fitting model (red curve in Fig. 1) indicates that the observed quasi-periodic (O-C) variability can be explained by the presence of a brown dwarf with the minimal mass of 15 Jupiter masses, orbiting the binary with the period of about 10 years.

Table 1. List of four new NSVS 14256825 eclipse times after November 3th, 2016.

Cycle	BJD	Error [days]	Ecl.Type	Telescope
31004.0	2457696.247527740	0.000012891	1	ADYU60
31212.0	2457719.205340000	0.000007616	1	TUG
32556.0	2457867.548033890	0.000012688	1	TUG
32773.0	2457891.499207930	0.000021861	1	ADYU60

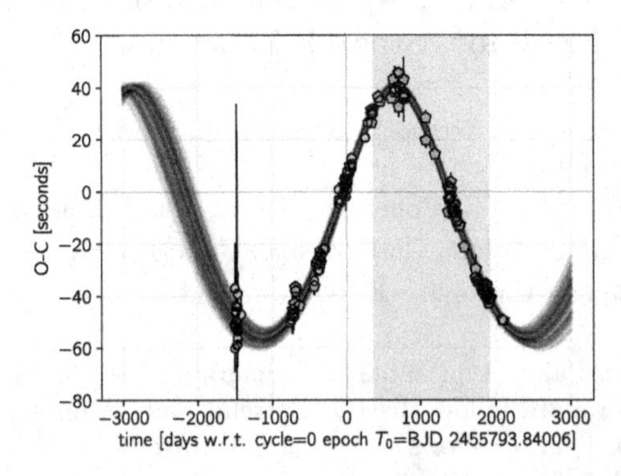

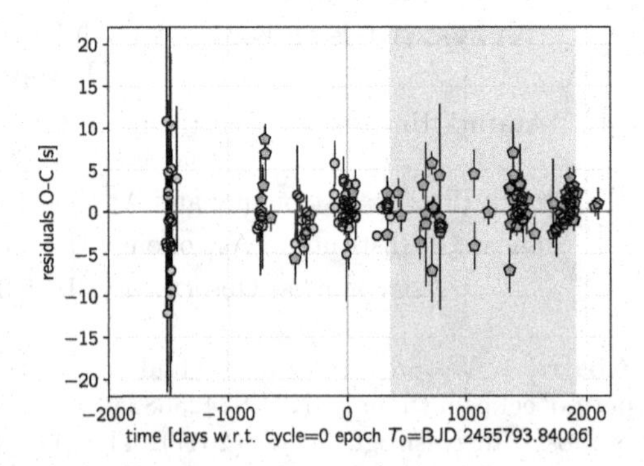

Figure 1. *The left panel:* the synthetic curve of the best-fitting model (red curve) to all data overplotted on 100 sample models from the MCMC posterior. Filled circles are for data in the up-to date literature and darker pentagons are for the new measurements presented in Nasiroglu *et al.* (2017), which contributed new data within the shaded box, and in this paper (after cycle number of 2000, see Tab. 1). *The right panel:* residuals to the best-fitting model.

A few compact, post-common envelope binaries exhibit similar, quasi-periodic variations of the (O-C) attributed to low-mass companions, out of which many are still uncertain. The GAIA mission has a great potential to confirm or dismiss the presence of such bodies. The eclipse timing will be useful to constrain astrometric models and to explain the nature of their observed (O-C) diagrams. Therefore we continue to monitor a sample of such binaries.

Acknowledgments

This work has been supported by The Scientific and Technological Research Council of Turkey (TUBITAK), through project 114F460 (IN, AS, HE). We thank the team of TUBITAK National Observatory for a partial support in using T100 telescope with project number TUG T100-631, and Adiyaman University Observatory.

References

Almeida, L. A., Jablonski, F., Tello, J., & Rodrigues, C. V. 2012, *MNRAS*, 423, 478
Almeida, L. A., Jablonski, F., & Rodrigues, C. V. 2013, *ApJ*, 766, 11
Beuermann, K., Breitenstein, P., Debski, B., *et al.* 2012, *A&A*, 40, 8
Irwin, J. B. 1952, *ApJ*, 116, 211
Nasiroglu, I., Goździewski, K., Słowikowska, A., *et al.* 2017, *AJ*, 153, 137
Goździewski, K., Nasiroglu, I., Słowikowska, A., *et al.* 2012, *MNRAS*, 425, 930
Goździewski, K., Słowikowska, A., Dimitrov, D., *et al.* 2015, *MNRAS*, 448, 1118
Hinse, T. C., Lee, J. W., Goździewski, K., *et al.* 2014, *MNRAS*, 438, 307
Wittenmyer, R. A., Horner, J., & Marshall, J. P. 2013, *MNRAS*, 431, 2150

Astronomy and Astrophysics in the Gaia sky
Proceedings IAU Symposium No. 330, 2017
A. Recio-Blanco, P. de Laverny, A.G.A. Brown
& T. Prusti, eds.

doi:10.1017/S1743921317005737

Observations of the satellites of the major planets at Pulkovo Observatory: history and present

N. A. Shakht, A. V. Devyatkin, D. L. Gorshanov and M. S. Chubey
Central (Pulkovo) Observatory RAS,
St- Petersburg 196140, Russia
email: natalia.shakht@yandex.ru

Abstract. In connection with long on-orbit European space satellite Gaia and the opportunity that now provides ESA, to use the results of observations of the space telescope, we would like to present some results of our long-term observations of the major planets satellites at Pulkovo Observatory. We hope to translate into reality these opportunities, namely the use of new observations and new ephemeris and a practical possibility of a new reduction for modern and old observations. The essential facilities can appear in the space, we give the shortest presentation of space project Orbital Stellar Stereoscopic Observatory.

Keywords. natural satellites of great planets, observations, space project OStSO

The astrometric positional observations of the major planets and their satellites started almost since the foundation of Pulkovo observatory. The first photographic observations made with Pulkovo Normal Astrograph (PNA) since 1894 yr and continued during one hundred years. Now we have about 6500 plates with the bodies of the Solar System from of total quantity 48 000 plates collected in Pulkovo glass library. The most positions and the list of publications took place in our database www.puldb.ru which is updated.

In the late nineteenth A. A. Belopolsky and S.K. Kostinsky received the first observations of the satellites of Mars on PNA . In 1899-1902 Kostinsky received positions of the satellite of Neptune, these observations were continued until 1930. In 1908-1922 Kostinsky also received the relative positions of six satellites of Saturn. Later the photographic observations of major planets and their satellites were continued at Pulkovo by means of the PNA and with 26-inch refractor that was established in 1956. In general, the bodies of the Solar System were observed in Pulkovo Observatory and in Pulkovo expeditions, on more than ten instruments, see www.puldb.ru.

Some of the most old plates of the late XIX and early XX century were lost during the Second World War, however, remained intact results of the measurements of planets and satellites and reference stars. Currently, observations are fulfilling also with automatic telescopes complexes ZA-320M at Pulkovo and MTM-500M at Pulkovo Mountain Station in Kislovodsk (North Caucasus).

The main telescopes that are now working at Pulkovo: 26-inch refractor (D=65 cm; F=10400 mm), PNA (D=330 mm, F=3467 mm), ZA-320 - complex automatic mirror astrograph ZA-320 (D=320 mm, F= 3200 mm) Now Pulkovo telescopes are completely automated and equipped with modern CCD-cameras. The time distribution of the observations of major planets and their satellites is presented in table 1 , by asterisk the first photographic observations are noted.

We represent the astrometric observations in 2004-2006 of Jupiter's and Saturn's satellites, table 2 , which were made in Pulkovo Observatory with ZA-320 Mirror Astrograph. Processing was done by means of APEX program system in USNO-A2.0 and USNO-B1.0

Table 1. Observations, distributed in time

Mars	1960-1973	Saturn, satellites	1908-1922, 1971-2009
Mars,satellites (old)*	1894-1909	Pluton	1930-2010, 2003-2006
Mars,satellites (modern)	1973, 1986, 1988	Neptune,satellites	1899-1955, 1990, 1993
Jupiter, satellites	1974-2015	Uranus, satellites	1919-2016

Table 2. Observations with ZA-320 in 2004-2006

Satellites	N	$(O-C)_{\alpha\cos\delta}$	$\sigma_{\alpha\cos\delta}$	$(O-C)_\delta$	σ_δ
Himalia	45	+0.13	±0.21	+0.24	±0.18
Elara	39	+0.11	±0.28	+0.17	±0.32
Pasiphae	33	+0.18	±0.30	+0.16	±0.32
Hyperion	56	+0.24	±0.24	+0.25	±0.22
Yapetus	62	+0.14	±0.22	+0.20	±0.23
Phoebe	91	+0.11	±0.26	+0.16	±0.23

catalogues system, see Devyatkin *et al.* 2006. Values of (O-C) and errors of observations are given in arcsecs.

We fulfilled the comparison some of old observations with the new ephemerides. As an example we give the results of Neptune's positions in 1899-1902.

We had $(O-C)$ for relative distances S and for positional angles P. The mean values for 1899-1901 are equal $(O-C)_s = -0''.25$ and $(O-C)_p = -0^o.48$ in comparison of observations with calculations by means of Hermann Struve ephemerides and $-0''.26$ and $-0^o.11$ which are calculated with new ephemerides of Emel'yanov, 2015. In 1902 correspondingly $(O-C)_s = -0''.51$ and $(O-C)_p = -0^o.91$ with old ephemerides and $(O-C)_s = -0''.53$; $(O-C)_p = -0^o.35$ with new ones. We used the database UAI/IMCCE/SAI.

The old photographic images of Saturnian Satellites 2-6 recorded with the 26-inch refractor and the PNA at Pulkovo from 1972 to 1974 have been recently digitized see Kiseleva *et al.* 2016. The observed position of the satellites and Saturn time were compared with the ephemeris using the software at MULTI-SAT Emel'yanov & Arlo 2008. Authors note the remarkable increasing of precision. The errors after new reduction reached to $0''.08$ and $0''.17$ for each instrument correspondingly. At Pulkovo the project of creation of the Orbital Stellar Stereoscopic Observatory (OStSO) has been developed, which is designed to meet the multi-program action for fundamental researches in space, see:
`mmg.tversu.ru/images/publications/2016-vol4-n3/Chubey–2016_12_08.pdf`

The authors thank the organizing Committee for the invitation and the opportunity to participate in the Symposium.

References

Devyatkin, A. V., Gorshanov, D. L., Kouprianov, V. V., Aleshkina, E. Y.u, Bekhteva, A. S., Baturina, G. D., Ibragimov, F. M., Vereschagina, I. A., Krakosevich, O. V., & Barshevich, K. V. 2006, *Izv. GAO* 218, 68

Kiseleva, T. P., Vasilieva, T. A., Izmailov, I. S., & Roshchina, E. A. 2015, *Sol.Syst. Res.* 49,1,72

Emel'yanov, N. V.& Arlot, J.-E. 2008, *A&A* 487, 759

Astronomy and Astrophysics in the Gaia sky
Proceedings IAU Symposium No. 330, 2017
A. Recio-Blanco, P. de Laverny, A.G.A. Brown & T. Prusti, eds.

doi:10.1017/S1743921317006238

Precise CCD positions of Triton in 2014-2016 from the Gaia DR1†

N. Wang[1], Q. Y. Peng[1], H. W. Peng[1,2] and Q. F. Zhang[1]

[1]Department of Computer Science, Jinan University,Guangzhou 510632, China,
email: `tpengqy@jnu.edu.cn`

[2]Yunnan Observatory, CAS, Kunming 650216, China

Abstract. 755 CCD observations during the years 2014-2016 have been reduced to derive the precise positions of Triton, the first satellite of Neptune. The observations were made by the 1 m telescope at Yunnan Observatory over 15 nights during the years 2014-2016. The theoretical position of Triton was retrieved from the Jet Propulsion Laboratory Horizons system. Our results show that when the newest Gaia catalogue (Gaia DR1) is referred to the mean O-Cs (observed minus computed) residuals are about 0.042 and -0.006 arcsec, the dispersions are 0.012 and 0.012 arcsec in right ascension and declination, respectively. The dispersions are improved very significantly when the Gaia DR1 is referred to. However, the agreement in right ascension is not so good as that in declination, the reason might come from the uncertainty of planet ephemeris. More observations are needed to confirm this.

Keywords. methods: observational, techniques: image processing, astrometry.

---

1. Introduction

Triton is the largest moon of Neptune and it has a retrograde, inclined and circular orbit. This unusual configuration has led to the belief that Triton originally orbited the Sun before being captured in orbit around Neptune (McKin 1984). Therefore Triton holds important clues to the evolution of the solar system. In recent years, generous observations of Triton have been made (Stone 2000; Vieira Martins *et al.* 2004; Qiao *et al.* 2014) and the precision of these observations is usually about 0.05 to 0.5 arcsec (Emelyanov & Samorodov 2015). As the newest Gaia DR1 star catalogue is available (Gaia Collaboration 2016a,b), the precision of Triton must be further improved and a quite great CCD field of view allows us to calibrate its geometry accurately (Peng *et al.* 2012). Furthermore, the high precision of Triton can help improve the precision of Neptune (Robert *et al.* 2011).

2. Observations

Since 2014, we have been engaged in a systematic observation of Triton. All the observations were made with the 1m telescope at Yunnan Observatory. The IAU site code is 286. A total of 755 frames of CCD images have been obtained for the satellite as well as 398 frames of CCD calibration images. The exposure time for each CCD frame ranged from 30 to 120 s, depending on the meteorological conditions.

† This research work is supported by the National Science Foundation of China (Grant Nos U1431227, 11273014). This work has made use of data from the European Space Agency (ESA) mission *Gaia* (https://www.cosmos.esa.int/gaia), processed by the *Gaia* Data Processing and Analysis Consortium (DPAC, https://www.cosmos.esa.int/web/gaia/dpac/consortium). Funding for the DPAC has been provided by national institutions, in particular the institutions participating in the *Gaia* Multilateral Agreement.

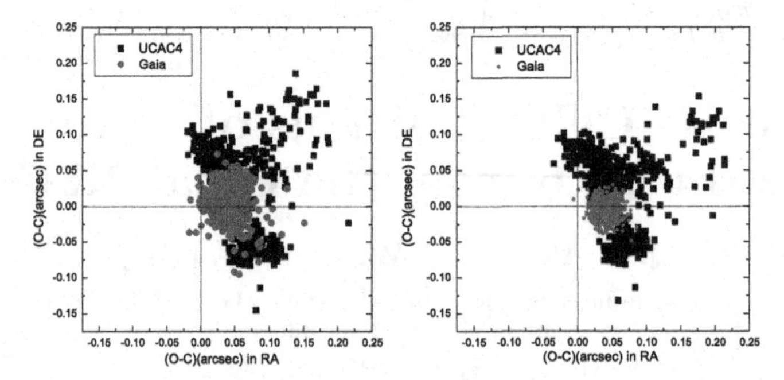

Figure 1. (O-C) residuals of Triton before and after Geometric Distortion Correction (GDC). The left panel shows the (O-C) residuals before GDC and the right panel after GDC.

Table 1. Statistics of (O-C) residuals for Triton by using different planetary and satellite ephemerides. Column 1 and 2 shows the planetary ephemeris and the satellite ephemerides respectively. The following columns list the mean (O-C) and its standard deviation (SD). Emelyanov15 means Emelyanov & Samorodov (2015). All units are arcsec.

planetary ephemeris	satellite ephemeris	<O-C> RA	SD RA	<O-C> DEC	SD DEC
INPOP10	Emelyanov15	0.083	0.013	-0.099	0.013
INPOP13C	Emelyanov15	0.062	0.013	-0.059	0.013
EPM2015	Emelyanov15	0.062	0.013	-0.014	0.013
DE405	Emelyanov15	0.149	0.016	-0.028	0.014
DE421	Emelyanov15	0.057	0.013	-0.018	0.013
DE431	Emelyanov15	0.042	0.013	-0.007	0.013
DE431	JPL nep081xl	0.042	0.012	-0.006	0.012

3. Results and conclusions

In order to analysis the observations, we compared our results using different referenced star catalogues, using different planetary ephemerides and satellite ephemerides.

The positional precision ($\sim$ 12 mas in each direction) of Triton has been greatly improved mainly due to the application of Gaia DR1 catalogue. The geometric distortion correction of a CCD frame has significant effect to the improvement of the positional precision of Triton. The ephemeris of Neptune even the new DE431 perhaps have a somewhat space ($\sim$ 40 mas in RA) to improve according to our present observations. On the other hand, the ephemerides of Triton from JPL and IMCCE have a very good agreement.

References

Emelyanov, N. V. & Samorodov, M. Y. 2015, *MNRAS*, 454, 2205
Gaia Collaboration, Prusti, T., de Bruijne, J. H. J., *et al.* 2016, *aap*, 595, A1
Gaia Collaboration, Brown, A. G. A., Vallenari, A., *et al.* 2016, *aap*, 595, A2
McKinnon, W. B. 1984, *nat*, 311, 355
Peng, Q. Y., Vienne, A., Zhang, Q. F., *et al.* 2012, *AJ*, 144, 170
Qiao, R. C., Zhang, H. Y., Dourneau, G., *et al.* 2014, *MNRAS*, 440, 3749
Robert, V., de Cuyper, J.-P., Arlot, J.-E., *et al.* 2011, *MNRAS*, 415, 701
Stone, R. C. 2000, *AJ*, 120, 2124
Vieira, M. R., Veiga, C. H., Bourget, P., *et al.* 2004, *aap*, 425, 1107

Astronomy and Astrophysics in the Gaia sky
Proceedings IAU Symposium No. 330, 2017
A. Recio-Blanco, P. de Laverny, A.G.A. Brown & T. Prusti, eds.

doi:10.1017/S1743921317005555

Astrometric Reduction of Cassini ISS Images of Enceladus in 2015 Based on Gaia DR1

Q. F. Zhang[1]*, V. Lainey[2], A. Vienne[2], N. J. Cooper[3], Q. Y. Peng[1] and N. Wang[1]

[1]Department of Computer Science, Jinan University, Guangzhou 510632, P.R. China

[2]IMCCE, Observatoire de Paris, UMR 8028 du CNRS, UPMC, Université de Lille 1, 77 av. Denfert-Rochereau, 75014 Paris, France

[3]Astronomy Unit, School of Mathematical Sciences, Queen Mary University of London, Mile End Road, London E1 4NS, UK

* email: tqfz@jnu.edu.cn

Abstract. The Gaia DR1 catalogue stars are taken as reference ones to reduce the Cassini ISS images of Enceladus in 2015, and a total of 494 Cassini-centered astrometric observation are obtained in right ascension(α) and declination (δ) in the international Celestial Reference Frame(ICRF). Compared with JPL ephemerides SAT367, we derive that their mean residuals are a few tens meters in $\alpha^*\cos(\delta)$ and a few kilometers in δ, and their standard deviation is not over 2 kilometers. Compared with the results from UCAC4 catalogue stars, The Gaia DR1 has the equivalent precision of reduction.

Keywords. Astrometric Reduction, Enceladus, Gaia DR1, Cassini ISS

1. Introduction

During the past a few years, the Cassini ISS images have been routinely used to measure the astrometric positions of planetary satellites (Cooper *et al.* 2006, Cooper *et al.* 2014, Tajeddine *et al.* 2013 and Tajeddine *et al.* 2015). The soft package Caviar has been also implemented for the task (Cooper *et al.* 2016), which is convenient to reduce space images. In the research, the ones in 2015 are measured by Caviar, and some of them are selected for comparing the effects resulted from Gaia DR1 (Gaia Collaboration 2016a and Gaia Collaboration 2016b) and UCAC4 (Zacharias *et al.* 2013).

2. Method

To reduce one image by CAVIAR involves two successive steps: pointing correction and limb fitting. In the first step, some catalogue stars are taken as reference ones to correct camera's pointing. To study how much the Gaia DR1 benefits the reduction, we take the same stars in Gaia DR1 and in UCAC4 catalogue respectively as reference stars in one image, and keep the same operation in the second step. So, each image will has two results of reduction, one from Gaia DR1, and the other from UCAC4. Finally, we compare the result pairs of total 368 images to analyze the effects of Gaia DR1.

3. Data and Results

All the images of Enceladus in 2015 have been taken by CASSINI ISS NAC from 2015-151T to 2015-337T.

At first, 494 images of Enceladus have been measured by Caviar with Gaia DR1, and then the residuals of observed position relative to JPL ephemerides SAT367 are

Table 1. Mean values and standard deviations(SD) of residuals of these observed positions by using Gaia DR1 relative to the JPL SAT367 ephemeris.

	Mean	SD
Sample (pixel)	-0.0078	0.1466
Line (pixel)	0.3218	0.2525
$\alpha^*\cos\delta$ (arcsec)	0.0091	0.1802
δ (arcsec)	0.4101	0.3210
$\alpha^*\cos\delta$ (km)	0.0757	1.1174
δ (km)	2.8488	1.8862

Table 2. Mean values of (O-C)s in pixels, arcseconds and kilometers relative to the JPL SAT367 ephemeris, including standard deviations, when the Gaia DR1 and UCAC4 are used to obtain Enceladus' positions respectively.

	Gaia DR1		UCAC4	
	Mean	SD	Mean	SD
Sample (pixel)	-0.0044	0.1288	-0.0152	0.1459
Line (pixel)	0.3581	0.1656	0.3502	0.1728
$\alpha^*\cos\delta$ (arcsec)	0.0125	0.1618	-0.0047	0.1825
δ (arcsec)	0.4569	0.2060	0.4465	0.2158
$\alpha^*\cos\delta$ (km)	0.1013	1.1404	0.0187	1.2789
δ (km)	3.1460	1.5278	3.0676	1.5843

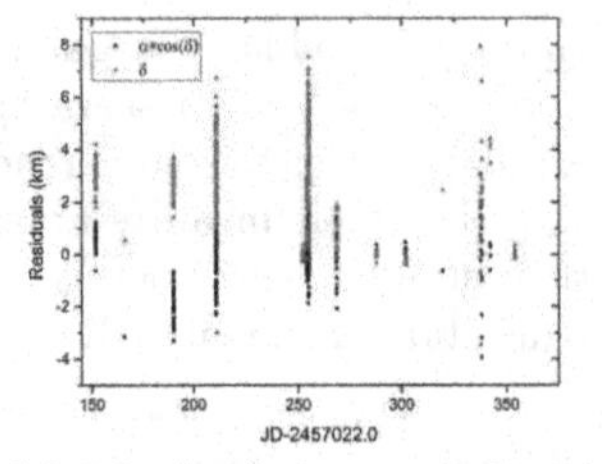

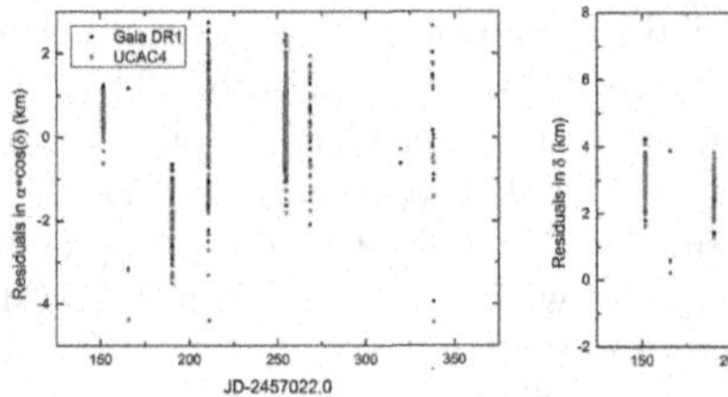

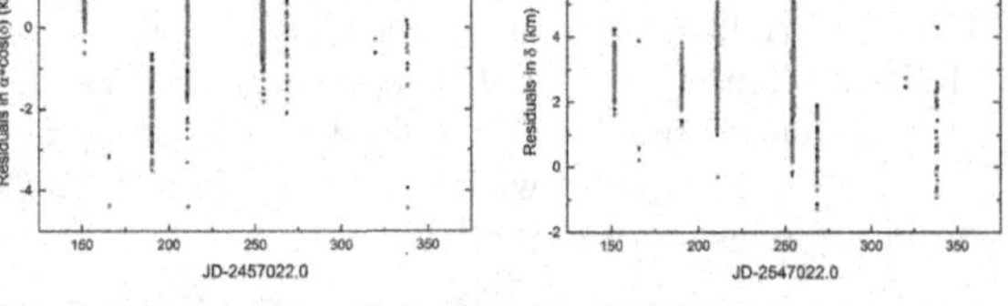

(a) The (o-c)s by using Gaia. (b) The (o-c)s by using Gaia DR1 and UCAC4 respectively.

Figure 1. The (O-C) residuals in km of observed positions relative to SAT367.

computed. The figure 1(a) displays the residuals and their mean and standard deviation values are listed in table 1. After that, 368 images of them have been reduced by Caviar with UCAC4 again. Figure 1(b) displays the (o-c)s of the position obtained by using Gaia DR1 and UCAC4 respectively relative to SAT367. Table 2 lists their means and standard deviations in different directions.From figure 1(b) and table 2, we can find that the two catalogues have equivalent effects. This is because the astrometric reduction's error comes from several sources that conceal the benefits from Gaia DR1's improvement.

4. Conclusion

A total of 494 ISS images of Enceladus in 2015 are reduced by using Gaia DR1, the best mean residual is $75.7m$ and their SD values are not over $2km$. Comparing the results from UCAC4 and Gaia DR1, we find they bring the equivalent precision of reduction.

References

Cooper, N. J., Murray, C. D., Porco, C. C., & Spitale, J. N. 2006, *Icarus*, 181, 223

Cooper, N. J., Murray, C. D., Lainey, V., Tajeddine, R., Evans, M. W., & Williams, G. A. 2014, *A&A*, 572, A43

Cooper, N. J., Evans, M. W., Meunier, L-E, *et al.* Caviar: a software package for the astrometric reduction of spacecraft images, 2016, *6th ICATT Conference*

Gaia Collaboration, T. Prusti, J. H. Bruijne *et al.* 2016, *A&A*, 595, A1

Gaia Collaboration, A. G. Brown, A. Vallenari *et al.* 2016, *A&A*, 595, A2

Tajeddine, R., Cooper, N. J., Lainey, V., Charnoz, S., & Murray, C. D. 2013, *A&A*, 551, A129

Tajeddine, R., Lainey, V., Cooper, N. J., & Murray, C. D. 2015, *A&A*, 575, A73

Zacharias, N., Finch, C. T., Girard, T. M. *et al.* 2013, *AJ*, 145, 44

Astronomy and Astrophysics in the Gaia sky
Proceedings IAU Symposium No. 330, 2017
A. Recio-Blanco, P. de Laverny, A.G.A. Brown & T. Prusti, eds.

doi:10.1017/S174392131700624X

Astrometry and Spectra Classification of Near Earth Asteroids with Lijiang 2.4m Telescope†

X. L. Zhang[1,2], B. Yang[1,2], and J. M. Bai[1,2]

[1]Yunnan Observatories, Chinese Academy of Sciences, Kunming 650011, China
email: zhangxiliang@ynao.ac.cn

[2]Key Laboratory of the Structure and Evolution of Celestial Objects, Chinese Academy of Sciences, Kunming 650011, China

Abstract. The Lijiang 2.4m telescope of Yunnan Observatories is located at longitude E100°01′51″, latitude N26°42′32″ and height 3250m above sea level (IAU code O44). Because of low latitude of the site, long-focus system and planetary tracking mode of telescope, high accuracy positioning and spectral classification of the near Earth objects (NEAs) especially in the Southern Hemisphere can be studied with the Lijiang 2.4m telescope. As a set of observational campaigns organized by the GAIA-FUN-SSO, astrometry of several near Earth asteroids including (367943) Duende and (99942) Apophis were made with Lijiang 2.4m telescope during 2013. From December 12, 2015, spectra of three near earth asteroids were also observed with the YFOSC terminal attached to the Lijiang 2.4m telescope. This paper will give the detailed introduction of Lijiaing 2.4m telescope and observational results of near Earth asteroids obtained with it.

Keywords. methods: observational, techniques: astrometry, spectrum classification.

1. Introduction

Knowledge of asteroids and comets, the relic of planet building blocks, directly sheds light on the formation of terrestrial planets, and the origin and evolution of the Solar System. Modern asteroid surveys have greatly enhanced the detectability of small NEAs and distant comets. In the past two decades, the number of known NEAs and comets has increased drastically. Given that the observable windows of newly detected objects are often quite narrow (merely a few days), prompt follow-up observations are critical for orbit determinations well as for the interpretations of the physical properties.

Lijiang 2.4m telescope (Figure 1) is equipped with the Yunnan Faint Object Spectrograph and Camera(YFOSC), which is capable of conducting optical imaging and low-resolution spectroscopy from 300 to 1000 nm, the parameters of grisms often used are given in table 1. Aided with the planetary tracking mode, Lijiang 2.4m telescope is an ideal facility to carry out follow-up physical studies on newly discovered NEAs.

2. Astrometry

As parts of pre-launch training programs organized by the Gaia-FUN-SSO, several NEAs such as (99942) Apophis, (367943) Duende and 2013 TV135, were observed with the Lijiang 2.4m telescope (Zhang *et al.*, 2015; Wang *et al.*, 2015; Thuillot *et al.*, 2015).

† This research work is supported by the National Science Foundation of China (Grant Nos 11203070, U1431227, U1631124, 11103092, U1631127, 11473068 and 11573067).

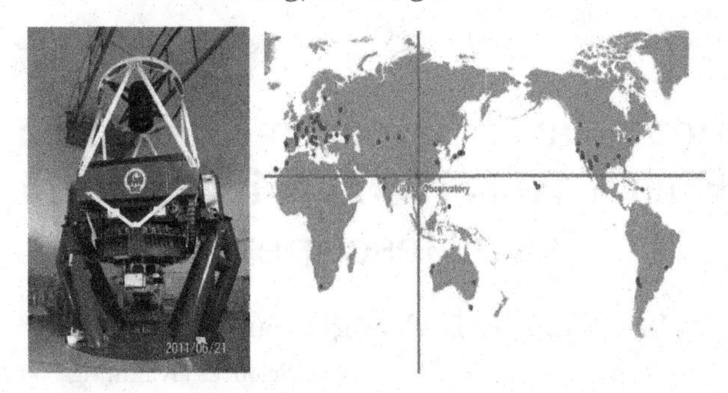

Figure 1. Lijiang 2.4m telescope and its location.

Table 1. Parameters of grisms of YFOSC.

Grisms NO.	λ_c (nm)	λ_{Blaze} (nm)	Grooves (nm/mm)	Dispersion (nm/piexl)	Resolution (@600nm)	Sp.Range (nm)
12	730	700	75	1.1	545	520-980
10	380	390	150	0.79	760	340-980
5	650	700	300	0.46	1300	496-980
15	586	527	300	0.39	1540	410-980
3	390	430	400	0.29	2068	340-910
14	463	428	600	0.17	3520	360-746
8	650	700	600	0.15	4000	510-960

Table 2. Detailed observational information of five newly found near Earth asteroids.

Objects	date (UT)	Δ (AU)	$r_\odot$ (AU)	Φ (Deg)	Airmass
163899	2015-12-19	0.076	0.965	102.5	1.43
363599	2016-04-09	0.056	1.022	67.5	1.05
154244	2016-07-26	0.071	1.044	64.7	1.11
2011 UU106	2016-12-08	0.155	1.070	53.1	1.11
2005 TF	2016-12-09	0.109	1.075	32.3	1.20

3. Optical Spectrum Classification

In table 2, we present the detailed observational information of five NEAs with the Lijiang 2.4m telescope during 2015-16, and their taxonomic types according to the feature-based Bus-taxonomy(Bus *et al.*, 2002) will be given in our follow-up work.

4. Acknowledgments

We acknowledge the support of the staff of the Lijiang 2.4m telescope. Funding for the telescope has been provided by Chinese Academy of Sciences and the People's Government of Yunnan Province.

References

Bus S. J., Vilas F., & Barucci M. A. 2002, *Asteroids III*, 169
Thuillot W., Bancelin D., Ivantsov A., *et al.* 2015, *A&A*, 583, A59
Wang N., Peng Q. Y., Zhang X. L., *et al.* 2015, *MNRAS*, 454, 3809
Zhang, X. L., Yu Y., Wang X. L., *et al.* 2015, *RAA*, 15, 3, 170

Author index

Abbas, U. – 79
Abreu, A. – 343
Adelman, S. J. – 362
Adibekyan, V. Zh. – 156
Adibekyano, V. – 391
Agüeros, M. A. – 297
Akhmetov, V. S. – 81, 100
Andreasen, D. T. – 271
Andrei, A. – 88
Andrei, A. H. – 75
Andrews, J. J. – 297
Anguiano, B. – 201
Apostolovska, G. – 393, 395
Aprilia, M. – 243
Arenou, F. – 313
Arifyanto, I. – 243
Arlot, J.-E. – 83
Ashley, R. – 181
Assafin, M. – 382, 397

Babusiaux, C. – 313
Bagdonas, V. – 241, 283
Bai, J. M. – 413
Bailer-Jones, C. A. L. – 144, 189
Baines, D. – 35, 277
Banda-Huarca, M. V. – 397
Barache, C. – 75
Barstow, M. A. – 301
Basak, N. – 331
Bebekovska, E. V. – 393, 395
Bellini, A. – 261
Belmonte, M. T. – 203
Bendjoya, P. – 399
Benedetti-Rossi, G. – 382, 397
Bensby, T. – 218
Bérard, D. – 382, 397
Bernard, E. J. – 148
Bertone, G. – 255
Besla, G. – 261
Bhardwaj, A. – 337
Binney, J. – 111, 152
Bisterzo, S. – 331
Boffin, H. M. J. – 323, 339
Bond, H. E. – 301
Borgniet, S. – 305
Borisov, G. – 395
Boubert, D. – 321
Bovy, J. – 210
Bragaglia, A. – 119
Braga-Ribas, F. – 382, 397
Brandner, W. – 214
Breddels, M. A. – 275
Brown, A. G. A. – 13, 181, 197, 245, 269
Bruzual, G. – 271
Bucciarelli, B. – 79
Busso, G. – 30

Cacciari, C. – 30
Camargo, J. – 382
Camargo, J. I. B. – 397
Carrasco, J. M. – 30
Carrasco-Kind, M. – 397
Casagrande, L. – 206
Casetti-Dinescu, D. I. – 85
Casewell, S. L. – 301
Cellino, A. – 399
Chanamé, J. – 297
Chayer, P. – 98
Chen, B. – 193
Chen, B.-Q. – 208, 220
Chen, H.-C. – 243
Chubey, M. S. – 407
Clear, C. – 203
Coelho, B. – 75
Cole, D. R. – 152
Cooper, N. J. – 411
Côté, B. – 331
Crosta, M. – 79, 231
Crowley, C. – 343
Cummings, J. – 317
Curir, A. – 263

da Costa, L. A. N. – 397
Damljanović, G. – 88
Danielski, C. – 313
David, P. – 63
De Angeli, F. – 30, 399
de Laverny, P. – 216, 267
Deason, A. – 261
Delbo, M. – 399
Delchambre, L. – 59
Delgado-Mena, E. – 271, 391
Derekas, A. – 214
Desmars, J. – 382, 397
Devyatkin, A. V. – 407
Donchev, Z. – 393, 395
Drazdauskas, A. – 241, 283
Drimmel, R. – 185
Drlica-Wagner, A. – 397
Ducourant, C. – 59, 265
Durán, J. – 35

Eggl, S. – 386

Er, H. – 405
Erece, O. – 94
Escorza, A. – 323
Evans, D. W. – 30
Evans, N. R. – 325
Evans, N. W. – 321
Eyer, L. – 30

Fabricius, C. – 30
Fedorov, P. N. – 81, 100
Figueira, P. – 391
Finet, F. – 59
Fink, M. – 343
Fouesneau, M. – 189
Fraser, M. – 321
Freeman, K. – 201
Frenk, C. – 253
Fritz, T. K. – 210

Gai, M. – 79
Gallenne, A. – 305, 325
Galli, P. A. B. – 265
Galluccio, L. – 59, 399
Gänsicke, B. T. – 317
García-Berro, E. – 201
Gattano, C. – 75
Gentile-Fusillo, N. – 317
Giammaria, M. – 263
Gibson, B. K. – 331
Gieren, W. – 305
Gilmore, G. – 23
Girard, T. M. – 85
Glamazda, D. – 403
Godunova, V. – 401
Goldman, B. – 214
Gonoretzky, E. R. – 265
González Hernández, J. I. – 156
González-Núñez, J. – 35
Gorbaneva, T. – 331
Goriely, S. – 352
Gorshanov, D. L. – 407
Gouda, N. – 90, 164
Goźiewski, K. – 405
Groenewegen, M. A. T. – 287
Guiglion, G. – 216, 267
Gutiérrez-Sanchéz, R. – 35

Hagen, J. H. J. – 160
Hanson, R. – 189
Hattori, K. – 164
Hayden, M. – 267
Hees, A. – 63
Helmi, A. – 160, 229, 275
Henning, T. – 214
Herwig, F. – 331
Hestroffer, D. – 63, 386
Hobbs, D. – 67

Høg, E. – 67, 92
Holberg, J. B. – 301
Howes, L. M. – 218
Huang, Y. – 193, 208, 220
Hunt, J. A. S. – 222

Irfan, M. – 243
Israelian, G. – 156
Ivantsov, A. – 386

Jableka, D. – 364
Janík, J. – 273
Janulis, R. – 241, 283
Jiang, I.-G. – 251
Jiménez-Esteban, F. – 225
Jordan, C. – 331
Jordan, S. – 317
Jordi, C. – 30
Jorissen, A. – 323, 329, 345, 350, 352
Joshi, Y. C. – 227
Joyce, S. R. G. – 301
Just, A. – 168

Kaiser, G. – 403
Kalirai, J. S. – 317
Kallivayalil, N. – 210
Kanbur, S. M. – 337
Kaplan, M. – 94
Karinkuzhi, D. – 352
Kervella, P. – 305, 325
Kharchenko, N. V. – 281
Kılıç, Y. – 94
Klebonas, L. – 241, 283
Klioner, S. – 71
Koppelman, H. H. – 229
Kordopatis, G. – 172, 181, 279
Korotin, S. A. – 331
Kostov, A. – 393
Koutsouridou, I. – 168
Kovalevsky, J. – 1
Kovtyukh, V. V. – 331
Kreiner, J. M. – 364
Krone-Martins, A. – 59
Krone-Martins, A. G. O. – 265
Krzeszowski, K. – 405
Kunder, A. – 176
Kuzmanovska, O. – 393
Kuznetsov, E. – 403

Lainey, V. – 83, 411
Lallement, R. – 243
Lallo, M. – 98
Lattanzi, M. G. – 79, 185, 231, 263
Le Campion, J. F. – 59
Le Poncin-Lafitte, C. – 63, 96
Liao, S. – 231
Liggings, F. – 203

Lin, C.-C. – 233
Lindegren, L. – 41
Linden, S. T. – 210
Liu, X. – 193
Liu, X.-W. – 208, 220
López, J. A. – 235, 237

Maia, M. A. G. – 397
Malasan, H. L. – 243
Manara, C. F. – 309
Marchetti, T. – 181
Marco, F. J. – 235, 237
Martínez, M. J. – 235, 237
Martín-Fleitas, J. M. – 343
Mashonkina, L. – 327
McLean, B. – 98
McMillan, P. J. – 239, 279
Mena, E. D. – 156
Mérand, A. – 305, 325
Merle, T. – 329, 350, 352
Mignard, F. – 59, 71
Mikolaitis, Š. – 241, 283
Minchev, I. – 127
Mishenina, T. – 331
Monreal-Ibero, A. – 243
Montegriffo, P. – 30
Moor, A. – 214
Mora, A. – 35, 343
Murante, G. – 263

Naagaya, T. – 333
Nardetto, N. – 305, 335
Nasiroglu, I. – 405
Neiner, C. – 83
Nelan, E. G. – 98
Ngeow, C.-C. – 337
Nishi, R. – 104
Ogando, R. L. – 397
Ogloza, W. – 364
Oudmaijer, R. D. – 277

Pakhomov, Y. – 327
Pakštienė, E. – 241, 283
Pancino, E. – 30
Pang, X.-Y. – 233
Paul, A. – 331
Peng, H. W. – 409
Peng, Q. Y. – 409, 411
Pickering, J. C. – 203
Pietrzynski, G. – 305
Pignatari, M. – 331
Pihlström, Y. M. – 245
Piskunov, A. E. – 281
Plez, B. – 352
Poggio, E. – 185
Posti, L. – 275
Pourbaix, D. – 323, 339
Proffitt, C. – 325
Prusti, T. – 7, 309
Puspitarini, L. – 243

Qi, Z. – 231
Quiroga-Nuñez, L. H. – 245

Racero, E. – 35
Re Fiorentin, P. – 263
Read, J. – 255
Rebassa-Mansergas, A. – 201
Recio-Blanco, A. – 216, 267
Reddy, B. E. – 348
Reffert, S. – 281
Reshetnyk, V. – 401
Rezaei Kh., S. – 189
Riello, M. – 30
Rimoldi, A. – 181
Ritter, C. – 331
Robert, V. – 83, 96
Röser, S. – 214, 281
Rossi, E. M. – 181
Ruiz-Dern, L. – 313
Russeil, D. – 341
Sahlmann, J. – 98, 249, 343
Sakai, N. – 164
Salgado, J. – 35
Santos, N. C. – 156, 271, 369, 391
Sariya, D. P. – 251
Sartoretti, P. – 313
Schilbach, E. – 214, 281
Schöfer, P. – 214
Schriefer, M. – 85
Segovia, J. C. – 35
Sementsov, V. – 354
Sergeev, O. – 401
Shakht, N. A. – 407
Sharma, M. – 253
Shetye, S. – 323, 345, 352
Shi, J. – 193
Shirasaki, Y. – 104
Sicardy, B. – 377, 382, 397
Siess, L. – 323, 345, 352
Silva de Souza, R. – 259
Silverwood, H. – 255
Simon, A. – 401
Singh, R. – 348
Sitnova, T. – 327
Sivertsson, S. – 255
Sjouwerman, L. O. – 245
Slezak, E. – 59
Słowikowska, A. – 405
Smart, R. – 79
Smart, R. L. – 185
Smiljanic, R. – 259
Sohn, S. T. – 261

Solano, E. – 225
Soubiran, C. – 331
Souchay, J. – 75
Sousa, S. G. – 156, 271, 391
Sozzetti, A. – 79
Spagna, A. – 185, 263
Starkenburg, E. – 181
Steger, P. – 255
Steinmetz, M. – 279
Surdej, J. – 59
Sysoliatina, K. – 168

Tagawa, H. – 164
Tang, Z. – 231
Tanga, P. – 399
Taradii, V. – 401
Taris, F. – 75, 88
Tautvaišienė, G. – 241, 283
Teixeira, G. D. C. – 271
Teixeira, R. – 59, 265
Theuns, T. – 253
Thielemann, F.-K. – 331
Thorne, A. P. – 203
Thouvenin, N. – 83
Tian, Z.-J. – 208
Titarenko, A. – 267
Torres, S. – 201, 269
Trahin, B. – 305
Travaglio, C. – 331
Traven, G. – 329
Tremblay, P.-E. – 317
Tsantaki, M. – 156, 271
Turon, C. – 313

van der Marel, R. – 249
van der Marel, R. P. – 261
Van der Swaelmen, M. – 329, 350
Van Eck, S. – 323, 329, 345, 350, 352
van Langevelde, H. J. – 245
van Leeuwen, F. – 30
Van Winckel, H. – 323, 345
Vecchiato, A. – 79, 231
Velčovský, J. –273
Velichko, A. B. – 81, 100
Velichko, S. – 401
Veljanoski, J. – 275
Vieira-Martins, R. – 382, 397
Vienne, A. – 411
Vioque, M. – 277
Voirin, J. – 309
Voloshina, I. – 354

Wang, C. – 193, 208, 220, 409, 411
Watkins, L. – 261
Wehmeyer, B. – 331
Weiler, M. – 30
Wertz, O. – 59
Wiebe, Y. – 403
Wojno, J. – 279
Worley, C. – 267
Worley, C. C. – 216
Wyse, R. F. G. – 136

Xia, F. – 356
Xiang, M. – 193
Xiang, M.-S. – 208, 220
Xiaoli, W. – 358

Yadav, R. K. S. – 251
Yamada, Y. – 104
Yang, B. – 413
Yano, T. – 164, 360
Yen, S. X. – 281
Youakim, K. – 181
Yuan, H. – 193
Yuan, H.-B. – 208, 220
Yüce, K. – 362

Zacharias, N. – 49
Zakrzewski, B. – 364
Zari, E. – 197, 309
Żejmo, M. – 405
Ženovienė, R. – 241, 283
Zhang, H.-W. – 208, 220
Zhang, Q. F. – 409, 411
Zhang, X. L. – 413
Zivick, P. – 210
Zola, S. – 364, 405
Zschocke, S. – 106
Zwart, S. P. – 269
Zwitter, T. – 201, 329, 350

Printed in the United States
by Baker & Taylor Publisher Services